D0235946

WITHDRAWN

INTERNATIONAL ATOMIC WEIGHTS* BASED ON $^{12}C=12$

Element	Symbol	Atomic Number	Atomic Weight
Actinium	Ac	89	(227)
Aluminum	Al	13	26.9815
Americium	Am	95	(243)
Antimony	Sb	51	121.75
Argon	Ar	18	39.948
Arsenic	As	33	74.9216
Astatine	At	85	(210)
Barium	Ba	56	137.34
Berkelium	Bk	97	(247)
Beryllium	Be	4	9.0122
Bismuth	Bi	83	208.980
Boron	B	5	10.811
Bromine	Br	35	79.909
Cadmium	Cd	48	112.40
Calcium	Ca	20	40.08
Californium	Cf	98	(251)
Carbon	C	6	12.01115
Cerium	Ce	58	140.12
Cesium	Cs	55	132.905
Chlorine	Cl	17	35.453
Chromium	Cr	24	51.996
Cobalt	Co	27	58.9332
Copper	Cu	29	63.54
Curium	Cm	96	(247)
Dysprosium	Dy	66	162.50
Einsteinium	Es	99	(254)
Erbium	Er	68	167.26
Europium	Eu	63	151.96
Fermium	Fm	100	(257)
Fluorine	F	9	18.9984
Francium	Fr	87	(223)
Gadolinium	Gd	64	157.25
Gallium	Ga	31	69.72
Germanium	Ge	32	72.59
Gold	Au	79	196.967
Hafnium	Hf	72	178.49
Helium	He	2	4.0026
Holmium	Ho	67	164.930
Hydrogen	H	1	1.00797
Indium	In	49	114.82
Iodine	I	53	126.9044
Iridium	Ir	77	192.2
Iron	Fe	26	55.847
Krypton	Kr	36	83.80
Lanthanum	La	57	138.91
Lawrencium	Lw	103	(256)
Lead	Pb	82	207.19
Lithium	Li	3	6.9417
Lutetium	Lu	71	174.967
Magnesium	Mg	12	24.312
Manganese	Mn	25	54.9380
Mendelevium	Md	101	(258)
Mercury	Hg	80	200.59
Molybdenum	Mo	42	95.94
Neodymium	Nd	60	144.24
Neon	Ne	10	20.183
Neptunium	Np	93	(237)
Nickel	Ni	28	58.71
Niobium	Nb	41	92.906
Nitrogen	N	7	14.0067
Nobelium	No	102	(255)
Osmium	Os	76	190.2
Oxygen	O	8	15.9994
Palladium	Pd	46	106.4
Phosphorus	P	15	30.9738
Platinum	Pt	78	195.09
Plutonium	Pu	94	(244)
Polonium	Po	84	(209)
Potassium	K	19	39.102
Praseodymium	Pr	59	140.907
Promethium	Pm	61	(145)
Protactinium	Pa	91	(231)
Radium	Ra	88	(266)
Radon	Rn	86	(222)
Rhenium	Re	75	186.2
Rhodium	Rh	45	102.905
Rubidium	Rb	37	85.47
Ruthenium	Ru	44	101.07
Samarium	Sm	62	150.35
Scandium	Sc	21	44.956
Selenium	Se	34	78.96
Silicon	Si	14	28.086
Silver	Ag	47	107.870
Sodium	Na	11	22.9898
Strontium	Sr	38	87.62
Sulfur	S	16	32.064
Tantalum	Ta	73	180.948
Technetium	Tc	43	(97)
Tellurium	Te	52	127.60
Terbium	Tb	65	158.924
Thallium	Tl	81	204.37
Thorium	Th	90	232.038
Thulium	Tm	69	163.934
Tin	Sn	50	118.69
Titanium	Ti	22	47.90
Tungsten	W	74	183.85
Uranium	U	92	238.03
Vanadium	V	23	50.9415
Xenon	Xe	54	131.30
Ytterbium	Yb	70	173.04
Yttrium	Y	39	88.905
Zinc	Zn	30	65.37
Zirconium	Zr	40	91.22

*Parentheses indicate the atomic weight of the most stable isotope.

Second Edition

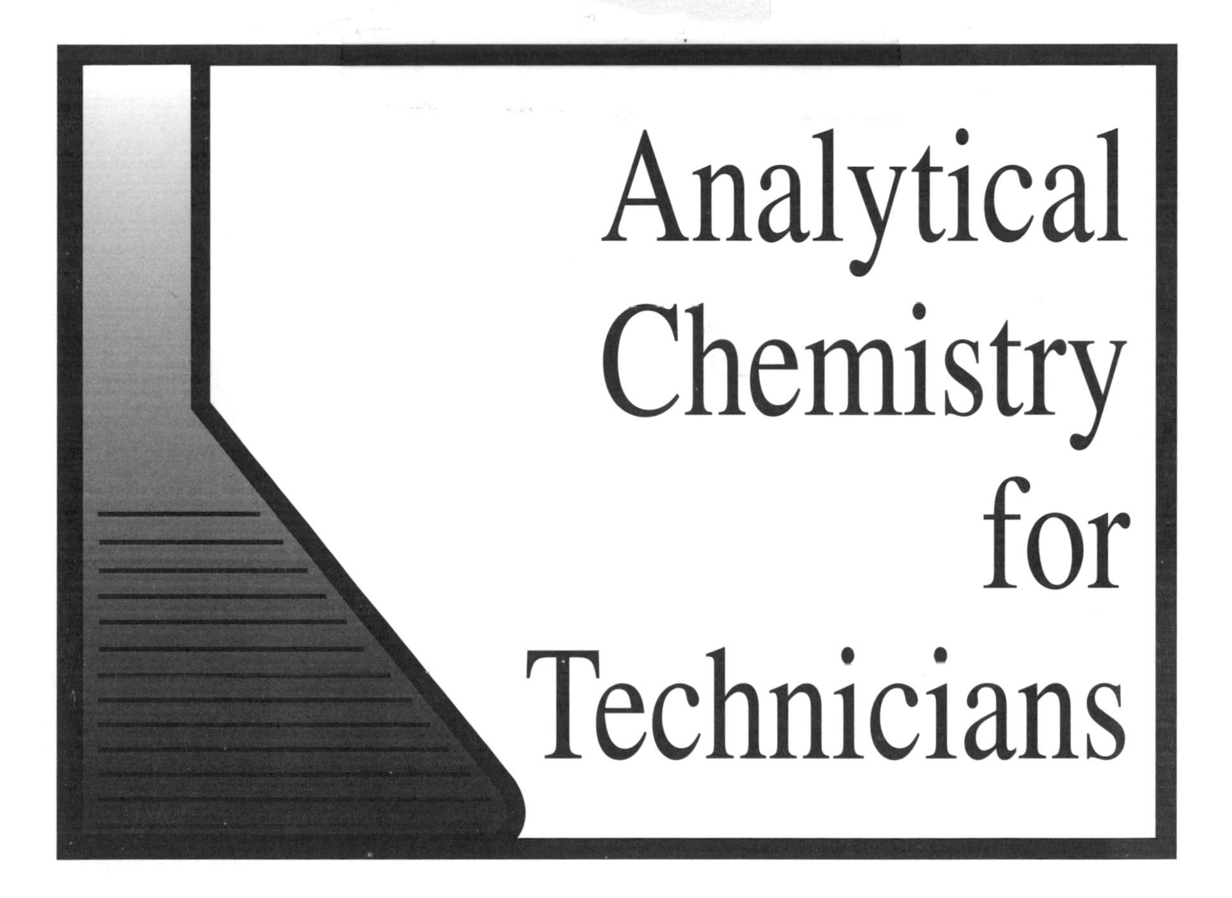

Analytical Chemistry for Technicians

John Kenkel

CRC LEWIS PUBLISHERS
Boca Raton New York

Library of Congress Cataloging-in-Publication Data

Kenkel, John.
Analytical chemistry for technicians / John Kenkel. – 2nd ed.
p. cm.
Includes bibliographical references and index.
ISBN 0-87371-966-2
1. Chemistry, Analytic. I. Title.
QD75.2.K445 1994
543 – dc20 94-1435
CIP

Lewis Publishers is an imprint of CRC Press LLC

International Standard Book Number 1-87371-966-2
Library of Congress Card Number 94-1435
Printed in the United States of America 3 4 5 6 7 8 9 0
Printed on acid-free paper

Preface

Industrial laboratory supervisors frequently express the view that persons with bachelor's degrees hired as laboratory technicians require extensive training in practical analytical chemistry before they can become productive employees. Indeed, it is not uncommon to hear one of my 2-year college chemistry technician graduates express some degree of shock and surprise when they determine that the 4-year college graduate that they have been asked to train has not acquired the basic skills of pipetting, extracting, titrating, and preparing solutions, to name a few. Such shock has also been expressed by university professors. Risley*, for example, states " . . . when upper division undergraduate students and graduate students first begin research, few have acquired the skill of preparing solutions . . . ". A recent editorial in *Spectroscopy*** makes a plea for more practical training in the bachelor degree programs. Dr. Ciurczak states: "the arcane points covered in most classes are often of little practical use and seldom serve to prepare the students for applying their knowledge while applying their trade. The sad fact is that most colleges use the undergraduate curriculum to prepare students for graduate school . . .". He also says: ". . . the classroom work is totally theoretical and seldom has any connection with the lab." Dr. Ciurczak then proposes the obvious solution, "My plea is

*Risley, J. M., *J. Chem. Ed.*, 68(12), 1054, 1991.
**Ciurczak, E. W., *Spectroscopy*, 8(3), 12–14, 1993.

to mix in a little 'nuts and bolts' science with the quantum theory – after all, the real world is where most of us are forced to live."

My textbook, *Analytical Chemistry For Technicians*, has been successfully providing the "nuts and bolts" of analytical chemistry to 2-year college chemistry technician trainees since 1988. While this new edition retains all the qualities that made that possible, it has undergone significant revision such that 4-year colleges heeding the call of more practical course work in the area of analytical chemistry may also find it attractive.

The first edition of *Analytical Chemistry For Technicians* has enjoyed overwhelming success. My original premise, that there was a need for an analytical chemistry textbook taking a traditional approach to the subject while stressing the practical aspects for the training of technician level laboratory workers, has been affirmed. This new edition represents my desire to expand the text, including subjects that were missing in the first edition, rewriting certain sections so as to take a fresh approach, and raising the level of the text slightly in order to present a greater challenge to the technician trainee and, at the same time, make it more adaptable to the 4-year college programs.

The instructor-user of this edition will notice many changes in nearly every chapter. While the order of coverage is roughly the same as in the first edition, the depth of coverage has changed for almost every topic. In some cases, where the analytical method discussed has been determined to be less important, such as gravimetric analysis, fluorometry, and polarography, the emphasis has either shifted or has decreased. In other cases, where the method discussed has been determined to be of greater importance, such as infrared (IR) spectrometry, analytical separations, NMR and mass spectrometry, entire new chapters have been written. Changes in other chapters can be described as ranging from subtle to major. Probably the most noticeable change, however, is in the number of end-of-chapter questions and problems. These have been dramatically increased in every chapter. There have been some experiments added and others have been modified.

In this new edition, the introduction to the subjects of analytical chemistry and gravimetric analysis, which spanned the first two chapters in the first edition, are now combined into one chapter, Chapter 1. Also, the discussion of analytical balances, which before was relegated to an appendix, has also become part of Chapter 1. Accompanying the discussion of laboratory acids in Chapter 2 is a discussion of common organic solvents and fusion. The statistics chapter, now Chapter 3, has been modified in response to several users, including a very substantial increase in the end-of-chapter questions and problems. The chemical equilibrium chapter, Chapter 4, has been expanded to include a comprehensive treatment of buffer solutions.

Changes in the titrimetric analysis chapters covering glassware and acid-base, complex ion formation, and oxidation-reduction reactions are more subtle. These include discussions of when to be accurate and when not (in both Chapter 1 and Chapter 5), an expanded discussion of the Kjeldahl method, and an emphasis on molarity and moles where EDTA and water hardness calculations are concerned, rather than normality and equivalents.

One major change has occurred with respect to the potentiometry chapter. I have merged this subject with the subject of polarography/voltammetry and have moved it to near the end of the book where the polarography chapter was previously located. While the aim here was to bring greater continuity to the subject of electroanalytical chemistry, instructors should still find it possible to follow-up the oxidation-reduction chapter with the subject of potentiometry if desired. Discussions of the Ag/AgCl and combination pH electrodes have been added and the discussion of ion-selective electrodes has been expanded. The discussion of polarography has been condensed somewhat.

Instructors will find that the chapter introducing instrumental analysis and computers has been expanded to include more on the subject of "$y = mx + b$", more on the method of least squares, and a new section summarizing serial dilution, and internal standard and standard additions methods. There is also a discussion of control samples and reagent and sample blanks and a new section entitled "Modern Data Handling". The increase in the number of end-of-chapter questions and problems mentioned above is probably most noticeable here.

More subtle changes have occurred in the chapter introducing light and light absorption. However, the chapters which cover spectroscopy techniques have been substantially rearranged. This subject was covered in three chapters in the first edition, but with fluorometry combined with UV/vis spectrophotometry and separate chapters now covering infrared and NMR/mass spectrometry, this subject, including a rewritten chapter on atomic absorption, is now covered in four chapters. This represents a very important change in that infrared spectrometry now is covered in a separate chapter and includes FTIR, and the topics of NMR and mass spectrometry, which were totally absent in the first edition, now comprise an entire chapter.

The subject of chromatography is still covered in three chapters, but there has been considerable expansion. The chapter introducing the subject is now called "Analytical Separations" and includes a comprehensive discussion of extraction and also introduces recrystallization and distillation. The electrophoresis section has been expanded.

The chapters covering GC and HPLC have been rewritten. The most noticeable changes include more on detectors in both chapters, and also a more comprehensive treatment of chromatogram parameters, such as resolution, the number of theoretical plates, HETP, capacity factor, and selectivity. The subject of troubleshooting is covered in both chapters.

In the final chapter, the application summary, flow-injection analysis, and robotics have been added.

Since the first edition of *Analytical Chemistry For Technicians* was published in 1988, I have written a companion manual entitled *Analytical Chemistry Refresher Manual*, also published by Lewis Publishers (1992). This manual was written for employed technicians who needed a refresher in the techniques of modern analytical chemistry. Some of the text and figures in this new edition of *Analytical Chemistry For Technicians* are taken from *Analytical Chemistry Refresher Manual.*

Many of the points made in the preface to the first edition continue to be valid. These include the following:

1. That a student with a limited background in chemistry can follow and comprehend the text.

2. That a new analytical psychology needs to be instilled in the beginning analytical chemistry student and that Experiments 1 and 2 are designed to do that.
3. That the lecture material presented supports and coincides with the laboratory experiences.
4. That it is organized in a manner similar to texts designed for baccalaureate programs.
5. That the experiments included represent a significant mix of "prepared" and "real-world" sample analyses that allow for the evaluation of the students' abilities and achievements while also providing the important real-world experiences.

It is my hope that this book continues to fulfill the analytical chemistry needs of instructors and students in modern chemistry technician training programs and that 4-year programs will now also want to take a look. I feel, as Dr. Ciurczak does, that there is a serious need for more practical training in the 4-year curriculum. As always, I hope that as students and instructors use the book that they will provide me with feedback, positive or negative, so that the text can continue to improve.

John Kenkel
Southeast Community College
Lincoln, Nebraska

Acknowledgments

I would first like to express my gratitude to those instructors who have adopted the first edition of *Analytical Chemistry For Technicians* for their analytical chemistry courses. Your positive comments about the book's quality have kept me going. I would also like to express my appreciation to the analytical chemistry community as a whole. Those of you who have purchased the first edition as a reference book for yourself or for the technicians working in your labs have given me unexpected satisfaction. I appreciate your support very much and hope that the book has served your needs.

A number of people at Lewis Publishers deserve a large dose of thanks. These include Jon Lewis, who supported the development of the first edition and who now encouraged this revision. In addition, I would like to acknowledge the help of Sharon Ray, Editorial Assistant, for her help, and also Mr. Brian Lewis for his continuous enthusiasm for my work. I appreciate it very much. I would also like to acknowledge Julie Haydu, Project Editor, for her hard work.

My artist, Lana Johnson of Lincoln, Nebraska, has been terrific. I greatly appreciate the quality of her work and her patience. I have enjoyed working with her.

I would like to acknowledge the following past and current users of the first edition of *Analytical Chemistry For Technicians* who either responded to a questionnaire asking for their recommendations for this new edition, or who were otherwise asked to comment.

Mary Clare Lambden, Centennial College, Scarborough, Ontario
Graham Sparrow, Centennial College, Scarborough, Ontario
Gunay Ozkan, Central Community College, Hastings, Nebraska
Bruce Schuhmacher, Albert Lea Technical College, Albert Lea, Minnesota
Carol White, Athens Area Technical Institute, Athens, Georgia
Gary Adams, Thames Valley State Technical College, Norwich, Connecticut
Ben L. Owen, State Technical Institute at Memphis, Memphis, Tennessee
Martha Brosz, Cincinnati Technical College, Cincinnati, Ohio
Kathy Riesterer, Northwest Wisconsin Technical College, Green Bay, Wisconsin
Bill Killian, Ferris State University, Big Rapids, Michigan
Alex Bell, Trident Technical College, Charleston, South Carolina
John Spille, University of Cincinnati, Cincinnati, Ohio
Gundega Michel, Truman College, Chicago, Illinois
Ildy Boer, County College of Morris, Randolph, New Jersey
Mike Sorrentino, Erie Community College, Buffalo, New York
Ken Hokanson, Aiken Technical College, Aiken, South Carolina

There have been many people during my 16 years at Southeast Community College that have played a major role in my success and in the enjoyment I've received from my teaching career. I would like to acknowledge the numerous students that have passed through the halls of Southeast Community College and have touched my life. I would also like to acknowledge my campus president, Dr. Jack Huck, my division dean, Mr. Richard Ross, and our division secretary, Ms. Marcia Blender, for their smiling faces and continued support. In addition, my colleagues, Mr. Don Mumm, Mr. Robin Monroe, Mr. Steven Bassett, Ms. Cindy Martin, and Dr. Reena Roy, have been simply outstanding. It has been a joy working with them.

The task of book writing often takes me away from my family during evenings and weekends. I cannot say enough about the love and understanding of my wife and daughters, to whom this book is once again dedicated. Thanks to you from the bottom of my heart.

Finally, to my God and Savior, thank you for your many blessings, especially for life and happiness. I know that nothing I do is possible without your help.

List of Experiments

1. The Percent of Water in Hydrated Barium Chloride 15
2. The Gravimetric Determination of Sulfate in a Commercial Unknown 18
3. The Gravimetric Determination of Iron in a Commercial Unknown 25
4. A Study of the Dissolving Properties of Water, Some Common Organic Liquids and Laboratory Acids 41
5. Statistical Analysis of Weighing 58
6. Perfecting the Art of Pipetting...............................99
7. Preparation and Standardization of HCl and NaOH Solutions ...128
8. Titrimetric Analysis of a Commercial Soda Ash Unknown for Sodium Carbonate...130
9. Titrimetric Analysis of a Commercial KHP Unknown for KHP 130
10. Analysis of a Grain Sample for Protein Content by the Kjeldahl Method 131
11. Preparation and Standardization of an EDTA Solution 150
12. Titrimetric Determination of Total Hardness in a Synthetic or "Real" Water Sample.......................................151

13. Titrimetric Determination of Temporary and Permanent Hardness in a Water Sample 152
14. Preparation and Standardization of a $KMnO_4$ Solution 167
15. Titrimetric Determination of Sodium Oxalate in a Commercial Unknown 168
16. The Preparation and Standardization of a Sodium Thiosulfate Solution 168
17. The Determination of Residual Chlorine in Wastewater Plant Effluent by Iodometry 169
18. Titrimetric Determination of Calcium in a Commercial Limestone Unknown 170
19. The Determination of Chloride in a Commercial Unknown by the Mohr Method 170
20. Electrical Connections Between Instruments, Recorders, and Computers 193
21. The Measurement of a Strong Acid-Strong Base Titration Curve by Data Acquisition with a Microcomputer 195
22. Colorimetric Analysis of Prepared or Real Water Samples for Iron 235
23. Spectrophotometric Analysis of a Prepared Sample for Toluene 236
24. Determination of Nitrate in Water by UV Spectrophotometry 236
25. The Colorimetric Determination of Ozone in Air 237
26. The Colorimetric Determination of Manganese in Steel 238
27. The Determination of Phosphorus in Water and Soil 239
28. Fluorometric Analysis of a Prepared Sample for Riboflavin 241
29. Fluorometric Analysis of Vitamin Tablets for Riboflavin 241
30. Qualitative Analysis by Infrared Spectrometry – Liquid Sampling 265
31. Qualitative Analysis by Infrared Spectrometry – Solid Sampling 266
32. Quantitative Infrared Analysis of Isopropyl Alcohol in Toluene 267
33. Verifying Optimum Instrument Parameters for Flame AA 309
34. Quantitative Atomic Absorption Analysis of a Prepared Sample 312
35. The Analysis of Soil Samples for Potassium by Atomic Absorption 313
36. The Analysis of Soil Samples for Iron Using Atomic Absorption 313
37. The Analysis of Snack Chips for Sodium by Atomic Absorption or Flame Photometry 314
38. The Atomic Absorption Analysis of Water Samples for Iron Using the Standard Additions Method 315

39. The Atomic Absorption or Flame Photometric Determination of Sodium in Soda Pop 315
40. Titrimetric Determination of the Effectiveness of an Extraction .345
41. The Thin-Layer Chromatography Analysis of Cough Syrups for Dyes 346
42. The Extraction and Analysis of a Pain Relief Tablet for Caffeine 347
43. The Extraction of Capsaicin from Chile Peppers 348
44. A Qualitative Gas Chromatography Analysis of a Prepared Sample 383
45. A Study of the Effect of the Changing of GC Instrument Parameters on Resolution 383
46. The Quantitative Gas Chromatography Analysis of a Prepared Sample for Benzene by the Internal Standard Method 384
47. The Gas Chromatography Determination of a Gasoline Component by the Method of Standard Additions and the Internal Standard Method. 385
48. The Determination of Dichloromethane in Commercial Paint Strippers by Gas Chromatography and the Internal Standard Method. 385
49. The Determination of Ethanol in Wine by Gas Chromatography and the Internal Standard Method. 386
50. A Study of the Effects of Mobile-Phase Composition on Resolution in Reverse-Phase HPLC 412
51. The Quantitative Determination of Methyl Paraben in a Prepared Sample by HPLC 413
52. HPLC Determination of Caffeine and Sodium Benzoate in Soda Pop 414
53. The Quantitative Analysis of Chile Pepper Extract for Capsaicin 415
54. The Analysis of Mouthwash by HPLC – A Research Experiment 415
55. Determination of the pH of Soil Samples .451
56. Determination of Fluoride in Municipal Water (or Other Sample) with the Use of an Ion-Selective Electrode 451
57. Familiarization with Polarographic Techniques 452
58. Polarographic Analysis of a Prepared Sample for Cadmium. 452

Safety in the Analytical Laboratory

The analytical chemistry laboratory is a very safe place to work. However, that is not to say that the laboratory is free of hazards. The dangers associated with contact with hazardous chemicals or with flames, etc., are very well documented and as a result, laboratories are constructed and procedures are carried out with these dangers in mind. Hazardous chemical fumes are, for example, vented into the outdoor atmosphere with the use of fume hoods. Safety showers for diluting spills of concentrated acids on clothing are now commonplace. Eye wash stations are strategically located for the immediate washing of one's eyes in the event of accidental contact with a hazardous chemical. Fire blankets, extinguishers, and sprinkler systems are also located in and around analytical laboratories for immediate extinguishing of flames and fires. Also, a variety of safety gear, such as safety glasses, aprons, and shields are available. There is never a good excuse for personal injury in well-equipped laboratories where well-informed analysts are working.

While the pieces of equipment mentioned above are now commonplace, it remains for the analysts to be well-informed of potential dangers and of appropriate safety measures. To this end, we list below some safety tips of which any laboratory worker must be aware. This list should be studied carefully by all students who have chosen to enroll in an analytical chemistry course. *This is not intended to be a complete list, however.* Students should consult with their course instructor in order to establish safety ground rules for the particular laboratory in which they will be working. Total awareness of hazards and dangers and what to do in case of an accident are the responsibility of the student and the instructor.

1. Safety glasses must be worn at all times by students and instructors. Visitors to the lab must be appropriately warned and safety glasses made available to them.
2. Fume hoods must be used when working with chemicals which may produce hazardous fumes.
3. The location of fire extinguishers, safety showers, and eye wash stations must be known.
4. All laboratory workers must know how and when to use the items listed in #3.
5. There must be no unsupervised or unauthorized work going on in the laboratory.
6. A laboratory is never a place for practical jokes or pranks.
7. The toxicity of all the chemicals you will be working with must be known. Consult the instructor, safety charts, and container labels for safety information about specific chemicals. Recently, many common organic chemicals, such as benzene, carbon tetrachloride, and chloroform, have been deemed unsafe.
8. Eating, drinking, or smoking in the laboratory is never allowed. Never use laboratory containers (beakers or flasks) for drinking coffee or tea.
9. Shoes (and not open-toed sandals) must always be worn—hazardous chemicals may be spilled on the floor or feet.
10. Long hair should always be tied back.
11. Mouth-pipetting is *never* allowed.
12. Cuts and burns must be immediately treated. Use ice on new burns and consult a doctor for serious cuts.
13. In the event of acid spills on one's person, flush thoroughly with water immediately. Be aware that acid/water mixtures will produce heat. Removing clothing from the affected area while water flushing may be important so as to not trap hot acid/water mixtures against the skin. Acids or acid/water mixtures can cause very serious burns if left in contact with skin only for a very short period of time.
14. Weak acids (such as citric acid) should be used to neutralize base spills and weak bases (such as sodium carbonate) should be used to neutralize acid spills. Solutions of these should be readily available in the lab in case of emergency.
15. Dispose of all waste chemicals from the experiments according to your instructor's directions.
16. In the event of an accident, report immediately to your instructor, regardless of how minor you perceive it to be.
17. Always be watchful and considerate of others working in the laboratory. It is important not to jeopardize their safety, nor yours.

18. Always use equipment that is in good condition. Any piece of glassware that is cracked or chipped should be discarded and replaced.

It is impossible to foresee all possible hazards that may manifest themselves in an analytical laboratory. Therefore, it is very important for all students to listen closely to their instructor and obey the rules of their particular laboratory in order to avoid injury. Neither the author of this text, nor its publisher, assumes any responsibility whatsoever in the event of injury.

Table of Contents

1 INTRODUCTION TO CHEMICAL ANALYSIS AND GRAVIMETRIC ANALYSIS 1
1.1 Introduction 1
1.2 Chemical Analysis 2
1.3 Qualitative vs. Quantitative, Wet vs. Instrumental 2
1.4 General Laboratory Considerations 3
1.5 The Laboratory Notebook 3
1.6 Gravimetric Analysis 4
1.7 Analytical Calculations 11
1.8 The Need for Stoichiometry 12
1.9 Gravimetric Factors 13
Experiments 15
Questions and Problems 27

2 SAMPLING AND SAMPLE PREPARATION 31
2.1 The Total Analysis 31
2.2 Step One: Obtaining the Sample 32
2.3 Step Two: Handling the Sample 33
2.4 Step Three: Preparing the Sample for Analysis 34
2.5 Quality Assurance of Reagents 39
Experiments 41
Questions and Problems 42

3 **STATISTICS IN CHEMICAL ANALYSIS** 47
3.1 Introduction to Error Analysis 47
3.2 Terminology 48
3.3 Distribution of Random Errors 53
3.4 Rejection of Data – The Q Test 54
3.5 Student's *t* Test 55
3.6 Statistics in Quality Control 57
3.7 Application to Experiments in This Text 58
Experiments 58
Questions and Problems 59

4 **CHEMICAL EQUILIBRIUM** 63
4.1 Review of Basic Concepts 63
4.2 The Solubility of Precipitates 64
4.3 The Solubility Product Constant 64
4.4 Solubility and K_{sp} 65
4.5 Addition of Ions from an External Source 67
4.6 Chemical Equilibrium and Weak Acids 68
4.7 Weak Bases 71
4.8 Buffer Solutions 72
Questions and Problems 75

5 **INTRODUCTION TO TITRIMETRIC ANALYSIS** 81
5.1 Background 81
5.2 Titration Defined 82
5.3 Preparation of Molar Solutions 84
5.4 Preparation from Pure Solid Chemicals 84
5.5 Preparation by Dilution 85
5.6 Volumetric Glassware 86
5.7 Solution Standardization and Terminology 97
5.8 When to be Accurate and When Not – Revisited 98
Experiments 99
Questions and Problems 100

6 **ACID-BASE TITRATIONS AND CALCULATIONS** 105
6.1 Introduction 105
6.2 Standardization Calculations 106
6.3 Equivalent Weights, Equivalents, and Normality 108
6.4 Examples of Acid-Base Primary Standards and Calculations 113
6.5 Percent Constituent Calculations 115
6.6 Back Titrations 116
6.7 The Kjeldahl Titration 117
6.8 Millimoles and Milliequivalents 119
6.9 Introduction to Titration Curves 120
6.10 The Role of Indicators 125
Experiments 128
Questions and Problems 133

7 COMPLEXOMETRIC TITRATIONS AND CALCULATIONS 139
7.1 Introduction 139
7.2 Complex Ion Terminology 139
7.3 The EDTA Ligand 142
7.4 The Determination of Water Hardness 144
7.5 Parts per Million 145
7.6 Standardization and Calculations 148
Experiments 150
Questions and Problems 153

8 OXIDATION-REDUCTION AND OTHER TITRATIONS 157
8.1 Introduction 157
8.2 Oxidation Number 158
8.3 Oxidation-Reduction Terminology 160
8.4 The Ion-Electron Method for Balancing Equations 160
8.5 Equivalent Weights in Redox Reactions 162
8.6 Potassium Permanganate 163
8.7 Iodometry – An Indirect Method 164
8.8 Summary of Redox Applications 165
8.9 Non-Redox Precipitation Titrations 167
Experiments 167
Questions and Problems 171

9 ANALYSIS WITH INSTRUMENTS AND COMPUTERS 177
9.1 Introduction 177
9.2 Instrumental Data and Readout 177
9.3 Methods for Quantitative Analysis 184
9.4 Blanks and Controls 186
9.5 Effects of Sample Pretreatment on Calculations 188
9.6 Use of Computers 190
9.7 Modern Data Handling 192
Experiments 193
Questions and Problems 196

10 FUNDAMENTALS OF LIGHT AND LIGHT ABSORPTION 201
10.1 Introduction 201
10.2 Nature and Parameters of Light 201
10.3 Converting and Calculating Light Parameters 205
10.4 Light Absorption and Emission 207
10.5 Atomic vs. Molecular 209
10.6 Names of Techniques and Instruments 212
Questions and Problems 212

11 UV-VIS SPECTROPHOTOMETRY 217
11.1 General Description 217
11.2 Instrument Design 218
11.3 Qualitative and Quantitative Analysis 224

11.4 The Question of Matched Cuvettes 231
11.5 The Absorption Spectrum and λ_{MAX} 232
11.6 Fluorometry 232
Experiments 235
Questions and Problems 242

12 INFRARED SPECTROMETRY 249
12.1 Introduction 249
12.2 Liquid Sampling 250
12.3 Solid Sampling 254
12.4 Instrument Design 256
12.5 Qualitative and Quantitative Analysis 259
Experiments 265
Questions and Problems 267

13 NUCLEAR MAGNETIC RESONANCE SPECTROSCOPY AND MASS SPECTROMETRY 275
13.1 Nuclear Magnetic Resonance Spectroscopy 275
13.2 Mass Spectrometry 280
Questions and Problems 284

14 ATOMIC SPECTROSCOPY 289
14.1 Introduction 289
14.2 Atomization and Excitation 289
14.3 Flame Emission 294
14.4 Flame Atomic Absorption 297
14.5 Graphite Furnace Atomic Absorption 304
14.6 Vapor Generation Methods 306
14.7 Inductively Coupled Plasma 307
14.8 Other Atomic Emission Techniques 307
14.9 Summary of Atomic Techniques 309
Experiments 309
Questions and Problems 316

15 ANALYTICAL SEPARATIONS 321
15.1 Introduction 321
15.2 Recrystallization 322
15.3 Distillation 322
15.4 Liquid-Liquid Extraction 324
15.5 Liquid-Solid Extraction 329
15.6 Chromatography 330
15.7 Electrophoresis 341
Experiments 345
Questions and Problems 349

16 GAS CHROMATOGRAPHY 355
16.1 Introduction 355
16.2 Instrument Design 356
16.3 Sample Injection 358

16.4 Columns 360
16.5 Other Variable Parameters 363
16.6 The Chromatogram 366
16.7 Detectors 369
16.8 Qualitative Analysis 375
16.9 Quantitative Analysis 376
16.10 Troubleshooting 381
Experiments 383
Questions and Problems 387

17 HIGH-PERFORMANCE LIQUID CHROMATOGRAPHY 393
17.1 Introduction 393
17.2 Solvent Delivery 397
17.3 Sample Injection 398
17.4 Column Selection 399
17.5 The Chromatogram 402
17.6 Detectors 404
17.7 Qualitative and Quantitative Analysis 410
17.8 Troubleshooting 410
Experiments 412
Questions and Problems 416

18 ELECTROANALYTICAL METHODS 421
18.1 Introduction 421
18.2 Transfer of Electrons via Electrodes 422
18.3 Transfer Tendencies – Standard Reduction Potentials 423
18.4 Determination of Overall Redox Reaction Tendency: E^0_{cell} 426
18.5 The Nernst Equation 427
18.6 Potentiometry 429
18.7 Polarography and Voltammetry 438
Experiments 451
Questions and Problems 453

19 APPLICATIONS SUMMARY 459
19.1 Introduction 459
19.2 Wet Methods 460
19.3 Instrumental Methods 461
19.4 Reference Sources 464
19.5 Introduction to Automation 464
Questions and Problems 471

APPENDICES 475
Appendix 1. Significant Figures 475
Appendix 2. Table of Some K_{sp} Values 479
Appendix 3. Table of K_a and K_b Values 481
Appendix 4. Compositions of Some Concentrated Commercial Acids and Bases 483
Appendix 5. Summary of Calculations for Titrimetric Analysis 485
Appendix 6. Recipes for Selected Acid-Base Indicator Solutions 487

Appendix 7. Basic Electrical Terminology 489
Appendix 8. Common Metric Prefixes 493
Appendix 9. Answers to Selected Questions and Problems 495
Appendix 10. Reference Sources 523

INDEX 527

The Lord continues to bless me with many special gifts. The most special gift of all is my family. Once again I dedicate this book to my wife Lois and my daughters, Angie, Jeanie, and Laura. Thank you for your understanding and love. It means more to me than you know.

Chapter 1

Introduction to Chemical Analysis and Gravimetric Analysis

1.1 INTRODUCTION

A large percentage of the chemical laboratory work that exists in the modern world is "analytical"; that is, the accurate "chemical analysis" of some substance or material is directly involved. For example, an industry that manufactures a product for consumer use cannot hope to be in business long if its product does not pass the consumer's demand for quality. In many cases, a chemical analysis provides this needed certification of quality. As another example, consider the immense array of environments into which we human beings place our bodies just as a matter of daily routine. Air enters, interacts with, and leaves our lungs continuously every second of every day. Is not the chemical makeup of the air important to our health? The same perspective exists with the water we drink and with the food we eat. It is obvious that the reliable chemical analysis of environmental samples is important.

A third example is the pharmaceutical industry. Serving to improve the overall health of the world community and to eradicate sickness and disease, this industry is continuously seeking new ways to fight health problems through research and development, which frequently involves a chemical analysis of raw materials, manufactured products, clinical samples, etc. The list of examples is endless. The importance of a study of the techniques involved in a chemical analysis by those who hope to be involved in this work someday is obvious.

1.2 CHEMICAL ANALYSIS

Exactly what do we mean by a "chemical analysis"? The chemical analysis of a sample of matter is the determination of the chemical composition or chemical makeup of that sample. The chemical analysis of a sample of steel would be the determination of its carbon content, its iron content, its manganese content, or all of these substances taken collectively. The chemical analysis of a sample of water would be the determination of its mineral content (hardness), its pollutant content, or perhaps its dissolved oxygen content. The chemical analysis of a blood sample would involve determining its iron content, its alcohol content, or perhaps its drug content. Chemical analysis consists of a set of chemistry laboratory operations designed to reveal the chemical makeup of a material sample.

1.3 QUALITATIVE VS. QUANTITATIVE, WET VS. INSTRUMENTAL

Chemical analysis procedures are usually classified in two ways: first, in terms of the goal of the analysis, and second, in terms of the nature of the method used. In terms of the goal of the analysis, the classification is based on whether the analysis is "qualitative" or "quantitative". The analysis of a sample for *what* constituents are present – the identification of the sample components – is referred to as qualitative analysis. The analysis of a sample for *how much* of these constituents is present – the determination of quantity of sample components – is referred to as quantitative analysis (see Figure 1.1). The major emphasis of this text is on quantitative analysis, although some qualitative applications will be discussed for some techniques.

Analysis procedures can be additionally classified into those that are "wet" procedures and those that are "instrumental" procedures. Wet techniques are those that employ chemical reactions and/or classical reaction stoichiometry as the sole basis for arriving at the results. They are reputed to have a high degree of inherent accuracy, but suffer from disadvantages, the major ones being time required and the tedium involved. Instrumental techniques are appropriately described as "modern" techniques. They make use of the wonders of electronics and the wonders of high technology as an integral part of the technique. These add speed and frequently a degree of sophistication to the procedure, which can offer a much greater scope and practicality to the analysis. A disadvantage is that sometimes accuracy suffers in the process. Most of the time the advantages of an instrumental

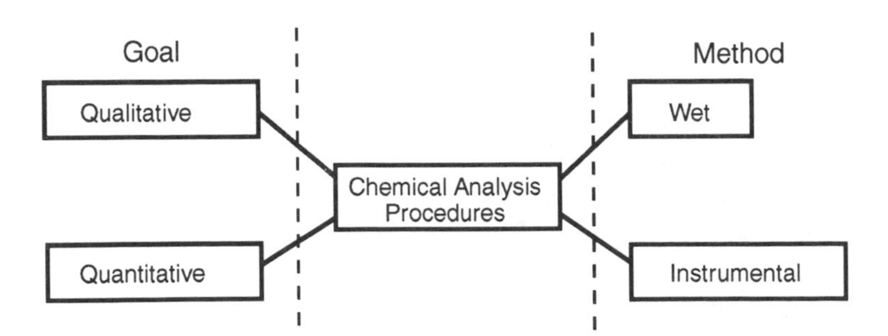

Figure 1.1. A depiction of the quantitative vs. qualitative and the wet vs. instrumental descriptions of chemical analysis.

technique outweigh this disadvantage. In addition, instrumental methods are generally used to determine the minor constituents, rather than the major constituents, of a sample. Approximately the first half of this text is concerned with wet methods, while the second half involves instrumental methods. A general discussion of instrumental methods is given in Chapter 9.

1.4 GENERAL LABORATORY CONSIDERATIONS

Because of the value and importance that are usually riding on the results of a chemical analysis, great care must be exercised in the lab when handling the sample and all peripheral materials. Contamination and/or loss of sample through avoidable accidental means cannot be tolerated. The results of a chemical analysis could affect such ominous decisions as the freedom or incarceration of a prisoner on trial, whether to proceed with a move that could mean the loss of a million dollars for an industrial company, or the life or death of a hospital patient.

Students should develop a kind of psychology for operating in a chemical analysis laboratory. They should always stop and think before proceeding with a new step in the procedure. What might happen in this step that would cause contamination or loss of the sample? An example would be when stirring a solution in a beaker with a stirring rod. Students might wish to remove the stirring rod from the beaker when going on to the next step. If they were to stop and think in advance, they would recognize that they need to rinse the rod into the beaker as they remove it. This would prevent the loss of that part of the solution adhering to the rod. Such a loss might result in a significant error in the determination.

In order to introduce and study the basic equipment and techniques involved in simple quantitative determinations, three experiments are presented at the end of this chapter which represent three classical examples of wet chemical, gravimetric type experiments – the analysis of a common hydrated salt for water content, the analysis of a solid, powdered sample for sulfate content, and the analysis of a powdered iron ore sample for iron.

1.5 THE LABORATORY NOTEBOOK

There are a few very simple rules to follow when keeping a laboratory notebook:

1. The notebook should be permanently bound, with a spiral binding or other suitable binding method. The reason for this is to avoid the possibility of loose pieces of paper which can become lost or misplaced.
2. Laboratory data should be recorded directly in the notebook as they are taken and before any calculations are performed. This gives the analyst the capability of checking calculations, in the event that an error is suspected, without having to search for "intermediate" recordings which may easily be misplaced.
3. Errors made in the data entry or in the experimental procedure should not prompt one to erase data already recorded in the notebook. Such data may prove to be valuable later. "Lining out" the data should

indicate that the data were perceived to be bad, but allows one to still see it.

4. The notebook should be organized. This includes a table of contents and numbered pages. A suggested format is as follows:

   (a) *Experiment name (and/or number) and the date it was begun.* It may also be appropriate to record dates in the data section if the data are recorded over a period of days.

   (b) *Objective.* The objective of the experiment – what it is one hopes to prove or determine – should be clearly but briefly stated.

   (c) *Data.* All data, raw or calculated, should be recorded here. This includes unknown numbers, weight and volume measurements, chart recordings and computer printouts, and measurements taken or derived from chart recordings and printouts. It is important for this section to be as organized as possible, given the fact that all data are recorded directly, with none erased. A well-prepared analyst is also well organized.

   (d) *Calculations and Results.* This section includes the calculations (or at least a sample) required to compute the results of an experiment as well as the results (answers to the analysis) themselves. Graphs and report forms are appropriately included here. If a number of similar results are to be reported, perhaps for a series of tests, it may be appropriate to list these in the form of a table.

   (e) *Conclusion.* All experiments should be drawn to a close in any notebook at the appropriate point. A simple statement indicating whether the objective has been achieved serves nicely for this. Basically, such a statement represents a "line of demarcation" between where one experiment ends and the next one begins.

Two sample pages from the notebook of a student performing Experiment 1 are shown in Figure 1.2. Notice that each page is signed with the student's name and also is dated. This is often required in industry for patent protection.

1.6 GRAVIMETRIC ANALYSIS

Introduction

The first of the two wet chemical methods we will consider is the gravimetric method of analysis. Gravimetric analysis is characterized by the fact that only one kind of measurement, that of weight, is made on the sample, its constituents, and reaction products. Only measurements of weight are thus used in the calculation. We will be concerned with the essential procedures involved in a gravimetric experiment and subsequent calculations. However, since the measurement of weight is such an integral part of this and indeed all chemical analyses, let us first describe the common weighing devices that may be encountered.

Weighing Devices

The most fundamental and possibly the most frequent measurement made in an analysis laboratory is that of weight (or mass). While we speak of

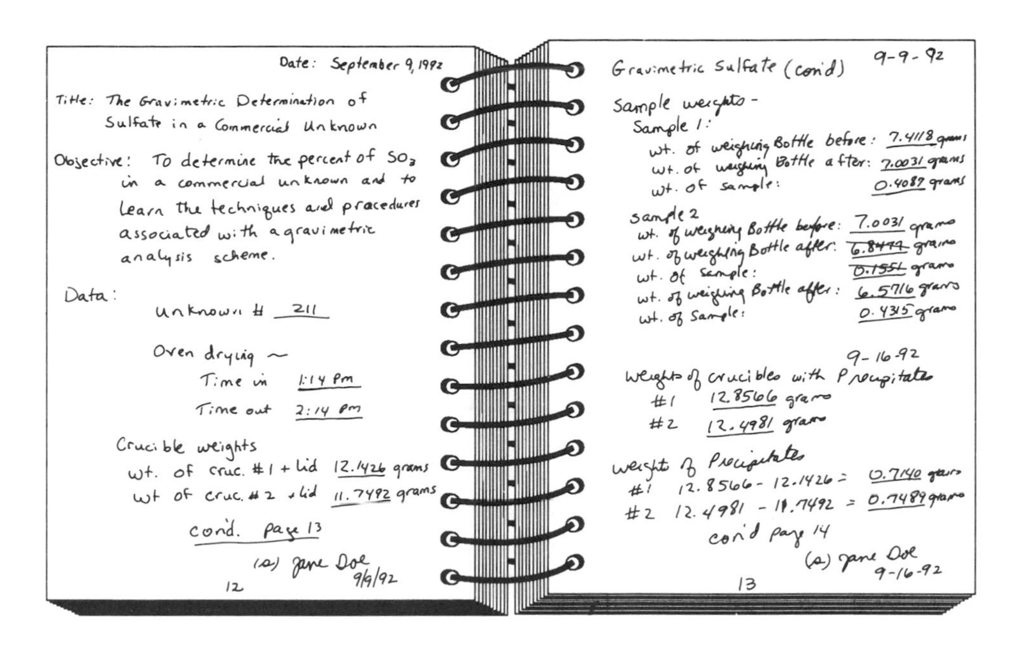

Figure 1.2. Sample pages from a laboratory notebook that a student is using for Experiment 2 in this text.

mass and weight often in the same breath, it is of some importance to recognize that they are not the same. Mass is the "quantity" or "amount" of a substance being measured. This quantity is the same no matter where the measurement is made – on the surface of the earth, in a spaceship speeding toward the moon, or on the surface of Mars. Weight is a measure of the earth's gravitational force exerted on a quantity of matter. Weight is one way to measure mass. In other words, we can measure the quantity of a substance by measuring the earth's gravitational effect on it. Since nearly 100% of all weight measurements made in any analysis laboratory are made on the surface of the earth, where the gravitational effect is nearly constant, weight has become the normal method of measuring mass. Thus, mass and weight have come to be interchangeable quantities even to the point of the unit involved. Weighing devices are calibrated in grams, which is defined as the basic unit of mass in the metric system.

The laboratory instrument built for measuring mass is called the "balance". A balance is a device in which the weight of an object is determined by balancing the object, usually across a knife-edge fulcrum with a series of known weights on the other side, as shown in Figure 1.3. Older balances formerly used in analytical laboratories closely resembled this basic design, having a pointer in the center to indicate when a balance between the two pans was achieved. The sum of the weights on the "known" pan was then calculated and the answer was taken to be the weight of the object.

The modern laboratory is of course equipped with balances which reflect the technological advances that have taken place over the years. Nearly all modern laboratory balances are of the single-pan variety. Older single-pan balances based on the principle of the double pan described above are still found in some laboratories. The basic principle of such a balance is shown in Figure 1.4. In this design, a permanent, constant counterbalancing weight

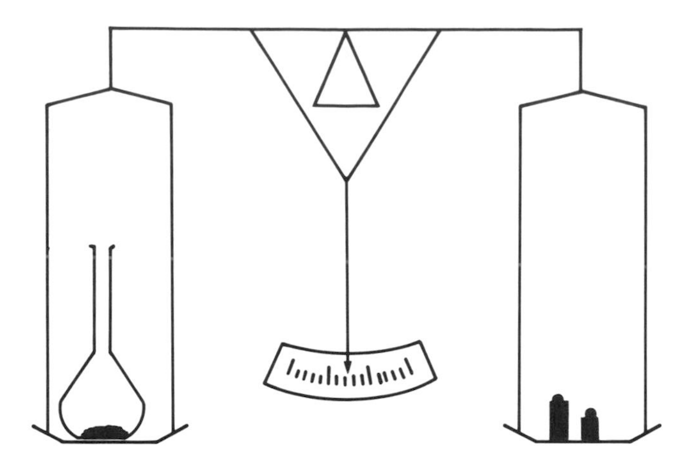

Figure 1.3. The basic concept of the device known as "balance". See text for a description. (From Kenkel, J., *Analytical Chemistry Refresher Manual*, Lewis Publishers, Chelsea, MI, 1992. With permission.)

is in place on a fulcrum across from the pan. The object to be weighed is placed on the pan and, along with a series of removable weights on the same side, is made to balance the constant weight on the other side. When a heavy object is placed on the pan, some of the removable weights are lifted via an external control so as to achieve a balance with the counterbalancing weight. We will discuss these in more detail in the next subsection.

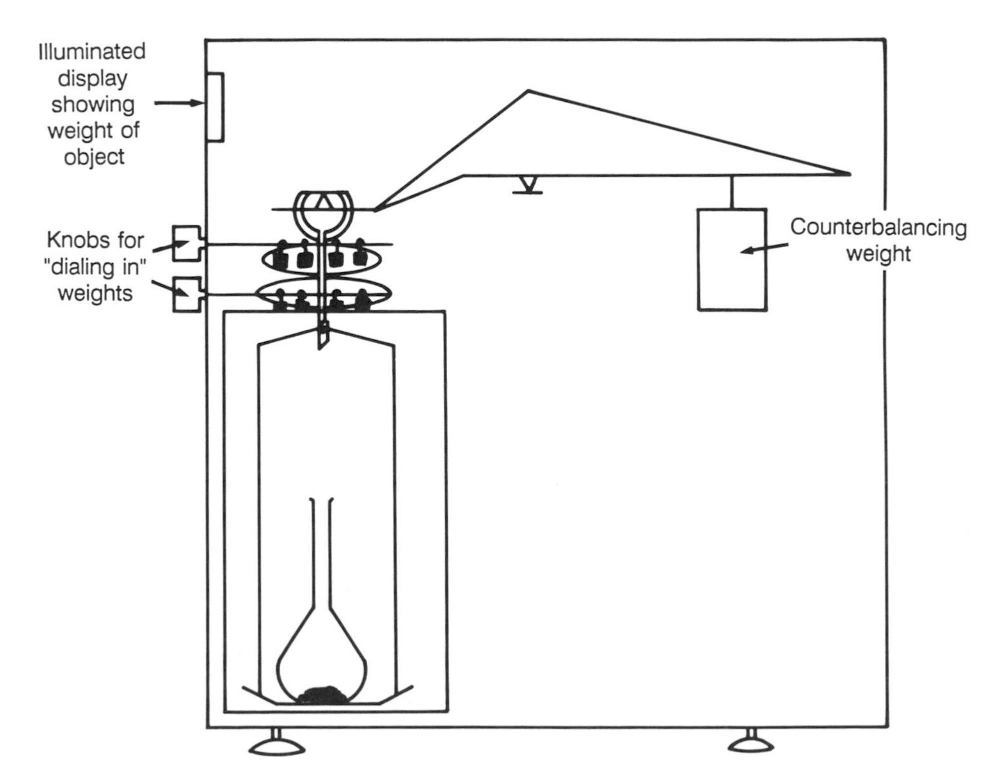

Figure 1.4. A diagram of a single-pan balance. (From Kenkel, J., *Analytical Chemistry Refresher Manual*, Lewis Publishers, Chelsea, MI, 1992. With permission.)

Different examples of laboratory work require different degrees of accuracy. Thus, there are a variety of balance designs available which reflect this need for varying accuracy. Some balances are accurate to the nearest gram, some are accurate to the nearest tenth or hundredth of a gram, some are accurate to the nearest milligram, and still others are accurate to the nearest tenth and hundredth of a milligram. We now proceed to describe some of the more popular designs, and our discussions will begin with a shorter treatment of some of the less accurate balances and end with a detailed treatment of the important electronic analytical balances, which are very accurate measuring devices.

Figure 1.5 depicts two common balances that can be described as "less accurate" or "auxiliary". These are used when the weight being measured does not necessarily need to be accurate. The multiple-beam balance (Figure 1.5A) is accurate to the nearest 0.01 g and resembles a balance commonly found in a doctor's office to measure a person's weight. The basic design here is more like the double-pan variety described briefly above in the sense that weights are added on a counterbalancing arm to balance the object on the pan, rather than removing weights from the same side as the pan as described above.

Electronic top-loading balances (Figure 1.5B) have variable accuracies depending on the model purchased. The one illustrated is an Ohaus Precision Standard balance, which is accurate to the nearest 0.01 g. Other such balances are accurate to the nearest 0.1 and 0.001 g. The electronic top-loaders often have a "tare" feature. A chemical can be conveniently weighed on a piece of weighing paper, for example, without having to determine the weight of the paper. "Taring" means that the balance is simply zeroed with the weighing paper on the pan. These balances are not based on balance/counterbalance, but rather utilize a torsion mechanism.

Analytical Balances

The balances that are used to achieve the highest degree of accuracy in the analytical laboratory are called "analytical" balances and are accurate to either 0.1 mg or 0.01 mg. The modern laboratory utilizes single-pan analytical balances almost exclusively for such accuracy. Such balances can be either of the balance/counterbalance variety, or the torsion variety.

Figure 1.6 shows a typical modern analytical balance. Notice that it is a single-pan balance with the pan enclosed. The chamber housing the pan has transparent walls for easy viewing. Sliding doors on the right and left sides make the pan accessible for loading and handling samples. The actual weight measurement is displayed via a digital readout on the face of the instrument. The counterbalancing mechanism, if it has one, is hidden in the upper portion of the instrument and is usually accessible by removing the cover.

The design of the readout and the exact weighing procedure vary with the age of the balance and the manufacturer. The specific techniques, such as the use of a "pan arrest" lever for obtaining the readout, are easy to learn, however, through demonstration and practice and therefore will not be discussed in detail here.

A

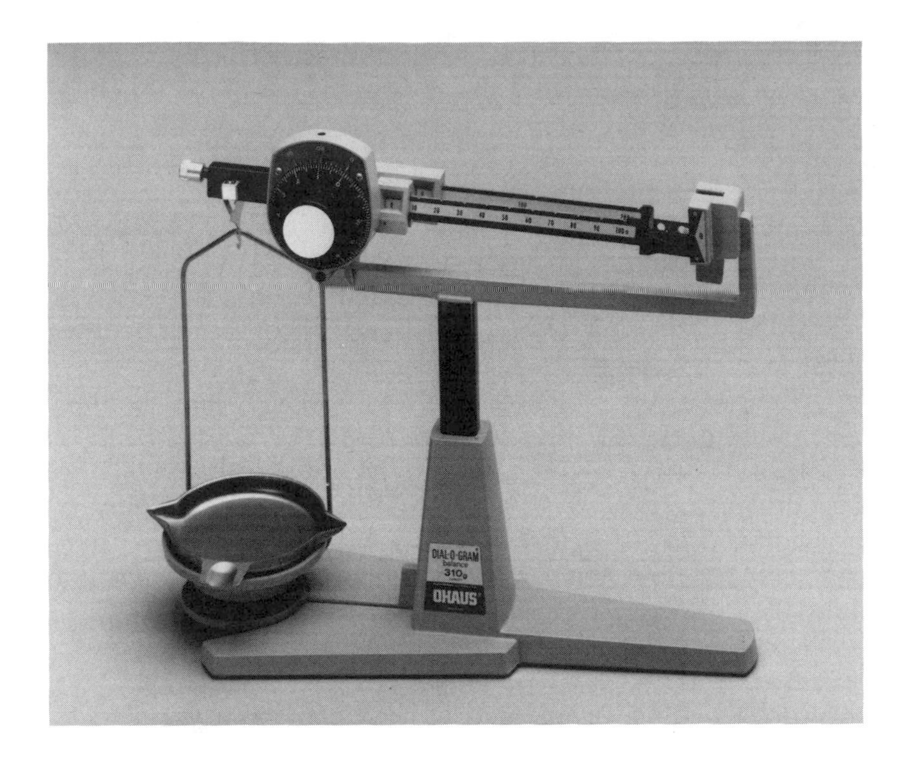

B

Figure 1.5. (A) A multiple-beam balance. (B) An electronic top-loading balance. (Courtesy of Ohaus Corporation, Florham Park, NJ.)

The step-by-step instructions which follow are written to be as generic as possible. The most important point is that the operator keep in mind that the analytical balance is extremely sensitive and extremely accurate and therefore must be handled carefully and correctly. Also, this discussion presumes that, if necessary, the sample has been dried (such as in an oven)

Figure 1.6. A typical modern analytical balance. (Courtesy of Ohaus Corporation, Florham Park, NJ.)

and has been kept dry (by storing in a desiccator—see Experiment 1) prior to weighing.

STEP 1: Some preliminary considerations are important. First, the sample to be weighed must be at room temperature. If it is not at room temperature, it must first be allowed to cool. The reason for this is that a warm or hot object placed on the pan in the enclosed chamber can create air currents that can buoy up the pan and cause an erroneous weight reading. Second, whenever a weight is being read on the readout the sliding doors must be closed. Again, this is to avoid the effect of air currents on the weight measurement. Third, the pan and the floor of the chamber must be swept free of chemical debris from prior spills. A camel-hair brush should be kept near the balance for this purpose. Fourth, care should be taken to avoid bumping the balance or its support while the measurement is taken. This would almost always cause the reading to change or vibrate and could also be damaging to the internal instrument parts. Fifth, the balance should be level. With most balances, a leveling method is built in, complete with a leveling bubble and vertically positionable legs.

STEP 2: If the absolute weight of an object on the pan is needed, such as a crucible weight, the balance must be zeroed. This is a calibration step in which, when there is nothing on the pan, the readout is made to read zero. All designs of analytical balances have a "zeroing control" for this purpose. With nothing on the pan, the zeroing control is adjusted such that the readout reads zero even to the fourth or fifth decimal place. If the absolute weight of an object on the pan is not needed, such as when weighing by difference or in taring, the zeroing step is not absolutely necessary. Zeroing (and taring) of modern torsion-style analytical balances, however, is as easy as pressing a button.

STEP 3: The object placed on the pan must be kept free of fingerprints or other interfering substance that could add weight and give an erroneous result. The use of gloves or finger cots is recommended. When weighing an object to the nearest tenth of a milligram, seemingly insignificant fingerprints can cause a significant error. Tongs can also be helpful here.

STEP 4: The object is placed on the pan and the weight obtained. As indicated earlier, balances vary in the mechanism for doing this. Some older balances have a "preweigh" mechanism. In addition, older balances require pan arrest while dialing in weights or when adding or removing objects from the pan. These require the operator to "dial in" the weights so as to achieve a null of some type. Modern torsion-type balances have a digital readout that automatically displays the weight without "dialing it in".

STEP 5: When finished, dial all weight controls to zero if necessary, clean up any spills, arrest the pan if required, and turn off any switches.

When To Be Accurate And When Not

The question of which balance to use for a given procedure, the less accurate multiple-beam or top-loader or the more accurate analytical, remains. An analyst must be able to recognize when a high degree of accuracy is important and when it isn't so that the analytical balance is used when accuracy is important and the less accurate balances are used when accuracy is not important.

First, analytical balances should *not* be used for any weight measurement when the overall objective is strictly qualitative or when quantitative results are to be reported to two significant figures or less. Whether the weight measurements are for preparing solutions to be used in such a procedure or for obtaining an appropriate weight of sample for such an analysis, etc., they need not be accurate since the outcome will be either only qualitative or not necessarily accurate.

Second, if the results of a quantitative analysis are to be reported to three or more significant figures, then weight measurements that enter directly into the calculation of the results should be made on an analytical balance so that the accuracy of the analysis is not diminished when the calculation is performed.

Third, if a weight is only incidental to the overall result of an accurate quantitative analysis, then it need not be accurate. This means that if the weight measurement to be performed has no bearing whatsoever on the quantity of the analyte tested, but is only needed to support the chemistry or some other factor of the experiment, then it need not be accurate.

Fourth, if a weight measurement does directly affect the numerical result of an accurate quantitative analysis (in a way other than entering directly into the calculation of the results), then it must be performed on an analytical balance.

These last two points require that the analyst carefully consider the purpose of the weight measurement and whether it will directly affect the quantity of analyte tested, the numerical value to be reported. It is often true that weight measurements taken during such a procedure need not be accurate even though the results are to be accurate.

1.7 ANALYTICAL CALCULATIONS

In any quantitative chemical analysis procedure, the final desired result is the amount of the constituent of interest present in the sample. This can be expressed in the form of percent (parts per hundred), parts per thousand (ppt), parts per million (ppm) or some other concentration unit. In this section, we shall deal with percent—the others will be discussed in later chapters.

The weight percent of a constituent in a sample is calculated using the definition of weight percent:

$$\text{weight \%} = \frac{\text{weight of part}}{\text{weight of whole}} \times 100 \tag{1.1}$$

$$\text{\% constituent} = \frac{\text{weight of constituent ("C")}}{\text{weight of sample ("S")}} \times 100 \tag{1.2}$$

The two variables in Equation 1.2 are the weight of the constituent (C) found in the sample and the weight of the sample (S) used. Usually, the value of "S" is directly measured in the procedure—it is usually the first measurement made. For example, in the measurement of the percent of moisture in a soil sample, the weight of the soil is first determined on a laboratory balance before most other laboratory operations are performed. Thus, "S" is transferred directly from the laboratory notebook to the calculation.

"C" is a different story. Since the constituent to be determined is physically mixed and/or combined chemically with at least one other constituent in the sample, its weight cannot be determined as readily. A separation procedure must first be executed in order to directly weigh this constituent so that we have a value for "C" that we can plug into the equation for percent. In some determinations this separation is easy to accomplish, in others it is not.

Consider again the analysis of the soil sample for moisture. The moisture is easily separated from the other constituents in the soil by simply heating the weighed sample in a drying oven. If the temperature of that oven is carefully controlled and there is plenty of air circulation, only the water (moisture) in the sample will be appreciably removed—it will evaporate. Following this drying step, the soil sample is cooled in a desiccator and then weighed again. The amount of weight lost in the drying process is the weight of the water contained in the original sample. "C" is thus determined in this example by subtracting the dry weight from the original weight.

$$\text{\% water} = \frac{\text{"S"} - \text{dried weight}}{\text{"S"}} \times 100 \tag{1.3}$$

Example 1

What is the percent of water in a sample that weighed 4.5027 g before drying and 3.0381 g after drying?

Solution

Utilizing Equation 1.3, we have

$$\% \text{ water} = \frac{(4.5027 - 3.0381)}{4.5027} \times 100$$

$$\% \text{ water} = 32.527\%$$

Example 2

A sample of grain is heated to determine the percent of volatile organics. The following data were obtained:

Weight of evaporating dish with sample .... 32.8201 g
Weight of evaporating dish after heating .... 28.9840 g
Weight of empty evaporating dish .... 16.1271 g

What is the percent of volatile organics in the grain?

Solution

Again utilizing Equation 1.3, we have

$$\% \text{ volatile organics} = \frac{(32.8201 - 28.9840)}{(32.8201 - 16.1271)} \times 100$$

$$= 22.980\%$$

1.8 THE NEED FOR STOICHIOMETRY

Of course, "C" (the numerator) frequently cannot be determined in such a straightforward fashion as in the above examples. Many times a separation of the constituent from the sample is difficult, if not impossible, to accomplish by physical means, such as evaporating, selective dissolving, etc. In that case, a gravimetric procedure still might be used if a chemical reaction is employed to convert the constituent to another chemical form that is both able to be separated cleanly and able to be weighed accurately. The weight of the reaction product, if it is a precipitate that has been separated and dried, is then determined and converted back to the weight of the constituent via a stoichiometry calculation.

In a stoichiometry calculation, the weight of one substance involved in a chemical reaction (reactant or product) is converted to the weight of another substance (reactant or product) appearing in the same reaction by making use of the balanced equation. The formula weight of the reactant and product involved are needed.* In the following general example

*The term "formula weight" will be used in this text rather than "molecular weight" in order to accurately represent all compounds whether ionic or molecular.

$$aA + bB \longrightarrow cC + dD$$

if the weight of "D" is known and the weight of "A" is needed, this weight of "A" can be calculated using a stoichiometry calculation of the form

$$\text{weight of "A"} = \text{weight of "D"} \times \frac{1}{FW_D} \times \frac{\text{a moles of "A"}}{\text{d moles of "D"}} \times FW_A \tag{1.4}$$

in which "FW" represents "formula weight".

Dimensional analysis shows that the units in Equation 1.4 cancel appropriately:

$$\text{grams of "D"} \times \frac{\text{moles of "D"}}{\text{grams of "D"}} \times \frac{\text{moles of "A"}}{\text{moles of "D"}} \times \frac{\text{grams of "A"}}{\text{moles of "A"}} = \text{grams of "A"} \tag{1.5}$$

A stoichiometry calculation is thus essentially a three-step procedure in which (1) the weight of "D" is divided by its formula weight to get moles of "D"; (2) the moles of "D" are converted to moles of "A" by multiplying by the mole ratio a/d as found in the chemical equation; and (3) the moles of "A" are converted to grams of A by multiplying by the formula weight of "A".*

1.9 GRAVIMETRIC FACTORS

The calculation process just described can be simplified by observing that the three steps can be combined as follows:

$$\text{weight of "A"} = \text{weight of "D"} \times \frac{FW_A}{FW_D} \times \text{mole ratio } \frac{a}{d} \tag{1.6}$$

A further simplification involving the mole ratio can be accomplished by noting that such a ratio can usually be determined independent of the balanced chemical equation by observing the formulas of the two substances involved and simply extracting the ratio from the formulas. Consider the following reaction:

$$AgNO_3 + NaCl \rightarrow AgCl + NaNO_3$$

This equation represents a gravimetric analysis in which the percent of NaCl in a sample is determined by reacting the NaCl with $AgNO_3$ to form the insoluble precipitate AgCl. The normal stoichiometry calculation would be

$$\text{weight NaCl} = \text{weight AgCl (g)} \times \frac{\text{1 mol of AgCl}}{\text{143.32 g AgCl}} \times \frac{\text{1 mol of NaCl}}{\text{1 mol of AgCl}} \times \frac{\text{58.43 g NaCl}}{\text{1 mol NaCl}} \tag{1.7}$$

Let us simplify this calculation as indicated above (Equation 1.6). In this simplification let us use the molecular formulas of the substance involved to symbolize their formula weights. Thus we have

$$\text{weight of NaCl} = \text{weight of AgCl} \times \frac{NaCl}{AgCl} \times \frac{1}{1} \tag{1.8}$$

The ratio 1:1 is "Step 2" from Equation 1.4 and was obviously extracted from the chemical equation. Now examine the formulas of the two

*Students should refer to any introductory chemistry text for a more detailed discussion and worked examples.

substances involved, AgCl and NaCl. They have one element in common, Cl. There is one Cl in NaCl and one Cl in AgCl – a 1:1 ratio. The same ratio is thus conveniently found by examining the formulas of the substances involved. Consider an example in which we do not have a 1:1 ratio:

$$Pb(NO_3)_2 + 2NaCl \rightarrow PbCl_2 + 2NaNO_3$$

Equation 1.6 becomes

$$\text{weight of NaCl} = \text{weight of } PbCl_2 \times \frac{NaCl}{PbCl_2} \times \frac{2}{1} \qquad (1.9)$$

The 2:1 ratio is obviously derived both by balancing the chlorides in the formulas and by looking at the equation. It is apparent we do not need to extract the ratio from the chemical equation. We can derive the same end result by examining the formulas of the substances used in the calculations – the substance whose weight is known (hereafter referred to as "substance known") and the substance whose weight is needed (hereafter referred to as "substance sought"). With the ratio of formula weights symbolized by the ratio of molecular formulas, all one needs to do to determine the mole ratio is to "balance" the common element in the two formulas with the use of a coefficient or coefficients in the numerator or denominator or both, as required. The conversion of weight of substance known to weight of substance sought thus utilizes a ratio of formula weights of the substance sought to the substance known multiplied by a ratio of numbers which balances the common element. This total conversion factor is what is known as the "gravimetric factor". The following examples illustrate the concept:

Example A

substance sought	Cl
substance known	AgCl
gravimetric factor	Cl/AgCl = 35.453/143.32 = 0.24737

Example B

substance sought	Cl
substance known	$PbCl_2$
gravimetric factor	$2Cl/PbCl_2$ = 2(35.453)/278.1 = 0.2550

Example C

substance sought	Cl_2
substance known	$AlCl_3$
gravimetric factor	$3Cl_2/2AlCl_3$ = 3(70.906)/2(133.34) = 0.79765

In cases in which there are two common elements, one of which is oxygen, the oxygen is ignored and the other element is balanced.

Example D

substance sought	SO_3
substance known	$BaSO_4$
gravimetric factor	$SO_3/BaSO_4 = 80.06/233.40 = 0.3430$

The equation for the weight of the substance sought thus becomes

$$\text{weight of substance sought} = \text{weight of substance known} \times \text{g.f.} \qquad (1.10)$$

where g.f. is the gravimetric factor. We can also write the equation for the percentage of a constituent as follows:

$$\% \text{ constituent} = \frac{\text{weight of precipitate} \times \text{g.f.}}{\text{weight of sample}} \times 100 \qquad (1.11)$$

Example 3

A sample that weighed 0.8112 g is analyzed for phosphorus (P) content by precipitating the phosphorus as $Mg_2P_2O_7$. If the precipitate weighs 0.5261 g, what is the percent of P in the sample?

Solution

$$\% \text{ P} = \frac{\text{weight of precip.} \times 2P/Mg_2P_2O_7}{\text{weight of sample}} \times 100 = \frac{0.5261 \times 0.2783}{0.8112} \times 100 = 18.05\%$$

The use of gravimetric factors is not limited to gravimetric analysis situations. As we shall observe in Chapter 7, any time it is necessary to convert the weight of one chemical to the weight of another, a ratio of the formula weights with the accompanying balancing ratio is the conversion factor to be used. Gravimetric factors for a wide variety of such conversions are well cataloged in chemistry handbooks, such as the *CRC Handbook of Chemistry and Physics* (see Table 1.1).

EXPERIMENTS

EXPERIMENT 1: THE PERCENT OF WATER IN HYDRATED BARIUM CHLORIDE

Introduction

The water that is trapped within the crystal structure of some ionic compounds (the water of "hydration") can be removed easily by heating. The amount of the water present in a given sample can be determined by weighing the sample before and after this heating. The weight loss that occurs is the weight of the water in the sample. It is a simple matter to then calculate the percent of water in the hydrate.

In this experiment, the percent of water in $BaCl_2 \cdot 2H_2O$ will be determined. The number one priority will be *accuracy*. This means that you will be using an analytical balance for the weighing and taking pains to be sure that there are no experimental errors that will interfere with the intended accuracy. You will do the experiment in duplicate (run two determinations at the same time) and report two answers. Both will be graded according to how close you come to the right answer.

Table 1.1. A number of gravimetric factors for certain substances sought, given certain substances weighed.

Weighed	Sought	Factor
$FeCO_3$	CO_2	0.37986
	FeO	0.62013
	Fe_2O_3	0.68913
$Fe(HCO_3)_2$	CO_2	0.49482
	Fe	0.31396
	FeO	0.40390
	Fe_2O_3	0.44887
FeO	CO_2	0.61256
	Fe	0.77732
	$FeCO_3$	1.61256
	$Fe(HCO_3)_2$	2.47588
	Fe_2O_3	1.11134
	$FePO_4$	2.09918
	FeS	1.22360
	SO_3	1.11436
Fe_2O_3	Fe	0.69943
	$FeCl_3$	2.03149
	$FeCO_3$	1.45099
	$Fe(HCO_3)_2$	2.22781
	$Fe(HCO_3)_3$	2.99200
	FeO	0.89981
	Fe_3O_4	0.96660
	$FePO_4$	1.88886
	FeS	1.10101
	$FeSO_4$	1.90252
	$FeSO_4 \cdot 7H_2O$	3.48190
	$Fe_2(SO_4)_3$	2.50406
	$FeSO_4\ (NH_4)_2SO_4 \cdot 6H_2O$	4.91040

From David L. Lide, Ed., *CRC Handbook of Chemistry and Physics*, 60th ed., CRC Press, Boca Raton, FL, 1980. With permission.

The specific objective of the experiment is to accurately determine the percent of water in the sample. The general objectives, which can be considered more important, are (1) to become accustomed to the kind of careful technique required for accurate analytical work and (2) to become familiar with a simple form of "gravimetric" analysis.

One problem encountered in this procedure is that a small amount of the $BaCl_2 \cdot 2H_2O$ may be evaporated along with the water when the sample is heated as just described. This would obviously cause an error because the weight loss would include more than just the water. To ensure that this doesn't happen, the sample will not be heated directly with the flame of a burner, but rather the radiant heat obtained from a flame-heated ceramic-centered wire gauze will be used.

Procedure

NOTE: Safety glasses are required.

1. Enough hydrated barium chloride should be dried ahead of time to get rid of adsorbed moisture due to room humidity (not the water of

hydration). This can be accomplished by storing overnight in a desiccator. The drying time, however, should not exceed about 12–15 hr because the water of hydration also may begin to be removed by the desiccant.

NOTE: A desiccator is a storage container used to either dry samples such as this barium chloride or, more commonly, to keep samples and crucibles dry and protected from the laboratory environment once they have been dried by other means. A typical laboratory desiccator is shown in Figure 1.7; however, any container that can be sealed, such as an ordinary coffee can, can be used. A quantity of water-absorbing chemical, called the desiccant, is placed in the bottom of the container and will absorb all the moisture inside the sealed vessel, thus providing a dry environment. A good commercially available desiccant is called Drierite® (anhydrous $CaSO_4$). This substance can be purchased in the "indicating" form, in which case a color change (from blue to pink) will be observed when the material is saturated with moisture. If the analyst suspects a desiccant to be saturated, it should be replaced. Drierite can be recharged by heating in a vacuum oven.

2. Set up two ring stands, each with one ring and one clay triangle, and place a Bunsen burner under each ring. Place a porcelain crucible with lid (with both crucible and lid premarked as #1 and #2 with a high-temperature marker) in each clay triangle and adjust the height of the rings so as to obtain the maximum heat with the burners at the bottom of the crucibles.
3. Heat the crucibles, with lids ajar, directly with the Bunsen burners for at least 20 min. During this time, the bottoms of the crucibles should get hot enough so as to glow with a dull red color. The purpose of this step is to remove any volatile materials (such as fingerprints) from the

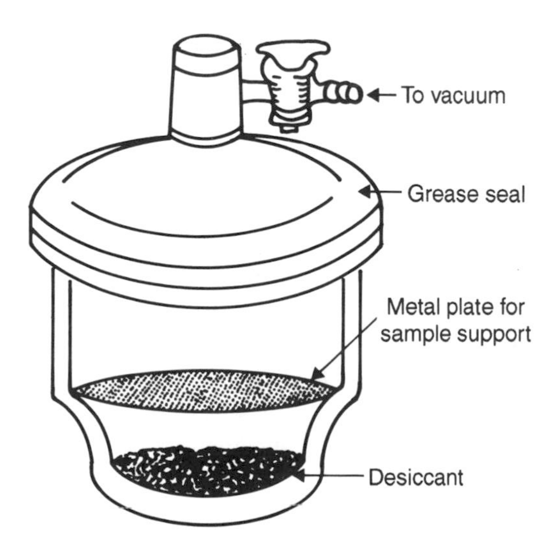

Figure 1.7. A drawing of a typical laboratory desiccant.

crucibles and lids in advance so that they will not be removed later and be included with the weight loss of the hydrate. At the end of the 20 min, allow the crucibles to cool in place for 5 min and then in your desiccator for an additional 15 min.

4. The next step is to weigh the samples of $BaCl_2 \cdot 2H_2O$ hydrate into the crucibles. Using tongs to handle the crucibles, weigh the first crucible, with lid, on the analytical balance and record the weight in your notebook. (Refer to Section 1.6 for suggestions concerning the notebook.) Then, with a spatula, add enough of the hydrate to the crucible so as to increase the weight by about 1 g. Also record this weight in your notebook. (Note – If your balance has a "tare" feature, you can use it, but be sure to have the weight of the empty crucible recorded.) Do the same with the second crucible. At this point, you can touch the crucibles with your fingers. Carry them back to your workspace.
5. Set up two ring stands with two rings and two Bunsen burners. In each setup, the lower ring should be 9 cm above the top of the burner and the upper ring should be placed 9 cm above the lower. Use a metric ruler for these measurements. Place a ceramic-centered wire gauze on each of the lower rings and clay triangles on the upper rings. Place the crucibles, with lids ajar, in the clay triangles. Proceed to heat the wire gauze with the full heat of the Bunsen burners for 30 min. The flame should be a tall, hot flame that covers the entire ceramic center of the gauze on the bottom side.
6. At the end of the 30 min, allow the crucibles to cool in place for 1 min and then in your desiccator for an additional 15 min. Handle them with tongs until after you've weighed them. Then weigh each, with lids, on the analytical balance. Record the weights in your notebook.
7. Calculate the percent of water in the hydrate for each sample and report to your instructor.

EXPERIMENT 2: THE GRAVIMETRIC DETERMINATION OF SULFATE IN A COMMERCIAL UNKNOWN

Introduction

This experiment represents an example of a gravimetric analysis based on stoichiometry rather than a simple physical separation, such as heating. A sample containing a soluble sulfate compound, such as Na_2SO_4 or K_2SO_4, is dissolved and the sulfate precipitated with $BaCl_2$ as $BaSO_4$. The percent of SO_3 is determined by a stoichiometry calculation as discussed in the text. Read the entire experiment completely before beginning the lab work. Safety glasses are required. The procedure calls for you to run the experiment in duplicate. Your instructor may want you to run it in triplicate.

Procedure

NOTE: Safety glasses are required.

Session 1

1. Obtain a sample of unknown sulfate from your instructor. Record the unknown number in your notebook.
2. Place the sample in a clean weighing bottle (Figure 1.8). Place the weighing bottle, with its lid ajar, in a small beaker, and place a watch glass on the beaker (Figure 1.9). Label the beaker with your initials or name. The beaker with its contents should then be placed in the drying oven and dried for at least 1 hr at 110°C.
3. While waiting for your sample to dry, prepare two porcelain crucibles by heating with Meker burners for 0.5 hr. This can be done with the use of two ring stands and rings with clay triangles to support the crucibles. The lids should be ajar during the heating and the tops of the burners should be about 0.5 in. from the bottom of the crucibles. Both the crucibles and lids should be labeled with heat-stable labels prior to heating. If only one Meker burner is available per student, use a Bunsen burner for one of the crucibles and alternate burners after 15 min so that each crucible is heated for 15 min with each burner.
4. After the 0.5 hr has expired, turn off the burners and allow the crucibles and lids to cool in place for 5 min. Then, using crucible tongs, transfer them to a desiccator (see Experiment 1, Step 1, for discussion) to allow them to cool to room temperature. The cooling should take approximately 15–20 min.
5. Continuing to handle the crucibles and lids with tongs, weigh each crucible with its lid accurately on an analytical balance and record the

Figure 1.8. Some examples of laboratory weighing bottles.

Figure 1.9. The correct configuration for drying the sample.

weights in your notebook. All measurements of weight should be made as accurately as possible; i.e., they should be made on an analytical balance. See Appendix 1 for a discussion of significant figure manipulation in measurements and calculations. Take special precautions to keep the original lids with the crucibles, since the lids and crucibles will be weighed together again later. Store in your desiccator until Session 2.

6. Remove the unknown from the oven, place the lid on the weighing bottle and place the weighing bottle in the desiccator to cool for 10 min. Thoroughly clean two 400-mL beakers. Weigh accurately by difference two samples of your unknowns, each weighing around 0.4 g, into the two 400-mL beakers labeled, for example, #1 and #2.

NOTE: Weighing a sample of the unknown by difference refers to first weighing the bottle containing the sample, then shaking out a portion of the contents into the beaker and weighing again. The difference in the two weights is the weight of the sample in the beaker. Care should obviously be taken not to get any of the sample anywhere but in the beaker. If such an error occurs, the sample in the beaker should be discarded and a new sample weighed (by difference). The advantage of weighing by difference is that no intermediate material contacts the sample between the weighing bottle and the beaker. Such materials as a spatula and/or weighing paper could entrap some of the sample so that not all of the weighed sample is contained in the beaker, thus creating an error. One potential minor problem with weighing by difference is that fingerprints deposited on the weighing bottle during the transfer may add sufficient weight to the bottle so as to give an incorrect weight for the second weighing. This can be eliminated by handling the weighing bottle with a rolled-up paper towel (Figure 1.10) so that the fingers don't contact the bottle between the two weighings. All of this reflects the kind of accuracy one frequently strives for. Given the use of the analytical balance and given the need for paying attention to significant figure rules (Appendix 1), seemingly insignificant things like fingerprints become likely sources of errors. The weights can be recorded in the DATA section of your notebook as follows:

Sample 1	Weight of weighing bottle before:	___ g
	Weight of weighing bottle after:	___ g
Sample 2	Weight of weighing bottle before:	___ g
	Weight of weighing bottle after:	___ g

Store the two beakers and the desiccator until Session 2, or proceed to Session 2 at the discretion of your instructor.

Session 2

7. Add to each of the two beakers 125 mL of water and 3 mL of concentrated HCl. Obtain two clean glass stirring rods and place one in each beaker. Stir the mixtures until the unknowns have dissolved. Leaving the stirring rods in the beakers, place watch glasses on the beakers and place both beakers on a hot plate. Place alongside these a third beaker containing about 130 mL of barium chloride stock solution

Figure 1.10. The use of rolled-up paper towel to handle a weighing bottle and avoid getting fingerprints on it.

(26 g/L concentration) prepared by your instructor. Place a watch glass on this beaker also.

8. Carefully heat all solutions to boiling, then remove from the hot plate and, while stirring, add 55 mL (use a 100-mL graduated cylinder) of the barium chloride to each of the unknowns. Use beaker tongs for handling hot beakers. The fine, white precipitate, barium sulfate, forms almost immediately. The amount of barium chloride used is in excess of the amount required. This is to ensure complete precipitation of the sulfate in your unknown. Continue to stir both solutions for at least 20 sec. Again, leave the stirring rods in the beakers.

9. Adjust the setting on the hot plate so that the two remaining solutions remain very hot (but not quite boiling). If either sample boils, cool it for a moment and place it back on the hot plate. Allow the precipitate to "digest" for 1 hr. Digestion refers to a procedure in which a sample is heated or stored for a period of time to allow a particular chemical or physical process to occur. In this case, a digestion step is required in order to change the small, finely divided particles of the barium sulfate precipitate into larger, filterable particles. Heating for a period of 1 hr accomplishes this goal.

   During this period, you can proceed to gather and assemble the equipment needed for the remainder of the experiment and prepare your notebook for the data to be recorded.

10. Assemble a filtration system consisting of a ring stand, a funnel rack (or two small rings), two clean long-stem funnels, and two beakers or flasks to catch the filtrate (Figure 1.11). Obtain two pieces of Whatman #40 filter paper, fold, and place into the funnels, moistening with a little distilled water so they adhere to the walls of the funnels (Figure 1.12). Label the funnels to correspond with the labels on your beakers.

11. After the 1-hr digestion period has expired, remove the beakers from the hot plate, being careful not to disturb the precipitate which has settled to the bottom. The filtration step will be completed in less time if the supernatant liquid is filtered first, followed by the transfer of the precipitate. For each solution, using the stirring rod as a guide for the solution (to avoid splashing), transfer the supernatant liquid to the funnel and allow it to filter. Be careful to transfer only solution #1 into funnel #1, etc. Then transfer the precipitate carefully, using the last 20 mL of supernatant to stir up the settled particles while transferring as with the supernatant.

    Add about 20 mL of distilled water and repeat. Repeat with one 10-mL portion and complete the transfer with the aid of a rubber policeman. Scrub the entire interior of the beaker with the policeman until no more white particles are visible. Rinse the policeman into the funnel with a stream of distilled water.

12. Rinse the precipitate with several portions of distilled water. To test for sufficient rinsing (elimination of all residual chloride from the ex-

Figure 1.11. The experimental setup for filtering the precipitates.

cess barium chloride, the HCl, and possibly from the sample) collect some of the most recent washings in a test tube and add two drops of silver nitrate solution. If a white precipitate (cloudy appearance) forms, more rinsing is required before proceeding to Step 13. Continue to rinse and test until the rinsings are clean.

13. After allowing the last rinsings to completely drain from the filter, carefully remove the filter paper from the funnel, fold, and place into the crucible with the same label. Place the lids on the crucibles. At this point, you will need a minimum of 60 min to complete the experiment. If time does not allow the completion of the experiment in the allotted time, store your crucibles outside your desiccator until the next lab period.

Session 3

14. Set up your ring stands with burners as before. Place the crucibles, with lids ajar, on the clay triangles in the rings and heat slowly with a Bunsen burner to evaporate the water in the crucibles. Do not allow

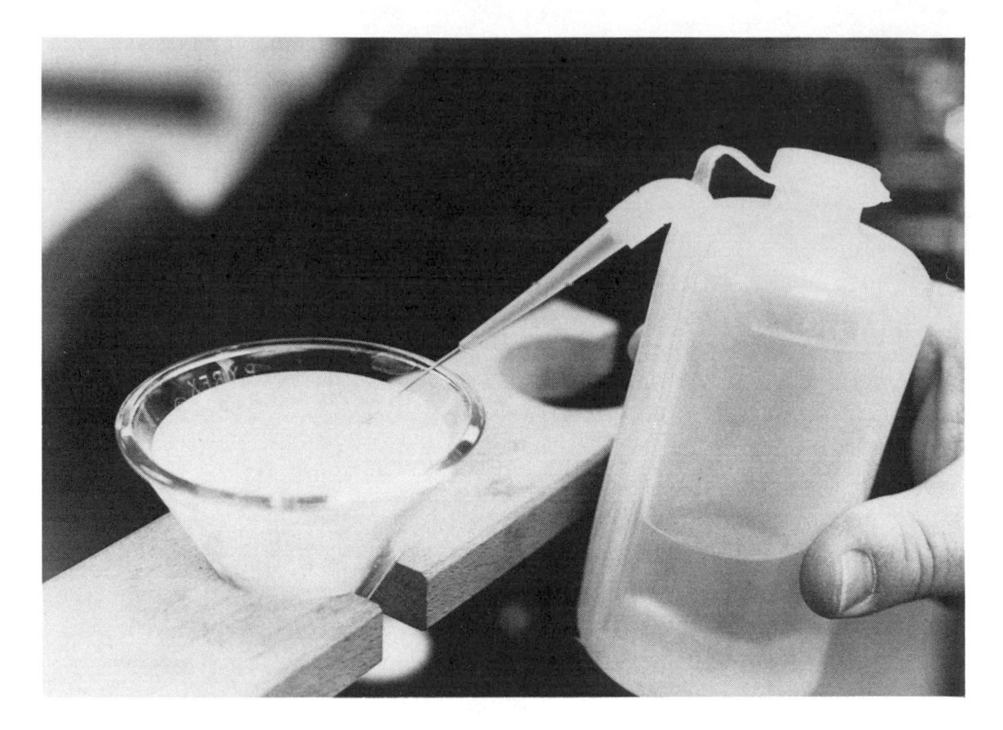

Figure 1.12. Moistening the filter with distilled water.

Figure 1.13. The correct way to position a crucible in the clay triangle for heating with the Meker burner.

the filter paper to catch fire. If it does catch fire, you can conveniently extinguish it by momentarily covering the crucible completely with the lid using tongs. The filter paper will slowly dry out and turn black, while releasing some smoke. Once the paper is completely charred (black), incline the crucible to one side as in Figure 1.13 and apply the full heat of the Meker burner to the bottom of the crucible for a minimum of 20 min. It is very important not to engulf the entire crucible in flame. Inclining the crucible to one side with lid positioned as in Figure 1.13 and directing the flame to the bottom of the crucible is important in order to allow oxygen to get to the sample so that the last of the paper can burn off. It is ashless, which means it burns completely, leaving no residue. This also prevents the reduction of the sulfate to sulfide, which would cause an error. If one were to leave the crucible upright and engulf the entire crucible in flame, no oxygen could get in.

If only one Meker burner is available, alternate with a Bunsen burner for a total of about 30 min (15 min per burner per sample). Allow to cool in place for 5 min and, using tongs, transfer to the desiccator to cool for an additional 15–20 min. Weigh the crucibles and lids and record the weights in your notebook.

15. Compute the weights of your sample and precipitates from the data in your notebook. The percentage of SO_3 in the unknown is calculated from these weights. Report the results to your instructor.

EXPERIMENT 3: THE GRAVIMETRIC DETERMINATION OF IRON IN A COMMERCIAL UNKNOWN

NOTE: Safety glasses are required.

Session 1

1. Obtain a sample from your instructor. Record the unknown number and dry as in Experiment 2. Also, obtain and prepare two porcelain crucibles and lids, as in Experiment 2. After the sample has cooled, accurately weigh by difference two samples, each around 0.4 g, into labeled 400-mL beakers. Store the beakers with your desiccator until Session 2.

NOTE: Session 1 here can coincide with a Session 3 for Experiment 2, if one was required.

Session 2

2. Add 5 mL of distilled water and 10 mL of concentrated HCl to each beaker and place stirring rods in the beakers and watch glasses on them.

3. Place the beakers on a hot plate in a good fume hood and bring to boiling. Allow to boil about 2 min to ensure complete dissolution.

4. When samples are completely dissolved, remove them from the hot plate and, while they are still in the fume hood, add 100 mL of water. Then remove from the hood and add 2 mL of concentrated HNO_3. (Be sure not to lose any yellow solution that might be adhering to the watch glass.) The purpose of the nitric acid is to ensure that all of the iron is present as Fe^{+3} and not Fe^{+2}.

5. Bring the solutions to boiling again (using a hot plate). While waiting for the solutions to boil, half-fill a squeeze bottle with 7 *M* NH_4OH and label this bottle.

6. When the solutions begin to boil, remove them from the hot plate and add the NH_4OH from the squeeze bottle with continuous stirring until the solutions become quite murky with the $Fe(OH)_3$ precipitate and the precipitate settles readily upon standing. Now test the solution with litmus paper (be sure to rinse the paper back into the beaker). If the paper does not turn a deep blue color, add more NH_4OH until it does. Be careful not to add too much. If the solution becomes too basic, other metals will precipitate, causing an error.

7. Boil gently for 5 min to digest the precipitate. Then begin filtering immediately through Whatman #41 filter paper prepared as in Experiment 2. This filtering step is identical to that in Experiment 2, except that rinsing is done with a 1% solution of NH_4NO_3 rather than distilled water. The reason for using NH_4NO_3 is that distilled water can cause "peptization" of the precipitate. This means that the precipitate particles can become dispersed throughout the solution and become small enough to be lost through the filter paper. The use of a dissolved salt such as NH_4NO_3 prevents this and, in addition, it completely volatilizes during the ignition step, thus not adding any weight to the precipitate. Once all of the precipitate has been transferred to the funnel, rinse it three more times with the NH_4NO_3 solution. Test for chloride in the filtrate by acidifying a few milliliters of the most recent rinsings with dilute HNO_3 and adding $AgNO_3$ dropwise as in Experiment 2. Continue to rinse until all chloride is gone.

8. Remove the filter paper and precipitate from the funnel and carefully fold as in Experiment 2 and place into the preweighed crucible. If you find the precipitate to be quite bulky and difficult to handle, place the folded filter paper "head first" into the crucible, so that if any precipitate oozes out it will go into the crucible. Be careful not to tear the filter paper at this point. It is wet and quite fragile.

Again, as with Experiment 2, you will need at least 60 min to finish. If you don't have this amount of time, store the crucibles outside your desiccator until the next laboratory period.

Session 3

9. Finish the experiment as you did Experiment 2. Slowly dry the filter paper using a Bunsen burner. Heat until black. Then incline the crucible at an angle, lid ajar, and direct the flame of a Meker burner to the bottom of the crucible, allowing air to get inside the crucible (Figure 1.13). Apply the full heat of the Meker burner for 20 min, or, if only one Meker burner is available, alternate, as before, with a Bunsen burner for 0.5 hr.

   Cool and weigh. Calculate the percent Fe in the sample via the gravimetric analysis calculation equation (Equation 1.11). The constituent is Fe and the precipitate is Fe_2O_3.

QUESTIONS AND PROBLEMS

1. Describe the differences between qualitative analysis and quantitative analysis.
2. Describe the differences between wet chemical methods and instrumental methods.
3. What is the difference between "weight" and "mass"?
4. Why is a weighing device called a "balance"?
5. What is a "single-pan balance"?
6. What does it mean to say that a balance has a "tare" feature?
7. What is an "analytical balance"?
8. The analytical balance is a much more accurate weighing device than the "ordinary" balances. Name three things that are important to remember when using the analytical balance that are not important when using an ordinary balance.
9. When a weight measurement enters directly into the calculation of a quantitative analysis result which is to be reported to four significant figures, is an ordinary top-loading balance satisfactory for this measurement? Explain.
10. A soil sample, as received by a laboratory, weighed 5.6165 g. After drying in an oven, this same sample weighed 2.7749 g. What is the percentage of moisture in this sample?
11. In an experiment in which the percentage of moisture in a sample was determined, the following data were obtained:
    Weight of crucible + sample before drying .... 11.9276 g
    Weight of crucible + sample after drying .... 10.7742 g
    Weight of empty crucible .... 7.6933 g
    What is the percent of moisture in the sample?
12. A sample of grain was analyzed for organics that volatilize at 500°C. Initially, a sample of this grain in an evaporating dish weighed 29.6464 g. After placing it in an oven at 500°C for 2 hr, this same sample in the dish weighed 20.9601 g. If the evaporating dish alone weighed 11.6626 g, what percentage of the grain was volatile organics?
13. An analyst performs an experiment to determine the percent of volatile organics in a grain sample. The grain is placed in a preweighed evaporating dish, the dish is weighed again, the organics are driven off by heating and the dish weighed a third time. Calculate the percent of volatile organics given the following data:
    Weight of empty evaporating dish .... 28.3015 g
    Weight of evaporating dish with the grain .... 39.4183 g
    Weight of dish with grain after heating .... 33.1938 g
14. Consider the analysis of the water from a lake for suspended solid particles. A sample of the water in a beaker is weighed and then filtered through a preweighed filter so as to separate the suspended solids from the water. The following data were recorded:

Weight of empty beaker .... 15.9201 g
Weight of beaker with water .... 37.2857 g
Weight of empty filter .... 11.6734 g
Weight of filter with solids .... 16.3758 g

What is the percent of suspended solids in the water?

15. Consider the analysis of a salt-sand mixture. If the mixture contained in a beaker is treated with sufficient water to dissolve the salt, what is the percent of both salt and sand in the mixture given the following data?

Weight of mixture .... 5.3502 g
Weight of sand isolated from mixture after filtering and drying .... 4.2034 g

16. What is the gravimetric factor in each of the following gravimetric analysis examples?

	Substance Sought	Substance Weighed
(a)	Ag	$AgBr$
(b)	SO_3	$BaSO_4$
(c)	Ag_2O	$AgCl$
(d)	Na_3PO_4	$Mg_2P_2O_7$
(e)	Pb_3O_4	$PbCrO_4$
(f)	SiF_6	CaF_2
(g)	Co_3O_4	$Co_2P_2O_7$
(h)	Bi_2S_3	Bi_2O_3

17. What is the gravimetric factor
    (a) when the weight of Al is to be converted to the weight of Al_2O_3?
    (b) when the precipitate is Li_2CO_3 and the percentage of Li_3PO_4 is to be calculated?

18. What is the gravimetric factor
    (a) for obtaining the weight of Cu from the weight of Cu_2O?
    (b) if one is calculating the percent of Mn_2O_3 in an ore when the weight of Mn_3O_4 is measured?
    (c) for obtaining the weight of Fe_3O_4 from the weight of Fe_2O_3?

19. What is the gravimetric factor
    (a) for obtaining the weight of Cr from the weight of Cr_2O_3?
    (b) if one is calculating the percent of Na_3PO_4 in a mixture when the weight of Na_2SO_4 is measured?
    (c) when converting the weight of Sb_2O_3 to the weight of Sb?

20. What is the gravimetric factor
    (a) for obtaining the weight of Ag_2CrO_4 from the weight of AgCl?
    (b) if one is calculating the percentage of Na_2SO_4 in a mixture when the weight of Na_3PO_4 is measured?
    (c) when converting the weight of $HgCl_2$ to the weight of Hg_2Cl_2?

21. What is the gravimetric factor that must be used in each of the following experiments?
    (a) The weight of $Mg_2P_2O_7$ is known and the weight of MgO is to be calculated.
    (b) The weight of Fe_3O_4 is to be converted to the weight of FeO.

(c) The weight of Mn_3O_4 is to be determined from the weight of Mn_2O_3.

22. What weight of Fe_2O_3 is equivalent to 0.2603 g of Fe_3O_4?
23. What weight of K_2SO_4 is equivalent to 0.6603 g of K_3PO_4?
24. What weight of P_2O_5 is equivalent to 0.6603 g of P?
25. What is the percent of K_2CrO_4 in a sample that weighed 0.7193 g if the weight of the Cr_2O_3 precipitate derived from the sample was 0.1384 g?
26. The gravimetric factor for converting the weight of $BaCO_3$ to Ba is 0.6959. If the weight of $BaCO_3$ derived from a sample was 0.2644 g, what weight of Ba was in this sample?
27. A sample containing sodium sulfate, Na_2SO_4, is analyzed gravimetrically by precipitation with barium chloride. If 0.4320 g of the sample gave 0.7446 g of $BaSO_4$, what is the percentage of SO_3 in this sample?
28. If 0.9110 g of a sample of silver ore yielded 0.4162 g of AgCl in a gravimetric experiment, what is the percentage of Ag in the ore?
29. Calculate the percent of Na_2SO_4 in a sample given the following data:
    Weight of weighing bottle before dispensing sample ..... 7.4834 g
    Weight of weighing bottle after dispensing sample ...... 7.0174 g
    Weight of empty crucible .... 19.0489 g
    Weight of crucible with $BaSO_4$ precipitate .... 19.7887 g
30. Given the following data, what is the percent S in the sample?
    Weight of weighing bottle before dispensing sample ..... 5.3403 g
    Weight of weighing bottle after dispensing sample ...... 4.8661 g
    Weight of crucible with $BaSO_4$ precipitate .... 19.3428 g
    Weight of crucible empty .... 18.7155 g
31. An iron ore sample is analyzed for iron content via a gravimetric analysis. The ore is placed in a weighing bottle and a sample is weighed by difference into a beaker. After obtaining the precipitate, it is filtered, placed into a preweighed crucible, ignited, and weighed. The following data were recorded:
    Weight of weighing bottle before dispensing sample ..... 6.9336 g
    Weight of weighing bottle after dispensing sample ...... 6.5002 g
    Weight of empty crucible .... 11.1435 g
    Weight of crucible with precipitate (Fe_2O_3) .... 11.7254 g
    What is the percentage of Fe in the ore?
32. Given the following data, what is the percent of Fe in the sample?
    Weight of weighing bottle before dispensing sample ..... 3.5719 g
    Weight of weighing bottle after dispensing sample ...... 3.3110 g
    Weight of crucible with Fe_2O_3 precipitate .... 18.1636 g
    Weight of empty crucible .... 18.0021 g
33. Nickel can be precipitated with dimethylglyoxime (DMG) according to the following reaction:

$$Ni_1^{+2} + 2HDMG \rightarrow Ni(DMG)_2 + 2H^{+1}$$

If 2.0116 g of a nickel-containing substance is dissolved and the nickel precipitated as above so that the $Ni(DMG)_2$ weighs 2.6642 g, what is the percentage of nickel in the substance? The formula weight of $Ni(DMG)_2$ = 288.92 g.

34. Imagine an experiment in which the percentage of manganese (Mn) in a manganese ore is to be determined by gravimetric analysis. If 0.8423 g of the ore yielded 0.3077 g of Mn_3O_4 precipitate, what is the percent of Mn in the ore?

35. Why should a stirring rod, which is being removed from a beaker containing a solution of the sample being analyzed, be rinsed back into that solution with distilled water?

36. What is meant by an "indicating" desiccant?

37. What is the advantage to weighing by difference as opposed to weighing using a spatula and weighing paper?

38. Why is the barium sulfate precipitate in Experiment 2 "digested"?

39. During what period of the "weighing by difference" process must fingerprints on the item being weighed be avoided? Explain.

40. When testing the precipitate rinsings for chloride to determine the completeness of the rinsing in Experiment 2, why is the white precipitate formed discarded and not added to the filter paper?

41. In Experiment 2, why must the heat-stable labels be placed on the crucibles before the first heating?

42. In Experiments 2 and 3, why is the desiccator not used to store the crucible and filter paper immediately after the filtration and before heating with a burner?

43. What does it mean to say that a filter paper is "ashless"? If the filter paper used in Experiment 2 were not ashless, what specific effect would it have on the calculated results?

44. In Experiment 2, if a student were to mark a crucible with the heat-stable marker *after* the first heating, rather than before, but before the first weighing, would his/her final results be higher or lower than the correct answer? Explain your reasoning.

45. In Experiment 2, consider an error in which a student accidentally poked a hole in the filter paper with the stirring rod as he/she was filtering and some precipitate was observed with the filtrate. If the student takes no steps to correct the error, would the percent of SO_3 calculated at the end be higher or lower than the true percent? Explain.

46. In Step 12 of Experiment 2, two drops of silver nitrate solution are added to a test tube containing rinsings from the funnel containing the precipitate. Suppose you were first asked to prepare this silver nitrate solution by weighing a given quantity of solid silver nitrate and dissolving it in water to make a given volume of solution. Would an analytical balance be required for this weight, or would an ordinary balance suffice? Explain.

Chapter 2

Sampling and Sample Preparation

2.1 THE TOTAL ANALYSIS

Chemical analysis procedures are not always limited to the kind of laboratory work that has been discussed thus far. A total analysis procedure would involve the manner in which the sample brought into the laboratory was obtained and handled and the manner in which the sample was prepared for analysis once it reached the laboratory, as well as what specific analysis technique was chosen and executed. Below is the total scheme by which an analysis is performed from beginning to end.

1. The sample is collected in a manner consistent with the goal of the analysis.
2. The sample, once collected, is handled in a manner so as to protect it in every way from contamination or alteration.
3. The sample is brought into the laboratory and prepared for the particular technique chosen.
4. The laboratory operations involved for the technique chosen are executed and the appropriate data obtained.
5. The data are worked up in such a way that the final desired result is determined.

The first three steps of this scheme are dealt with in this chapter. These are steps which are common to all analyses, regardless of the technique chosen, and thus are considered here as a group, separate from Steps 4 and 5. These latter steps are what could be termed "the heart" of the analysis.

The remainder of the text deals with the variety of techniques involved in these steps.

2.2 STEP ONE: OBTAINING THE SAMPLE

A laboratory analysis is almost always meant to give a result which is indicative of a concentration in a very large system. A farmer wants an analysis result to represent the concentration level in an entire 40-acre field. A pharmaceutical manufacturer wants an analysis result to represent the concentration of an active ingredient in 80 cases of its product, each case containing 3 dozen bottles of 100 tablets each. A governmental environmental control agency wants a single laboratory analysis to represent the concentration of a toxic chemical in every cubic inch of soil within 5 miles of a hazardous waste dump site.

An ideal analysis is one in which the entire system can be run through the analysis procedure. Obviously, this is not possible in most cases. The analyst is then faced with the serious problem of obtaining a sample from this system which is representative of the whole system. It is logical that an analytical result can only be as good as the sample itself.

Obviously, there are different degrees of difficulty and different sampling modes involved with obtaining samples for analysis, depending on the type of sample to be gathered, whether the source of the sample is homogeneous, the location of and access to the system, etc. For example, obtaining a sample of blood from a hospital patient is completely different from obtaining a sample of coal from a train car full of coal.

As far as blood is concerned, from what part of the body to take blood needs to be considered. Second, the time of day, along with a knowledge of the patient's recent diet habits, is important. Third, perhaps the patient is on some sort of medication which could affect the analysis.

With the coal sample it is important to recognize that the coal held in a train car may not be homogeneous, and a sampling scheme which takes this into account must be implemented.

The key word in any case is "representative". A laboratory analysis sample must be representative of the whole so that the final result of the chemical analysis represents the entire system that it is intended to represent. If there are variations, or at least suspected variations, in the system, small samples must be taken from all suspect locations. If results for the entire system are to be reported, these small samples are then mixed well to give the final sample to be tested. In some cases, analysis on the individual samples may be more appropriate. Some examples follow.

Consider the analysis of soil from a farmer's field. The farmer wants to know whether he needs to apply a nitrogen-containing fertilizer to his field. It is conceivable that different parts of the field could provide different types of samples in terms of nitrogen content, particularly if there is a cattle feedlot nearby, perhaps uphill from part of the field and downhill from another part of the field. Obviously the sample taken should include portions from all parts of the field which may be different so that it will truly represent the entire field. Alternatively, two samples could be taken, one from above the feedlot and one from below the feedlot, so that two analyses are performed and reported to the farmer. At any rate, one wants the results of the chemical analysis to be correct for the entire area for which the analysis is intended.

Consider the analysis of the leaves on a tree for pesticide residue. The tree grower wants to know if the level of pesticide residue on the leaves indicates whether the tree needs another pesticide application. Once again, the analyst must consider all parts of the tree that might be different. Leaves at the top, in the middle, and at the bottom should be sampled (one can imagine differences in application rates at the different heights); leaves on the outside and leaves close to the trunk should be sampled; and perhaps there should also be a difference between the shady side and the sunny side of the tree. All leaf samples are then combined and brought into the laboratory as a single sample.

Consider the analysis of a blood sample for alcohol content (imagine that a police officer suspects a motorist to be intoxicated). The problem here is not sampling different locations within a system, but rather a time factor. The blood must be sampled within a particular time frame which would demonstrate intoxication at the time the motorist was stopped.

The problems associated with sampling are unique to every individual situation. The analyst simply needs to take all possible variations into account when obtaining the sample so that the sample taken to the laboratory truly does represent what it is intended to represent. Figure 2.1 shows some specialized sampling devices that are helpful in some cases.

2.3 STEP TWO: HANDLING THE SAMPLE

How to get the sample from the sampling site to the laboratory without contamination or alteration is generally not as challenging, in most cases, as the problem of how to obtain a representative sample. There are basically two considerations associated with such sample preservation: (1) storage of the sample in a container which will not leach contaminants to a degree that would be damaging to the integrity of the sample, particularly if trace amounts of the constituent are to be determined; and (2) preservation of the sample from problems which may be internal, e.g., temperature effects or bacterial effects.

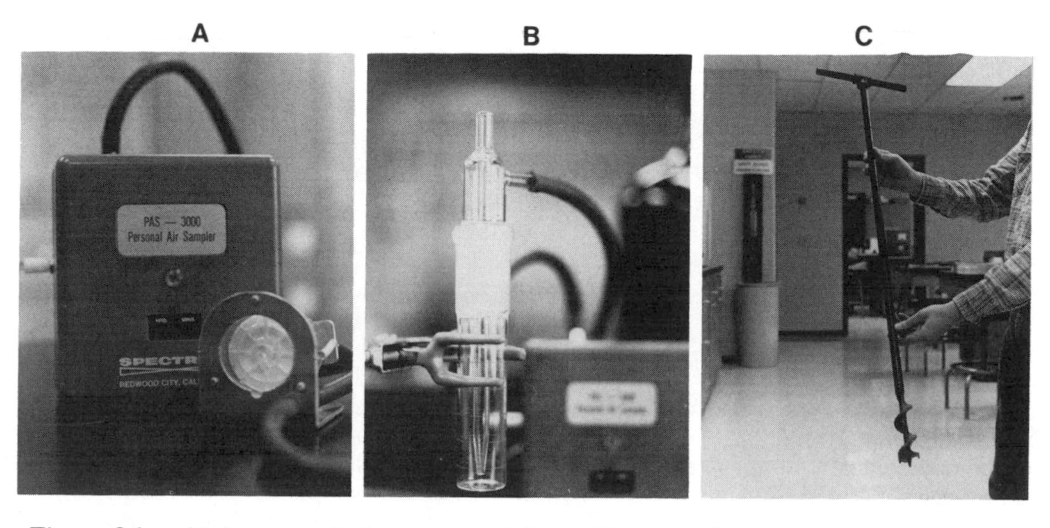

Figure 2.1. (A) A personal air sampler with air filter cartridge for particulates in air. (B) A midget impinger for chemicals in the air. (C) A soil sampler.

If trace amounts of metals are to be determined, one would not want to store the sample in a glass container, since glass can leach small amounts of metals. On the other hand, if trace organics are to be determined, plastic containers may be deemed inappropriate. Sometimes refrigeration may be important to avoid decomposition from bacterial sources. At any rate, proper sample handling methods must be employed to ensure sample integrity.

2.4 STEP THREE: PREPARING THE SAMPLE FOR ANALYSIS

Once sampling and sample preservation schemes have been properly executed, the sample is presumably in the laboratory ready for the analysis. Most of the time, however, it is still not in a state in which the chosen analytical technique, whatever it is, can be properly applied. The vast majority of all analysis procedures call for the sample to be in the dissolved state – a solution of the sample is what is required most of the time. Sometimes this will mean complete dissolution of the entire sample, while at other times it may mean only partial dissolution.

Reagents for Total Dissolution

Water. It should come as no surprise that ordinary water can be an excellent solvent for many samples. Due to its extremely polar nature, water will dissolve most substances of similar polar or ionic nature. Obviously, then, when samples are composed solely of ionic salts or polar substances, water would be an excellent choice. An example might be the analysis of a commercial table salt for sodium iodide content. A list of solubility rules for ionic compounds in water can be found in Table 2.1.

Table 2.1. Solubility rules for some common inorganic compounds in water.

Compound class	Soluble?	Exceptions
Nitrates	Yes	None
Acetates	Yes	Silver acetate is sparingly soluble.
Chlorides	Yes	Chlorides of Ag, Pb, and Hg are insoluble.
Sulfates	Yes	Sulfates of Ba and Pb are insoluble. Sulfates of Ag, Hg, and Ca are slightly soluble.
Carbonates	No	Carbonates of Na, K, and NH_4 are soluble.
Phosphates	No	Phosphates of Na, K, and NH_4 are soluble.
Chromates	No	Chromates of Na, K, NH_4, and Mg are soluble.
Hydroxides	No	Hydroxides of Na, K, and NH_4 are soluble. Hydroxides of Ba, Ca, and Sr are slightly soluble.
Sulfides	No	Sulfides of Na, K, NH_4, Ca, Mg, and Ba are soluble.
Sodium salts	Yes	Some rare exceptions
Potassium salts	Yes	Some rare exceptions
Ammonium salts	Yes	Some rare exceptions
Silver salts	No	Silver nitrate and perchlorate are soluble. Silver acetate and sulfate are sparingly soluble.

Hydrochloric Acid. Strong acids are used frequently for the purpose of sample dissolution when water won't do the job. One of these is hydrochloric acid, HCl. Concentrated HCl is actually a saturated solution of hydrogen chloride gas, fumes of which are very pungent. Such a solution is 38.0% HCl and about 12 *M*. Hydrochloric acid solutions are used especially for dissolving metals, metal oxides, and carbonates not ordinarily dissolved by water. Examples are iron and zinc metals, iron oxide ore, and the metal carbonates of which the scales in boilers and humidifiers are composed. Being a strong acid, it is very toxic and must be handled with care. It is stored in a blue color-coded container.

Sulfuric Acid. An acid that is considered a stronger acid than HCl in many respects is sulfuric acid, H_2SO_4. When sulfuric acid contacts clothing, paper, etc., one can see an almost instantaneous reaction – paper towels turn black and disintegrate; clothing fibers become weak and holes readily form. Concentrated sulfuric acid is about 96% H_2SO_4 (the remainder being water) or about 18 *M* and is a clear, colorless, syrupy, dense liquid. It reacts violently with water, evolving much heat, and so water solutions of sulfuric acid must be prepared cautiously, often to include a means of cooling the container. Its sample dissolution application is limited mostly to organic material, such as vegetable plants. It is not as useful for metals because many metals form insoluble sulfates. It is the solvent of choice for the Kjeldahl analysis (Chapter 4) for such materials as grains and products of grain processing. It is also used to dissolve aluminum and titanium oxides on airplane parts. It is stored in a yellow color-coded container.

Nitric Acid. Another acid that has significant application is nitric acid, HNO_3. This acid is also very dangerous and corrosive and is aptly referred to as an oxidizing acid. This means that a reaction other than hydrogen gas displacement (as with HCl) occurs when it contacts metals. Frequently, oxides of nitrogen form in such a reaction, and noxious brown, white, and colorless gases are evolved. Concentrated HNO_3 is 70% HNO_3 (16 *M*) and is used for applications where a strong acid with additional oxidizing power is needed. These include metals such as silver and copper, as well as organic materials such as in a wastewater sample. Nitric acid will turn skin yellow after only a few seconds of contact. It is stored in a red color-coded container.

Hydrofluoric Acid. An acid that has some very useful and specific applications, but is also very dangerous, is hydrofluoric acid, HF. This acid reacts with skin in a way that is not noticeable at first, but becomes quite serious if left in contact for a period of time. It has been known to be especially serious if trapped against the skin or after diffusing under fingernails. Treatment of this is difficult and painful. Concentrated HF is about 50% HF (26 *M*). It is an excellent solvent for silica (SiO_2)-based materials such as sand, rocks, and glass. It can also be used for stainless steel alloys. Since it dissolves glass, it must be stored in plastic containers. This is also true for low pH solutions of fluoride salts.

Perchloric Acid. Another important acid for sample preparation and dissolution is perchloric acid, $HClO_4$. It is an oxidizing acid like HNO_3, but is considered to be even more powerful in that regard when hot and concentrated. It can be used for metals, since metal perchlorates are highly soluble, but it is most useful for more difficult organic samples, such as leathers and rubbers, often in combination with nitric acid. It can also be used for stainless steels and other more stable alloys. Commercial $HClO_4$ is 72% (12 *M*). It is a very dangerous acid, especially when hot and in contact with organic matter, and should only be used in a fume hood designed for the collection of its vapors. Contact with alcohols, including polymeric alcohols such as cellulose, and other oxidizable materials should be avoided due to the potential for explosions.

"Aqua Regia". An acid mixture that is prepared by mixing one part concentrated HNO_3 with three parts concentrated HCl is called "aqua regia". This mixture is among the most powerful dissolving agents known. It will dissolve the very noble metals (gold and platinum) as well as the most stable of alloys.

Table 2.2 summarizes these acids and their properties.

Table 2.2 Various laboratory acids and their properties.

Chemical Substance	Formula	Description	Uses
Water	H_2O	Clear, colorless liquid with low vapor pressure; highly polar.	Dissolving polar and ionic compounds.
Hydrochloric acid	HCl	Commercially available as 38% water solution (concentrated HCl). Evolves irritating fumes and must be handled in fume hood. A strong and dangerous acid.	Dissolving metals and metal ores.
Sulfuric acid	H_2SO_4	Concentrated water solution is 96% H_2SO_4. A dense, syrupy liquid. Reacts on contact with skin and clothing. Evolves much heat when mixed with water.	Some organic sample dissolution (e.g., Kjeldahl – see Chapter 6). Also, oxides of Al and Ti.
Nitric acid	HNO_3	Available commercially as 70% water solution (concentrated HNO_3). Reacts with clothing and skin – turns skin yellow. Evolves thick white and brown fumes when in contact with most metals.	Dissolving more noble metals, e.g., Cu and Ag, and also some organic samples.
Hydrofluoric acid	HF	Concentrated HF is a 50% water solution. Must be stored in plastic containers, since it dissolves glass. Very damaging to skin.	Dissolving silica-based materials and stainless steel.
Perchloric acid	$HClO_4$	Commercially available as a 72% solution	Dissolving difficult organic samples and stable metal alloys.
Aqua regia		A 1:3 (by volume) mixture of concentrated HNO_3 and concentrated HCl.	Dissolving highly unreactive metals, such as gold.

Partial Dissolution – Extraction

In cases in which it is not practical or necessary to dissolve an entire sample or in which it is easy to effect a separation of the analyte by selective dissolution, an extraction procedure is used. This technique is described in some detail in Chapter 15.

Briefly, the sample can be either a solution (typically a water solution) of the analyte or a solid material, e.g., soil. When the sample is a water solution, "solvent extraction" refers to an experiment in which an organic solvent, immiscible with the sample, is brought into intimate contact with the sample (in a separatory funnel – see Chapter 15) in such a way that the substance of interest is at least partially removed (extracted) from the water and dissolved in the solvent.

When the sample is a solid, the extracting solvent is often an aqueous solution of an inorganic compound, such as an acid, although it can also be an organic solvent. Here again the sample is brought into intimate contact with the extracting solvent by shaking in the same container (usually after grinding to a fine powder) or by repeated contact with fresh solvent in an apparatus called a Soxhlet extractor (Chapter 15). With both of these methods, the analyte is at least partially separated from the sample and dissolved in this solvent, while most other materials remain undissolved. This method is acceptable since the undissolved materials play no role at all in the analysis and in fact may contain interfering substances which are best left undissolved. The analysis of the sample extract can then proceed as with any other sample solution. Please refer to Chapter 15 for more details of the extraction technique.

Following is a description of some organic solvents that are commonly used in the laboratory for dissolution, extraction, or other sample preparation schemes.

n-Hexane. *n*-Hexane is a non-polar alkane, C_6H_{14}. Its solubility in water is virtually nil. It is less dense than water (density = 0.66 g/mL), and thus it would be the top layer in a separatory funnel with a water solution. It is obviously a poor solvent for polar compounds, but is very good for extracting traces of non-polar solutes in water samples. It is used as a solvent for ultraviolet (UV) spectrophotometry (Chapter 11). It is highly flammable and has a low toxicity level. It can be used as the mobile phase in normal-phase chromatography (Chapter 17).

Acetonitrile. Acetonitrile is a polar solvent, $CH_3C{\equiv}N$. It is completely miscible with water. It is used for the extraction of polar components of solid samples, such as the extractions of chlorinated pesticide residues from vegetables, and also as a mobile phase for reverse-phase chromatography (Chapter 17). It can be used as a solvent for UV (Chapter 11) and nuclear magnetic resonance (NMR) (Chapter 13) work. It is flammable, toxic, and has an unpleasant odor. Breathing acetonitrile vapors should be avoided.

Methylene Chloride. This solvent is a somewhat polar solvent also known as dichloromethane, CH_2Cl_2. Its solubility in water is about 2%. It is denser than water (density = 1.33 g/mL), and thus it would be in the bottom layer when used with a water solution in a separatory funnel. It may form an emulsion when shaken in a separatory funnel with water solutions. It is used for extractions as well as for infrared (IR) (Chapter 12) and NMR

(Chapter 13) solvents. It is not flammable and is considered to have a low toxicity level.

Acetone. Acetone (CH_3COCH_3) is a volatile, highly flammable liquid that has intermediate polarity and is completely miscible with water. It is used as a solvent for both polar and non-polar substances, often as a cleaning solvent for glassware and as a degreaser. Its high volatility makes it useful for rapid drying of glassware. It can be used as a solvent for NMR work (Chapter 13). Acetone is occasionally used to extract components from solid samples. It is considered non-toxic, but prolonged breathing of vapors should be avoided.

Benzene. Benzene (C_6H_6) is a non-polar aromatic liquid insoluble in water. It can be used as a solvent in NMR work (Chapter 13). For extraction work, it has been used for extraction of non-polar components of water samples, but is prone to emulsion formation. Its density is 0.87 g/mL. It is highly flammable and toxic. It is especially noted as a carcinogen. Breathing of vapors and skin contact should be avoided. Due to its toxicity, it has been replaced in most laboratory applications with toluene (see below).

Toluene. Toluene ($C_6H_5{-}CH_3$) is also a non-polar aromatic liquid insoluble in water. It is flammable but less toxic than benzene. Its density is 0.87 g/mL, and it is slightly soluble in water (0.47 g/L). It is sometimes used as a mobile phase in normal-phase liquid chromatography (Chapter 17) and as an extraction solvent for non-polar sample components.

Methanol. Methanol ($CH_3{-}OH$) is a polar organic solvent completely miscible with water and most other solvents. It is a flammable and poisonous liquid. It is useful as a solvent for UV (Chapter 11) and NMR work (Chapter 13) as well as a mobile phase for reverse-phase liquid chromatography (Chapter 17). Like acetone, it is sometimes used to quick-dry laboratory glassware, since it is volatile. It is poisonous when ingested and inhalation should also be avoided.

Diethyl Ether. Diethyl ether ($CH_3CH_2{-}O{-}CH_2CH_3$), also frequently referred to as "ethyl ether" and as "ether", is a mostly non-polar organic liquid that is highly volatile and extremely flammable, a dangerous combination. Also, explosive peroxides form with time. Precautions discouraging peroxide formation include storage in metal containers in explosion-proof refrigerators. It should be disposed of after about 9 months of storage.

Ether is only slightly soluble (75 g/L) in water. Given its slight polarity, it is useful for extraction of solutes from water. Its density is 0.71 g/mL. Since it is volatile, it is easily evaporated from extraction fractions.

Chloroform. Chloroform (trichloromethane, $CHCl_3$) is a non-flammable organic liquid with low miscibility with water (10 g/L). It is useful as an extracting liquid for water samples, as a solvent for UV (Chapter 11), IR (Chapter 12), and NMR work (Chapter 13), and as a mobile phase for high-performance liquid chromatography (HPLC) (Chapter 17). It is very toxic and should be avoided when another solvent would do as well. Vapors should not be inhaled and contact with the skin should be avoided.

Fusion

For extremely difficult samples, a method called "fusion" may be employed. Fusion is the dissolving of a sample using a molten inorganic salt generally called a "flux". This flux dissolves the sample and, upon cooling, results in a solid mass that is then soluble in a liquid reagent. The dissolving action is mostly due to the extremely high temperatures (usually 300 to 1000°C) required to render most inorganic salts molten.

Additional problems arise within fusion methods, however. One is the fact that the flux must be present in a fairly large quantity in order to be successful. The measurement of the analyte must not be affected by this large quantity. Also, while a flux may be an excellent solvent for difficult samples, it will also dissolve the container to some extent, creating contamination problems. Platinum crucibles are commonly used, but nickel, gold, and porcelain have been successfully used for some applications.

Probably the most common flux material is sodium carbonate. It may be used by itself (melting point 851°C) or in combination with other compounds, such as oxidizing agents (nitrates, chlorates, and peroxides). These may be used to dissolve silicates and silica-based samples as well as samples containing alumina (Al_2O_3).

For dissolving particularly difficult metal oxides, the acidic flux potassium pyrosulfate ($K_2S_2O_7$) may be used. The required temperature for this flux is about 400°C.

2.5 QUALITY ASSURANCE OF REAGENTS

It is of utmost importance in an analytical laboratory to be able to trust all reagents used for both the sample preparation schemes, which have been outlined in this section, as well as for other analytical preparations. A stringent quality assurance program for such reagents is necessary. This means that the quality of all inventoried and newly purchased chemicals must be assured. It also means that the integrity of the various solutions prepared and samples gathered for analytical purposes in the laboratory must be maintained.

First, chemicals purchased and/or stored will have a designation of their purity on the label. This designation will likely be one of the following.*

Primary Standard. This is a specially manufactured analytical reagent of exceptional purity for standardizing solutions and preparing reference standards.

Reagent (ACS). Maximum limits of purity for most commonly used reagents (mostly inorganic) have been established by the Committee on Analytical Reagents of the American Chemical Society (ACS). Whenever a specification exists and a product is produced in conformity with it, the bottle is so labeled.

Reagent. When the ACS has not developed specifications for a specific reagent, the manufacturer establishes its own standards, and the maxi-

*These designations have been reprinted, in part, from the *Modern Chemical Technology Guidebook*, Revised Edition, page 157. Copyright 1972 by the American Chemical Society, Washington, DC. Used with permission.

mum limits of allowable impurities are shown on the labels of these reagents.

C.P. "Chemically Pure" grades of chemicals are offered by manufacturers. They meet or exceed U.S.P. or N.F. requirements, but are lower grade than "Reagent" or "Reagent ACS" chemicals.

U.S.P. Chemicals labeled U.S.P. meet the requirements of the U.S. Pharmacopeia. Generally of interest to the pharmaceutical profession, these specifications may not be adequate for reagent use. This designation and the N.F. designation are now the same.

N.F. Chemicals labeled N.F. meet the requirements of the National Formulary. Chemicals ordered with an N.F. label may not be useful for reagents. It will be necessary to check the National Formulary in each case. This designation and the U.S.P. designation are now the same.

Practical. This grade designates chemicals of sufficiently high quality to be suitable for use in some syntheses. Organic chemicals of practical grade may contain small amounts of intermediates, isomers, or homologs.

Purified. This is a grade of chemical for which care has been exercised by the manufacturer to offer a product that is physically clean and of good quality but not meeting Reagent ACS, Reagent, U.S.P., or C.P. standards.

Technical. This is a grade of chemical generally suitable for industrial use. Purity is not specified and is generally determined by on-site analysis.

Spectro Grade. This is a designation for organic solvents which have been prepared for use in ultraviolet or infrared spectroscopy without further purification. Many of these also conform to Reagent or Reagent ACS standards.

HPLC Grade. This is a designation for organic solvents which have been specifically prepared for use as HPLC mobile phases.

Second, the integrity of inventoried chemicals, or chemicals that have been on the shelf for a period of time, can be determined. This determination of integrity, or shelf-life, can involve proper labeling at the manufacturing plant, various reference sources, such as the *Merck Index*, and/or various laboratory tests to determine purity.

Third, samples and solutions gathered or prepared by laboratory personnel must also be properly labeled at the time of sampling or preparation. This means that the label must include the name of the sample or the chemical or chemicals dissolved and the date the sample was gathered or the solution prepared. In addition, the laboratory should maintain a notebook of all samplings and preparations, giving details of each sampling or preparation and the references which called for the particular chemicals or techniques used. This would allow tracking the reagent or sample in the event an error is suspected.

The point was made earlier that "an analysis is only as good as the sample" used. We can now extend this statement and say that an analysis is only as good as the sample *and the reagents* that are used.

EXPERIMENTS

EXPERIMENT 4: A STUDY OF THE DISSOLVING PROPERTIES OF WATER, SOME COMMON ORGANIC LIQUIDS, AND LABORATORY ACIDS

Introduction

In this experiment, the miscibility of water with some common organic solvents and the dissolving power of water, hydrochloric acid, sulfuric acid, and nitric acid will be confirmed. First, small amounts of water will be mixed with roughly equal amounts of the organic solvents and the miscibilities observed. Then small volumes of water and the acids will be allowed to contact granules of some selected metals and other compounds and their dissolving power will be noted. The objective is to confirm some of the statements made in Table 2.2 and accompanying text.

NOTE: Safety glasses are required.

Part A: A Study of the Miscibility of Water with Selected Organic Solvents

1. Obtain a test tube rack and nine small test tubes. Add distilled water to each tube such that each is about one fourth to one third full. Place a piece of labeling tape on each.
2. Obtain samples of the following nine organic liquids contained in individual small dropper bottles: *n*-hexane, acetonitrile, methylene chloride, acetone, benzene, toluene, methanol, diethyl ether, and chloroform. Then label each of the test tubes from Step 1 with the names, or an abbreviation of the names, of these liquids.
3. Add small amounts of each liquid indicated in Step 2 to the water in the test tubes with the corresponding label. Stopper, shake, and observe the miscibility. Record the miscibility ("miscible" or "immiscible") and the polarity ("polar" or "nonpolar") in a table, such as the following, in the DATA section of your notebook.

Solvent	Miscibility	Polarity
n-hexane		
acetonitrile		
methylene chloride		
acetone		
benzene		
toluene		
methanol		
diethyl ether		
chloroform		

   Compare the results in the table with what was stated in the text for miscibility with water and polarity.

4. Optional: Check out the liquids in the above table to see if they are miscible with each other rather than with water. You might expect those that are polar to be immiscible with those that are nonpolar. Check out, for example, the miscibility of acetone and methanol with

those you identified as non-polar. Place your observations in a table and draw some appropriate conclusions regarding polarities.

5. Dispose of the contents of each test tube according to the directions of your instructor.

Part B: A Study of the Solubility of Some Selected Inorganic Materials in Water, Hydrochloric Acid, Sulfuric Acid, and Nitric Acid

1. Obtain small samples of NaCl, $CaCO_3$, Fe_2O_3, Al_2O_3, and the metals zinc, iron, copper, silver, and gold (or platinum). Also obtain around 20 small test tubes and a test tube rack. Place a piece of labeling tape on each tube.
2. Make a table or "grid" in the DATA section of your notebook as follows:

Acid	NaCl	$CaCO_3$	Fe_2O_3	Al_2O_3	Zn	Fe	Cu	Ag	Au
Water									
Hydrochloric									
Sulfuric									
Nitric									

3. Place very small granules of each of the nine materials indicated in Step 1 individually into test tubes. Label each of the nine tubes with the chemical symbol of the contents.
4. Half-fill each of the test tubes in Step 3 with distilled water. Now also place a symbol for water (such as a "W" or "H_2O)" on each label. After 2 min, carefully observe the contents of each tube to determine whether there has been a reaction. If there has been no visible reaction, stopper and shake, then observe whether the material has dissolved. Place an "S" (for "soluble") in the grid if the material has either reacted or dissolved (or both) and an "I" (for "insoluble") if nothing happened.
5. Obtain dropper bottles containing 50% (by volume) solutions of hydrochloric acid, sulfuric acid, and nitric acid. Repeat Steps 3 and 4 using the hydrochloric acid in place of the water, testing all those for which water gave an "I". Then repeat with sulfuric acid. Repeat finally with nitric acid, but only for those for which hydrochloric and sulfuric acids gave an "I".
6. Compare the results with statements of uses of these acids in the text.

QUESTIONS AND PROBLEMS

1. What tasks relating to the chemical analysis of a material sample must be performed prior to beginning the actual laboratory work?
2. What is a representative sample? Do not use the words "represent" or "representative" in your answer.

3. How would you obtain a representative sample of each of the following?
   (a) the water in a creek
   (b) the grain stored in a bin
   (c) the aspirin in a bottle of aspirin tablets
   (d) the soil in your yard
   (e) old paint on a building
   (f) corn plants in a field of corn

4. If a water sample is to be analyzed for trace levels of metals, why is a *glass* container for sampling and storage inappropriate?

5. If a sample to be analyzed in a laboratory is subject to alteration due to the action of bacteria, what can be done to preserve its integrity?

6. Look up in a reference book exactly what is involved in taking and preserving a sample of soil to be analyzed for nitrate.

7. Identify water or a specific acid as the least drastic dissolving agent for each of the following:
   (a) a mixture of sodium chloride and sodium sulfate
   (b) the material from a silver mine
   (c) a sample of stainless steel
   (d) a sample of iron carbonate ore
   (e) the material from a gold mine
   (f) a sample of sand from a beach
   (g) the oxide on the surface of an airplane part

8. What concentrated acid is correctly described by each of the following?
   (a) a very dense, syrupy liquid
   (b) a liquid that can be absorbed through the skin
   (c) a liquid that turns skin yellow
   (d) a liquid that gives off unpleasant fumes
   (e) a liquid that is mixed with HCl to give "aqua regia"
   (f) a solvent for silver metal
   (g) a solvent for corrosion products on an airplane part
   (h) a solvent for sand
   (i) a solvent for stainless steel
   (j) a solvent for iron ore
   (k) a solvent for leather
   (l) a liquid that gets especially hot when mixed with water
   (m) an acid that turns paper towels black on contact

9. What is "aqua regia"?

10. Fill in the blanks with a specific example of the kind of sample dissolved by the acid indicated.
    (a) hydrofluoric acid ______________________________
    (b) nitric acid ______________________________
    (c) sulfuric acid ______________________________
    (d) hydrochloric acid ______________________________

(e) aqua regia ____________________
(f) perchloric acid ____________________

11. Fill in the blanks with the name of the acid described by the statement.
    (a) The bottle containing this acid has a red color code. __________
    (b) This acid, when mixed with nitric acid in certain proportions, gives aqua regia. __________
    (c) When an open bottle of this acid sits next to an open bottle of ammonium hydroxide, thick white fumes form. __________
    (d) This acid gets especially hot when mixed with water. __________
    (e) This acid diffuses through skin and is especially bad when it gets under the fingernails. __________
    (f) The bottle containing this acid has a yellow color code. __________
    (g) This acid is used to dissolve the sample for the Kjeldahl analysis. __________

12. From the following list, choose those samples that would be best dissolved by HCl and those that would be best dissolved by HNO_3. Not all samples will be used.
    (a) NaCl
    (b) iron ore
    (c) gold metal
    (d) lettuce leaves
    (e) copper metal
    (f) sand
    (g) aluminum oxide

13. List the organic solvents described in this chapter that (a) dissolve in water, and (b) do not dissolve in water.
14. Of those organic solvents that do not dissolve appreciably in water, which would be the top layer in a separatory funnel? Explain what the solvents' density has to do with your answer.
15. List the organic solvents described in this chapter that (a) are toxic, (b) are flammable, (c) are polar, (d) are nonpolar.
16. List the organic solvents described in this chapter that (a) are useful as solvents for UV spectrophotometry, (b) are useful as solvents for IR spectrophotometry, (c) are useful as solvents for NMR spectrometry.
17. List the organic solvents described in this chapter that (a) are useful as mobile phases for reverse phase chromatography, (b) are useful as mobile phases for normal phase chromatography.
18. What is it about benzene that causes particular concern when used in the laboratory?
19. Describe the safety hazards involving the use and storage of diethyl ether in the laboratory and also tell what specific precautions are taken.
20. Give two examples of when an "extraction" is more useful than total sample dissolution.

21. Describe a separatory funnel and describe its function and usage.
22. Define “extraction” as it is used in a chemical analysis laboratory.
23. In a reference book, look up and report the following:
    (a) the extracting solvent for analyzing pesticides residues in water.
    (b) a specific application of a Soxhlet extraction apparatus (see also Chapter 15).
    (c) what a “grab” sampler is and what it is used for.
24. What is meant by “fusion” as a method for sample dissolution?
25. What is meant by the “flux” in a fusion procedure?
26. What materials must be used as containers for the flux/sample mixture in a fusion procedure? Explain.
27. Why must the quality of all reagents used in the laboratory be assured before use?
28. Perhaps the most common grade of chemical used in the laboratory is “Reagent grade”. Explain what “Reagent grade” refers to.
29. Can chemicals labeled as “Practical” or “Technical” grade be used as standards in analytical procedures? Explain.
30. What designation of purity is used for solvents that are appropriate for UV or IR analysis?
31. What designation of purity is used for solvents that are appropriate for use as HPLC mobile phases?
32. How important is the label on a sample or reagent? What information should appear on the label of a sample to be analyzed?

Chapter 3

Statistics in Chemical Analysis

3.1 INTRODUCTION TO ERROR ANALYSIS

The most important aspect of the job of the chemical analyst is to assure that the data and results that are reported are of the maximum possible quality. This means that the analyst must be able to recognize when the testing instrument is breaking down and when a human error is suspected. The analyst must be as confident as he/she can be that the readout from an instrument, for example, does in fact indicate a true readout as much as possible. The analyst must be familiar with error analysis schemes that have been developed and be able to use them to the point where confidence and quality are assured.

Errors in the analytical laboratory are basically of two types: "determinate", also called "systematic", and "indeterminate", also called "random". Determinate errors are avoidable blunders that were known to have occurred, or at least were determined later to have occurred, in the procedures. They arise from such avoidable sources as contamination, wrongly calibrated instruments, reagent impurities, instrumental malfunctions, poor sampling techniques, incomplete dissolution of sample, errors in calculations, etc. Sometimes correction factors can adjust for their occurrence; at other times the procedures must be repeated so as to avoid the error.

Indeterminate errors, on the other hand, are impossible to avoid. They are random errors, human errors which were not known to have occurred, or errors inherent in measurements. Such errors are known to occur, but can neither be accounted for directly nor avoided. Examples include problems inherent in the manner in which an instrument operates, errors inherent in reading a meniscus or a meter, errors, such as sample loss, that occur in sample and solution handling, etc. With regard to errors inherent in a measurement, if measuring devices are used that are more accurate (give more significant figures) than the results that are to be reported, then such errors are of no consequence. However, this is often not the case, and unknown human error also occurs.

Since indeterminate errors are unavoidable and unknown, they are presumed to affect the result both positively and negatively and are dealt with by statistics. A given result can be rejected as being too inaccurate based on statistical analysis indicating too great a deviation from the established norm. The procedure to check for such errors involves running a series of identical tests on the same sample, using the same instrument or another piece of equipment, over and over. Those results that agree to within certain predetermined limits are averaged and the average is then considered the correct answer. Any results that fall outside these predetermined limits are "rejected" and are not used to compute the average. The "rejectability" parameters are the standards which a given laboratory must determine and adopt for the particular situation. Some methods for the determination of these parameters will be discussed in this chapter.

Thus the answer to the question "Can I really trust the result that I get to be accurate?" is "yes!"–assuming that determinate errors have been detected and accounted for and statistical methods are correctly applied to deal with random, indeterminate errors. Such a result would have a high probability of being accurate.

At this point, a distinction ought to be made between the term "accuracy" and the term "precision". These two terms are frequently misrepresented. "Accuracy" refers to the "correctness" of a given measurement or series of measurements. That is, does the measurement (or measurements) come close to what the correct answer is? Precision does not necessarily relate to accuracy, but rather to how closely several measurements relate to each other. Figure 3.1 should make this distinction clear. Precision is, however, usually taken to be an indication of accuracy, unless there is known to be a factor which inherently affects accuracy of a continuous and constant basis.

3.2 TERMINOLOGY

In working with the statistical methods used to establish the "correct" answer in an analysis, a number of terms appear which need to be defined. These are as follows:

1. *Mean*. In the case in which a given measurement on a sample is repeated a number of times, the average of all measurements is an important number and is called the "mean". It is calculated by adding together the numerical values of all measurements and dividing this sum by the number of measurements.
2. *Median*. For this same series of identical measurements on a sample, the "middle" value is sometimes important and is called the median. Thus the median is always one of the actual measurements, while the mean may not be. However, if the total number of measurements is an even number, there is no single "middle" value. In this case, the median is the average of two "middle" values.

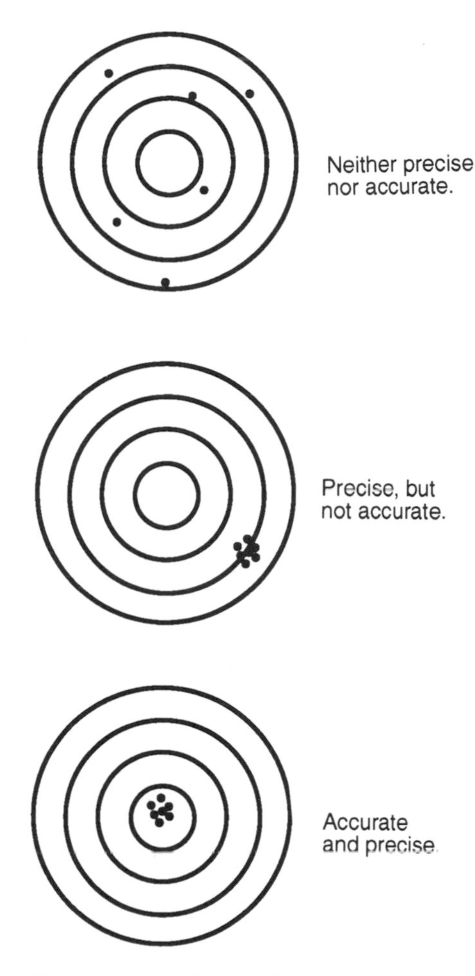

Figure 3.1. Illustration of precision and accuracy.

3. *Mode.* The value that occurs most frequently in the series is called the "mode". For a large number of identical measurements, the median and mode should be the same.
4. *Deviation.* How much each measurement differs from the mean is an important number and is called the deviation. A deviation is associated with each measurement, and if a given deviation is large compared to others in a series of identical measurements, this may signal a potentially rejectable measurement which will be tested by the statistical methods. Mathematically, the deviation is calculated as follows:

$$d = |m - e| \tag{3.1}$$

in which d is the deviation, m is the mean, and e represents the individual experimental measurement. [The bars (||) refer to "absolute value", which means the value of d is calculated without regard to sign; i.e., it is always a positive value.]

5. *Relative Deviation.* Perhaps of more practical significance is the "relative deviation". This quantity relates the deviation to the value of the mean. If the mean is a relatively large number, a deviation that appears to be large may not be out of line because the value from which it shows deviation is large also. Thus a relative deviation is more useful. Mathematically, it is defined as follows:

$$d_R = \frac{d}{m} \tag{3.2}$$

in which d_R is the relative deviation. d_R can also be expressed as a percent, or part per thousand, etc.:

$$\text{relative \% deviation} = d_R \times 100 \tag{3.3}$$

$$\text{relative ppt deviation} = d_R \times 1000 \tag{3.4}$$

in which ppt represents "parts per thousand".

Example 1

Calculate the relative ppt deviation for each measurement in the following series of measurements:

Measurement #	Value
1	0.1013
2	0.1011
3	0.1014
4	0.1009
5	0.1010
6	0.1019

Solution

$$\text{relative ppt deviation} = d_R \times 1000$$

$$= \frac{d}{m} \times 1000$$

$$m = \frac{0.1013 + 0.1011 + 0.1014 + 0.1009 + 0.1010 + 0.1019}{6} = 0.1013$$

Deviation:

$$d = |m - e|$$

$$\#1 : |0.1013 - 0.1013| = 0.0000$$

$$\#2 : |0.1013 - 0.1011| = 0.0002$$

$$\#3 : |0.1013 - 0.1014| = 0.0001$$

$$\#4 : |0.1013 - 0.1009| = 0.0004$$

$$\#5 : |0.1013 - 0.1010| = 0.0003$$

$$\#6 : |0.1013 - 0.1019| = 0.0006$$

Relative ppt deviation:

$$\#1: \quad \frac{0.0000 \times 1000}{0.1013} = 0$$

$$\#2: \quad \frac{0.0002 \times 1000}{0.1013} = 2$$

$$\#3: \quad \frac{0.0001 \times 1000}{0.1013} = 1$$

$$\#4: \quad \frac{0.0004 \times 1000}{0.1013} = 4$$

$$\#5: \quad \frac{0.0003 \times 1000}{0.1013} = 3$$

$$\#6: \quad \frac{0.0006 \times 1000}{0.1013} = 6$$

6. *Average Deviation.* An overall picture of the quality of data (in terms of precision) is the "average deviation". It is the average of all deviations for all measurements:

$$d_a = \frac{(d_1 + d_2 + d_3 + \ldots\ldots)}{n} \tag{3.5}$$

in which d_a is the average deviation and n is the number of deviations (corresponding to the number of measurements) considered.

7. *Standard Deviation.* Of more popular use for demonstrating data quality is the "standard deviation", which is similar to the average deviation:

$$s = \sqrt{\frac{(d_1^2 + d_2^2 + d_3^2 \ldots\ldots)}{(n - 1)}} \tag{3.6}$$

The term (n – 1) is referred to as the number of "degrees of freedom", and s represents the standard deviation. The significance of both d_a and s is that the smaller they are numerically, the more precise the data and thus presumably the more accurate the data. The standard deviation is used in most statistical methods in one way or another in evaluating reliability, that is, in establishing the level of confidence for arriving at the correct answer in the analysis.

8. *Variance.* A more statistically meaningful quantity for expressing data quality is the variance. It is defined as:

$$v = s^2 \tag{3.7}$$

It is considered to be more statistically meaningful because if the variation in the measurements is due to two or more causes the overall variance is the sum of the individual variances.

9. *Relative Standard Deviation.* One final deviation parameter is the relative standard deviation, s_R. It is analogous to d_R in the sense that d_R is obtained by dividing d by the mean and multiplying by 100 or

1000. In this case, s is divided by the mean and then multiplied by 100 or 1000:

$$s_R = \frac{s}{m} \tag{3.8}$$

and

$$\text{relative \% standard deviation} = s_R \times 100 \tag{3.9}$$

and

$$\text{relative ppt standard deviation} = s_R \times 1000 \tag{3.10}$$

The relative % standard deviation (Equation 3.9) is also called the coefficient of variance, c.v. Relative standard deviation relates the standard deviation to the value of the mean and represents a practical and popular expression of data quality.

Example 2

The following numerical results were obtained in a given laboratory experiment: 0.09376, 0.09358, 0.09385, and 0.09369. Calculate the relative ppt standard deviation.

Solution

We must calculate both the mean and the standard deviation in order to use Equation 3.8 and then 3.10.

First, the mean, m:

$$m = \frac{0.09376 + 0.09358 + 0.09385 + 0.09369}{4}$$

$$= 0.09372$$

Next, the deviations:

$$d_1 = 0.09376 - 0.09372 = 0.00004$$

$$d_2 = 0.09372 - 0.09358 = 0.00014$$

$$d_3 = 0.09385 - 0.09372 = 0.00013$$

$$d_4 = 0.09372 - 0.09369 = 0.00003$$

Then, the standard deviation:

$$s = \sqrt{\frac{(0.00004)^2 + (0.00014)^2 + (0.00013)^2 + (0.00003)^2}{(4 - 1)}}$$

$$= 0.000114$$

Finally, the relative ppt standard deviation:

$$s_R = \frac{s}{m} \times 1000 = \frac{0.000114}{0.09732} = 1.17$$

3.3 DISTRIBUTION OF RANDOM ERRORS

A graphical picture of the distribution of the results of an experiment, and thus a picture of the distribution of the random errors, is the "normal distribution curve". It is a plot of the frequency of occurrence of a result on the y-axis vs. the numerical value of the results on the x-axis. The normal (also called "Gaussian") distribution curve is bell-shaped, with the average of all results being the most frequently observed at the center of the bell shape and an equal drop-off in the frequency of occurrence in both directions away from the mean. (See Figure 3.2.) It is a picture of the precision (and thus presumably accuracy) of a given data set. The more points there are bunched around the mean and the sharper the drop-off away from the mean, the smaller the standard deviation, the more precise the data and the more confidence one has in any one result being correct. It can be shown that approximately 68% of the area under the curve falls within one standard deviation from the mean and approximately 95% of the area falls within two standard deviations from the mean. "Confidence limits" are sometimes defined based on the number of standard deviations from the mean. Typically, any measurement that is within predetermined confidence limits, and thus within the so-called confidence "interval", is deemed to be of sufficient accuracy to be reported or to be included in the calculation of the mean, which is then reported. Any measurements outside these confidence limits are of insufficient accuracy and are rejected.

Example 3

For the data in Example 1, a standard deviation of 3.0×10^{-4} is chosen as the confidence limit. Is the data acceptable? If not, try rejecting the value with the highest deviation and test it again.

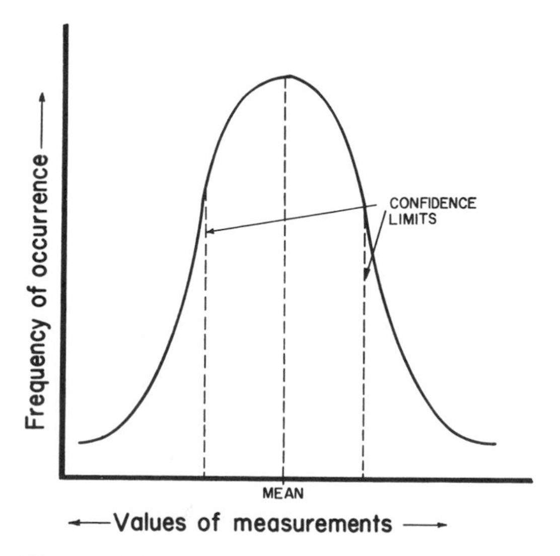

Figure 3.2. A normal distribution curve showing confidence limits that are a particular number of standard deviations from the mean, such as one or two.

Solution

$$s = \sqrt{\frac{(0.0000)^2 + (0.0002)^2 + (0.0001)^2 + (0.0004)^2 + (0.0003)^2 + (0.0006)^2}{(6 - 1)}}$$

$$= \sqrt{1.3 \times 10^{-7}} = 3.6 \times 10^{-4}$$

$3.6 \times 10^{-4} > 3.0 \times 10^{-4}$, therefore data are unacceptable.

Repeating, rejecting d = 0.0006:

$$s = \sqrt{\frac{(0.0000)^2 + (0.0002)^2 + (0.0001)^2 + (0.0004)^2 + (0.0003)^2}{(5 - 1)}}$$

$$= \sqrt{7.5 \times 10^{-8}} = 2.7 \times 10^{-4}$$

$2.7 \times 10^{-4} < 3.0 \times 10^{-4}$, therefore data are acceptable.

3.4 REJECTION OF DATA–THE Q TEST

In any set of data, those individual values that are the farthest from the mean are candidates for deletion ("rejection") from the data set. As discussed in Section 3.3, those values that lie outside the established confidence limits are rejected. For small data sets ($n < 10$), which a chemical analyst often encounters, a simple method to determine if the individual values that are farthest from the mean lie outside the confidence limits is the statistical test known as the Q test.

In this test, a value for "Q" is calculated and compared to a table of Q values that represent a certain percentage of confidence that the proposed rejection is valid. If the calculated Q value is greater than the value from the table, then the suspect value is rejected and the mean, which is what is to be reported as the answer to the analysis, is recalculated. If the Q value is less than or equal to the value from the table then the calculated mean is reported. "Q" is defined as follows:

$$Q = \frac{\text{gap}}{\text{range}} \tag{3.11}$$

where the "gap" is the difference between the suspect value and it nearest neighbor and the "range" is the difference between the lowest and highest values. A table of Q values for the 90% confidence level is given in Table 3.1.

Example 4

In a laboratory experiment in which the molarity of an acid solution was determined to four significant figures in five consecutive tests, the following results were obtained: 0.09589 *M*, 0.09534 *M*, 0.09597 *M*, 0.09581 *M*, 0.09612 *M*. Use the Q test at the 90% confidence level to determine what value is to be reported for this molarity.

Table 3.1. A table of Q values representing the 90% confidence level.

Number of Measurements	Q
3	0.94
4	0.76
5	0.64
6	0.56
7	0.51
8	0.47
9	0.44
10	0.41

Solution

$$\text{mean} = \frac{0.09589 + 0.09534 + 0.09597 + 0.09581 + 0.09612}{5}$$

$$= 0.09583$$

The suspect value is 0.09534, which is the value farthest from the mean. Using Equation 3.11,

$$Q = \frac{(0.09581 - 0.09534)}{(0.09612 - 0.09534)}$$

$$= 0.60$$

Since 0.64 (from Table 3.1) is greater than 0.60, the suspect value cannot be rejected and 0.09583 *M* is the value to be reported.

Despite the ease with which one can apply the Q test, its results are generally not considered useful unless n is greater than 4. Thus, for most laboratory work, unless the experiment is fast or the time required is not important, the Q test is not useful. It is considered more important to be able to look at the data and, from a common sense perspective, decide whether to repeat the analytical test or accept the data already found. This is certainly true if a recognizable error (a determinate error) is thought to have occurred in the case of the suspect value. Common sense would dictate that the suspect value be rejected and the analytical test repeated in that case.

3.5 STUDENT'S *t* TEST

From the discussion so far, it would seem that an inordinately large amount of time is needed to make enough measurements (n) to make the results meaningful each time an analysis is attempted so as to determine an appropriate confidence interval. A useful shorter method requiring a small number of measurements utilizes a parameter called the "probability factor", *t*. This parameter was the result of the work of W.S. Gossett, who published it under the pen name "Student" in 1908. It is often called "Student's *t*". The calculation establishes the confidence limits:

$$\text{analysis result} = m \pm \frac{t \times s}{\sqrt{n}} \tag{3.12}$$

and thus an analysis result, r, can be expressed as r ± c, where c is the confidence limit or interval.

The values of t depend on the number of measurements, n, or, more correctly, on the number of "degrees of freedom", n - 1, and are chosen based on the analyst's own probability or reliability requirements. That is, the values of t differ according to whether the analyst wants a 90% probability of being accurate, a 95% probability, a 99% probability, etc. Table 3.2 summarizes values for t for three different confidence limit levels and for a number of different degrees of freedom. The smaller the number of measurements, the less confidence the analyst has in his/her analysis result, as indicated by the larger values for t in Table 3.2. The larger the value of t the larger the confidence interval. The more measurements that are made the better. From a practical point of view, it is the analyst's decision as to how much time is spent making measurements weighed against a desired confidence in the results. An analyst can reject results that push the interval past a desired value.

Example 5

How would an analysis result be reported if the mean for a series of five measurements is 0.674, the standard deviation is 0.012, and a 90% confidence level is chosen?

Solution

From Table 3.2, for n - 1 = 4 and for a 90% confidence level, t = 2.13.

$$\text{analysis result} = m \pm \frac{t \times s}{\sqrt{n}}$$

Thus,

$$\text{analysis result} = 0.674 \pm \frac{2.13 \times 0.012}{\sqrt{5}}$$

$$= 0.674 \pm 0.011$$

Table 3.2. *t* values for calculating confidence limits.

n - 1	*t* (at 90% level)	*t* (at 95% level)	*t* (at 99% level)
1	6.3	12.7	63.7
2	2.9	4.3	9.9
3	2.35	3.2	5.8
4	2.13	2.78	4.6
5	2.02	2.57	4.03
6	1.94	2.45	3.71
7	1.90	2.37	3.50
8	1.86	2.31	3.36
9	1.83	2.26	3.25

3.6 STATISTICS IN QUALITY CONTROL

In addition to providing the basis for rejection of "bad" laboratory data, statistics can also be used as a basis for detecting unacceptable batches of raw materials or problems in a quality control laboratory. If a given analysis yields results that are quite different from what is expected, the "rejection" indicated by statistics results in the rejection of the batch or sample from being used or shipped, or whatever.

Confidence limits that are exceeded can also sometimes be indicated in an instrument reading that exceeds a statistically predetermined limit, rather than a final answer. An example is the gas chromatography (GC) analysis of cellophane extracts for formaldehyde. Injection of the extracts one after the other into a GC apparatus may give a series of formaldehyde peaks as shown in Figure 3.3. Sample number 5 in this chromatogram shows a peak that exceeds the pre-set confidence limit. Under these circumstances, the batch of cellophane from which this sample was acquired is discarded.

In addition, "quality control charts" are sometimes kept in a lab in order to detect problems or trends in analysis procedures or material analyzed. The idea is that over a period of time certain standard deviation values or confidence limit values are established which create a range of acceptable analysis results. A measurement falling outside the range or a trend away from

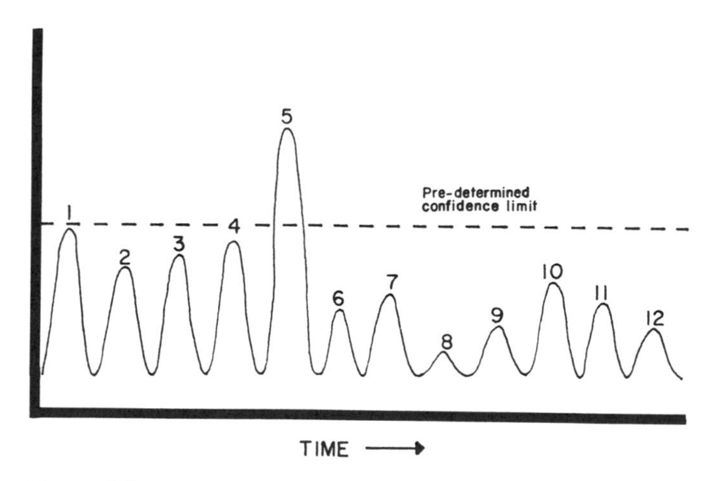

Figure 3.3. A hypothetical gas chromatogram of a series of cellophane extracts. See text for a brief description.

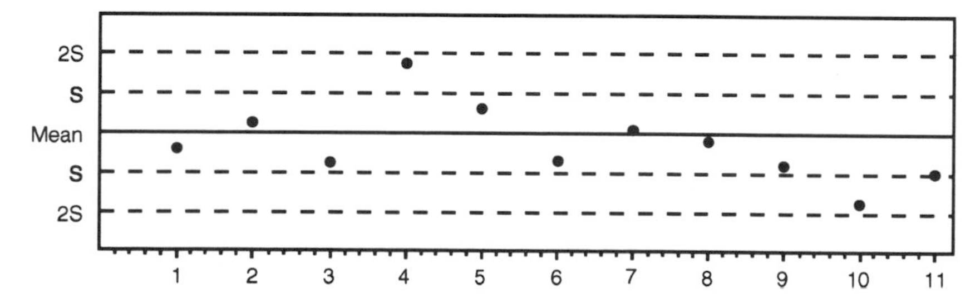

Figure 3.4. A typical quality control chart. Notice the measurement outside the range (defined by S) on day 4 and the trend away from the mean after about day 7.

the mean would then indicate a problem – possibly a determinate error, or a series of bad raw material batches, etc. An example of a quality control chart is shown in Figure 3.4.

3.7 APPLICATION TO EXPERIMENTS IN THIS TEXT

Experiments in this text which are analyses of "commercial" samples are designed to use preanalyzed samples for which the instructor has a "correct" answer which can then be compared to the students' results. In these cases, the instructor can calculate a relative deviation (or, more correctly, a relative "error") on which to base a student's grade:

$$E_R = \frac{|T - e|}{T} \times 1000 \qquad (3.13)$$

In this equation, T is the true value, e is the experimental determination (as before), and E_R is the parts per thousand error. A grading scheme based on the values of E_R for each measurement is then in effect.

Certain predetermined confidence limits are suggested in some experiments. (See, for example, Step 13, Experiment 7.) Any results outside these limits should then be rejected and the experiment repeated. In cases in which the correct answer is not known (such as Experiment 7) the instructor can base a grading scheme on the values of deviations rather than errors.

Other experiments are analyses of "synthetic" samples, or samples the instructor has prepared for the student to analyze. In this case a "correct" answer is again known and a grade can be based on parts per thousand error or percent error. Many experiments, especially instrumental analyses, however, are analyses of real samples, and relative errors cannot be calculated. Relative deviations, however, can be calculated if the run (or the "critical" measurement) is repeated several times, or deviations from the average results of all students can be calculated and perhaps the Q test can be used.

EXPERIMENTS

EXPERIMENT 5: STATISTICAL ANALYSIS OF WEIGHING

Introduction

This experiment has two objectives: (1) to demonstrate the concepts of laboratory statistics presented in this chapter, and (2) to check the precision of the analytical balances in your laboratory.

Procedure

1. Obtain the object to be weighed from your instructor. In the procedure that follows, be sure to protect the object from fingerprints, benchtop debris, and any other contamination or avoidable weighing error.
2. Proceed to weigh the object on five different analytical balances in your laboratory, first on one, then on the next, etc., until you've obtained the weight on each balance. Each time, before you obtain the weight on a given balance, dust the pan and level and zero the balance.

3. Repeat Step 2, cycling through the entire set of balances again so that you have two weights for the object from each balance.
4. Calculate the mean and determine the mode and the median. Then calculate the deviation and the relative ppt deviation for each measurement.
5. Calculate the standard deviation and the relative ppt standard deviation. Plot the normal distribution curve for your data.
6. Use the Q test at the 90% confidence level to determine if any data are rejectable on that basis. Then, use the Student's t test at the 90% confidence level and report the weight of the object with confidence limits included (weight ± ?).
7. Decide whether any of the balances you used are not functioning properly based on your results.

QUESTIONS AND PROBLEMS

1. Given the care with which laboratory equipment (balances, burets, instruments, etc.) are calibrated at the factory, why should the chemical analyst worry about errors?
2. Distinguish between determinate errors and indeterminate errors.
3. List five specific examples of determinate errors that are possible in the lab work you have performed up to this point.
4. An analyst determines that the analytical balance he/she used in a given analytical test is wrongly calibrated. Is this a determinate error or an indeterminate error? Explain.
5. While measuring the weight of an object on an analytical balance in the lab, a student realizes that there is a certain amount of error inherent in the measurement. Is such an error determinate or indeterminate? Explain.
6. Can a person really trust the results of a laboratory analysis to be accurate? Explain.
7. Distinguish clearly between "accuracy" and "precision".
8. A given analytical test was performed five times. The answers to the analysis are represented by the following values: 6.738, 6.738, 6.737, 6.739, 6.738. Would you say that these results are precise? Can you say that they are accurate? Explain both answers.
9. A given analytical test was performed five times. The answers to the analysis are represented by the following values: 37.23, 32.91, 45.38, 35.22, 41.81. Would you say that these results are precise? Would you say that they are accurate? Explain both answers.
10. Suppose the correct answer to the analysis represented in #8 above is 6.923. What can you say now about the precision and accuracy?
11. Discuss the differences between the mean, the median, and the mode.

12. Determine the mean, the median, and the mode for the data in #9 above.
13. What is the difference between "deviation" and "relative deviation"? Which is more useful and why?
14. What is meant by relative ppt deviation?
15. Calculate the relative ppt deviation for each of the following weight measurements:

Measurement #	Value
1	16.7724
2	16.7735
3	16.7722
4	16.7756
5	16.7729
6	16.7716
7	16.7720
8	16.7733

16. Calculate the average deviation, the standard deviation, and the relative standard deviation for the data in #15 above.
17. A series of eight absorbance measurements using an atomic absorption spectrophotometer are as follows: 0.855, 0.836, 0.848, 0.870, 0.859, 0.841, 0.861, 0.852. According to the instrument manufacturer, the precision of the absorbance measurements using this instrument should not exceed 1% relative standard deviation. Does it in this case?
18. Why is the relative standard deviation considered a popular and practical expression of data quality?
19. Explain this statement: "A relative standard deviation of 1% can be achieved in this experiment."
20. For the data in problem #15, a relative standard deviation of 0.050 ppt is chosen as the confidence limit. Are the data acceptable? If not, try rejecting values one at a time until the data are acceptable.
21. Plot the normal distribution curve for the following data set:
    5.69, 5.76, 5.65, 5.70, 5.71. 5.75, 5.69, 5.73, 5.71, 5.70, 5.72, 5.69, 5.68, 5.67, 5.74, 5.70, 5.68, 5.66, 5.70, 5.72, 5.69, 5.67, 5.71, 5.73, 5.72, 5.69, 5.68, 5.73, 5.70, 5.72, 5.66, 5.67, 5.68, 5.74.
22. Determine the range of the data in #21 above.
23. A given data set includes the first nine points in #21. If the value 5.59 is added to this data set, determine if this value is "rejectable" by the Q test at the 90% confidence level.
24. In a laboratory experiment in which the molarity of a base solution was determined, the following molarities were calculated after a series of six identical analytical determinations: 0.1033, 0.1031, 0.1034, 0.1030, 0.1025, 0.1033. Perform the Q test on these results at the 90% confidence level and determine what single answer is to be reported as the molarity of the base.

25. The following data are from a student's analysis of iron ore by gravimetric analysis, as in Experiment 3, but repeated numerous times. The true value, as given by the instructor, is also given. What is the (a) mean; (b) median; (c) mode; (d) deviation of each measurement; (e) relative ppt deviation of each measurement; (f) average deviation; (g) standard deviation; and (h)% standard deviation? Also, plot (i) the normal distribution curve, and calculate (j) ppt error of the mean.
    %Fe found: 54.21, 54.60, 54.07, 54.33, 54.11, 54.51, 54.33, 54.47, 54.29; true % = 54.28

26. Perform the Q test (90% confidence level) on the data set in #25 above and determine if any of the values should be rejected.

27. Based on a Student's t analysis at the 90% confidence level, how should the results of an instrument reading be reported if the mean for six measurements is 0.492 and the standard deviation is 0.013?

28. If the mean of nine analysis results is 14.28% and the standard deviation is 0.21, how should the results be reported according to a Student's t analysis at the 95% confidence level?

29. Calculate the confidence limits for the data in Example 1 for (a) 90% probability level, (b) 95% probability level, and (c) 99% probability level.

30. Calculate the confidence limits for the data in Problem #25 (above) for a 95% probability level for (a) the first five results, (b) for the first seven results, and (c) for all nine results.

31. Repeat Problem #20 (above), but rather than using the relative standard deviation to establish the confidence limits, calculate the formal confidence limits according to Equation 3.12 and a probability level of 95%.

32. What laboratory analysis results might cause a batch of raw material at a manufacturing plant to be rejected from potential use in the plant process? Explain.

33. How can a quality control chart signal a problem with a routine laboratory procedure?

34. In Experiment 7, Step 13, the three results of a student's normality determination were 0.09986, 0.09954, and 0.09995. Based on the criteria for rejection of data outlined there, must the student repeat the determination one more time? Explain.

35. A fourth determination of the normality in #34 (above) gave a result of 0.09964. Are the data now acceptable? What is considered now to be the accurate value of the normality?

Chapter 4

Chemical Equilibrium

4.1 REVIEW OF BASIC CONCEPTS

Chemical reactions relating to chemical analysis are frequently reversible reactions; that is, a "reverse" reaction competes with the "forward" reaction. After a time, these two opposing reactions occur at the same rate, and an equilibrium is established. Such a reaction system is written with the double arrow ($\rightleftarrows$) as in the following general reaction

$$aA + bB \rightleftarrows cC + dD \tag{4.1}$$

A and B represent reactant chemicals, C and D represent the products, and a, b, c, and d represent the coefficients required to balance the equation.

A mathematical expression called the equilibrium constant expression can be written as follows for Reaction 4.1:

$$K_{eq} = \frac{[C]^c\,[D]^d}{[A]^a\,[B]^b} \tag{4.2}$$

The brackets ([ ]) refer to the molar concentration of the chemical.* The value of K_{eq} is constant at a given temperature.

*More correctly, a quantity called the "activity" of the chemical should be used rather than the concentration. Activity is directly proportional to concentration, the "activity coefficient" being the proportionality constant. However, for most applications, especially at small concentrations, the activity coefficient is assumed to equal one, and the activity therefore is equal to the concentration. A discussion of activity and the activity coefficient is beyond the scope of this text, and these concepts will not be considered further.

The explanation of this constancy is based on Le Chatelier's principle. Le Chatelier's principle says that if a stress is placed on a chemical reaction system at equilibrium, the equilibrium shifts so as to relieve the stress. If the nature of the stress is the external addition of more reactant or product to the reaction system, a "concentration" stress, the equilibrium shifts so as to consume this added substance, and it will shift so as to maintain the constancy of K_{eq}. The value of K_{eq} remains the same regardless of concentration changes in the system.

For example, consider a solution of a weak acid, acetic acid, $HC_2H_3O_2$. The chemical equilibrium of this system is represented by

$$HC_2H_3O_2 \rightleftarrows H^+ + C_2H_3O_2^{-1} \tag{4.3}$$

The equilibrium constant expression, referred to here as K_a, the equilibrium constant for a weak acid ionization, is

$$K_a = \frac{[H^+]\,[C_2H_3O_2^{-1}]}{[HC_2H_3O_2]} \tag{4.4}$$

If more acetate ion, $C_2H_3O_2^{-1}$, is added from an external source, e.g., as $NaC_2H_3O_2$, the equilibrium in Reaction 4.3 shifts to the left, the added acetate and hydrogen ions are consumed, and more acetic acid is formed. Thus in Equation 4.4, $[H^+]$ decreases, $[C_2H_3O_2^{-1}]$ increases, $[HC_2H_3O_2]$ increases, and the overall result is that K_{eq} does not change.

4.2 THE SOLUBILITY OF PRECIPITATES

In gravimetric experiments involving reaction stoichiometry, such as in Experiments 2 and 3, the degree to which a precipitate dissolves is important. All precipitates dissolve to some extent, and for analyses such as the sulfate analysis, the iron analysis, and others, this dissolution is very slight – so slight that it should not affect the outcome. It is interesting and important to consider these facts, however, since, for one thing, the extent of the rinsing of the precipitate in the funnel could affect the outcome if it is overdone so that too much of the precipitate dissolves and is swept away. We will not address this problem directly, but the general considerations with respect to precipitate solubility will be discussed.

4.3 THE SOLUBILITY PRODUCT CONSTANT

The dissolving of a precipitate is an equilibrium process. For example, the dissolving of barium sulfate can be represented by

$$BaSO_4 \rightleftarrows Ba^{+2} + SO_4^{-2} \tag{4.5}$$

The dissolving of $BaSO_4$ to form the Ba^{+2} ion and the SO_4^{-2} ion occurs at the same rate as the recombination of the Ba^{+2} ion and the SO_4^{-2} ion, and thus we have equilibrium. A special form of the equilibrium constant is useful here. Because the amount of undissolved $BaSO_4$ does not affect the equilibrium system and it doesn't make sense to refer to the concentration of an undissolved chemical, the equilibrium expression is modified and only the product of the concentration of the two ions appears. This constant is referred to as the solubility product constant, K_{sp}. Thus, for the equilibrium system represented in Reaction 4.5 we have

$$K_{sp} = [Ba^{+2}]\,[SO_4^{-2}] \tag{4.6}$$

Some additional examples are the following:

$$AgCl \rightleftarrows Ag^{+1} + Cl^{-1} \qquad K_{sp} = [Ag^{+1}]\,[Cl^{-1}] \tag{4.7}$$

$$PbBr_2 \rightleftarrows Pb^{+2} + 2Br^{-1} \qquad K_{sp} = [Pb^{+2}]\,[Br^{-1}]^2 \tag{4.8}$$

$$Fe(OH)_3 \rightleftarrows Fe^{+3} + 3OH^{-} \qquad K_{sp} = [Fe^{+3}]\,[OH^{-}]^3 \tag{4.9}$$

The magnitude of the solubility product constant relates to the position of the equilibrium as it does for all equilibrium systems. For example, if the solubility of $BaSO_4$ is very slight, there will be very few ions present in the system represented by Reaction 4.5, and the concentrations multiplied together in the calculation of K_{sp} result in a very small number. Similarly, if the solubility of the precipitate is high (as with any soluble ionic compound), the ion concentrations are large, and the value of the solubility product constant is large. Some examples of the values of K_{sp} are listed in Appendix 2.

4.4 SOLUBILITY AND K_{sp}

Let us consider again Reaction 4.5:

$$BaSO_4 \rightleftarrows Ba^{+2} + SO_4^{-2} \tag{4.10}$$

The $BaSO_4$ that dissolves results in the formation of one Ba^{+2} ion and one SO_4^{-2} ion. Two things can be said regarding this process assuming that there is no Ba^{+2} or SO_4^{-2} from any other source.

1. The solubility of $BaSO_4$, which can be expressed as the number of moles of $BaSO_4$ that dissolve per liter, is equal to the concentration of the Ba^{+2} ion and also the concentration of the SO_4^{-2} ion in these saturated solutions.
2. The Ba^{+2} concentration and the SO_4^{-2} concentration are equal since for each $BaSO_4$ unit that dissolves one Ba^{+2} and one SO_4^{-2} ion form.

These two points can be summarized as follows:

$$\text{solubility} = s = [Ba^{+2}] = [SO_4^{-2}] \tag{4.11}$$

The expression 4.6 can be rewritten as follows:

$$K_{sp} = s \times s = s^2 \tag{4.12}$$

Thus, given the solubility of a solid ionic compound, the K_{sp} can be calculated. Conversely, given the K_{sp}, the solubility can be calculated.

Example 1

What is the solubility product constant, K_{sp}, of AgCl if the solubility of AgCl is 1.33×10^{-5} *M*?

Solution

$$AgCl \rightleftarrows Ag^{+1} + Cl^{-1}$$

$$K_{sp} = [Ag^{+1}]\,[Cl^{-1}] = (s) \times (s) = s^2$$

$$= (1.33 \times 10^{-5})^2 = 1.77 \times 10^{-10}$$

Example 2

What is the solubility of CuS if the K_{sp} is 1.27×10^{-46}?

Solution

$$CuS \rightleftarrows Cu^{+2} + S^{-2}$$

$$K_{sp} = [Cu^{+2}]\,[S^{-2}] = (s) \times (s) = s^2 = 9.0 \times 10^{-45}$$

$$s = \sqrt{1.27 \times 10^{-46}} = 1.13 \times 10^{-23}\ M$$

A slightly more complicated problem is one in which there are more than two ions formed each time one species dissolves. For example, consider the silver chromate precipitate, Ag_2CrO_4.

$$Ag_2CrO_4 \rightleftarrows 2Ag^{+1} + CrO_4^{-2} \qquad (4.13)$$

Two silver ions and one chromate ion form when one Ag_2CrO_4 species dissolves. This means that

$$[CrO_4^{-2}] = s \qquad (4.14)$$

but

$$[Ag^{+1}] = 2s \qquad (4.15)$$

and thus

$$K_{sp} = [Ag^{+1}]^2\,[CrO_4^{-2}] = (2s)^2(s) \qquad (4.16)$$

and

$$K_{sp} = 4s^3 \qquad (4.17)$$

Thus, Equation 4.17 is used whenever three ions result from a precipitate unit dissolving.

Example 3

What is the solubility product constant, K_{sp}, of CaF_2 if the solubility of CaF_2 is $3.32 \times 10^{-4}\ M$?

Solution

$$CaF_2 \rightleftarrows Ca^{+2} + 2F^{-1}$$

$$K_{sp} = 4s^3 = 4(3.32 \times 10^{-4})^3$$

$$K_{sp} = 1.46 \times 10^{-10}$$

When four ions result from a precipitate unit dissolving, the following applies:

$$K_{sp} = 27s^4 \tag{4.18}$$

as is seen in the following example.

Example 4

If the solubility product constant for $Fe(OH)_3$ is 2.64×10^{-39}, what is the solubility? (Assume for this example that there are no OH^{-1} ions present due to water ionization. The presence of ions from another source is briefly discussed in the next section.)

Solution

$$Fe(OH)_3 \rightleftarrows Fe^{+3} + 3OH^{-1}$$

$$K_{sp} = [Fe^{+3}]\,[OH^{-1}]^3 = (s)(3s)^3$$

$$K_{sp} = 27s^4 = 2.64 \times 10^{-39}$$

$$s^4 = \frac{2.64 \times 10^{-39}}{27} = 9.78 \times 10^{-41}$$

$$s = \sqrt[4]{9.78 \times 10^{-41}} = 9.94 \times 10^{-11}$$

Table 4.1 summarizes the information required for the above examples.

4.5 ADDITION OF IONS FROM AN EXTERNAL SOURCE

The discussion in the previous section is based on the premise that there are no ions present except those which result from the ionization of the precipitate. As pointed out earlier, if sulfate ions from, say, Na_2SO_4 are added to the equilibrium mixture containing $BaSO_4$, Ba^{+2}, and SO_4^{-2}, Le Chatelier's principle is in effect, and K_{sp} remains the same. What have changed are the concentrations of the two ions. $[Ba^{+2}]$ is decreased and $[SO_4^{-2}]$ is increased. If we add Na_2SO_4 so that the sulfate concentration is 0.10 M, the $[Ba^{+2}]$, although changed, can still be determined:

$$K_{sp} = [Ba^{+2}]\,[SO_4^{-2}] \tag{4.19}$$

$$[Ba^{+2}] = \frac{K_{sp}}{[SO_4^{-2}]} = \frac{K_{sp}}{0.10} \tag{4.20}$$

Table 4.1. A summary of the relationships between K_{sp} and s according to the number of ions that form from one precipitate unit, assuming that no such ions are present from another source.

2 ions	$K_{sp} = s^2$
3 ions	$K_{sp} = 4s^3$
4 ions	$K_{sp} = 27s^4$

4.6 CHEMICAL EQUILIBRIUM AND WEAK ACIDS

When weak acids dissolve in water, we again have a chemical equilibrium system in which a species is in equilibrium with its ions. An example, as indicated previously, is acetic acid, $HC_2H_3O_2$.

$$HC_2H_3O_2 \rightleftarrows H^{+1} + C_2H_3O_2^{-1} \tag{4.21}$$

The equilibrium constant for this reaction is

$$K_a = \frac{[H^{+1}]\,[C_2H_3O_2^{-1}]}{[HC_2H_3O_2]} \tag{4.22}$$

The symbol "K_a", or K with "a" as the subscript rather than "eq", is a symbol reserved for the equilibrium constant for weak acid ionization. The "a" refers to "acid" just as "sp" referred to solubility product.

The concentration of the nonionized species $[HC_2H_3O_2]$ is retained in the K_a expression because the $HC_2H_3O_2$ is dissolved even though it isn't ionized. If it weren't dissolved, and this was the case with precipitates, it would not appear in the K_a expression. This places a different perspective on the calculations involved and results in a much more complicated absolute solution. As we'll see, however, there is an important approximation that can be made because the acid is weak, and this greatly simplifies the problem.

The question that is to be answered is: "What is the pH of a solution of a weak acid in which the total concentration (ionized + nonionized) is given?" For the purposes of this discussion let us use acetic acid as the example, and let us say that the total acid concentration is 0.10 *M*. In other words, "What is the pH of a 0.10 *M* solution of acetic acid?" The solution to this problem is to determine $[H^{+1}]$ since $pH = -\log[H^{+1}]$.

The solution relies on the use of K_a (Equation 4.22). The first observation to make is that for every $HC_2H_3O_2$ molecule that dissociates, one H^{+1} and one $C_2H_3O_2^{-1}$ are formed. Thus we can say

$$[C_2H_3O_2^{-1}] = [H^{+1}] \tag{4.23}$$

and the K_a expression is simplified as follows:

$$K_a = \frac{[H^{+1}]^2}{[HC_2H_3O_2]} \tag{4.24}$$

Next, we need to know K_a and $[HC_2H_3O_2]$. K_a can be found quite simply by referring to a table of K_a values in a handbook or reference book. (See Appendix 3.) For acetic acid, $K_a = 1.8 \times 10^{-5}$. The concentration of undissociated but dissolved acid $HC_2H_3O_2$ is another story. It is only a part of the total concentration, with the other part being that which is dissociated.

For this example, which is a "monoprotic" acid – it has only one "acidic" hydrogen – the number that do dissociate is the same as the number of H^{+1} or $C_2H_3O_2^{-1}$ ions that form, since for every $HC_2H_3O_2$ that dissociates one H^{+1} ion and one $C_2H_3O_2^{-1}$ ion form. Thus, if we designate the total acid concentration (dissociated + undissociated) as C_{HAc} (HAc meaning acetic acid), we can then say

$$C_{HAc} = [HC_2H_3O_2] + [H^{+1}] \tag{4.25}$$

and therefore

$$[HC_2H_3O_2] = C_{HAc} - [H^{+1}] \tag{4.26}$$

Substituting this back into Equation 4.24 gives us

$$K_a = \frac{[H^{+1}]^2}{C_{HAc} - [H^{+1}]} \tag{4.27}$$

Since C_{HAc} was given in the problem as 0.10 *M* we have only one unknown in Equation 4.27, but solving for it is not so easy:

$$[H^{+1}]^2 = K_a(C_{HAc} - [H^{+1}]) \tag{4.28}$$

$$[H^{+1}]^2 = K_aC_{HAc} - K_a[H^{+1}] \tag{4.29}$$

$$[H^{+2}]^2 + K_a[H^{+1}] - K_aC_{HAc} = 0 \tag{4.30}$$

The problem with obtaining a solution (a value for $[H^{+1}]$) here is that there is a squared term ($[H^{+1}]^2$) as well as an unsquared term ($Ka[H^{+1}]$) for the same variable ($[H^{+1}]$). The equation is called a quadratic equation, which has the general form

$$ax^2 + bx + c = 0 \tag{4.31}$$

In our example, $a = 1$, $x = [H^{+1}]$, $b = K_a$, and $c = -K_aC_{HAc}$.

Quadratic equations are conveniently solved with the use of the so-called quadratic formula, which is

$$x = \frac{-b \pm \sqrt{b^2 - 4ac}}{2a} \tag{4.32}$$

and thus an exact solution can be obtained if desired. There is, however, an approximation that can be made in Equation 4.26 which greatly simplifies this kind of problem. Since the discussion involves a weak acid, the number of acid molecules that dissociate is quite small compared to those that do not. Thus, C_{HAc} is much greater than $[H^{+1}]$. Subtracting a very small number from a significantly larger number, as we would do in Equation 4.26, results in a number not too different from the larger number. In other words,

$$C_{HAc} - [H^{+1}] = C_{HAc} \tag{4.33}$$

or

$$[HC_2H_3O_2] = C_{HAc} \tag{4.34}$$

Thus, Equation 4.27 becomes

$$K_a = \frac{[H^{+1}]^2}{C_{HAc}} \tag{4.35}$$

and the $[H^{+1}]$ is found by rearranging this equation:

$$[H^{+1}]^2 = K_aC_{HAc} \tag{4.36}$$

$$[H^{+1}] = \sqrt{K_aC_{HAc}} \tag{4.37}$$

Thus, for our example,

$$[H^{+1}] = \sqrt{(1.8 \times 10^{-5})(0.10)} \quad (4.38)$$

$$[H^{+1}] = 1.3 \times 10^{-3}\,M \quad (4.39)$$

$$pH = -\log[H^{+1}] = 2.89 \quad (4.40)$$

(See Appendix 1 for the rule regarding significant figures for logarithms.)

Example 5

What is the pH of a 0.10 *M* solution of benzoic acid (HBn), $HC_7H_5O_2$, the K_a for which is 6.6×10^{-5}?

Solution

$$HC_7H_5O_2 \rightleftarrows H^{+1} + C_7H_5O_2^{-1}$$

$$K_a = \frac{[H^{+1}]\,[C_7H_5O_2^{-1}]}{[HC_7H_5O_2]}$$

$$K_a = \frac{[H^{+1}]^2}{C_{HBn}}$$

$$[H^{+1}]^2 = K_a \times C_{HBn} = 6.6 \times 10^{-5} \times 0.10$$

$$[H^{+1}] = 2.6 \times 10^{-3}\,M; \qquad pH = 2.59$$

In addition to these examples of monoprotic acids, one can also consider polyprotic acids—those that have more than one acidic hydrogen. Such acids present an additional challenge to the student of analytical chemistry because the loss of hydrogen ions occurs in steps, and there is an equilibrium constant associated with each step:

$$H_3A \rightleftarrows H^{+1} + H_2A^{-1} \quad (4.41)$$

$$H_2A^{-1} \rightleftarrows H^{+1} + HA^{-2} \quad (4.42)$$

$$HA^{-2} \rightleftarrows H^{+1} + A^{-3} \quad (4.43)$$

and

$$K_1 = \frac{[H^{+1}]\,[H_2A^{-1}]}{[H_3A]} \quad (4.44)$$

$$K_2 = \frac{[H^{+1}]\,[HA^{-2}]}{[H_2A^{-1}]} \quad (4.45)$$

$$K_3 = \frac{[H^{+1}]\,[A^{-3}]}{[HA^{-2}]} \quad (4.46)$$

In most cases, the problem is simplified by working with only the first ionization step, since the other two usually contribute a negligible number of hydrogen ions to the system. Sometimes, however, it is important to know which ionic species predominates when the pH is made more basic through addition of a strong base. (See, for example, the EDTA example in Section 7.3). The more basic the pH becomes, the more important the additional ionization steps become. As OH^{-1} ions are added, H^{+1} ions are removed ($H^{+1} + OH^{-1} \rightarrow H_2O$), and, by Le Chatelier's principle, Steps 2 and 3 ultimately become more involved.

4.7 WEAK BASES

The principles applied to weak acid equilibria in the last section also apply to weak bases:

$$B + H_2O \rightleftarrows BH^{+1} + OH^{-1} \tag{4.47}$$

$$K_b = \frac{[BH^{+1}][OH^{-1}]}{[B]} \tag{4.48}$$

$$K_b = \frac{[OH^{-1}]^2}{C_B} \tag{4.49}$$

$$[OH^{-1}]^2 = K_bC_B \tag{4.50}$$

$$[OH^{-1}] = \sqrt{K_bC_B} \tag{4.51}$$

The calculation of the pH here involves the use of the ion-product constant for water, which is symbolized K_w and defined according to the ionization of the water molecule:

$$H_2O \rightleftarrows H^{+1} + OH^{-1} \tag{4.52}$$

In all water solutions, regardless of pH, this ion-product constant is

$$K_w = [H^{+1}][OH^{-1}] = 1.00 \times 10^{-14} \tag{4.53}$$

If we define $pK_w = -\log K_w$ and $pOH = -\log[OH^{-1}]$ we can write

$$pK_w = pH + pOH = 14.00 \tag{4.54}$$

Thus, given pOH, we can calculate pH by subtracting from 14.00:

$$pH = 14.00 - pOH \tag{4.55}$$

Example 6

What is the pH of a 0.10 *M* solution of ammonia (NH_3), $K_b = 1.7 \times 10^{-4}$?

Solution

$$NH_3 + H_2O \rightleftarrows NH_4^{+1} + OH^{-1}$$

$$K_b = \frac{[NH_4^{+1}][OH^{-1}]}{[NH_3]} = 1.7 \times 10^{-4}$$

$$K_b = \frac{[OH^{-1}]^2}{0.10} = 1.7 \times 10^{-4}$$

$$[OH^{-1}]^2 = C_{NH_3} \times 1.7 \times 10^{-4} = 0.10 \times 1.7 \times 10^{-4}$$

$$[OH^{-1}] = \sqrt{0.10 \times 1.7 \times 10^{-4}} = 4.1 \times 10^{-3}$$

$$pOH = 2.39; \quad pH = 11.61$$

4.8 BUFFER SOLUTIONS

An important direct laboratory application of chemical equilibrium processes and Le Chatelier's principle involves the use of buffer solutions. Buffer solutions are solutions that resist changes in pH even when strong acids or strong bases are added. The explanation for this resistance lies in the fact that the added hydrogen ions or hydroxide ions are immediately consumed by shifting equilibrium processes in the solution.

Consider as an example a solution of acetic acid and sodium acetate in appropriate concentrations for a pH of 4.70. The following chemical equilibrium process occurs:

$$HC_2H_3O_2 \rightleftarrows H^{+1} + HC_2H_3O_2^{-1} \qquad (4.56)$$

and there are a large number of acetate ions present because of the presence of dissolved sodium acetate. When a strong acid is added, a large number of H^{+1} is added, which one would expect would cause a drastic change in the pH. Actually, because of Le Chatelier's principle, the hydrogen ions from the strong acid are consumed by combination with the acetate ions present, the equilibrium shifts to the left in Equation 4.56, and thus there is no significant change in the pH.

A similar reaction occurs in the case of a strong base. Upon the addition of OH^{-1} ions, a reaction with the H^{+1} ions in Equation 4.56 occurs:

$$H^{+1} + OH^{-1} \rightleftarrows H_2O \qquad (4.57)$$

and the H^{+1} is removed from the equilibrium. The pH, however, does not change because the H^{+1} is replenished by the shifting of the equilibrium in Equation 4.56 to the right.

Another example of a buffer solution is a solution of ammonium hydroxide and ammonium chloride (NH_4Cl):

$$NH_4OH \rightleftarrows NH_4^{+1} + OH^{-1} \qquad (4.58)$$

In this case, there is a reservoir of NH_4^{+1} ions from the NH_4Cl which will consume OH^{-1} ions when the strong base is added. When H^{+1} ions are added, the OH^{-1} ions in Equation 4.58 will consume them, and once again the pH doesn't change in either case. (Refer to Section 7.3 and Experiments 11 and 12 for further discussion and application of this buffer solution.)

Buffer solutions can be prepared quite simply by appropriate combination of weak acids or bases and their salts, as in the above examples. Although commercially prepared buffer solutions are available, these are most often utilized solely for pH meter calibration and not for adjusting or maintaining a chemical reaction system at a given pH. It is not surprising, therefore, that the analyst often needs to prepare his/her own solutions for this purpose. It then becomes a question of what proportions of the acid, or base, and its salt should be mixed to give the desired pH.

The answer is in the expression for the ionization constant, K_a or K_b, where the ratio of the salt concentration to the acid concentration is found. In the case of a weak acid:

$$HA \rightleftarrows H^{+} + A^{-} \qquad (4.59)$$

$$K_a = \frac{[H^+][A^-]}{[HA]} \tag{4.60}$$

and in the case of a weak base:

$$B + H_2O \rightleftarrows BH^+ + OH^- \tag{4.61}$$

$$K_b = \frac{[BH^+][OH^-]}{[B]} \tag{4.62}$$

Knowing the value of K_a or K_b for a given weak acid or base, and knowing the desired pH value, one can calculate the ratio of salt concentration to acid (or base) concentration that will produce the given pH. Rearranging Equation 4.60, for example, would give

$$[H^+] = K_a \times \frac{[HA]}{[A^-]} \tag{4.63}$$

Taking the negative logarithm of both sides would give

$$pH = pK_a - \log \frac{[HA]}{[A^-]} \tag{4.64}$$

or

$$pH = pK_a + \log \frac{[A^-]}{[HA]} \tag{4.65}$$

The equivalent expression derived for a weak base would be

$$pOH = pK_b + \log \frac{[BH^+]}{[B]} \tag{4.66}$$

Thus, for a weak acid with a given K_a (or pK_a) and a given ratio of salt concentration to acid concentration, the pH may be calculated. Or, given the desired pH and the K_a (pK_a), the ratio of salt concentration to acid concentration can be calculated and the buffer subsequently prepared. These equations (4.63, 4.64, 4.65, and 4.66) are each a form of the Henderson-Hasselbalch equation for dealing with buffer solutions.

Example 7

What is the pH of an acetic acid/sodium acetate solution if the acid concentration is 0.10 *M* and the sodium acetate concentration is 0.20 *M*? (The K_a of acetic acid is 1.8×10^{-5}.)

Solution

Utilizing Equation 4.65, we have

$$pH = -\log(1.8 \times 10^{-5}) + \log \frac{0.20}{0.10}$$

$$pH = 5.04$$

Example 8

What is the pH of a solution of THAM and THAM hydrochloride (THAM is also sometimes called TRIS – see Table 4.2) if the base concentration is 0.20 *M* and the hydrochloride concentration is 0.25 *M*? (The K_b for THAM is 1.2×10^{-6}.)

Solution

Utilizing Equation 4.66, we have

$$pOH = -\log(1.2 \times 10^{-6}) + \log \frac{0.25}{0.20}$$

$$pOH = 6.02$$

$$pH = 14.00 - 6.02 = 7.98$$

Example 9

What ratio of the concentration of sodium trichloroacetate to trichloroacetic acid is needed to prepare a buffer solution of pH = 2.0 using these chemicals? (The K_a of trichloroacetic acid is 2.0×10^{-1}.)

Solution

Rearranging Equation 4.65, we have

$$\log \frac{[A^-]}{[HA]} = pH - pK_a = 2.0 - 0.7 = 1.3$$

$$\frac{[A^-]}{[HA]} = 20.0$$

It should be stressed that since the K_a, or K_b, enters into the calculation, how weak the acid or base is dictates what is a workable pH range for that acid or base. Table 4.2 gives commonly used examples of acid/salt and base/salt combinations and notes the applicable pH range for each. It should also be stressed that the pH value of an actual buffer solution prepared by mixing quantities of the weak acid or base and its salt based on the calcu-

Table 4.2. Commonly used examples of acid/salt and base/salt combinations and corresponding useful pH ranges.

Combination	pH Range
Trichloroacetic acid + sodium trichloroacetate	1.8–3.8
Acetic acid + sodium acetate	3.7–5.7
Sodium dihydrogen phosphate + sodium monohydrogen phosphate	6.1–8.1
THAM[a] (also called TRIS) + THAM hydrochloride	7.5–9.0
Ammonium hydroxide + ammonium chloride	8.3–10.3

[a] Tris-(hydroxymethyl)amino methane

lated ratio will likely be different from what was calculated. The reason for this is the use of approximations in the calculations, such as the use of molar concentration rather than activity and the fact that the values for the non-ionized acid and base concentrations in the denominators in Equations 4.60 and 4.62 are actually approximations (C_{HA} rather than [HA]). A better method of preparing a buffer solution would be to prepare a solution of the salt (the concentration isn't important) and then add a solution of a strong acid (or base, if the salt is a salt of a weak base) until the pH, as measured by a pH meter, is the desired pH. The strong acid (or base), in combination with the salt, creates the equilibrium required for the buffering action.

For example, to prepare a pH = 9 buffer solution, one would prepare a solution of ammonium chloride (refer to Table 4.2) and then add a solution of sodium hydroxide while stirring and monitoring the pH with a pH meter. The preparation is complete when the pH reaches 9. The equilibrium created would be

$$NH_4OH \rightleftarrows NH_4^+ + OH^-$$

and the solution would be a solution containing a weak base (NH_4OH) and its salt (NH_4Cl).

Recipes for standard buffer solutions can be useful, however. Table 4.3 gives specific directions for preparing some popular buffer solutions.

Many applications for buffer solutions are found in the analytical laboratory. It is frequently necessary to have solutions which do not change pH during the course of an experiment.

Table 4.3. Recipes for some of the more popular buffer solutions.

pH = 4.0 phthalate buffer	Dissolve 10.12 g of dried potassium hydrogen phthalate (KHP) in 1 L of solution.
pH = 6.9 phosphate buffer	Dissolve 3.39 g of dried potassium dihydrogen phosphate and 3.53 g of dried sodium monohydrogen phosphate in 1 L of solution.
pH = 10.0 ammonia buffer	Dissolve 70.0 g of dried ammonium chloride and 570 mL of concentrated ammonium hydroxide in 1 L of solution.

QUESTIONS AND PROBLEMS

1. What is the difference between "concentration" and "activity"? What is an "activity coefficient"?
2. State Le Chatelier's principle.
3. Write the reaction representing the chemical equilibrium occurring between NaCl and its ions in a saturated solution of NaCl.
4. Write the solubility product constant expression for the chemical equilibrium between $PbBr_2$ and its ions.
5. The solubility of PbI_2 in water is 1.29×10^{-3} *M*. What is the K_{sp} of PbI_2?
6. The K_{sp} of $BaCO_3$ is 2.58×10^{-9}. What is the solubility of $BaCO_3$?

7. The solubility product constant for Ag_2SO_4 is 4.35×10^{-13}. What is the solubility of Ag_2SO_4?

8. In problem #7 above, the expression $K_{sp} = [Ag^+]^2[SO_4^{-2}]$ is used. The substitution $[Ag^+] = 2s$ is made and this is eventually followed by a squaring of the quantity 2s. Explain why $[Ag^+] = 2s$ is valid, and also explain why the quantity 2s is squared after the substitution is made.

9. What is the solubility of the precipitate $Pb(OH)_2$ if its solubility product constant is 7.4×10^{-12}?

10. What is the solubility of the precipitate BaF_2 given that the solubility product constant is 1.84×10^{-7}?

11. What is the solubility product constant of the precipitate Ag_2CO_3 if its solubility is 1.28×10^{-4} *M*?

12. The solubility of $Co(CN)_2$ in water is 4.18×10^{-3} g per 100 mL. What is the K_{sp}?

13. What is the concentration of silver ions in a saturated solution of Ag_2CO_3 if the carbonate ion concentration is 0.10 *M*? (The K_{sp} for Ag_2CO_3 is 8.45×10^{-12}.)

14. What is the pH of a 0.50 *M* solution of chloroacetic acid ($ClCH_2COOH$) if the K_a is 1.40×10^{-3}?

15. What is the pH of a 0.25 *M* solution of HCN with $K_a = 7.2 \times 10^{-10}$?

16. The pH of a 0.10 *M* solution of a weak acid is 4.6. What is the K_a for this acid?

17. What is the pH of a 0.30 *M* solution of iodoacetic acid, which has a K_a of 7.5×10^{-4}?

18. What is the pH of a 0.25 *M* solution of bromoacetic acid, which has a K_a of 2.05×10^{-3}?

19. What is the pH of a 0.35 *M* solution of ascorbic acid, given that the K_a for this acid is 7.94×10^{-5}?

20. What is the pH of a 0.20 *M* solution of iodoacetic acid, which has a K_a of 7.5×10^{-4}?

21. State in words and with an equation what the approximation is when performing the calculation required in #15, for example.

22. What is the pH of a 0.10 *M* solution of H_3A, where $K_1 = 7.2 \times 10^{-3}$, $K_2 = 2.46 \times 10^{-8}$, and $K_3 = 8.1 \times 10^{-13}$? (Hint: Assume that the second and third ionization steps do not contribute significantly to the pH.)

23. What is the pH of a 0.15 *M* NH_4OH solution ($K_b = 1.7 \times 10^{-4}$)?

24. In the process of deriving the mathematical equation needed for Problem #14 (for example), the expression for the K_a of a weak acid changes from

$$K_a = \frac{[H^+][A^-]}{[HA]} \qquad \text{(equation 1)}$$

to

$$K_a = \frac{[H^+]^2}{C_{HA}} \qquad \text{(equation 2)}$$

Explain the reasoning used to change the numerator and denominator in going from equation 1 to equation 2.

25. Concerning the equilibrium system of the weak acid, HA, dissolved in water, and also the equilibrium system of the precipitate, AgCl, in water, indicate which of the following are correct, which are incorrect, and which are approximations.

(a) $K_a = \frac{[H^+][A^-]}{[HA]}$

(b) $[H^+] = [HA]$

(c) $C_{HA} = [H^+] + [HA]$

(d) $[H^+] = [A^-]$

(e) $C_{HA} = [H^+]$

(f) $K_{sp} = s^2$

(g) $K_{sp} = 4s^3$

(h) $[Ag^+] = [Cl^-]$

(i) $K_{sp} = \frac{[Ag^{+1}][Cl^-]}{[AgCl]}$

(j) $C_{HA} = [HA]$

26. Consider the following equilibrium system for a precipitate, AgBr:

$$AgBr \rightleftarrows Ag^+ + Br^-$$

and also the following equilibrium system for a weak acid, HA:

$$HA \rightleftarrows H^+ + A^-$$

Some of the following statements concerning these systems are absolutely true, some are approximations, and some are false. On the line to the left of each place either a "T" (for true), an "A" (for approximation), or an "O" (for false). (Assume that no ions from any external source are present.)

(a) ______ $[Ag^+] = [Br^-]$

(b) ______ $K_{sp} = 4s^3$

(c) ______ $[H^+] = [A^-]$

(d) ______ $[HA] = C_{HA}$

(e) ______ $[Ag^+] = s$

(f) ______ $C_{HA} - [H^+] = [HA]$

(g) ______ $C_{HA} - [H^+]$

(h) ______ $K_{sp} = [Ag^+][Br^-]$

(i) ______ $[Br^-] = s$

(j) ______ $[A^-] = [HA]$

27. (a) What is the hydrogen ion concentration in a 0.20 *M* solution of lactic acid (HL) which has a K_a of 8.4×10^{-4}?
    (b) Comment on the validity of the following when calculating the result in (a)
    (i) $[H^+] = [L^-]$
    (ii) $C_{HA} = [HA]$

28. Lead sulfate, $PbSO_4$, has a solubility product constant of 1.6×10^{-8}.
    (a) What is the solubility of $PbSO_4$ in water? (Show all steps.)
    (b) Comment on the validity of the following when calculating the result in (a):
    (i) $[Pb^{+2}] = [SO_4^{-2}]$
    (ii) $[Pb^{+2}]$ = solubility

The next six questions are multiple choice.

29. Which of the following is a correct expression concerning the solubility equilibrium of the precipitate Ag_2O?
    (a) $K_{sp} = s^2$
    (b) $K_{sp} = 4s^3$
    (c) $K_{sp} = 27s^4$
    (d) $K_{sp} = \sqrt[3]{s/4}$

30. Which of the following is not a correct expression relating to the solubility equilibrium of $PbSO_4$?
    (a) $[Pb^{+2}] = [SO_4^{-2}]$
    (b) $[Pb^{+2}] = s$
    (c) $[SO_4^{-2}] = s$
    (d) $[PbSO_4] = s$

31. Which of the following is a correct way to calculate the solubility of PbI_2?
    (a) $s = 2[I^-]$
    (b) $s = K_{sp}$
    (c) $s = \sqrt[3]{K_{sp}/4}$
    (d) $s = \frac{[Pb^{+2}]}{2}$

32. Which of the following is an approximation that is made when calculating the pH of a 0.10 *M* solution of acetic acid, HAc?

    (a) $C_{HAc} = [HAc]$

    (b) $[H^+] = [Ac^-]$

    (c) $C_{HAc} = [H^+]$

    (d) $C_{HAc} - [H^+] = [HA]$

33. Which of the following is the correct way to calculate the $[H^+]$ of a 0.10 *M* solution of HAc once the approximation is made?

    (a) $[H^+] = [Ac^-]$
    (b) $[H^+] = s^2/C_{HAc}$
    (c) $[H^+] = \sqrt{K_aC_{HAc}}$
    (d) $[H^+] = [H^+]^2/C_{HAc}$

34. Which of the following is an incorrect expression concerning the equilibrium of HAc and its ions, H^+ and Ac^-?

    (a) $[Ac^-] = [HAc]$

    (b) $K_a = [H^+]^2/[HAc]$

    (c) $[H^+] = [Ac^-]$

    (d) $C_{HAc} = [H^+] + [HAc]$

35. Consider the NH_4OH/NH_4Cl buffer system.
    (a) Write the chemical equation representing the equilibrium involved in the buffering action.
    (b) Tell specifically what occurs when a strong acid is added and why the pH doesn't change appreciably.

36. What is the Henderson/Hasselbalch equation? Tell how it is useful in the preparation of buffer solutions.

37. What is the pH of a solution of chloroacetic acid (0.25 *M*) and sodium chloroacetate (0.20 *M*)?

38. What is the pH of a solution that is 0.30 *M* in iodoacetic acid and 0.40 *M* in sodium iodoacetate?

39. What is the pH of a solution of THAM and THAM hydrochloride if the THAM concentration is 0.15 *M* and the THAM hydrochloride concentration is 0.40 *M*?

40. What ratio of the concentrations of acetic acid to sodium acetate are needed to prepare a solution of pH = 4.0?

41. A buffer solution of pH 3.0 is needed. From Table 4.2, select a weak acid/salt combination that would give that pH and calculate the ratio of acid to salt concentrations that would give that pH.

Chapter 5

Introduction to Titrimetric Analysis

5.1 BACKGROUND

The second wet chemical method of analysis to be discussed is the method known as "titrimetric" or "volumetric" analysis. The difference between this method and gravimetric analysis is that the measurement of volume as well as the measurement of weight is an integral part of the procedure. Let us first develop this method in terms of the stoichiometry involved.

As with gravimetric analysis, the sample is weighed on an analytical balance and dissolved, but the constituent is not converted to a form that is isolated and weighed, as in the gravimetric procedure. In the titrimetric procedure, a substance that reacts with the constituent is added, as before, but rather than measuring the amount of a precipitate formed, one measures the exact amount of the added reagent needed to consume the constituent. The question then becomes: "How does one get back to the constituent if the amount of a product formed is not determined?" Recall that in stoichiometry one can measure the amount of any substance involved in a chemical reaction, reactant or product and then calculate the corresponding stoichiometric amount of any other substance involved, be it a reactant or product. In titrimetric analysis, the amount of the added reagent required to react with the constituent is what is measured and converted back to the constituent. Figure 5.1 should help clarify this basic difference between the gravimetric and titrimetric techniques.

How then does one measure the amount of the added reagent and how does one know at what point the constituent has been exactly consumed? The answers to these two questions are the key points of a titrimetric experiment and will be addressed in quite some detail in the sections and

Gravimetric Analysis

C + AR ⟶ (P1) + P2

The weight of the product of the reaction (a precipitate) is measured and converted to the weight of the constituent via a stoichiometric calculation.

Titrimetric Analysis

C + (AR) ⟶ P1 + P2

The amount of the added reagent needed to consume all of the constituent is measured and converted to the weight of the constituent via a stoichiometric calculation.

Figure 5.1. Comparison of gravimetric analysis to titrimetric analysis with respect to the stoichiometric calculation involved. In the reactions, C = constituent, AR = added reagent, P1 = first product, P2 = second product.

chapters to follow. For now, suffice it to say that (1) the amount of added reagent is measured by measuring the volume of a solution of the reagent as it is added, and (2) the total consumption of the constituent is signaled via some visual sign, such as a color change, in the reaction solution.

5.2 TITRATION DEFINED

At this point, some definitions of terms are useful. First, the volume of the added reagent is typically measured with a piece of glassware called a "buret" and the color change is caused by a substance known as an "indicator". An experiment in which a solution of a reactant is added to a reaction flask with the use of a buret is called a "titration". The solution in the buret, as well as the substance dissolved in the solution in the buret, is called the "titrant". The substance in the reaction flask with which the titrant reacts is called the "titrand", or "substance titrated". The experimental setup for a titration is shown in Figure 5.2. The buret is a sort of specialized graduated cylinder. There are two major characteristics which differentiate it from an ordinary graduated cylinder. First, rather than having to invert it to release the solution into a receiving vessel, a buret has a "stopcock" at the bottom, which is a valve that can be opened to release solution into a receiving vessel. Second, because of the use of the stopcock, it is convenient to be able to measure the solution, via the graduation lines, from the top down. The zero line is at the top of the cylinder and, as the solution is dispensed via the stopcock, the volume added can be determined by directly observing the position of the meniscus using the top-to-bottom graduations. Figure 5.3 gives a close-up view to clarify this concept.

Another important point with regard to correctly reading a buret is that the number of significant figures obtainable from a standard 50-mL buret is

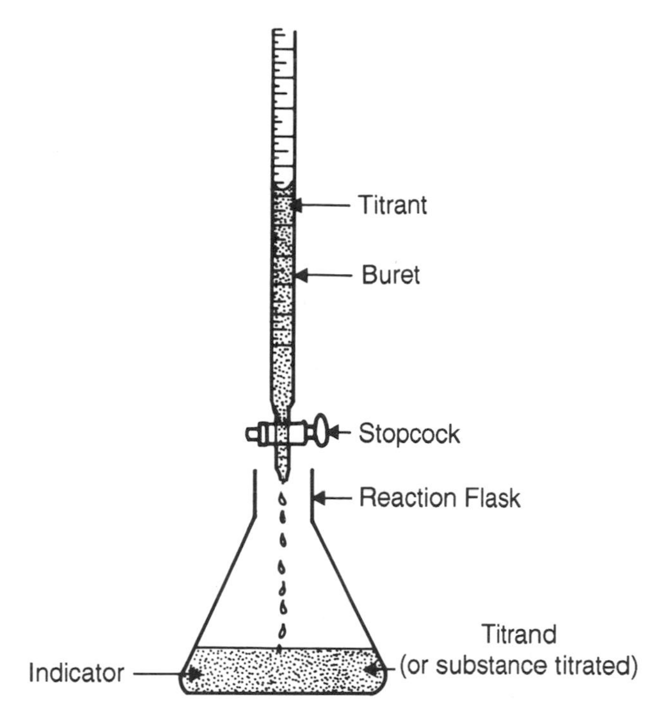

Figure 5.2. The experimental setup and terminology for a titration.

four – there are two digits to the right of the decimal point in the correct reading. The key is to always read the number of significant figures known with certainty and then estimate one more. (See Appendix 1.) On a buret, this estimated digit is the second digit to the right of the decimal point. Refer to Figure 5.3 for an example. The fact that four significant figures are obtainable on a buret also means that the accuracy of the weight of the sample obtained on an analytical balance is not diminished in the course of the calculation since these sample weights are also known to four significant

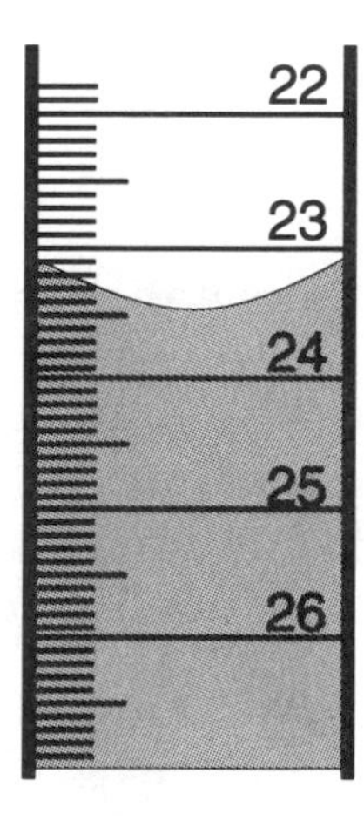

Figure 5.3. The concept of reading the set of graduations on a buret from the top down. The correct reading is 23.48 mL.

figures. (Refer to Appendix 1 for a thorough discussion of the rules for significant figures.)

5.3 PREPARATION OF MOLAR SOLUTIONS

The calculation of the percent of the constituent in a titrimetric analysis involves, as step one of the three-step stoichiometry calculation, the conversion of the volume of the titrant to moles of the titrant. The conversion factor for doing this is the concentration of the solution in moles/L.

$$\text{liters of titrant} \times \frac{\text{moles of titrant}}{\text{liter of titrant}} = \text{moles of titrant} \quad (5.1)$$

$$\cancel{\text{L}} \times \frac{\text{moles}}{\cancel{\text{L}}} = \text{moles} \quad (5.2)$$

$$\text{L} \times \text{M} = \text{moles} \quad (5.3)$$

Thus the molarity (moles/L) of the titrant must be known and, in addition, it must be known to four significant figures in order to avoid diminishing the accuracy of the volume and weight measurements in the course of the calculation. Thus, for titrimetric analysis, it is very important to prepare solutions of known concentration with a high degree of accuracy, or at least prepare them so that their concentrations can be determined accurately. Following, then, is a review of molar solution preparation procedures and calculations involving both preparation from a pure solid chemical and preparation by dilution.

5.4 PREPARATION FROM PURE SOLID CHEMICALS

Many solutions are prepared from chemicals which are pure and in the solid state. The weight of the chemical needed for the solution is conveniently calculated using the formula

$$L_D \times M_D \times FW_{SOL} = \text{grams to be weighed} \quad (5.4)$$

in which L_D is the liters desired, M_D is the molarity desired, and FW_{SOL} is the formula weight of the solute.

A check of the units reveals that grams of the solute do indeed result from this calculation:

$$\cancel{\text{liters}} \times \frac{\cancel{\text{moles}}}{\cancel{\text{liter}}} \times \frac{\text{grams}}{\cancel{\text{mole}}} = \text{grams} \quad (5.5)$$

The solution is then simply prepared by weighing the chemical according to the grams found in the above calculation, placing this weight in a container of appropriate calibration, dissolving it in water, adding water up to the calibration line for the volume required, and then by shaking the solution to make it homogeneous.

Example 1

How would you prepare 500 mL of a 0.15 *M* solution of NaOH from pure solid NaOH?

Solution

$$\text{grams to be weighed} = 0.500\,\cancel{L} \times \frac{0.15\,\cancel{\text{mole}}}{\cancel{L}} \times \frac{40.0\text{ g}}{\cancel{\text{mole}}} = 3.0\text{ g}$$

3.0 g of NaOH are placed in a container with a 500-mL calibration line, dissolved in water, diluted to the calibration line with additional water, and then made homogeneous by shaking.

5.5 PREPARATION BY DILUTION

Many solutions are prepared by diluting more concentrated solutions. This is especially true for solutions of acids, since many common acids are purchased as concentrated solutions. The key to solution preparation is to determine that volume of the more concentrated solution which must be measured and diluted with water. This can be easily found by keeping in mind that the number of moles of solute does not change in the dilution process—one just adds water to this quantity of moles. Thus,

$$\text{moles of solute before dilution} = \text{moles of solute after dilution} \tag{5.6}$$

$$\text{moles}_B = \text{moles}_A \tag{5.7}$$

where B refers to before dilution and A refers to after dilution.

The number of moles of a solute in any solution can be calculated by multiplying the molarity of the solution by the number of liters:

$$\cancel{L} \times \frac{\text{moles}}{\cancel{L}} = \text{moles} \tag{5.8}$$

Thus,

$$M_B \times V_B = M_A \times V_A \tag{5.9}$$

where V is the volume in liters. V can also be in milliliters or any volume unit, since the conversion factor to convert it to liters would be the same on both sides and thus would cancel:

$$\frac{\text{moles}}{\cancel{L}} \times \cancel{\text{mL}} \times \frac{1\,\cancel{L}}{1000\,\cancel{\text{mL}}} = \frac{\text{moles}}{\cancel{L}} \times \cancel{\text{mL}} \times \frac{1\,\cancel{L}}{1000\,\cancel{\text{mL}}} \tag{5.10}$$

Example 2

How would you prepare 500 mL of 0.15 *M* HCl from concentrated HCl which is 12.0 *M*? (Concentration values for the various concentrated acid and base solutions are given in Appendix 4.)

Solution

$$M_B \times V_B = M_A \times V_A$$

$$12.0 \times V_B = 0.15 \times 500$$

$$V_B = \frac{0.15 \times 500}{12} = 6.3\text{ mL}$$

One would measure out 6.3 mL of concentrated HCl and dilute it with distilled water to 500 mL.

5.6 VOLUMETRIC GLASSWARE

The accurate measurement of volumes of solutions is very important in titrimetric analysis. First, as indicated at the beginning of the last section, the volume of the titrant is used directly in the titrimetric percent constituent calculation. Second, volumes of solutions are important measurements when preparing solutions for titrimetric analysis, whether such preparation is accomplished by dilution or by weighing a pure solid chemical. Third, a variety of reasons exist for accurately transferring a volume of a solution from one vessel to another. It should not be surprising that an analytical chemist needs to be well versed in the selection and proper use of volumetric glassware.

There are basically three types of accurate volume measuring devices in common use. These are the volumetric flask, the pipet, and the buret. Let us study the characteristics of each type individually.

The Volumetric Flask

Let's first deal with the container that is typically used for accurate solution preparation. First, this container must have a calibration line corresponding to the volume of the solution being prepared. There are four different pieces of glassware that fulfill this requirement – a beaker, an Erlenmeyer flask, a graduated cylinder, and a volumetric flask. All of these are shown in Figure 5.4. Second, the calibration line must be affixed to the container so that the volume of the solution is of such accuracy that the accuracy of the planned experiment is not diminished beyond the required limit. In other words, if the planned use of the solution requires that its concentration be known accurately directly through its preparation, then a piece of glassware that is calibrated to reflect this accuracy for the total

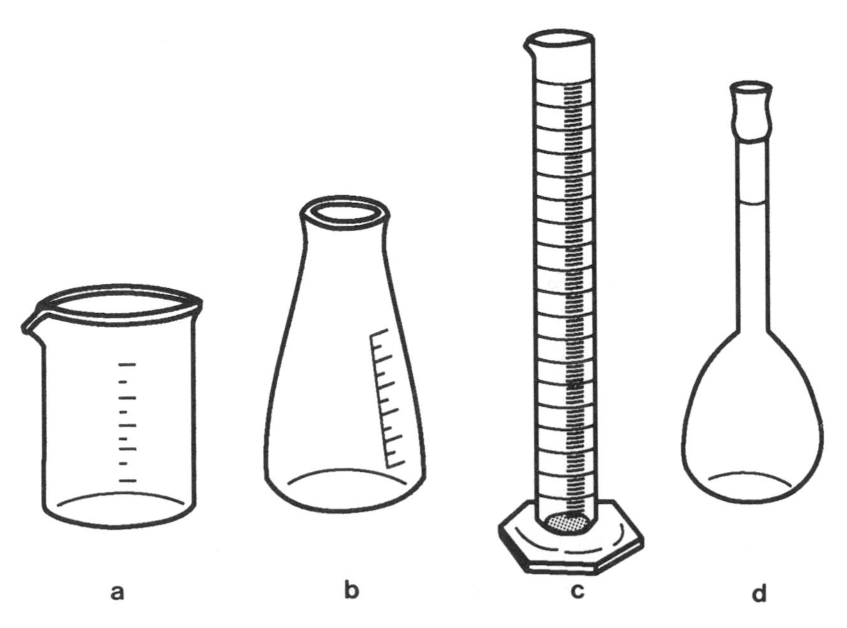

Figure 5.4. Four pieces of glassware having calibration lines for solution volumes. (a) Beaker, (b) Erlenmeyer flask, (c) graduated cylinder, (d) volumetric flask.

volume of the solution prepared must be used. For this reason, and also for convenience, the beaker and the Erlenmeyer flask shown in Figure 5.4 are never used for accurate solution preparation. The calibration lines reflect such poor accuracy on these two containers that they can only be taken to mean very rough indications of liquid volume. Even the graduated cylinder is of insufficient accuracy most of the time, especially since it is usually meant to deliver the specified volume and not to contain the volume.

One important clue as to the accuracy of a volume measured in any piece of glassware is the narrowness of the tube on which the calibration line is imprinted. The narrower this diameter, the more accurate the measurement. Consider, for example, bringing a solution volume up to the final resting point of the bottom of the meniscus – i.e., bringing it up to the desired calibration line during the solution preparation process. Refer to Figure 5.5. If you imagine doing this in the 500-mL beaker, it would take perhaps as much as 5 mL of solvent to make any noticeable change in the position of the meniscus (an accuracy of 1.0%). In the 500-mL graduated cylinder, where the calibration line is on a narrower diameter tube, it may take as much as 1 mL for this noticeable change (an accuracy of 0.2%). In the volumetric flask (right in Figure 5.4), a single calibration line is imprinted on a very narrow (1 cm) diameter neck and just one drop (0.05 mL) of solution will make a visible change. That's an accuracy of ±0.01%! It is no wonder, then, that a volumetric flask is the container to choose when preparing a solution so that its concentration is known accurately. Conveniently, it can be purchased in a variety of sizes, from 5 mL up to several liters. Figure 5.6 shows some of the various sizes.

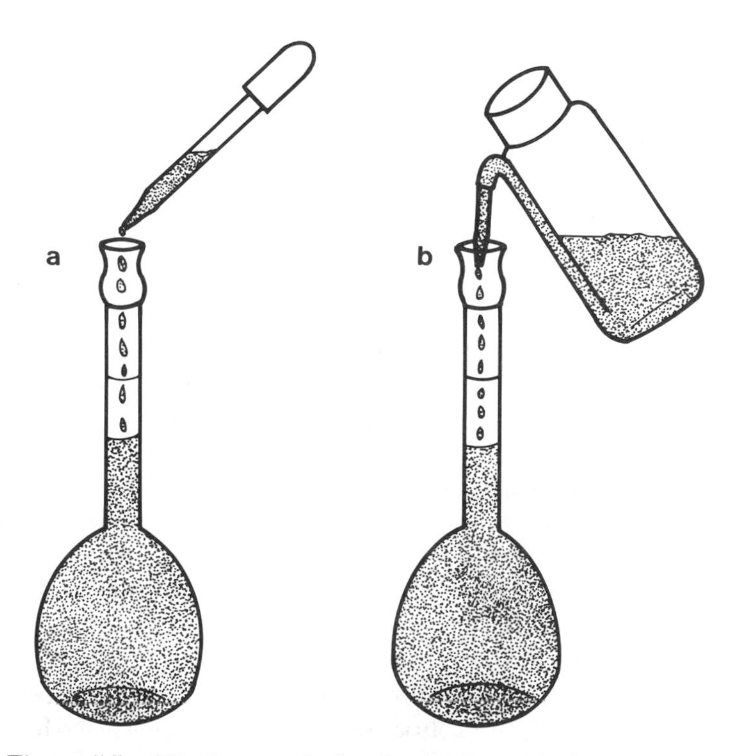

Figure 5.5. Diluting a solution in a volumetric flask to the mark (a) with an eye dropper and (b) with a squeeze bottle.

Figure 5.6. A variety of sizes of volumetric flasks.

The volumetric flask is characterized by a large base that tapers into a very narrow neck onto which the single calibration line is affixed. This calibration line is affixed so that the indicated volume is *contained* rather than delivered. Accordingly, the legend "TC" is imprinted on the base of the flask, thus marking the flask as a vessel "to contain" the volume indicated as opposed to "to deliver" a volume. The reason this imprint is important is that a contained volume is different from a delivered volume, since a small volume of solution remains adhering to the inside wall of the vessel and is not delivered when the vessel is drained. If a piece of glassware is intended to deliver a specified volume, the calibration obviously must take this small volume into account in the sense that it *will not* be part of the *delivered* volume. On the other hand, if a piece of glassware is not intended to deliver a specified volume, but rather to contain the volume, the calibration must take this small volume into account in the sense that it *will* be part of the *contained* volume. Other pieces of glassware, namely most pipets and all burets, are "TD" vessels, meaning they are calibrated "to deliver". Figure 5.7 shows a close-up of the base of a volumetric flask clearly showing the "TC" imprint (just below the flask's capacity – 500 mL).

Notice the other markings on the base of the flask in Figure 5.7. The imprint "A" refers to flasks (and pipets and burets as well) that have undergone more stringent calibration procedures at the factory (class "A" flasks). Such flasks are more accurate (and also more expensive) than flasks that do not have this marking. The imprint "20 °C" indicates that the flask is calibrated to contain the indicated volume when the temperature is 20 °C, which is a standard temperature of calibration. This marking is needed since the volume of liquids and liquid solutions changes slightly with temperature. For highest accuracy, the temperature of the contained fluid should be adjusted to 20 °C.

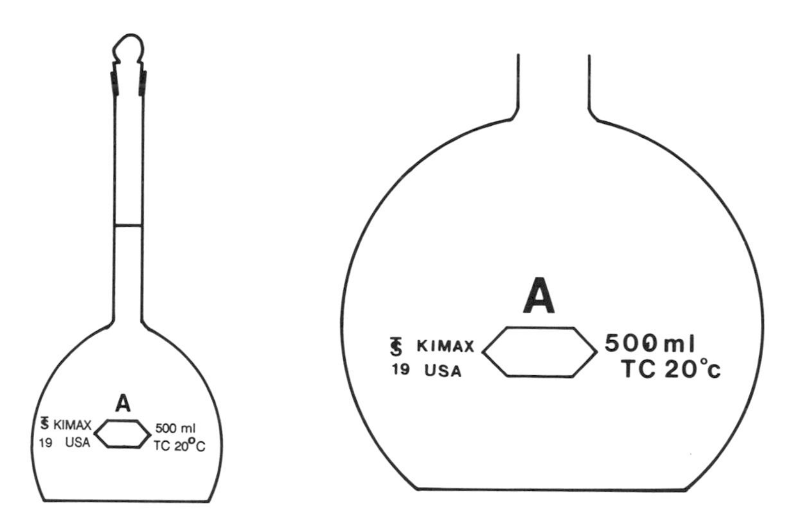

Figure 5.7. The base of a volumetric flask showing the markings imprinted there. (From Kenkel, J., *Analytical Chemistry Refresher Manual*, Lewis Publishers, Chelsea, MI, 1992. With permission.)

One final marking on the base of the flask is the "19" designation. This refers to the size of the tapered top on the flask. The stopper that is used either can be a ground-glass stopper, to match the flask opening, or it can simply be a plastic tapered stopper, which can also be used in a ground-glass opening. It has an identical numerical imprint on it ("19" in this case), and such number designations on these two items must match to indicate that the stopper is the correct size. The stopper is not necessarily a tapered stopper as in this example. It is fairly common for the stopper to be a "snap cap". In this case, the top of the flask is not tapered, nor is it a ground-glass opening. Rather, it has an unusually large lip around the opening, over which the snap cap is designed to seal. A number designation, however, is used as with the tapered opening to indicate the size of the opening and the size of the cap required. This number is found on both the flask base and the cap as in the other design. The technician should always be aware of the type and size of cap required for a given volumetric flask. If any amount of solution should leak out due to an improperly sealed cap while shaking before the solution is homogeneous, its concentration cannot be trusted.

One problem with the volumetric flask exists because of its unique shape, and that is the difficulty in making prepared solutions homogeneous. When the flask is inverted and shaken, the solution in the neck of the flask is not agitated. Only when the flask is set upright again is the solution drained from the neck and mixed. A good practice is to invert and shake at least a dozen times to ensure homogeneity.

Volumetric flasks should *not* be used to prepare solutions of reagents that can etch glass (such as sodium hydroxide and hydrofluoric acid), since if the glass is etched, its accurate calibration is lost. Volumetric flasks should *not* be used for storing solutions. Their purpose is to prepare solutions accurately. If they are used for storage then they are not available for their intended purpose. Finally, volumetric flasks should *not* be used to contain solutions when heating or performing other tasks for which their accurate calibration serves no useful purpose. There are plenty of other glass vessels to perform these functions.

The Pipet

As indicated above, most pipets are pieces of glassware that are designed to deliver (TD) the indicated volume. Pipets come in a variety of sizes and shapes. The most common is probably the "volumetric" or "transfer" pipet shown in Figure 5.8. This pipet, like the volumetric flask, has a single calibration line. It can thus be used in delivering only rather common volumes – 5, 10, 15 mL, etc. The correct use of a volumetric pipet is outlined in Table 5.1. The sequence of events is described with less detail as follows. The bottom tip of the pipet is placed into the solution to be transferred. A pipet bulb is evacuated, placed over the top of the pipet, and slowly released. The solution is drawn up into the pipet as a result of the vacuum. When the level of the solution has risen above the calibration line, the bulb is quickly removed from the pipet and replaced with the index finger. Next, the tip of the pipet is removed from the solution and wiped with a towel. The pressure exerted with the index finger is then used to adjust the bottom of the meniscus to coincide with the calibration line. The tip of the pipet is placed into the receiving vessel and the finger released. The solution is thus drained into the receiving vessel. When the solution has completely drained from the pipet, the bottom tip should be placed against the wall of the receiving vessel so that the last bit of solution will drain out. A small drop of the liquid will remain in the pipet at this point. The pipet should be given a half-turn (twist) and then removed. *Under no circumstances should this last drop in a volumetric pipet be blown out with the bulb.* The volumetric pipet is not calibrated for blow-out.

For precise work, volumetric pipets that are labeled as class "A" have a certain time in seconds imprinted near the top (see Figure 5.9) which is the

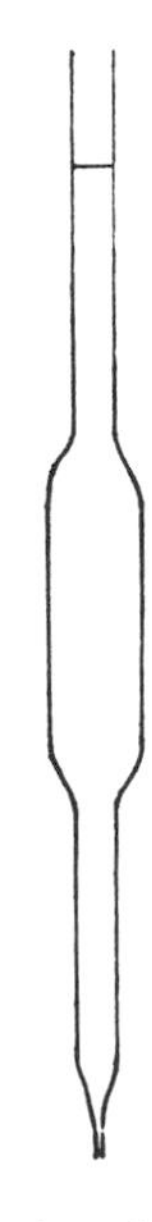

Figure 5.8. A volumetric pipet.

Table 5.1. The sequence of events involved in transferring a volume of solution with a volumetric pipet.

Step 1.	If the outside of the pipet is wet, dry it with a paper towel first, especially the tip.
Step 2.	Evacuate the pipet bulb (by squeezing).
Step 3.	Seat the bulb opening over the top opening of the pipet.
Step 4.	If the pipet is wet on the inside with a foreign liquid, immerse the tip of the pipet into the solution to be delivered while simultaneously releasing the bulb slightly to immediately draw the solution in to about half the pipet's capacity. Empty by inverting and draining into a sink through the top. Repeat this rinsing step at least three times. If the inside of the pipet is dry, proceed with step 5.
Step 5.	Fill the pipet to well past the calibration line by releasing the squeezing pressure, as in Step 4. Re-evacuate the bulb if necessary.
Step 6.	Quickly remove the bulb and seal the top of the pipet with the index finger.
Step 7.	Keeping the index finger in place, remove the tip from the solution and wipe with a towel. To avoid contamination from the towel, tilt the pipet to a 45° angle so that a small volume of air is drawn into the tip before wiping.
Step 8.	Slowly release the finger to adjust the meniscus to the calibration line.
Step 9.	Touch the tip to the *outside* of the receiving vessel so as to remove a drop of the solution that may be suspended there.
Step 10.	Place the tip into the mouth of the receiving vessel and completely release the finger, being careful not to shake the pipet so that some solution is lost and an air bubble appears in the tip before the release.
Step 11.	When the draining is complete, touch the tip to the inside wall of the vessel and give it a half-twist. You should see the solution level inside the tip drop slightly when this is done.

Note: See text for further discussion.

time that should be allowed to elapse from the time the finger is released until the pipet is given the half-turn and removed. The reason for this is that the film of solution adhering to the inner walls will continue to slowly run down with time, and the length of time one waits to terminate the delivery thus becomes important. The intent with class "A" pipets, then, is to take this "run-down" time into account by terminating the delivery in the specified time. After this specified time has elapsed, the pipet is touched to the wall of the receiving flask, given the half-turn, and removed.

Several additional styles of pipets other than the volumetric pipet are in common use. These are shown in Figure 5.10. Pipets that have graduations lines, much like a buret, are called "measuring" pipets. They are used whenever odd volumes are needed. There are two types of measuring pipets – the Mohr pipet and the serological pipet. The difference is whether or not the calibration lines stop short of the tip (Mohr pipet) or go all the way to the tip (serological pipet). The serological pipet is better in the sense that the menis-

Figure 5.9. The top of a class A volumetric pipet (positioned horizontally). Note the symbol "A" and the delivery time (35 sec).

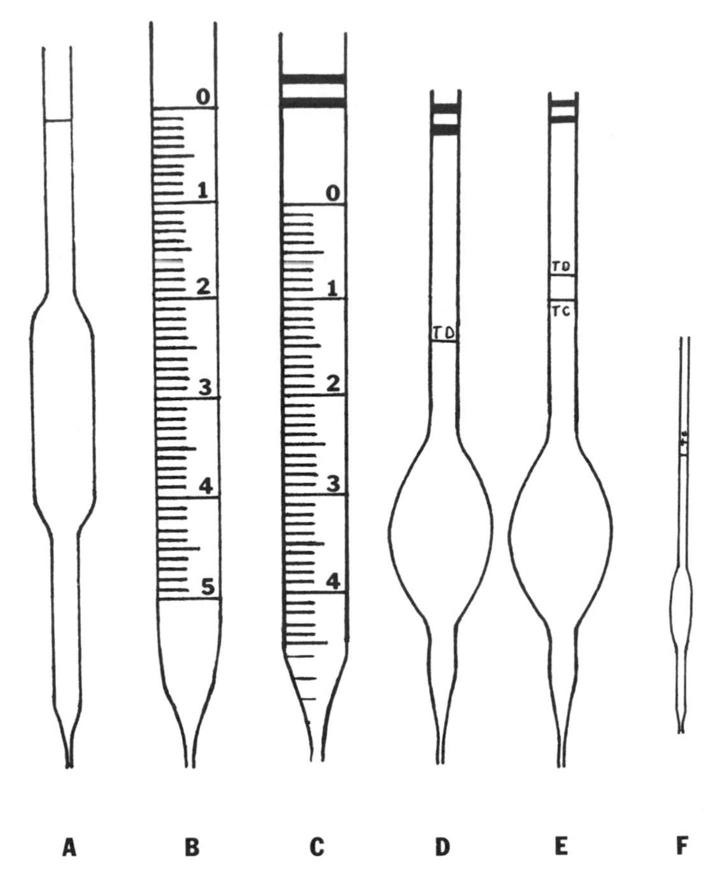

Figure 5.10. Some types of pipets. (A) Volumetric, (B) Mohr, (C) serological, (D) Ostwald-Folin, (E) duopette, and (F) lambda.

cus need be read only once since the solution can be allowed to drain completely out. In this case, the last drop of solution is blown out with the pipet bulb. With the Mohr pipet, however, the meniscus must be read twice – once before the delivery and again after the delivery is complete. The solution flow out of the pipet must be halted at the correct calibration line, and the error associated with reading a meniscus is thus doubled. The delivery of 4.62 mL, for example, is done as in Figure 5.11a with a Mohr pipet but as in Figure 5.11b with a serological pipet.

It should be stressed that with the serological pipet every last trace of solution capable of being blown out must end up inside the receiving vessel. Some analysts find that this is more difficult and perhaps introduces more error than reading the meniscus twice with a Mohr pipet. For this reason, these analysts prefer to use a serological pipet as if it were a Mohr pipet. It is really a matter of personal preference. A double or single frosted ring circumscribing the top of the pipet (above the top graduation line) indicates the pipet is calibrated for blow-out. Disposable pipets are most often of the serological blow-out type. They are termed disposable because the calibration lines are not necessarily permanently affixed to the outside wall of the pipet. The calibration process is thus less expensive, resulting in a less expensive product which can be discarded after use.

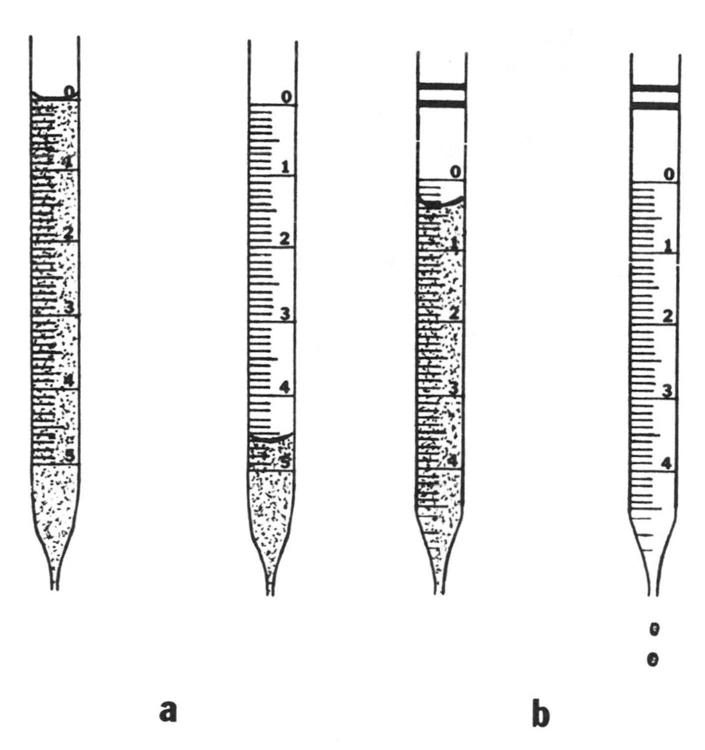

Figure 5.11. The delivery of 4.62 mL of solution (a) with a Mohr pipet (the meniscus is read twice) and (b) with a serological pipet (the meniscus is read once [at 0.38] and the solution is blown out).

One of two pipets that are calibrated "TC" is called the lambda pipet (Figure 5.10). Such a pipet is used to transfer unusually viscous solutions such as syrups, blood, etc. With such solutions, the thin film remaining inside would represent a significant non-transferred volume which translates into a significant error by normal TD standards. With the lambda pipet, the calibration line is affixed at the factory so that every trace of solution contained within is transferred by flushing the solution out with a suitable solvent. Thus, the pipetted volume is contained within and then quantitatively flushed out. Such a procedure would actually be acceptable with any TC glassware, including the volumetric flask. Obviously, diluting the solution in the transfer process must not adversely affect the experiment. Some pipets have both a TC line and a TD line and are called "duopettes" (Figure 5.10). Ostwald-Folin pipets (Figure 5.10) have a single line but are calibrated for blow-out.

Pipetting Devices

In addition to the several designs of pipets just described, various pipetting "devices" for measuring solution volumes from 0 to 1 mL have been invented for laboratory use and have become very popular. These "pipettors" employ a bulb concealed within a plastic fabricated body, a spring-loaded push button at the top, and a nozzle at the bottom for accepting a plastic disposable tip. They may be fabricated for either single or variable volumes. In the latter case, a ratchet-like device with a digital volume scale is present below the push button. The desired volume may be "dialed in" prior to use. An example is shown in Figure 5.12.

The Buret

The buret has some unique attributes and uses. One could call it a specialized graduated cylinder, having graduation lines that increase from top to bottom, with a usual accuracy of ±0.01 mL and a stopcock at the bottom for dispensing a solution. There are some variations in the type of stopcock that warrant some discussion. The stopcock itself, as well as the barrel into which it fits, can be made of either glass or Teflon®. Some burets have an all-glass arrangement, some have a Teflon® stopcock and a glass barrel, and some have the entire system made of Teflon®. The three types are pictured in Figure 5.13.

When the all-glass system is used, the stopcock needs to be lubricated so it will turn with ease in the barrel. There are a number of greases on the market for this purpose. Of course, the grease must be inert to chemical

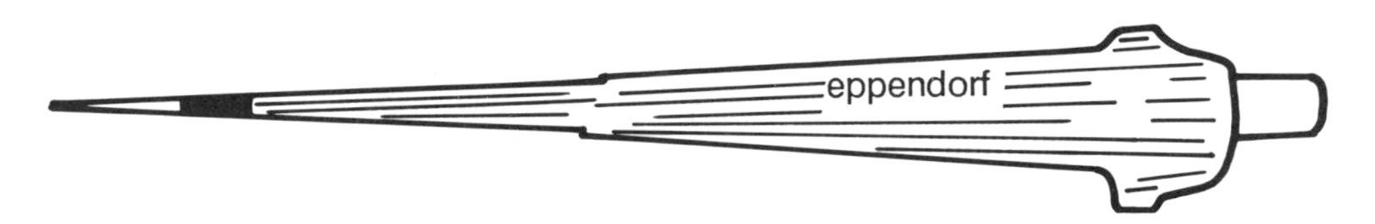

Figure 5.12. A drawing of an Eppendorf pipettor with variable volume capability. (From Kenkel, J., *Analytical Chemistry Refresher Manual*, Lewis Publishers, Chelsea, MI, 1992. With permission.)

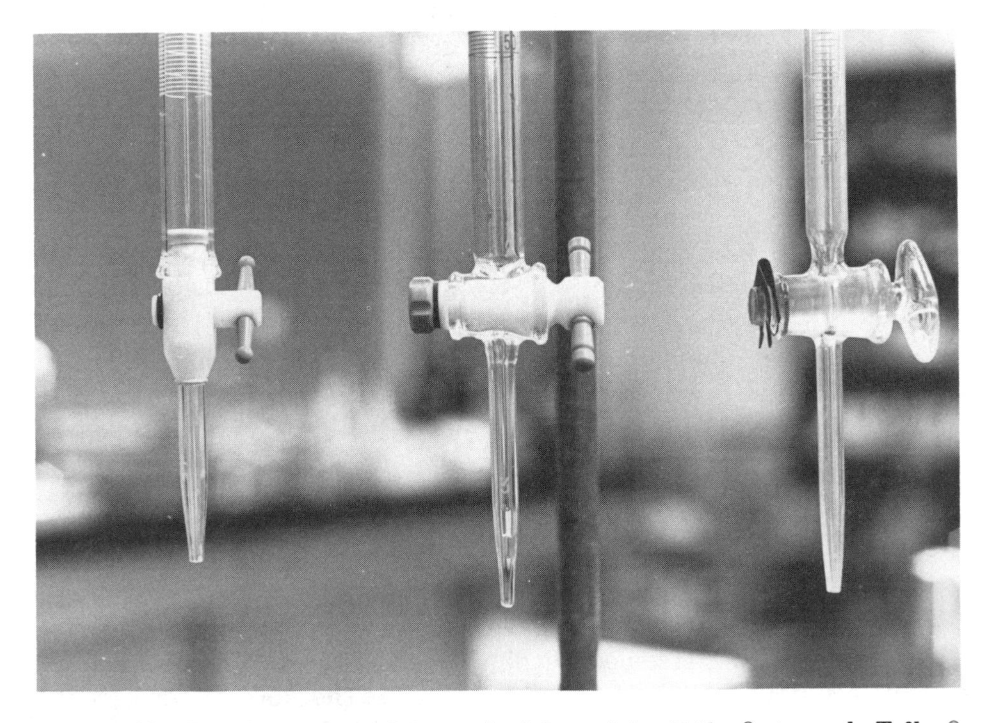

Figure 5.13. The three types of stopcocks left to right: Teflon® stopcock, Teflon® barrel; Teflon® stopcock, glass barrel; glass stopcock, glass barrel.

attack by the solution to be dispensed. Also, the amount of grease used should be carefully limited so that excess grease does not pass through the stopcock and plug the tip of the buret. Any material stuck in a buret tip can usually be dislodged with a fine wire inserted from the bottom when the stopcock is open and the buret full of solution. The Teflon® stopcocks are free of the lubrication problem. The only disadvantage is that the Teflon® can become deformed, and this can cause leakage.

The correct way to position one's hands to turn the buret's stopcock during a titration is shown in Figure 5.14. The natural tendency with this positioning is to pull the stopcock in as it is turned. This will prevent the stopcock from being pulled out, causing the titrant to bypass the stopcock. The other hand is free to swirl the flask as shown.

Cleaning and Storing Procedures

The use of clean glassware is of utmost importance when doing a chemical analysis. In addition to the obvious need for keeping the solution free of contaminants, the walls of the vessels, particularly the transfer vessels (burets and pipets), must be cleaned so the solution will flow freely and not "bead up" on the wall as the transfer is performed. If the solution beads up, it is obvious that the pipet or buret is not delivering the volume of solution that one intends to deliver. It also means that there is a film of grease on the wall which could introduce contaminants. The analyst should examine, clean, and re-examine his/her glassware in advance so that the free flow of

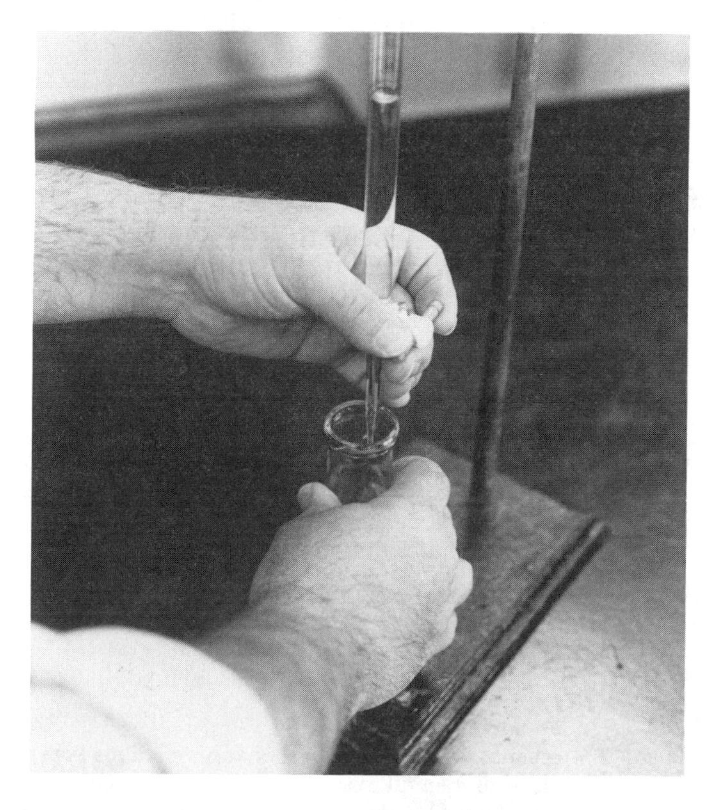

Figure 5.14. The correct way to position one's hands for a titration.

solution down the inside of the glassware is observed. For the volumetric flask, at least the neck must be cleaned in this manner so as to ensure a well-formed meniscus.

Of course, the next question is: What cleaning procedures are used? For most cleaning requirements, ordinary soap and water used with a brush where possible is sufficient. A commercial laboratory soap called Alconox® is one product often used. Other excellent phosphate-free detergents are available. With burets, a cylindrical brush with a long handle (buret brush) is used to scrub the inner wall. With flasks, a bottle or test tube brush is used to clean the neck. Also, there are special bent brushes available to contact and scrub the inside of the base of the flask.

Pipets pose a special problem. Brushes cannot be used because of the shape of some pipets and the narrowness of the openings. In this case, if soap is to be used, one must resort to soaking with a warm soapy water solution for a period of time proportional to the severity of the particular cleaning problem. Commercial soaking and washing units are available for this latter technique. Soap tablets are manufactured for such units and are easy to use.

For pipets, and for difficult cleaning problems for other pieces of glassware, special cleaning solutions, which chemically break down greasy films through soaking for a period of time, are used. One commonly used cleaning solution is chromic acid. This is a solution of concentrated sulfuric acid and potassium dichromate. Another is a solution of potassium hydroxide in ethanol or propanol. Recipes for preparing these solutions are given in Table 5.2. These are very tough on serious cleaning problems.

Safety should be stressed in the use of these cleaning solutions. The highly corrosive nature of the ingredients makes it imperative to prevent spills and splashes on one's person. In addition to normal safety gear (safety glasses, eye wash, shower, first aid kit), one should also have solutions of weak acids (e.g., acetic acid, citric acid) and weak bases (e.g., sodium carbonate) handy to neutralize spills quickly. It is highly recommended that use of these cleaning solutions be restricted to a good fume hood. Spent cleaning solutions should *not* be poured down the drain, but should be disposed of like any other hazardous waste.

Table 5.2. Recipes for preparing glassware cleaning solutions.

Chromic Acid, Method 1	Dissolve 90–100 g of sodium or potassium dichromate in 450 mL of water and add 80 mL of concentrated sulfuric acid. The solution will become a semisolid red mass. Add just enough sulfuric acid to dissolve this mass.
Chromic Acid, Method 2	Add the contents of a 25-mL bottle of Monostat Chromerge® (available from chemical products suppliers) to a standard 9-lb bottle of concentrated sulfuric acid. Add approximately 5 mL at a time, and shake well after each addition.
Alcoholic KOH	Dissolve 105 g of KOH in 120 mL of water. Add 1 L of 95% ethanol (or propanol) and shake well. This solution will etch glass to a small extent and therefore should not be left in contact with volumetric glassware or ground-glass joints for more than about 15 min at a time.

Once the glassware has been cleaned (by whatever method), one should also take steps to keep it clean. One technique is to rinse the items thoroughly with distilled water and then dry them in an oven. Following this, they are cooled and stored in a drawer. For shorter time periods, it may be convenient to store them in a soaker under distilled water. This prevents their possible recontamination during dry storage. When attempting to use a soaker-stored pipet or buret, however, it must be remembered that a thin film of water is present on inner walls and this must be removed by rinsing with the solution to be transferred, being careful not to contaminate the solution in the process.

5.7 SOLUTION STANDARDIZATION AND TERMINOLOGY

It was indicated previously that the concentration of solutions, especially titrants, used in titrimetric analysis needs to be known very accurately so that the number of significant figures in the percent is not diminished. In many cases, it is possible to obtain the desired accuracy by simply weighing the substance with an analytical balance and making up the solution in a volumetric flask. In just as many cases, however, it is not possible to obtain the desired accuracy in this manner. One example of this latter situation is when the solid material weighed into the flask is hygroscopic (e.g., sodium hydroxide); i.e., it readily absorbs water from humid air and thus cannot be weighed accurately. Another example is when diluting a concentrated acid when the accurate concentration of this acid is not known. In this case, the concentration of the prepared solution must be determined by a separate experiment.

Let us first define some of the terms relating to this subject. First, a "standard solution" is a solution which has a known concentration, meaning that it is known to the accuracy required for a given experiment. "Standardization" refers to an experiment in which the concentration of a solution is determined to the desired accuracy. To "standardize" a solution means to determine the concentration of a solution via a standardization experiment. A "primary standard" is the substance to which a solution is compared which allows standardization to take place.

This last definition warrants some additional comment. Obviously the quality of the primary standard substance is ultimately the basis for a successful standardization. This means that it must meet some special requirements with respect to purity, etc., and these are enumerated below.

1. It must be 100% pure, or at least its purity must be known.
2. If it is impure, the impurity must be inert.
3. It should be stable at drying oven temperatures.
4. It should not be hygroscopic—it should not absorb water when exposed to laboratory air.
5. The reaction in which it takes part must be quantitative and preferably fast.
6. A high molecular weight is desirable.

Most substances used as a primary standard can be purchased as primary standard grade (see Chapter 2), and this is appropriate and sufficient for a standardization experiment.

Additionally, there are two frequently used terms relating to any titration experiment. These are the terms "end point" and "equivalence point". As mentioned earlier, the function of an indicator is to signal the precise moment of the complete consumption of the substance titrated. It is rare that an indicator is able to do this at that exact moment. The term "equivalence point" refers to that point at which a titration reaction is complete. The term "end point" refers to that point at which a titration reaction is complete as indicated by an indicator. While there frequently is a difference between the two, the accuracy of an experiment usually does not suffer appreciably if the end point and the equivalence point do not exactly coincide. If there is an appreciable difference, some other equivalence point detection method must be used. One such method is with the use of electrodes, and this will be discussed in Chapter 18.

5.8 WHEN TO BE ACCURATE AND WHEN NOT–REVISITED

In Chapter 1 (near the end of Section 1.7) we discussed the need to know when a weight measurement must be accurate and when it doesn't have to be accurate so that the appropriate balance is chosen for the measurement. We now repeat many of our comments made there, but for volume measurements rather than weight.

The question of which piece of glassware to use for a given procedure, such as for solution preparation (pipet or graduated cylinder, beaker or volumetric flask), remains. An analyst must be able to recognize when a high degree of accuracy is important and when it isn't so that accurate volumetric glassware – pipets, burets, and volumetric flasks – is used when accuracy is important and regular glassware – graduated cylinders and beakers – is used when accuracy is not important.

First, volumetric glassware should *not* be used for any volume measurement when the overall objective is strictly qualitative or when quantitative results are to be reported to two significant figures or less. Whether the volume measurements are for preparing solutions to be used in such a procedure or for transferring an appropriate volume of solution or solvent for such an analysis, etc., they need not be accurate since the outcome will be either only qualitative or not necessarily accurate.

Second, if the results of a quantitative analysis are to be reported to three or more significant figures, then volume measurements that enter directly into the calculation of the results should be made with volumetric glassware so that the accuracy of the analysis is not diminished when the calculation is performed.

Third, if a volume measurement is only incidental to the overall result of an accurate quantitative analysis, then it need not be accurate. This means that if the volume measurement to be performed has no bearing whatsoever on the quantity of the analyte tested, but is only needed to support the chemistry or some other factor of the experiment, then it need not be accurate.

Fourth, if a volume measurement does directly affect the numerical result of an accurate quantitative analysis (in a way other than entering directly into the calculation of the results), then volumetric glassware must be used.

These last two points require that the analyst carefully consider the purpose of the volume measurement and whether it will directly affect the quantity of analyte tested, the numerical value to be reported. It is often true that volume measurements taken during such a procedure need not be accurate even though the results are to be accurate.

EXPERIMENTS

EXPERIMENT 6: PERFECTING THE ART OF PIPETTING

Introduction

The technique of accurate pipetting is an art. It is something that an analyst becomes proficient at only by practice and experience. There are many analytical procedures that require accurate pipetting. It is essential, then, that analytical laboratory workers practice the technique over and over in order to become proficient so that they can be successful in the lab. As with other analytical techniques, such as in weighing with the analytical balance, the ability to be accurate is important. The analytical balance is a very accurate measuring device, but does not require much manual dexterity. The volumetric pipet is also a very accurate measuring device, but in addition requires the ability of the analyst to manipulate his/her hands and fingers in order to be successful – hence the need for practice.

One way to practice and to determine whether your pipetting technique is sufficiently good after or during a period of practicing is to try to continuously reproduce the weight of a pipetted volume on an analytical balance. You've already mastered the use of the analytical balance and so you can use it to see if you have mastered the use of the pipet.

There is another variable involved in the use of a pipet, however, and that is the temperature of the solution pipetted. The density of water, and therefore the density of water solutions, changes with temperature – water expands when heated and contracts when cooled. Because of this, the weight of the pipetted volume will change if the temperature changes. Thus, the temperature of the water being pipetted must remain constant throughout the period of practice so that the weight doesn't change due to the temperature change. For this reason, we will be monitoring the temperature during this practice session.

The attempt to reproduce a measurement over and over is an attempt at good precision. We can say that if our results are reproduced over and over again, then we have good precision. Precision doesn't necessarily assure accuracy, but it is often used as an indication of accuracy. See Chapter 3 for additional information regarding the difference between accuracy and precision.

In this experiment, you will use a 25-mL volumetric pipet to practice pipetting. You will weigh the pipetted volume on an analytical balance after each attempt. If, after a number of attempts, the weights of the pipetted volumes are consistent (to within about ± 0.02 g for a 25-mL volumetric pipet), then you can say that your technique is good and you can proceed confidently to a real analytical experiment.

Procedure

1. Obtain a 400-mL beaker and fill it about 3/4 full with tap water. Immerse a thermometer into the water and let stand for a few minutes. Record the temperature. Also obtain a 125-mL Erlenmeyer flask and a clean 25-mL volumetric pipet. Practice using the pipet (refer to Table 5.1) to accurately deliver 25.00 mL of the water to the flask. Once you feel comfortable with your technique, proceed to Step 2.
2. Dry the outside of the flask thoroughly with a paper towel. Weigh the flask carefully with an analytical balance. Now carefully pipet 25.00 mL of the water into the flask and weigh it again. Record the weight in your notebook and subtract the first weight from the second. The weight of the water should be something less than 25.00 g since the density of water is less than 1 at room temperature.
3. Repeat over and over again until the measured weight is reproduced within ± 0.02 g each time. If you experience difficulty getting good precision, check the temperature again. If the temperature is different from your first recording, it's possible that this is the reason for your lack of precision. Once your weight readings are precise, your pipetting technique is good enough for you to proceed to Experiment 7.

QUESTIONS AND PROBLEMS

1. How is titrimetric analysis different from gravimetric analysis?
2. Define buret, indicator, titration, titrant, titrand, stopcock.
3. Tell how you would prepare each of the following:
   (a) 500 mL of a 0.10 *M* solution of KOH from pure solid KOH.
   (b) 250 mL of a 0.15 *M* solution of NaCl from pure solid NaCl.
   (c) 100 ml of a 2.0 *M* solution of glucose from pure solid glucose ($C_6H_{12}O_6$).
   (d) 500 mL of a 0.10 *M* solution of HCl from concentrated HCl which is 12.0 *M*.
   (e) 100 mL of a 0.25 *M* solution of NaOH from a solution of NaOH that is 2.0 *M*.
   (f) 2.0 L of a 0.50 *M* solution of sulfuric acid from concentrated sulfuric acid, which is 18.0 *M*.
4. How would you prepare 250 mL of a 0.35 *M* solution of NaOH using
   (a) a bottle of pure solid NaOH?
   (b) a solution of NaOH that is 12.0 *M*?
5. How many grams of NH_4Cl are required to prepare 350 mL of a 0.25 *M* solution?
6. How many milligrams of Na_3PO_4 are required to prepare 70 mL of a 0.0030 *M* solution?

7. How many grams of Na_2CO_3 are needed to prepare 1.2 L of a 0.045 *M* solution?
8. How many milliliters of a concentrated H_3PO_4 solution are required to prepare 350 mL of a 0.20 *M* solution?
9. How many milliliters of concentrated HNO_3 solution are required to prepare 1.20 L of a 0.55 *M* solution?
10. How many milliliters of a KOH solution, prepared by dissolving 60.0 g of KOH in 100 mL of solution, are needed to prepare 450 mL of a 0.70 *M* solution?
11. How many milliliters of a KCl solution, prepared by dissolving 45 g of KCl in 500 mL of solution, are needed to prepare 750 mL of a 0.15 *M* solution?
12. 500 mL of a 0.35 *M* solution of KNO_3 are needed. Tell how you would prepare this solution
    (a) from pure solid KNO_3.
    (b) from a solution of KNO_3 that is 4.5 *M*.
13. How many milliliters of 12 *M* KOH are needed to prepare 150 mL of a 0.30 *M* solution?
14. How many grams of solid KOH are needed to prepare the same solution requested in #13?
15. To prepare a buffer solution (Chapter 4), it is determined that acetic acid must be present at 0.17 *M* and that sodium acetate must be present at 0.29 *M*, both in the same solution. If the sodium acetate is a pure solid chemical and the acetic acid to be measured out is a concentrated solution (see Appendix 4), how would you prepare 500 mL of this buffer solution?
16. Tell how you would prepare 500 mL each of the two buffer solutions in Questions 39 and 40 in Chapter 4.
17. Tell which statements are TRUE and which are FALSE.
    (a) In a titrimetric analysis, the weight of a reaction product is determined and converted to the weight of a constituent.
    (b) The "titrant" is the solution in the buret in a titrimetric analysis.
    (c) The buret's graduation lines are read from the top down.
    (d) "TC" means "to control".
    (e) The volumetric pipet has graduation lines on it, much like a buret.
    (f) The volumetric flask is never used for delivering an accurate volume of solution to another vessel.
    (g) Volumetric pipets are not calibrated for blow-out.
    (h) Two frosted rings found near the top of a pipet mean that the pipet is a Mohr pipet.
    (i) The volumetric flask has the letters "TD" imprinted on it.
    (j) The volumetric pipet is a type of "measuring" pipet.
    (k) "Standardization" refers to an experiment in which the percent of a constituent in a sample is determined by experiment.
    (l) The serological pipet is a type of volumetric pipet.
    (m) The key ingredients in chromic acid cleaning solution are $K_2Cr_2O_7$ and ethyl alcohol.

18. Completion
    (a) The two key ingredients in chromic acid cleaning solutions are ____________ and ____________.
    (b) Two pipets that have the legend TC imprinted are the ____________ and ____________.
    (c) The kind of pipet that has graduations on it much like a buret is called the ____________ pipet.
    (d) To say that a given pipet is not calibrated for blow-out means that ____________.
    (e) NaOH cannot be used as a primary standard because ____________.
    (f) Burets and pipets that are not dry should be rinsed first with the solution to be used because ____________.
19. Which is more accurate, a Mohr pipet or a volumetric pipet? Why?
20. Should a 100-mL volumetric flask be used to measure out 100 mL of solution to be added to another vessel? Why or why not?
21. What does a frosted ring near the top of a pipet indicate?
22. What is a primary standard?
23. How does a volumetric flask differ from an Erlenmeyer flask?
24. What is the difference between the end point and the equivalence point in a titration?
25. Explain the reasons for rinsing a pipet as directed in Step 4 of Table 5.1.
26. With a Mohr pipet the meniscus must be read twice, but with a serological pipet the meniscus may be read only once, if desired. Explain this.
27. A student is observed using a 50-mL volumetric flask to "accurately" transfer 50 mL of a solution from one container to another. What would you tell the student (a) to explain his/her error and (b) to help him/her do the "accurate" transfer correctly?
28. What are the two key ingredients in a chromic acid cleaning solution?
29. Why is sodium hydroxide not useful as a primary standard?
30. Why must a pipet that has been stored in distilled water be thoroughly rinsed with the solution to be transferred before use?
31. A technician is directed to prepare accurately to four significant figures 100 mL of a solution of Na_2CO_3 that is approximately 0.25 *M*. The laboratory supervisor provides a solution of Na_2CO_3 that is 4.021 *M* and also some pure solid Na_2CO_3 and tells the technician to proceed by whichever method is easier. Give specific details as to how you would prepare the solution by two different methods, one by dilution and one by weighing the pure chemical, including how many grams or milliliters to measure (show calculation), what type of glassware is used (include the size and kind of pipets and flasks, if applicable), and how it is used.

32. For one of the methods in #31, you needed a pipet. Is the pipet you would choose calibrated for blow-out? How can you tell by looking at the markings on the pipet? What is the name of the pipet you would choose? Give specific details as to how the volume is delivered using this pipet.

33. The concentration of solutions can be known accurately to four significant figures either directly through their preparation or by standardization.
    (a) If the solute is a pure solid, give specific instructions which would ensure such accuracy directly through its preparation.
    (b) What is meant by "standardization"?

34. Compare a volumetric pipet with a serological measuring pipet in terms of
    (a) the number of graduation lines on the pipet.
    (b) whether it is calibrated for blow-out.
    (c) which one to select if you need to deliver 3.72 mL.
    (d) whether it is calibrated "TC" or "TD".

35. What are three requirements of a primary standard?

36. Some pipets are calibrated "TC". Which pipets are they and why would one ever want a pipet calibrated "TC" rather than "TD"?

37. What does it mean to "standardize" a solution?

38. What is a "quantitative transfer"?

39. Why must a pipet that has been stored in distilled water be thoroughly rinsed with the solution to be transferred before use?

40. Consider an experiment in which a pure solid chemical is weighed on an analytical balance into an Erlenmeyer flask. The nature of the experiment is such that this substance must be dissolved before proceeding. The experiment calls for you to dissolve it in 75 mL of water. In the subsequent calculation of the results, which are to be reported to four significant figures, the weight measurement is needed but the 75 mL measurement is not. Should the 75 mL of water be measured with a pipet, or is a graduated cylinder good enough? Explain.

Chapter 6

Acid-Base Titrations and Calculations

6.1 INTRODUCTION

A very common reaction system utilized in titrimetric analysis is the reaction of an acid with a base. This includes all combinations involving strong and weak acids and strong and weak bases. The reaction of an acid with a base is given the general term "neutralization". We say that an acid can be "neutralized" with a base and vice versa. The reaction has the general form

$$\text{acid} + \text{base} \rightarrow \text{salt} + \text{water} \qquad (6.1)$$

Some specific examples are

$$HCl + NaOH \rightarrow NaCl + H_2O$$

$$H_2SO_4 + 2NaOH \rightarrow Na_2SO_4 + 2H_2O$$

$$HC_2H_3O_2 + NaOH \rightarrow NaC_2H_3O_2 + H_2O$$

$$HCl + Na_2CO_3 \rightarrow 2NaCl + H_2CO_3$$

$$\rightarrow 2NaCl + CO_2 + H_2O$$

Note that one product, water, is common to all the examples given. The reason for this, of course, is that neutralization, in most cases, is really the reaction of H^+ with OH^-:

$$H^+ + OH^- \rightarrow H_2O \qquad (6.2)$$

It is the consumption of H^+ by OH^- (or vice versa) that really constitutes neutralization since H^+ ions give a solution its acid properties and OH^- ions give a solution its base properties.

One of the above reactions does not fit this description. It is the reaction of the Bronsted-Lowry base, Na_2CO_3, with HCl. The product analogous to H_2O in this case is H_2CO_3 (all neutralization reactions are double replacement reactions). This weak acid, carbonic acid, decomposes, as it does in a freshly opened can of soda pop, into carbon dioxide and water. This is, however, a neutralization reaction because it consists of the consumption of H^+ by a base. It happens in this case that the base is a Bronsted-Lowry base.

The neutralization reaction lends itself well to titrimetric analysis because (1) the reactions are usually quantitative and fast, (2) primary standard-grade chemicals are available, and (3) there are a large number of substances available which are suitable for use as indicators.

6.2 STANDARDIZATION CALCULATIONS

As indicated in the previous chapter, it is frequently necessary to standardize the titrant prior to calculating the percent of the constituent in a sample. Let us first focus our attention on the calculations involved in such an experiment.

Consider first the simplest (stoichiometrically) sort of reaction that is encountered – the reaction in which the reactants combine on a one-to-one basis. The common example of this for neutralization is

$$HCl + NaOH \rightarrow NaCl + H_2O \tag{6.3}$$

The reaction is one-to-one, meaning that 1 mole of HCl reacts with 1 mole of NaOH. This also means that the number of moles of HCl that react equals the number of moles of NaOH that react:

$$\text{moles of acid} = \text{moles of base} \tag{6.4}$$

$$\text{moles}_A = \text{moles}_B \tag{6.5}$$

In more general terms, in which either A or B can be the titrant, we write:

$$\text{moles of titrant} = \text{moles of substance titrated} \tag{6.6}$$

$$\text{moles}_T = \text{moles}_{ST} \tag{6.7}$$

Let us elaborate on the form of the calculation using this more general terminology. The moles of titrant (moles_T) is always given by multiplying liters of titrant by the molarity of the titrant:

$$\cancel{L_T} \times \frac{\text{moles}_T}{\cancel{L_T}} = \text{moles}_T \tag{6.8}$$

Thus, Equation 6.7 becomes

$$L_T \times M_T = \text{moles}_{ST} \tag{6.9}$$

L_T is the buret reading at the end point converted to liters, and M_T is the molarity of the titrant. It may be the object of the standardization.

Standardization with a Standard Solution

The moles of substance titrated ($moles_{ST}$) is calculated in either of two ways, depending on whether the standardization is accomplished with the use of a primary standard or another standard solution. If a standard solution of the substance titrated is available, the moles of substance titrated is also calculated by multiplying the number of liters used by the molarity,

$$L_{ST} \times M_{ST} = moles_{ST} \tag{6.10}$$

and thus Equation 6.9 becomes

$$L_T \times M_T = L_{ST} \times M_{ST} \tag{6.11}$$

The volume of the substance titrated is known (presumably the solution was carefully pipetted into the reaction flask) and, if the titrant's molarity is to be determined, the concentration of the substance titrated is accurately known. The only unknown in Equation 6.11 is, therefore, M_T, which is the object of the experiment. The equation can be rearranged as follows:

$$M_T = \frac{L_{ST} \times M_{ST}}{L_T} \tag{6.12}$$

and the molarity computed. Again, as with solution preparation, the volumes can both be in milliliters since the conversion factors cancel.

If the molarity of the substance titrated (M_{ST}) is to be calculated, M_T is known and Equation 6.11 is solved for M_{ST}.

Standardization with a Primary Standard

If a primary standard is used to standardize the titrant, the number of moles of the substance titrated, the right side of Equation 6.9, is computed by dividing the weight of the primary standard by its molecular weight:

$$\cancel{grams_{ST}} \times \frac{moles_{ST}}{\cancel{grams_{ST}}} = moles_{ST} \tag{6.13}$$

$$\frac{grams_{ST}}{FW_{ST}} = moles_{ST} \tag{6.14}$$

Now Equation 6.9 becomes

$$L_T \times M_T = \frac{grams_{ST}}{FW_{ST}} \tag{6.15}$$

and can be rearranged to give M_T:

$$M_T = \frac{grams_{ST}}{FW_{ST} \times L_T} \tag{6.16}$$

In this case, the volume of the titrant must be in liters, since there is no cancellation of the conversion factor.

Example 1

What is the molarity of an NaOH solution that has been standardized with a 0.1012 *M* HCl solution 25.00 mL of which required 28.76 mL of the NaOH?

Solution

$$L_T \times M_T = L_{ST} \times M_{ST}$$

$$28.76 \times M_T = 25.00 \times 0.1012$$

$$M_T = \frac{25.00 \times 0.1012}{28.76} = 0.08797\ M$$

Example 2

What is the molarity of an NaOH solution that has been standardized with a primary standard acid (FW = 204.23), 0.4119 g requiring 21.66 mL of the NaOH? (Assume a one-to-one reaction.)

Solution

$$L_T \times M_T = \frac{\text{grams}_{ST}}{FW_{ST}}$$

$$0.02166 \times M_T = \frac{0.4119}{204.23} \rightarrow M_T = \frac{0.4119}{0.02166 \times 204.23} = 0.09311\ M$$

A more complicated problem is one in which the reaction is not one-to-one. In this case, our initial equation

$$\text{moles}_T = \text{moles}_{ST} \tag{6.17}$$

must be modified. This can be done by extracting the mole ratio from the equation much like one would do in an ordinary stoichiometry calculation when converting the substance known to the substance sought. Thus, we would have

$$\text{moles}_T = \text{moles}_{ST} \times \text{ratio} \tag{6.18}$$

or

$$\text{moles}_T \times \text{ratio} = \text{moles}_{ST} \tag{6.19}$$

The ratio is typically 2:1 (or 1:2) or 3:1 (or 1:3), etc. Some confusion and error may result here in determining whether the ratio is 1:2 or 2:1, etc., and this can be serious, especially when the calculation for percent constituent involves a similar ratio. Scientists recognized this problem many years ago and solved it by devising an entirely new system for working with quantities of chemical in laboratory reactions. This is the system of equivalent weights, equivalents, and normality, which we abbreviate in this text as "EEN".

6.3 EQUIVALENT WEIGHTS, EQUIVALENTS, AND NORMALITY

The EEN System

The scheme alluded to in the previous section is based on the fact that all reactions are one-to-one if we think in terms of a single unit of one substance reacting with a single unit of another substance. The question is what part of each substance is considered to be this "single unit". It obviously can be equal to 1 mole (in a "real" one-to-one reaction) or it can equal some fraction of a mole.

Consider two neutralization reactions, one that is one-to-one (in terms of moles) and one that is not:

$$NaOH + HCl \rightarrow NaCl + H_2O \tag{6.20}$$

$$2NaOH + H_2SO_4 \rightarrow Na_2SO_4 + 2H_2O \tag{6.21}$$

Reaction 6.20 is one-to-one; Reaction 6.21 is two-to-one. NaOH reacts with HCl as follows:

$$\square + \square \rightarrow NaCl + H_2O \tag{6.22}$$

and it reacts with H_2SO_4 as follows:

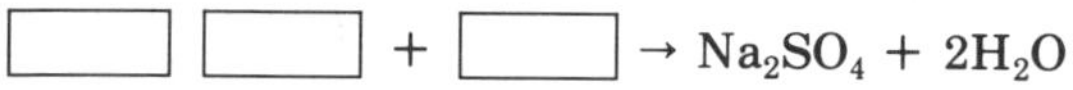

$$\square\ \square + \square \rightarrow Na_2SO_4 + 2H_2O \tag{6.23}$$

where the rectangles indicate 1 mole of each substance.

If we were to split the 1 mole of H_2SO_4 into two "pieces", 1 mole of NaOH would react with each "piece", and our reaction would be one-to-one – one "piece" of NaOH reacting with one "piece" of H_2SO_4. A "piece" of NaOH would be 1 mole; one "piece" of H_2SO_4 would be 0.5 mole:

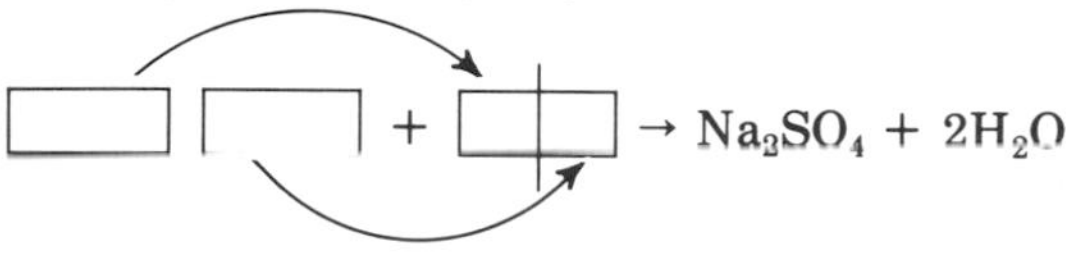

$$\square\ \square + \boxplus \rightarrow Na_2SO_4 + 2H_2O \tag{6.24}$$

This "piece" of a reactant is an "equivalent" of that reactant. It is the amount of one substance that will react with an equivalent of another substance. In other words, one equivalent of one substance reacts with an equivalent of another substance – the reaction is one-to-one. All reactions can be viewed in this manner, and thus all reactions are considered to be one-to-one under this scheme.

The weight of one of these "pieces" is useful information just as the formula weight, the weight of 1 mole, is useful in the scheme of formula weight, moles, and molarity ("FMM"). The weight of one equivalent is called the "equivalent weight". It is used for all of the same types of calculations that the formula weight was used for in the FMM system. This includes converting grams to equivalents and vice versa.

Concentrations of solutions are expressed differently in this EEN system. Under the FMM system, molarity (moles per liter) was the important concentration unit. In this EEN system, equivalents per liter is the important concentration unit. It is called "normality" (as opposed to "molarity"), and solutions are referred to as "normal" solutions (as opposed to "molar" solutions). Table 6.1 summarizes these comparisons.

In the example of NaOH reacting with H_2SO_4, it was obvious after writing the equation that 1 mole of NaOH reacts with 0.5 mole of H_2SO_4. Since one equivalent reacts with one equivalent, it is apparent that 0.5 mole of H_2SO_4 is one equivalent of H_2SO_4, and the equivalent weight is one half the molecular weight. There is a simpler way of reaching this conclusion and this simpler method is true for all acids. *The equivalent weight of an acid is the*

Table 6.1. Comparison of the two systems for expressing quantities of reactants.

FMM System	EEN System
The mole	The equivalent
The formula weight	The equivalent weight
Molarity	Normality
0.10 *M*	0.10 *N*
Molar	Normal

molecular weight divided by the number of hydrogens donated by one molecule of acid in the reaction. With H_2SO_4, the number of hydrogens available for donating is two and they are indeed both donated in the reaction

$$H_2SO_4 + 2NaOH \rightarrow Na_2SO_4 + H_2O \quad (6.25)$$

H_2SO_4 has become Na_2SO_4. Both hydrogens have been donated, as is evidenced by the fact that both hydrogens are gone in the formula Na_2SO_4. This and some additional examples are shown in Table 6.2.

We can use similar reasoning when determining the equivalent weight of a base (an H^+ acceptor). *The equivalent weight of a base is the molecular weight divided by the number of H^+ accepted per molecule in the reaction.* Some examples are shown in Table 6.3.

Table 6.2. Some examples of neutralization reactions and the determination of the equivalent weight (EW) of the acid involved.

Reaction	H's Donated	EW
$H_2SO_4 + 2NaOH \rightarrow Na_2SO_4 + 2H_2O$	2	$\frac{H_2SO_4}{2}$
$H_3PO_4 + 3KOH \rightarrow K_3PO_4 + 3H_2O$	3	$\frac{H_3PO_4}{3}$
$H_3PO_4 + 2NaOH \rightarrow Na_2HPO_4 + 2H_2O$	2	$\frac{H_3PO_4}{2}$
$NaH_2PO_4 + NaOH \rightarrow Na_2HPO_4 + H_2O$	1	$\frac{NaH_2PO_4}{1}$

Table 6.3. Some examples of neutralization reactions and the determination of the equivalent weight (EW) of the base involved.

Reaction	H's Accepted	EW
$HCl + NaOH \rightarrow NaCl + H_2O$	1	$\frac{NaOH}{1}$
$2HCl + Ba(OH)_2 \rightarrow BaCl_2 + 2H_2O$	2	$\frac{Ba(OH)_2}{2}$
$H_2SO_4 + 2KOH \rightarrow K_2SO_4 + 2H_2O$	1	$\frac{KOH}{1}$
$2HCl + Na_2CO_3 \rightarrow 2NaCl + H_2CO_3$	2	$\frac{Na_2CO_3}{2}$

Solution Preparation

Solution preparation in the EEN system is a straightforward extension of that in the FMM system. Normality is substituted for molarity, and equivalent weight is substituted for formula weight.

For preparing solutions of a solute that is a pure solid chemical the following formula was useful before (compare with Equation 5.4):

$$L_D \times M_D \times FW_{SOL} = \text{grams to be weighed} \quad (6.26)$$

in which L_D is liters desired, M_D is molarity desired, and FW_{SOL} is the formula weight of the solute. In the EEN system, the equation becomes

$$L_D \times N_D \times EW_{SOL} = \text{grams to be weighed} \quad (6.27)$$

in which N_D represents normality desired and EW_{SOL} represents the equivalent weight of the solute. A check of the units reveals appropriate cancellation:

$$\cancel{L_D} \times \frac{\cancel{\text{equiv.}}}{\cancel{L}} \times \frac{\text{grams}}{\cancel{\text{equiv.}}} = \text{grams} \quad (6.28)$$

In cases in which molecular weight and equivalent weight are the same (one hydrogen donated or accepted), such as with NaOH, KOH, etc., the calculation is identical with that of the FMM scheme, and molarity and normality are the same. Following are two examples in which the formula weight and equivalent weight are not the same.

Example 3

How would you prepare 500 mL of a 0.10 *N* solution of $Ba(OH)_2$ from pure solid $Ba(OH)_2$? (Refer to Table 6.3 for the reaction involved.)

Solution

In Table 6.3 we see that the equivalent weight is the formula weight divided by 2. Thus

$$0.500\,\cancel{L} \times \frac{0.10\ \cancel{\text{equiv.}}}{\cancel{L}} \times \frac{85.68\ \text{g}}{\cancel{\text{equiv.}}} = 4.3\ \text{g}$$

4.3 g of $Ba(OH)_2$ are weighed, placed in a 500-mL container, dissolved, and diluted to the mark with water.

Example 4

How would you prepare 1.0 L of 0.35 *N* phthalic acid from pure solid phthalic acid, FW = 166.14, if two hydrogens are donated in the reaction?

Solution

$$1.0\,\cancel{L} \times \frac{0.35\ \cancel{\text{equiv.}}}{\cancel{L}} \times \frac{83.07\ \text{g}}{\cancel{\text{equiv.}}} = 29\ \text{g}$$

One would weigh 29 g of phthalic acid, place it in a 1-L container, dissolve in water, and dilute to the mark.

For preparing solutions by dilution, the normality of the solution before and after is used rather than the molarity (compare with Equation 5.9):

$$N_B \times V_B = N_A \times V_A \tag{6.29}$$

Obviously, if the normality and molarity are the same, the calculation is identical with that in Chapter 5. If they are not the same, the form of the calculation is the same, but the numbers are different.

Example 5

(Compare with Problem #3f at the end of Chapter 5.)

How would you prepare 2.0 L of a 0.50 *N* solution of sulfuric acid, H_2SO_4, from concentrated H_2SO_4, which is 36.0 *N*?

Solution

$$N_B \times V_B = N_A \times V_A$$

$$36.0 \times V_B = 0.50 \times 2.0$$

$$V_B = \frac{0.50 \times 2.0}{36.0} = 0.028 \text{ L}$$

$$= 28 \text{ mL}$$

One would dilute 28 mL of the 36.0 *N* to 2.0 L.

Occasionally it may be necessary to compare the normality of a solution with its molarity. To determine the molarity, given the normality, or vice versa, the number of equivalents contained in a mole is useful information. Such a number can be used as the conversion factor for converting normality to molarity or vice versa:

$$\frac{\cancel{\text{equiv.}}}{\text{L}} \times \frac{\text{moles}}{\cancel{\text{equiv.}}} = \frac{\text{moles}}{\text{L}} \tag{6.30}$$

Example 6

The normality of concentrated H_2SO_4 is 36.0 *N*. What is its molarity (assume both hydrogens are neutralized)?

Solution

$$\frac{36.0\ \cancel{\text{equiv.}}}{\text{L}} \times \frac{1\ \text{mol}}{2\ \cancel{\text{equiv.}}} = 18.0\ M$$

Standardization Calculations

We can now extend the concepts of the system of equivalents, equivalent weights, and normality to standardization calculations. Earlier in this chapter we discussed the form and gave examples of standardization calculations in the FMM system. We discussed briefly the need to use a ratio in the calculation in order to account for a reaction condition that was other than one-to-one. (See Equation 6.18 and accompanying discussion.) Since in this EEN system all reactions are one-to-one, the ratio is eliminated, and Equation 6.18 becomes

$$\text{equiv.}_T = \text{equiv.}_{ST} \tag{6.31}$$

Standardization calculations, then, in the EEN system look very much like those in the FMM system in which the mole ratio is one-to-one. For a standardization with another standard solution the following is useful (compare with Equation 6.11):

$$L_T \times N_T = L_{ST} \times N_{ST} \tag{6.32}$$

For a standardization with a primary standard, we use the following (compare with Equation 6.15):

$$L_T \times N_T = \frac{\text{grams}_{ST}}{EW_{ST}} \tag{6.33}$$

Some examples are presented in the next section.

6.4 EXAMPLES OF ACID-BASE PRIMARY STANDARDS AND CALCULATIONS

For standardizing a base solution, primary standard grade potassium hydrogen phthalate is the popular choice. Also called potassium acid phthalate, or simply KHP, it is the salt representing partially neutralized phthalic acid. Figure 6.1 shows the chemical structure of phthalic acid and KHP. KHP is a white crystalline substance of high formula weight (204.23), stable at oven drying temperatures, and available in a very pure form.

For standardizing acid solutions, either primary standard sodium carbonate or primary standard tris-(hydroxymethyl)amino methane, or "THAM" (also sometimes referred to as "TRIS"), is used. These, too, possess all the qualities of a good primary standard. As mentioned earlier, when titrating sodium carbonate, carbonic acid (H_2CO_3) is one of the products and must be decomposed with heat to push the reaction to completion to the right.

$$Na_2CO_3 + 2HCl \rightleftharpoons 2NaCl + H_2CO_3$$

$$H_2CO_3 \rightleftharpoons CO_2 + H_2O$$

Heat also will eliminate CO_2 from the solution, which aids further in this "completion push". A further discussion of this appears later in this chapter.

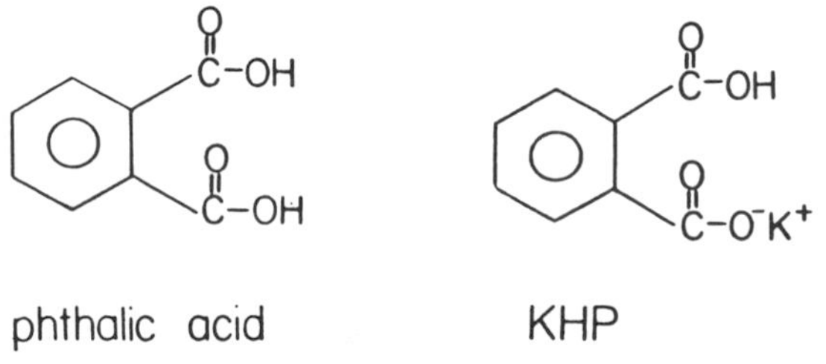

phthalic acid KHP

Figure 6.1. The molecular structures of phthalic acid and potassium hyrdrogen phthalate (KHP).

The $H_2CO_3 \rightleftharpoons CO_2$ equilibrium can be a problem in all base solutions due to CO_2 from the air dissolving in the solution, forming carbonic acid, resulting in the formation of carbonate:

$$H_2CO_3 \rightleftharpoons 2H^{+1} + CO_3^{-2}$$

which would then also react with whatever acid may be used in titrimetric procedures, thus creating an error. The water used to prepare such solutions is boiled in advance (or freshly distilled) so that all CO_2 is eliminated and no carbonate forms.

THAM, or TRIS, is also a popular primary standard base. Its formula is

$$(HOCH_2)_3CNH_2$$

It has a high formula weight (121.14), is stable at moderate drying oven temperatures, and is available in pure form. The reaction with an acid involves the acceptance of one hydrogen by the amine group, $-NH_2$:

$$(HOCH_2)_3CNH_2 + HCl \rightarrow (HOCH_2)_3CNH_3^{+}Cl^{-1}$$

The end point occurs at a pH between 4.5 and 5.

Example 7

In a standardization experiment, 0.5067 g of primary standard sodium carbonate (Na_2CO_3) was exactly neutralized by 27.86 mL of a hydrochloric acid solution. What is the normality of the HCl solution?

$$Na_2CO_3 + 2HCl \rightarrow 2NaCl + CO_2 + H_2O$$

Solution

$$\text{equiv.(HCl)} = \text{equiv.}(Na_2CO_3)$$

$$\text{L(HCl)} \times \text{N(HCl)} = \frac{\text{grams}(Na_2CO_3)}{\text{EW }(Na_2CO_3)}$$

$$0.02786 \times \text{N(HCl)} = \frac{0.5067}{52.995}$$

$$\text{N(HCl)} = \frac{0.5067}{0.02786 \times 52.995} = 0.3432\ N$$

Example 8

Standardization of a solution of sulfuric acid required 28.50 mL of 0.1077 *N* NaOH when exactly 25.00 mL of the H_2SO_4 was used. What is the normality of the H_2SO_4?

Solution

$$\text{equiv.}(H_2SO_4) = \text{equiv.(NaOH)}$$

$$\text{L}(H_2SO_4) \times \text{N}(H_2SO_4) = \text{L(NaOH)} \times \text{N(NaOH)}$$

or

$$mL(H_2SO_4) \times N(H_2SO_4) = mL(NaOH) \times N(NaOH)$$

$$25.00 \times N(H_2SO_4) = 28.50 \times 0.1077$$

$$N(H_2SO_4) = \frac{0.02860 \times 0.1077}{0.02500} = 0.1228\ N$$

6.5 PERCENT CONSTITUENT CALCULATIONS

The ultimate goal of any titrimetric analysis is to determine the percentage of some constituent in a sample. As mentioned at the beginning of Chapter 5 (see Figure 5.1 and accompanying discussion), the amount of the titrant needed to exactly consume the constituent is measured and converted to the amount of constituent via a stoichiometric calculation. In this case, this stoichiometry calculation consists of converting the liters of titrant to grams of constituent, and since the ratio is one-to-one in the EEN system, only two steps are required – the conversion of liters to equivalents and the conversion of equivalents to grams – as can be seen from the following derivation:

$$\cancel{L_T} \times \frac{\cancel{equiv}_T}{\cancel{L_T}} \times \frac{\cancel{equiv}_{ST}}{\cancel{equiv}_T} \times \frac{grams_{ST}}{\cancel{equiv}_{ST}} = grams_{ST} \quad (6.34)$$

Since the ratio represented by

$$\frac{equiv._{ST}}{equiv._T}$$

is one-to-one, it can be eliminated and Equation 6.34 rewritten:

$$L_T \times N_T \times EW_{ST} = grams_{ST} \quad (6.35)$$

Since the substance titrated (ST) is the constituent, we can write the formula for percent constituent as follows:

$$\%\ Const = \frac{L_T \times N_T \times EW_{const}}{sample\ weight} \times 100 \quad (6.36)$$

Example 9

In the analysis of a soda ash (impure Na_2CO_3) sample for sodium carbonate content, 0.5203 g of the soda ash required 36.42 mL of 0.1167 *N* HCl for titration. What is the percent of Na_2CO_3 in this sample? (See Example 7 for the reaction involved.)

Solution

$$\%\ Na_2CO_3 = \frac{L(HCl) \times N(HCl) \times EW(Na_2CO_3)}{sample\ weight} \times 100$$

$$\%\ Na_2CO_3 = \frac{0.03642 \times 0.1167 \times 52.995}{0.5203} \times 100$$

$$\%\ Na_2CO_3 = 43.29\%$$

6.6 BACK TITRATIONS

Sometimes a special kind of titrimetric procedure known as a "back titration" is required. In this procedure, the constituent is consumed using excess titrant (the end point is intentionally overshot) and the end point is then determined by titrating the excess with a second titrant. Thus, two titrants are used, and the exact concentration of each is needed for the calculation.

Figure 6.2 shows a pictorial representation of a back titration. The first buret reading represents the equivalents of the first titrant added. This amount actually exceeds the equivalents of the substance titrated. The second buret reading (the buret is inverted to show a titration back to the end point) represents the amount of the second titrant (the so-called "back titrant") used to titrate the excess amount of the initial titrant. The difference between the total equivalents of the first titrant and the total equivalents of the second titrant is the number of equivalents that actually reacted with the constituent, and this is the number that is needed for the calculation. Thus

$$\% \text{ Const} = \frac{(L_T \times N_T - L_{BT} \times N_{BT}) \times EW_{ST}}{\text{sample weight}} \times 100 \qquad (6.37)$$

in which "BT" symbolizes the back titrant.

It may seem strange that we would ever want to perform an experiment of this kind. First of all, it would be used in the event of a slow reaction taking place in the reaction flask. Perhaps the sample is not dissolved completely, and addition of the titrant causes dissolution to take place over a period of minutes or hours. Adding an excess of the titrant and back titrating it later would seem an appropriate course of action in a case of this kind.

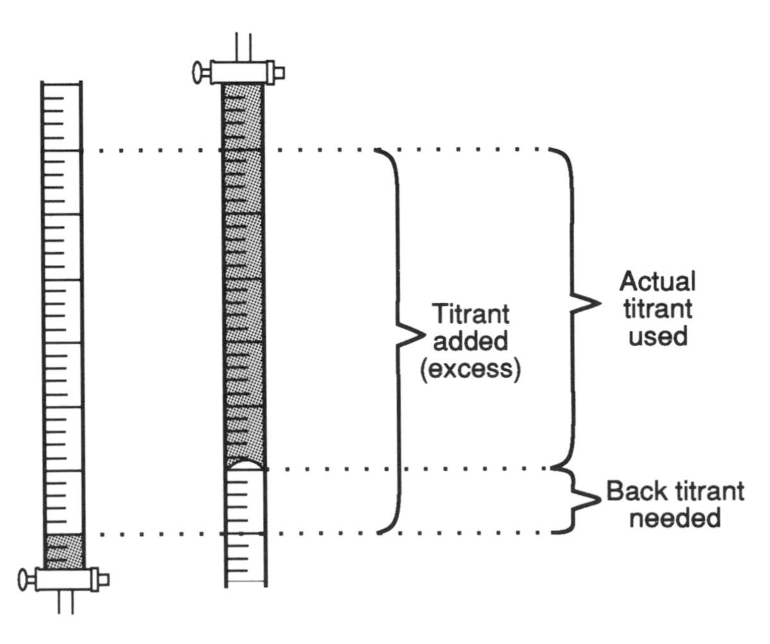

Figure 6.2. A pictorial representation of a back titration.

Second, perhaps the constituent must be calculated from a gaseous product of another reaction. In this case, one would want the gas to react with the titrant as soon as it is formed, since it could escape into the air because of a high vapor pressure. Thus, an excess of the titrant would be present in a solution through which the gas is bubbled. After the gas-forming reaction has stopped, the excess titrant in this "bubble flask" could be titrated with the back titrant and the results calculated. A possible application of this latter description is the Kjeldahl titration.

6.7 THE KJELDAHL TITRATION

A titrimetric method that has been used for many years for the determination of nitrogen and/or protein in a sample is the Kjeldahl method. Examples of samples include grain, protein supplements for animal feed, fertilizers, and food products. It is a method that often makes use of the "back titration" concept. We will now describe this technique in detail.

The method consists of three parts: the digestion, the distillation, and the titration. The digestion step is in essence the dissolving step. The sample is weighed and placed in a Kjeldahl flask, which is a round-bottom flask with a long neck, similar in appearance to a volumetric flask except for the round bottom and the lack of a calibration line. A fairly small volume of concentrated sulfuric acid along with a quantity of K_2SO_4 (to raise the boiling point of the sulfuric acid) and a catalyst (typically an amount of $CuSO_4$, selenium, or a selenium compound) are added, and the flask is placed in a heating mantle and heated. The sulfuric acid boils and the sample digests for a period of time until it is evident that the sample is dissolved and a clear solution is contained in the flask. The digestion must be carried out in a fume hood, since thick SO_3 fumes evolve from the flask until the sample is dissolved. At this point, the contents of the flask are diluted with water and an amount of fairly concentrated sodium hydroxide is added to neutralize the acid. Immediately upon neutralization, the nitrogen originally present in the sample is converted to ammonia. At this point the distillation step begins.

A laboratory that runs Kjeldahl analyses routinely would likely have a special apparatus set up for the distillation. The essence of this apparatus is shown in Figure 6.3. A baffle is placed on the top of the Kjeldahl flask and subsequently connected to a condenser which in turn guides the distillate into a receiving flask as shown. The ammonia is then distilled into the receiving flask. The receiving flask contains an acid for reaction with the ammonia.

The acid in the receiving flask can either be a dilute (perhaps 0.10 *N*) standardized solution of a strong acid, such as sulfuric acid, or a solution of boric acid. If it is the former, it is an example of a back titration. If it is the latter, it is an example of an indirect titration.

In the back titration method, an excess, but carefully measured, amount of the standardized acid is contained in the flask such that, after the ammonia bubbles through it and is consumed, an excess remains. The flask is then removed from the apparatus and the excess acid titrated with a standardized NaOH solution. The analyte in this procedure is the nitrogen (in the form of ammonia as it enters the flask), and thus the amount of acid consumed is the important measurement. The amount of acid consumed is the

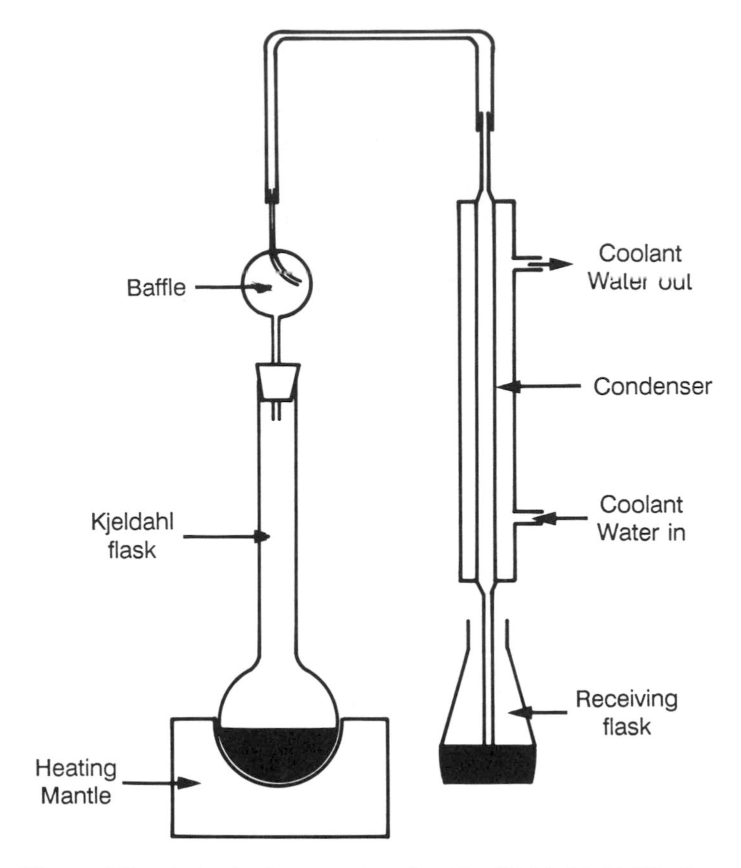

Figure 6.3. A typical apparatus for the Kjeldahl distillation step. (From Kenkel, J., *Analytical Chemistry Refresher Manual*, Lewis Publishers, Chelsea, MI, 1992. With permission.)

difference between the total amount present and the amount that was in excess. It is called a back titration because the amount of acid in the flask is in excess and in essence goes beyond the end point for the reaction with the ammonia. Thus, the analyst must come *back* to the end point with the sodium hydroxide. The calculation is

$$\% \text{ N} = \frac{(L_T \times N_T - L_{BT} \times N_{BT}) \times 14.00}{\text{sample weight}} \times 100 \qquad (6.38)$$

in which the titrant is the dilute sulfuric acid and the back titrant (BT) is the NaOH. The equivalent weight shown is the atomic weight of nitrogen. The percent of protein may also be calculated, in which case the equivalent weight of the protein is substituted for the 14.00.

In the indirect method using boric acid, the ammonia reacts with the boric acid, producing a partially neutralized salt of boric acid:

$$NH_3 + H_3BO_3 \rightarrow H_2BO_3^{-1} + NH_4^+$$

which can then be titrated with a standardized acid:

$$H_2BO_3^{-1} + H^{+1} \rightarrow H_3BO_3$$

The amount of standardized acid needed is proportional to the amount of ammonia that bubbled through. It is an called an indirect method because the ammonia is determined but is not titrated. It is determined indirectly by titration of the $H_2BO_3^{-1}$. In a direct titration, the analyte would be reacted directly with the titrant. Equation 6.36 is used for both direct and indirect methods. The concentration of the boric acid in the receiving flask does not enter into the calculation and need not be known. Experiment 10 in this chapter is a Kjeldahl titration experiment, a back titration experiment as written.

Example 10

In a Kjeldahl analysis, a flour sample weighing 0.9857 g was digested in concentrated H_2SO_4 for 45 min. A concentrated solution of NaOH was added such that all of the nitrogen was converted to NH_3. The NH_3 was then distilled into a flask containing 50.00 mL of 0.1011 *N* H_2SO_4. The excess required 5.12 ml of 0.1266 *N* NaOH for titration. What is the percent of nitrogen in the sample?

Solution

$$\% \text{ N} = \frac{\text{L}(H_2SO_4) \times \text{N}(H_2SO_4) - \text{L(NaOH)} \times \text{N(NaOH)]} \times 14.00}{\text{sample weight}} \times 100$$

$$= \frac{[0.050000 \times 0.1011 - 0.00512 \times 0.1266] \times 14.00}{0.9857} \times 100$$

$$= 6.259\% \ N$$

Example 11

In a Kjeldahl analysis, a grain sample weighing 1.1033 g was digested in concentrated H_2SO_4 for 40 min. A concentrated solution of NaOH was added such that all of the nitrogen was converted to NH_3. The NH_3 was then distilled into a flask containing a solution of boric acid. Following this, the solution in the receiving flask was titrated with 0.1011 *N* HCl, requiring 24.61 mL. What is the percent of nitrogen in the sample?

Solution

$$\% \text{ N} = \frac{L_{HCl} \times N_{HCl} \times 14.00 \times 100}{\text{sample weight}}$$

$$= \frac{0.02461 \times 0.1011 \times 14.00 \times 100}{1.1033}$$

$$= 3.157\%$$

Appendix 5 summarizes all the above calculations for both the FMM and the EEN systems.

6.8 MILLIMOLES AND MILLIEQUIVALENTS

Besides moles and equivalents, it is fairly common to express amounts of chemicals as millimoles or as milliequivalents. The metric prefix "milli" has the same meaning here as in other examples we have seen. Thus, a millimole, mmol, is one-thousandth of a mole and a milliequivalent, meq, is one-thousandth of an equivalent just as one milligram is one-thousandth of a gram and one milliliter is one-thousandth of a liter. In order to convert millimoles to moles, we divide by 1000; or in order to convert moles to millimoles, we multiply by 1000, etc.

All the calculations discussed previously in this chapter, and summarized in Appendix 5, can accomodate these smaller units without conversion to the larger units. Wherever "L" appears, we may use "mL" (such as for the volume desired for solution preparation, and buret readings) and wherever grams appears, we may use milligrams, as long as mL are used only when mg are used, etc. Milligrams per millimole, or milligrams per milliequivalent, are alternate units for formula weight and equivalent weight respectively, so conversion to smaller units is not an issue for these. The same can be said for molarity and normality. The number of millimoles per milliliter are alternate units for molarity and milliequivalents per milliliter are alternate units for normality.

Thus, for example, when preparing a molar solution of a pure solid chemical, the following formula can be used:

$$mL_D \times M_D \times FW_{SOL} = \text{mg to be weighed} \tag{6.39}$$

Compare this equation with Equation 5.4.

When standardizing with a primary standard, the following formula can be used:

$$mL_T \times N_T = \frac{mg_{ST}}{EW_{ST}} \tag{6.40}$$

Compare this with Equation 6.33.

When calculating the percent of a constituent, the following can be used:

$$\%\ \text{Const} = \frac{mL_T \times N_T \times EW_{const}}{\text{mg of sample}} \times 100 \tag{6.41}$$

Compare this equation with Equation 6.36.

6.9 INTRODUCTION TO TITRATION CURVES

A graphic picture of what happens during an acid-base titration is easily produced in the laboratory. Consider again what is happening as a titration proceeds. Consider, specifically, NaOH as the titrant and HCl as the substance titrated. In the titration flask, the following reaction occurs when titrant is added (see Figure 6.4):

$$H^{+1} + OH^{-1} \rightarrow H_2O$$

Obviously, as H^{+1} ions are consumed in the flask by the OH^{-1}, the pH will change since pH = –log $[H^{+1}]$. In fact, the pH should increase since the number of H^{+1} ions decreases. The magnitude of this increase can be measured with the use of a pH meter, an electronic instrument designed to measure the pH and display it on a meter (see Chapter 18). Thus, if we were to measure the pH in this manner after each addition of NaOH and graph the pH vs. milliliters of NaOH added we would have a graphical display of

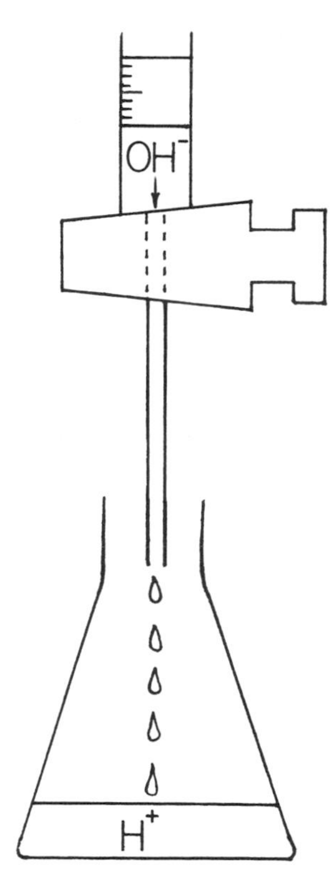

Figure 6.4. Acid-base titration – an acid (represented by H^+) in the reaction flask and a base (represented by OH^-) in the buret.

the experiment. Figure 6.5 shows the results of such an experiment for the case in which 0.10 *M* HCl is titrated with 0.10 *M* NaOH.

An initial response to this "titration curve" might be: "Why the strange shape? Why the gradual increase in the beginning? Why the sharp change in the middle? Why the gradual increase at the end?" An acid-base titration curve is not simply a mysterious result as measured by a mysterious pH meter. The entire curve can also be arrived at independent of the pH meter experiment by calculating the $[H^{+1}]$ and the pH after each addition and once

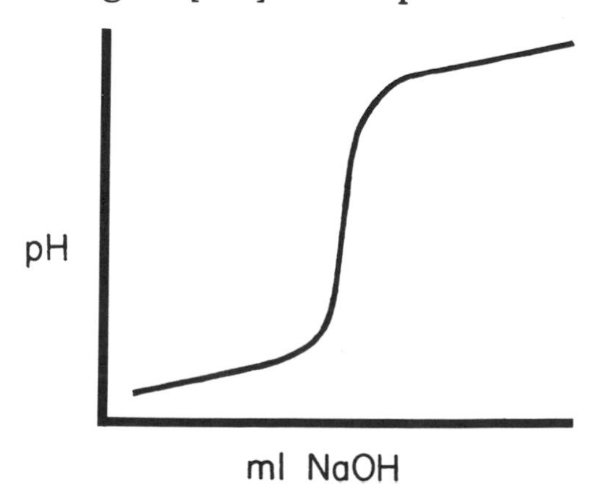

Figure 6.5. A titration curve resulting from the measurement of the pH after adding quantities of titrant.

again plotting the results. Let us illustrate the idea by performing several sample calculations.

Let us imagine that the experiment consists of adding 0.10 *M* NaOH in the buret to 25.00 mL of 0.10 *M* HCl in the reaction flask. The following steps may be used to obtain the titration curve:

A. At 0.00 mL NaOH added:

$$[H^{+1}] = 0.10\ M$$

$$pH = -\log[H^{+1}] = -\log(0.10) = 1.00$$

B. After adding 5.00 mL NaOH (30.00 mL of solution in the flask):

$$[H^{+1}] = \frac{0.02500\ L \times \frac{0.10\ mol}{L} - 0.00500\ L \times \frac{0.10\ mol}{L}}{0.03000\ L}$$

$$= \frac{0.002000\ mol}{0.03000\ L} = 0.0667\ M$$

$$pH = 1.18$$

C. After adding 10.00 mL of NaOH (35.00 mL in the flask):

$$[H^{+1}] = \frac{0.02500\ L \times \frac{0.10\ mol}{L} - 0.01000\ L \times \frac{0.10\ mol}{L}}{0.03500\ L}$$

$$= \frac{0.00150\ mol}{0.03500\ L} = 0.0429\ M$$

$$pH = 1.37$$

D. After adding 20.00 mL NaOH (45.00 mL in the reaction flask):

$$[H^{+1}] = \frac{0.02500\ L \times \frac{0.10\ mol}{L} - 0.02000\ L \times \frac{0.10\ mol}{L}}{0.04500\ L}$$

$$= \frac{0.00050\ mol}{0.04500\ L} = 0.0111\ M$$

$$pH = 1.95$$

E. At the equivalence point (25.00 mL of NaOH added) – all H^{+1} due to HCl are gone:

$$[H^{+1}] = 1.0 \times 10^{-7}\ M \text{ (due to water)}$$

$$pH = 7.00$$

F. After the equivalence point – 35.00 mL NaOH added:

$$[OH^{-1}] = \frac{(0.03500 - 0.02500)L \times 0.10 \frac{mol}{L}}{0.06000\ L}$$

$$= \frac{0.00100\ mol}{0.0600\ L} = 0.0167\ M$$

$$pOH = 1.78$$

$$pH = 12.22$$

G. 50.00 mL of NaOH added:

$$[OH^{-1}] = \frac{(0.05000 - 0.02500)L \times 0.10 \frac{mol}{L}}{0.07500\ L}$$

$$= \frac{0.0025\ mol}{0.07500\ L} = 0.0333\ M$$

$$pOH = 1.48$$

$$pH = 12.52$$

Summarizing the data to be plotted, we have

	pH	mL NaOH
A	1.00	0.00
B	1.18	5.00
C	1.37	10.00
D	1.95	20.00
E	7.00	25.00
F	12.22	35.00
G	12.52	50.00

Figure 6.6 represents the graph of this data. The results show the exact curve measured experimentally with the use of a pH meter (Figure 6.5).

We should emphasize here that the concentrations of the acid and base are important to consider. These concentrations in the discussions thus far have been 0.10 M. The lower the concentration of the acid, the fewer H^{+1} ions are present, the higher the initial pH, and the initial flat portion of the titration curve would occur at a higher pH level.

It is interesting to compare this with the titration of a weak acid, such as acetic acid, with NaOH. The difference between this and the strong acid-strong base case just discussed is that the pH starts and continues to the equivalence point at the same higher level as with the lower concentration of HCl. This should not be unexpected, since a solution of a weak acid is being measured, and once again there are fewer H^{+1} ions in the solution. Still weaker acids start at even higher pH values.

The weak acid curves can also be calculated. This involves the manipulation of the equilibrium constant expression for a weak acid and is beyond the scope of this text. However, we can correlate the value of the K_a with the position of the titration curve. As discussed in Chapter 4, the weaker the acid, the smaller the K_a. Figure 6.7 represents a series of four acids at a concentration of 0.10 M titrated with a strong base. The curves for HCl and

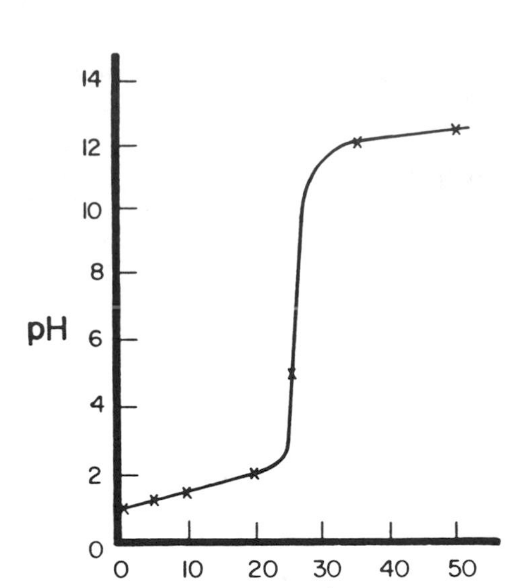

Figure 6.6. Titration curve of the HCl titrated with NaOH as calculated in the text.

acetic acid are shown as well as two curves for two acids even weaker than acetic acid. (The K_a's are indicated.)

If the titration of a 0.10 *M* base with a 0.10 *M* acid is considered (Figure 6.8; compare with Figure 6.4), we have a curve that starts at a high pH value (a solution of a base has a high pH) and ends at a low pH value (just the opposite to that observed when titrating an acid with a base). Likewise, the

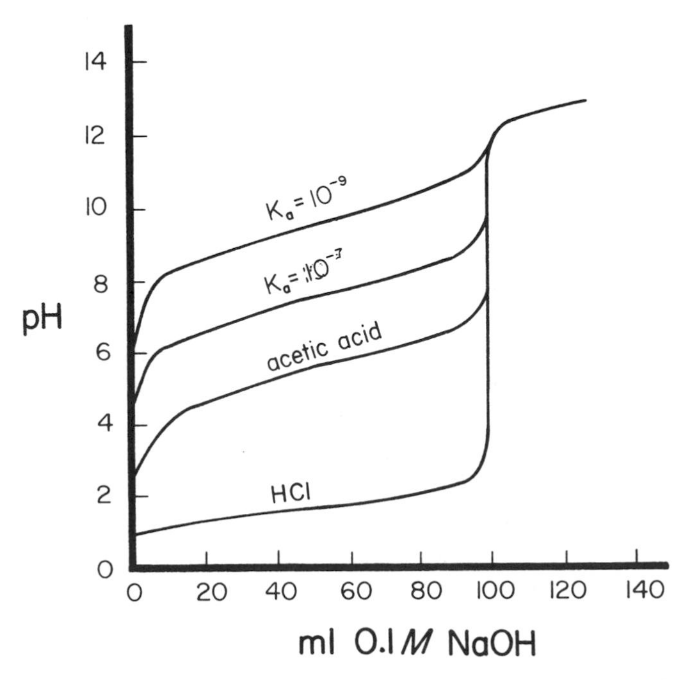

Figure 6.7. The curves of several weak acids compared with 0.1 *M* HCl.

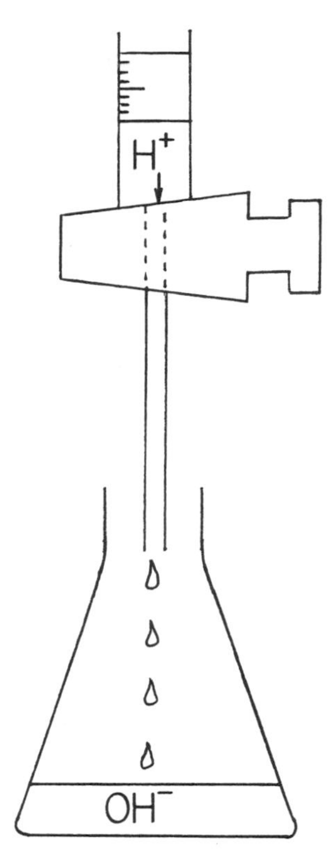

Figure 6.8. The titration of a base with an acid.

curves for 0.10 *M* weak bases start out at a lower pH, just as those for 0.10 *M* weak acids started out at a high pH.

Figure 6.9 shows the family of curves for titration of 0.10 *M* bases with a 0.10 *M* strong acid. These curves, too, can be calculated as we have done for the HCl with NaOH case. (See Problem #53 at the end of the chapter.) An example of a weak base in Figure 6.9 is ammonium hydroxide, NH_4OH, the analogy to acetic acid in the previous discussion.

6.10 THE ROLE OF INDICATORS

The sharp rise (or decline) in the pH in the middle of these curves occurs at the equivalence point, the exact point at which all of the substance titrated has been consumed by the titrant. The exact position for this in the strong acid with strong base case, and also the strong base with strong acid case, is at pH = 7 (see previous calculation), exactly in the middle of the sharp rise (or decline). For the weak acid with strong base case, the equivalence point is again in the middle of the sharp rise, but this is at a pH higher than pH = 7 (see Figure 6.7). In fact, the weaker the acid, the higher the pH value that corresponds to the equivalence point. The opposite is observed in the case of the weak base with strong acid cases (see Figure 6.9). The equivalence point occurs at progressively lower pH values the weaker the base. The exact pH at these equivalent points can be calculated as indicated previously.

The problem the analyst has is to choose indicators that change color close enough to an equivalence point so that the accuracy of the experiment is not diminished, which really means at any point during the sharp rise or

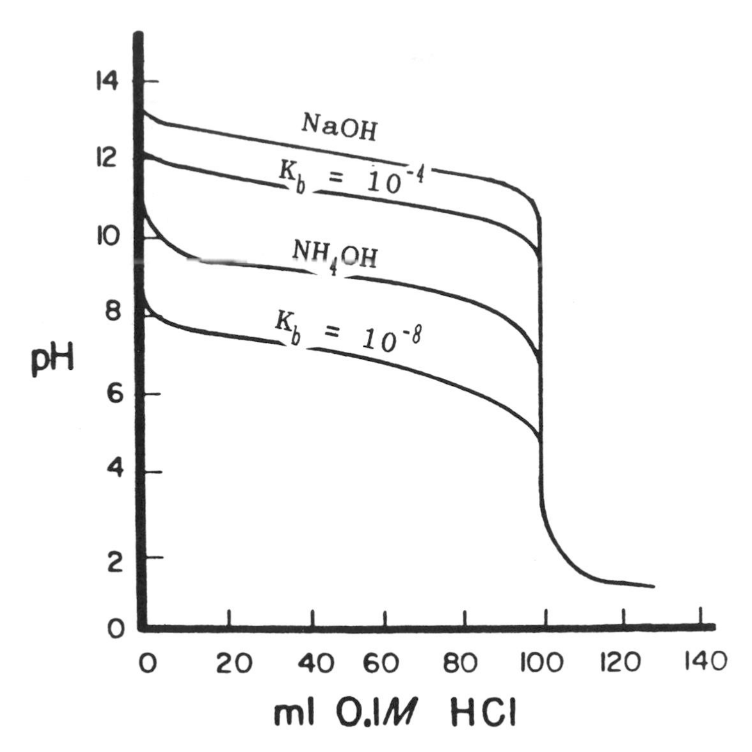

Figure 6.9. Titration curves for bases titrated with a strong acid.

decline. (Refer to Section 5.7 for a related discussion of equivalence points vs. end points.) It almost seems like an impossible task, since there must be an indicator for each possible acid or base to be titrated. Fortunately, there are a large number of indicators available, and there is at least one available for every acid and base with the exception of only the extremely weak acids and bases. Figure 6.10 lists some of these indicators and shows the pH ranges over which they change color.

Thus, referring to Figure 6.10, in the HCl with NaOH case (Figure 6.7) the following indicators would work: phenolphthalein, thymolphthalein, methyl red, methyl orange, etc.—virtually all of them in the list except for thymol blue and cresol red. As the acid becomes weaker and weaker, the pH range available for the indicator to change color becomes narrower and narrower, and a smaller number of indicators are useful. The color change must take place during the sharp rise in the pH. The same observation is made when titrating bases with acids. For the titration of acetic acid, phenolphthalein, thymolphthalein, and perhaps bromthymol blue are useful, while for ammonium hydroxide (see Figure 6.9), methyl red and bromcresol green, etc., are useful.

Finally, let us consider briefly polyprotic acids and polybasic bases and their titration curves. As discussed in Chapter 4, these acids and bases have two or more hydrogens ions to be donated or accepted, with each individual loss or gain having a K_a associated with it. For this reason such acids and bases have two or more sharp pH changes (inflection points) during the course of the titration. See Figure 6.11a for an idea of the appearance of these curves.

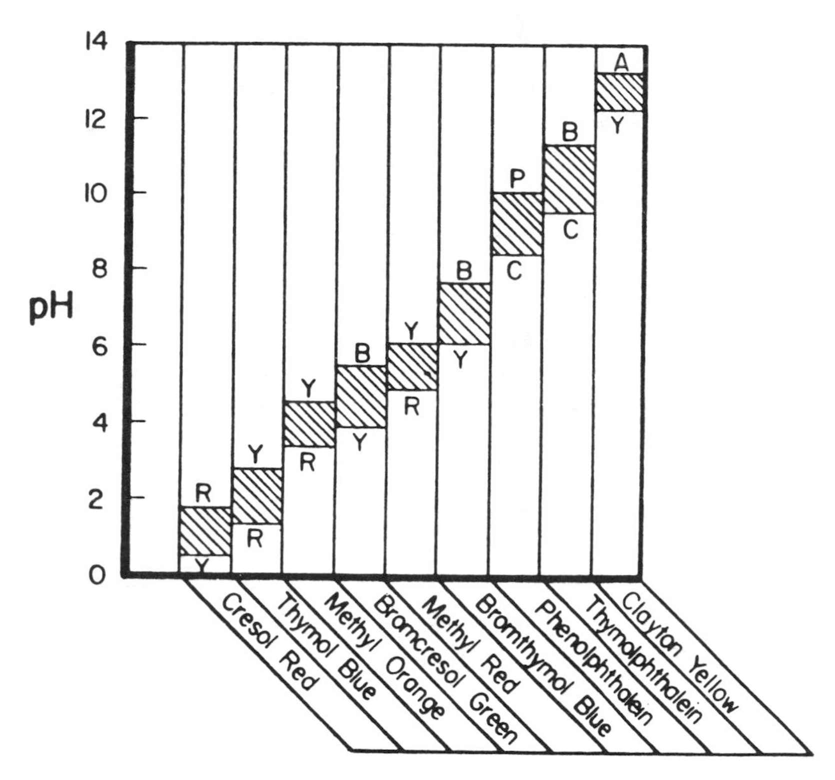

Figure 6.10. Some acid-base indicators and their color change ranges. R = red, Y = yellow, B = blue, C = colorless, A = amber.

Choice of indicators here obviously depends on the inflection point at which one wants to terminate the experiment. Again the indicator must change color during the sharp rise or decline.

Figure 6.11b represents the titration curve of a dibasic base, an example of which is sodium carbonate:

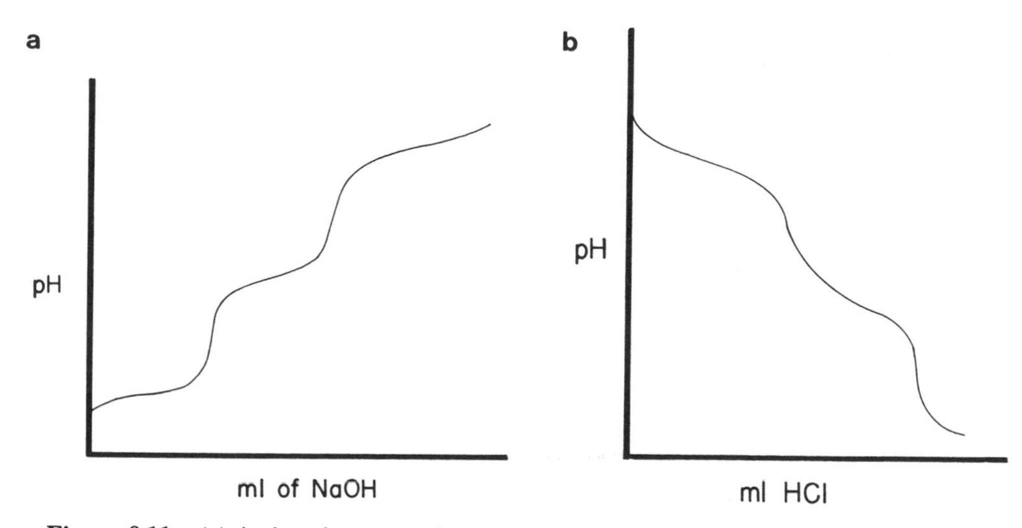

Figure 6.11. (a) A titration curve for a triprotic acid titrated with a strong base. (b) A titration curve for a dibasic base titrated with a strong acid.

$$Na_2CO_3 + 2H^{+1} \rightarrow H_2CO_3 + 2Na^{+1}$$

This is the curve for the example discussed earlier in which the following two equilibria played a role as well:

$$H_2CO_3 \rightleftharpoons H_2O + CO_2$$

$$H_2CO_3 \rightleftharpoons 2H^{+1} + CO_3^{-2}$$

The actual titration of Na_2CO_3 (see Experiment 8) uses the second inflection point, at which bromcresol green (also methyl red) is useful. The end point, however, is not sharp (the drop in pH is not sharp) because of a buffering effect due to the presence of a large concentration of H_2CO_3, the product of the titration to the second inflection point. If the solution is heated (boiled) close to the end point, however, the $H_2CO_3 \rightleftharpoons CO_2$ equilibrium is pushed to the right (recall the previous discussion), consuming the H_2CO_3 and thus greatly diminishing this buffering effect. This end point then becomes sharp and very satisfactory. Boiling at the end point is therefore what is done in Experiment 8 and also in acid standardization experiments in which sodium carbonate is the primary standard.

EXPERIMENTS

EXPERIMENT 7: PREPARATION AND STANDARDIZATION OF HCl AND NaOH SOLUTIONS

NOTE: Safety glasses are required.

1. For this experiment you will need a minimum of 3–4 g of primary standard KHP for three titrations. Place at least 6 g of it in a weighing bottle and dry in a drying oven for 2 hr. Your instructor may choose to dispense this to you.
2. Prepare CO_2-free water for a 0.10 *M* NaOH solution by boiling 1000 mL of distilled water in a covered beaker on a hot plate. While waiting for this water to boil continue with Step 3.
3. Prepare 1 L of 0.10 *M* HCl by diluting the appropriate volume of concentrated HCl (12.0 *M*). Use a 1-L glass bottle and half-fill it with water before adding the concentrated acid. Use a graduated cylinder to measure the acid. Add more water to have about 1 L, shake well, and label.
4. Once the water from Step 2 has boiled, remove the beaker from the heat and cool so that it is only warm to the touch. This can be accomplished by immersing the beaker in cold water (such as in a stoppered sink).
5. Using this freshly boiled water, prepare 1000 mL of a 0.10 *M* NaOH solution. Weigh out the appropriate number of grams of NaOH, and place it in a 1000-mL plastic bottle. Add the water to a level approximately equal to 1000 mL. Shake well to completely dissolve the solid. Allow to cool completely to room temperature before proceeding. Label the bottle.

6. Assemble the apparatus for a titration. The buret should be a 50-mL buret and should be washed thoroughly with a buret brush and soapy water. Clamp the buret to a ring stand with either a buret clamp or an ordinary ring stand clamp. The receiving flask should be a 250-mL Erlenmeyer flask. You should clean and prepare three such flasks. Place a piece of white paper (a page from your notebook will do) on the base of the ring stand. This will help you see the end point better.
7. Give both your acid and base solutions one final shake at this point to ensure their homogeneity. Rinse the buret with 5–10 mL of NaOH twice, then fill it to the top. Open the stopcock wide open to force trapped air bubbles from the stopcock and tip. Allow this excess solution to drain into a waste flask. Bring the bottom of the meniscus to the 0.00 mL line. Using a clean 25-mL pipet (volumetric) carefully place 25.00 mL of the acid solution into each of the three flasks. Add three drops of phenolphthalein indicator to each of the three flasks.
8. Titrate each acid sample with the base, one at a time. For the first sample, you can very easily make a good estimate of how much titrant should be required. Since both solutions are the same molar concentration and since the reaction is one-to-one in terms of moles (see Equation 6.3), the volume of NaOH required should be approximately the same as the volume of HCl pipetted, 25.00 mL. You should be able to open the stopcock and allow about 20 mL of the base to enter the flask without having to worry about overshooting the end point. From that point on, however, you should proceed with caution so that the indicator changes color (from clear to pink) upon the addition of the smallest amount of titrant possible – a fraction of a drop. At the point that you think the indicator will change color in this manner, rinse down the walls of the flask with water from a squeeze bottle. This will ensure that all of the titrant has had a chance to react with all of the substance titrated. This step can be performed more than once – addition of more water has no effect on the location of the end point. The fraction of a drop can then be added if necessary by (1) a very rapid rotation of the stopcock or (2) slowly opening the stopcock, allowing the fraction of a drop to hang on the tip of the buret, and then washing it into the flask with water from a squeeze bottle. At the end point, a faint pink color will persist in the flask for at least 20 sec. Read the buret to four significant figures, record the reading, and repeat with the next sample.
9. All three of your titrations should agree to within at least 0.05 mL in order to be acceptable. This is a relative deviation of less than 2 ppt – see Chapter 3. If they do not agree in this manner, more such titrations must be performed in order to have three good answers to rely on or until you have a precision satisfactory to your instructor.
10. After the 2-hr drying period for the KHP has expired, remove from the drying oven and allow to cool to room temperature in your desiccator.
11. Prepare three solutions of the KHP for titration. To do this, again clean three 250-mL Erlenmeyer flasks and give each a final thorough

rinse with distilled water. Now, weigh into each flask, by difference, on the analytical balance a sample of the KHP weighing between 0.7 and 0.9 g. Add approximately 50 mL of distilled water to each and swirl to dissolve.

12. Add three drops of phenolphthalein indicator to each flask and titrate as before.
13. Calculate the exact normality of the NaOH solution from the KHP data. If your three normalities do not agree to within 2 ppt relative deviation from the average (see Chapter 3), repeat until you have three that do or until you have a precision that is satisfactory to your instructor.
14. Calculate the exact normality of the HCl solution using the three volume readings from Step 8 and the average NaOH normality from Step 13. Compute the average of these three results.
15. Record in your notebook at least a sample calculation for each of these standardizations and all results.

EXPERIMENT 8: TITRIMETRIC ANALYSIS OF A COMMERCIAL SODA ASH UNKNOWN FOR SODIUM CARBONATE

NOTE: Safety glasses are required.

1. Obtain a soda ash sample from your instructor and dry, as before, for 2 hr. Allow to cool and store in your desiccator.
2. Prepare the flasks and weigh by difference, as usual, three samples of the soda ash, each weighing between 0.4 and 0.5 g. Dissolve in 75 mL of water.
3. Perform this on each of your samples one at a time, starting with the one of least weight. Add three drops of bromcresol green indicator and titrate with your standard HCl solution from Experiment 7. The color will change slowly from blue to a light green. When the light green color is apparent, stop the titration, place a watch glass on the flask, and bring to boil on a hot plate. Boil for 2 min. The color of the solution will turn back to blue. Cool to room temperature (you can use a cold water bath) and resume the titration. (Do not refill the buret!) The color change now should be sharp, from blue to greenish-yellow. Record the buret reading next to the corresponding weight.
4. Calculate the percentage of Na_2CO_3 in the sample, and record at least a sample calculation in your notebook along with all results and the average. If the calculated percentages do not agree to within 10 ppt relative deviation from the average, you should titrate additional samples until you have three that do agree in this manner or until you have a precision that is satisfactory to your instructor.

EXPERIMENT 9: TITRIMETRIC ANALYSIS OF A COMMERCIAL KHP UNKNOWN FOR KHP

NOTE: This experiment calls for you to use a pH meter and a combination pH electrode (Chapter 18) to detect the end point of a titration.

Your instructor may choose to have you use an indicator instead. Safety glasses are required.

1. Obtain an unknown KHP sample from your instructor and dry, as usual, for 2 hr. Allow to cool and store in your desiccator.
2. Prepare three 250-mL beakers and weigh, by difference, three samples of the unknown KHP, each weighing between 1.0 to 1.3 g, into the beakers. Dissolve in 75 mL of distilled water. Place watch glasses on the beakers.
3. Prepare and standardize a pH meter with a combination probe for pH measurement. Your instructor may provide special instructions for the pH meter you are using. You will use an automatic stirrer with magnetic stirring bar to stir the solution in the beaker while you are titrating. Mount the electrode in a ring stand clamp on a ring stand so that it is just immersed in the solution in one of the beakers. The beaker should be positioned on the stirrer with the stirring bar in the center of the beaker and the pH probe off to one side so as to not contact the stirring bar. The stirring speed should be slow enough so as not to splash the solution but fast enough to thoroughly mix the added titrant quickly.
4. The titrant is the standardized 0.10 *M* NaOH from Experiment 7. The procedure is to monitor the pH as the titrant is added. A sharp increase in the pH will signal the end point. You will want to determine the midpoint of the sharp increase as closely as possible, since that will be the end point. The titrant is added very slowly when the pH begins to rise, since only a very small volume will be required at that point to reach the end point. You can add the titrant rapidly at first, but slow to a fraction of a drop when the pH begins to rise. The pH at the end point will be in the range of 8–10.
5. Record the buret readings at the end points for all three beakers. Calculate the percent of KHP in the sample for all three titrations and include at least a sample calculation in your notebook along with all results and the average. If the calculated percentages do not agree to within 10 ppt relative deviation from the average, you should titrate additional samples until you have three that do agree in this manner or until you have a precision satisfactory to your instructor.

EXPERIMENT 10: ANALYSIS OF A GRAIN SAMPLE FOR PROTEIN CONTENT BY THE KJELDAHL METHOD

NOTE: It is very important to wear safety glasses for this experiment. Your instructor may require additional safety gear, such as a shield and an apron.

Part A – Solution Preparation and Standardization

NOTE: Solution standardization can be performed while the sample is digesting and distilling in order to save time.

1. Prepare 500 mL (use a 500-mL plastic bottle) of 0.10 *N* NaOH. Use freshly boiled water so that it is CO_2-free. Standardize with KHP as in Experiment 7.

2. Prepare about 500 mL of 0.10 *N* H_2SO_4 (concentrated H_2SO_4 is 36 *N*) and standardize with the NaOH prepared in Step 1. (See Experiment 7.)

Part B – Digestion, Distillation, and Back Titration

1. Weigh approximately 1 g of a sample of grain, flour, or whatever your instructor requests on an analytical balance, wrap in a piece of filter paper, and place (filter paper and all) into an 800-mL Kjeldahl flask.

2. Add 20 mL of concentrated H_2SO_4 and 1½ packets of commercial Kjeldahl catalyst to the flask.

3. Digest on the Kjeldahl apparatus in a good fume hood for 45 min (or less, if the sample appears to be totally dissolved). Let cool, but not enough to cause crystallization (no more than about 10 min).

4. Pipet 25 mL of the 0.1 *N* H_2SO_4 into a 250-mL Erlenmeyer flask. Add 25 mL of water and three drops of methyl red indicator. Set in place in the apparatus so as to receive the condensing liquid from the distillation in Step 5. Place three or four glass beads in the Kjeldahl flask and then slowly add 275 mL of distilled water to it. Carefully add 50 mL of a 50% NaOH solution to the Kjeldahl flask, and immediately connect this flask to the rest of the apparatus. Swirl the flask.

5. Distill until about 150 mL of distillate is collected in the receiving flask. This distillation will require 30 to 60 min. After this, remove the receiving flask, and then turn off the heat to the Kjeldahl flask. It is important to remove the receiving flask first. Otherwise, cooling could create a vacuum inside which would draw the solution up out of the receiving flask, thus making impossible the important back titration in Step 6.

6. Titrate this solution with the 0.10 *M* NaOH until there is a color change from pink to yellow.

7. Calculate the percent protein as follows:

$$\%\ \text{protein} = \frac{[L(H_2SO_4) \times N(H_2SO_4) - L(NaOH) \times N(NaOH)] \times 14.007 \times F}{\text{sample weight}} \times 100$$

F = 5.7 for bread, wheat, or wheat flour

F = 6.38 for milk products

F = 6.25 for all others

QUESTIONS AND PROBLEMS

1. What is the molarity of a NaOH solution if 22.43 mL of it exactly neutralized 25.00 mL of a 0.1006 *M* HCl solution?
2. If 25.00 mL of a HCl solution required 23.04 mL of 0.1032 *M* NaOH for titration, what is the molarity of the HCl?
3. The molarity of a HCl solution is 0.1103 *M*. If 25.00 mL of this solution required 23.97 mL of NaOH for titration, what is the molarity of the NaOH?
4. The concentration of a solution of an acid (FW = 121.82 g/mol) was 0.1202 *M* and 25.00 mL of it exactly reacted with 27.93 mL of a solution of a base. What is the concentration of the base?
5. What is the molarity of a NaOH solution if 37.43 mL of it exactly neutralized 0.7211 g of KHP?
6. In standardizing a solution of NaOH, the following data were collected: 0.7269 g of KHP were neutralized by 34.01 mL of the NaOH. What is the molarity of the NaOH?
7. If 0.4853 g of a primary standard acid (FW = 194.22) were exactly neutralized by 23.09 mL of a solution of a base (FW = 54.99) in a one-to-one reaction, what is the molarity of the base?
8. In a standardization experiment using a primary standard base (FW = 354.80), 29.66 mL of an acid solution were required to neutralize 0.9911 g of the base in a one-to-one reaction. What is the molarity of the acid?
9. What is the equivalent weight of both reactants in each of the following?

   (a) $NaOH + HCl \rightarrow NaCl + H_2O$

   (b) $2NaOH + H_2SO_4 \rightarrow Na_2SO_4 + 2H_2O$

   (c) $2HCl + Ba(OH)_2 \rightarrow BaCl_2 + 3H_2O$

   (d) $3NaOH + H_3PO_4 \rightarrow Na_3PO_4 + 3H_2O$

   (e) $2HCl + Mg(OH)_2 \rightarrow MgCl_2 + 2H_2O$

   (f) $2NaOH + H_3PO_4 \rightarrow Na_2HPO_4 + 2H_2O$

   (g) $NaOH + Na_2HPO_4 \rightarrow Na_3PO_4 + H_2O$

   (h) $NaOH + H_3PO_4 \rightarrow NaH_2PO_4 + H_2O$

   (i) $Na_2CO_3 + 2HCl \rightarrow 2NaCl + H_2CO_3$
10. Consider the following two reactions:

    (a) $NaOH + H_3PO_4 \rightarrow NaH_2PO_4 + H_2O$

    (b) $KH_2PO_4 + Ba(OH)_2 \rightarrow KBaPO_4 + 2H_2O$

    What are the equivalent weights of all four reactants?
11. Calculate the equivalent weights of the acids and bases from any six equations chosen from Table 6.4.

Table 6.4. Acid-base equations for use in selected problems in the text.

(A)	$HBr + NaOH \rightarrow NaBr + H_2O$
(B)	$HBr + KOH \rightarrow KBr + H_2O$
(C)	$2HCl + Ca(OH)_2 \rightarrow CaCl_2 + 2H_2O$
(D)	$2HBr + Ba(OH)_2 \rightarrow BaBr_2 + 2H_2O$
(E)	$3HCl + Al(OH)_3 \rightarrow AlCl_3 + 3H_2O$
(F)	$H_2SO_4 + 2KOH \rightarrow K_2SO_4 + 2H_2O$
(G)	$H_2SO_4 + Ca(OH)_2 \rightarrow CaSO_4 + 2H_2O$
(H)	$3H_2SO_4 + 2Al(OH)_3 \rightarrow Al_2(SO_4)_3 + 6H_2O$
(I)	$H_3PO_4 + 3KOH \rightarrow K_3PO_4 + 3H_2O$
(J)	$2H_3PO_4 + 3Ba(OH)_2 \rightarrow Ba_3(PO_4)_2 + 6H_2O$
(K)	$H_3PO_4 + Al(OH)_3 \rightarrow AlPO_4 + 3H_2O$
(L)	$NaH_2PO_4 + 2KOH \rightarrow K_2NaPO_4 + 2H_2O$
(M)	$NaH_2PO_4 + Ca(OH)_2 \rightarrow CaNaPO_4 + 2H_2O$
(N)	$3NaH_2PO_4 + 2Al(OH)_3 \rightarrow Al_2(NaPO_4)_3 + 6H_2O$
(O)	$Na_2HPO_4 + KOH \rightarrow Na_2KPO_4 + H_2O$
(P)	$2Na_2HPO_4 + Ba(OH)_2 \rightarrow Ba(Na_2PO_4)_2 + 2H_2O$
(Q)	$3Na_2HPO_4 + Al(OH)_3 \rightarrow Al(Na_2PO_4)_3 + 3H_2O$
(R)	$KH_2PO_4 + NaOH \rightarrow KNa_2PO_4 + 2H_2O$
(S)	$KH_2PO_4 + Ba(OH)_2 \rightarrow KBaPO_4 + 2H_2O$
(T)	$3KH_2PO_4 + 2Al(OH)_3 \rightarrow Al_2(KPO_4)_3 + 6H_2O$
(U)	$K_2HPO_4 + KOH \rightarrow K_3PO_4 + H_2O$
(V)	$2K_2HPO_4 + Ca(OH)_2 \rightarrow Ca(K_2PO_4)_2 + 2H_2O$
(W)	$3K_2HPO_4 + Al(OH)_3 \rightarrow Al(K_2PO_4)_3 + 3H_2O$
(X)	$H_3PO_4 + KOH \rightarrow KH_2PO_4 + H_2O$
(Y)	$2H_3PO_4 + Ba(OH)_2 \rightarrow Ba(H_2PO_4)_2 + 2H_2O$
(Z)	$3H_3PO_4 + Al(OH)_3 \rightarrow Al(H_2PO_4)_3 + 3H_2O$
(AA)	$H_3PO_4 + 2KOH \rightarrow K_2HPO_4 + 2H_2O$
(BB)	$H_3PO_4 + Ca(OH)_2 \rightarrow CaHPO_4 + 2H_2O$
(CC)	$H_3PO_4 + Al(OH)_3 \rightarrow Al_2(HPO_4)_3 + 3H_2O$
(DD)	$NaH_2PO_4 + KOH \rightarrow KNaHPO_4 + H_2O$
(EE)	$NaH_2PO_4 + Ba(OH)_2 \rightarrow Ba(NaHPO_4)_2 + H_2O$
(FF)	$NaH_2PO_4 + Al(OH)_3 \rightarrow Al(NaHPO_4)_3 + H_2O$
(GG)	$KH_2PO_4 + KOH \rightarrow K_2HPO_4 + H_2O$
(HH)	$2KH_2PO_4 + Ca(OH)_2 \rightarrow Ca(KHPO_4)_2 + 2H_2O$
(II)	$3KH_2PO_4 + Al(OH)_3 \rightarrow Al(KHPO_4)_3 + 3H_2O$
(JJ)	$2HBr + Na_2CO_3 \rightarrow 2NaBr + H_2O + CO_2$
(KK)	$H_2SO_4 + Na_2CO_3 \rightarrow Na_2SO_4 + H_2O + CO_2$
(LL)	$2H_3PO_4 + 3Na_2CO_3 \rightarrow 2Na_3PO_4 + 3H_2O + 3CO_2$
(MM)	$2H_3PO_4 + Na_2CO_3 \rightarrow 2NaH_2PO_4 + H_2O + CO_2$
(NN)	$H_3PO_4 + Na_2CO_3 \rightarrow Na_2HPO_4 + H_2O + CO_2$
(OO)	$2NaHCO_3 + Mg(OH)_2 \rightarrow Mg(NaCO_3)_2 + 2H_2O$
(PP)	$HSO_3NH_2 + NaOH \rightarrow NaSO_3NH_2 + H_2O$
(QQ)	$HC_2H_3O_2 + KOH \rightarrow KC_2H_3O_2 + H_2O$
(RR)	$H_2C_2O_4 + 2NaOH \rightarrow Na_2C_2O_4 + 2H_2O$
(SS)	$H_2C_2O_4 + KOH \rightarrow KHC_2O_4 + H_2O$
(TT)	$H_2SO_3 + Ca(OH)_2 \rightarrow CaSO_3 + 2H_2O$
(UU)	$2H_2SO_3 + Mg(OH)_2 \rightarrow Mg(HSO_3)_2 + 2H_2O$
(VV)	$2KHC_2O_4 + Ba(OH)_2 \rightarrow Ba(KC_2O_4)_2 + 2H_2O$
(WW)	$HCl + K_2CO_3 \rightarrow 2KCl + H_2O + CO_2$

12. What is the equivalent weight of the acid in each of the equations in Table 6.4?

13. What is the equivalent weight of the base in each of the equations in Table 6.4?

14. How many grams are needed to prepare 500 mL of 0.20 *N* KH_2PO_4 if it is to be used as in the equation from Problem 10b above?

15. How would you prepare 500 mL of 0.11 *N* H_2SO_4 for use in the reaction in Problem 9b from concentrated H_2SO_4 (18.0 *M*)?

16. How would you prepare 750 mL of 0.11 *N* $Ba(OH)_2$ for use in the reaction from Problem 9c from pure solid $Ba(OH)_2$?

17. How many grams of the base in Equation B of Table 6.4 are needed to prepare 200 mL of a 0.15 *N* solution?

18. How many grams of the base in Equation D of Table 6.4 are needed to prepare 700 mL of a 0.25 *N* solution?

19. How many grams of the acid in Equation L of Table 6.4 are needed to prepare 400 mL of a 0.35 *N* solution?

20. How many grams of the acid in Equation O of Table 6.4 are needed to prepare 550 mL of a 0.25 *N* solution?

21. How many milliliters of a 15.0 *N* solution of $Ba(OH)_2$ are required to prepare 700 mL of a 0.30 *N* solution if it is to be used as in Equation Y of Table 6.4?

22. How many milliliters of conc. phosphoric acid are required to prepare 300 mL of a 0.15 *N* solution if it is to be used as in Equation J of Table 6.4?

23. How many milliliters of a 7.0 *N* solution of KOH are needed to prepare 900 mL of a 0.10 *N* solution if it is to be used as in Equation AA of Table 6.4?

24. How many milliliters of concentrated phosphoric acid are required to prepare 650 mL of a 0.65 *N* solution if it is to be used as in Equation MM of Table 6.4?

25. An H_3PO_4 solution is to be used to titrate an NaOH solution as in the equation in Problem 9d. If the normality of the H_3PO_4 solution is 0.2411 *N*, what is the molarity?

26. 0.7114 g of KHP was used to standardize a $Mg(OH)_2$ solution as in the following reaction:

$$Mg(OH)_2 + 2KHC_8H_4O_4 \rightarrow MgK_2(C_8H_4O_4)_2 + 2H_2O$$

If 31.18 mL of the $Mg(OH)_2$ were needed, what is the normality of the $Mg(OH)_2$?

27. A NaOH solution has standardized against a H_3PO_4 solution as in the equation in Problem 9f. If 25.00 mL of 0.1427 *N* H_3PO_4 required 40.07 mL of the NaOH, what is the normality of the NaOH?

28. A solution of KOH is standardized with primary standard KHP ($KHC_8H_4O_4$). If 0.5480 g of the KHP exactly reacted with 25.41 mL of the KOH solution, what is the normality of the KOH?

$$KOH + KHP \rightarrow K_2P + H_2O$$

29. What is the normality of a solution of HCl, 35.12 mL of which were required to titrate 0.4188 g of primary standard Na_2CO_3?

$$2HCl + Na_2CO_3 \rightarrow 2NaCl + CO_2 + H_2O$$

30. What is the normality of a solution of sulfuric acid that was used to titrate a 0.1022 *N* solution of KOH as in Equation F of Table 6.4 if 25.00 mL of the base was exactly neutralized by 29.04 mL of the acid?

31. If, when a solution of H_3PO_4 is standardized with a 0.1329 *N* solution of $Ca(OH)_2$ as in Equation BB of Table 6.4, 25.00 mL of the acid requires 26.77 mL of the base, what is the normality of the acid?

32. What is the normality of a solution of NaOH if 0.5022 g of KH_2PO_4 were exactly neutralized by 29.04 mL of the NaOH according to Equation R of Table 6.4?

33. If 39.05 mL of an HBr solution reacts exactly with 0.7744 g of Na_2CO_3 according to Equation JJ of Table 6.4, what is the normality of the HBr?

34. Primary standard THAM, or TRIS, is used to standardize a hydrochloric acid solution. If 0.4922 g of THAM are used and 23.45 mL of the HCl are needed, what is the normality of the HCl?

$$(HOCH_2)_3CNH_2 + HCl \rightarrow (HOCH_2)_3CNH_3^+ \, Cl^-$$

35. Suppose a sulfuric acid solution, rather than the hydrochloric acid solution as in #34, is standardized with primary standard THAM. Does the calculation change in any way? Explain.

36. What is the percent of K_2HPO_4 in a sample when 46.79 mL of 0.09223 *N* $Ca(OH)_2$ exactly neutralizes 0.9073 g of the sample according to Equation V of Table 6.4?

37. What is the percent of $Al(OH)_3$ in a sample when 0.6792 g of the sample is exactly neutralized by 23.45 mL of 0.1320 *N* H_3PO_4 according to Equation Z of Table 6.4?

38. What is the percent of KH_2PO_4 in a sample if after adding 50.00 mL of 0.1257 *N* $Al(OH)_3$ to 0.8744 g of the sample, according to Equation T of Table 6.4, the excess base required 10.45 mL of 0.1101 *N* standard acid?

39. What is the percent of $Ca(OH)_2$ in a sample weighing 0.5677 g if after adding 50.00 mL of 0.1239 *N* KH_2PO_4, as in Equation HH in Table 6.4, the excess acid required 5.16 mL of 0.1009 *N* NaOH according to Equation R in this same table.

40. What is the percent of NaH_2PO_4 in a sample if 24.18 mL of 0.1032 *N* NaOH was used to titrate 0.3902 g of the sample according to the following?

$$NaH_2PO_4 + 2NaOH \rightarrow Na_3PO_4 + 2H_2O$$

41. A 0.1057 *N* HCl solution was used to titrate a sample containing $Ba(OH)_2$ (see 9c). If 35.78 mL of HCl were required to exactly react with 0.8772 g of the sample, what is the percent of $Ba(OH)_2$ in the sample?

42. A grain sample was analyzed for nitrogen content by the Kjeldahl method. If 1.2880 g of the grain were used and 50.00 mL of 0.1009 *N* HCl were used in the receiving flask, what is the percent of *N* in the sample when 5.49 mL of 0.1096 *N* NaOH were required for back titration?

43. A flour sample was analyzed for nitrogen content by the Kjeldahl method. If 0.9819 g of the flour were used and 35.10 mL of 0.1009 *N* HCl were used to titrate the boric acid solution in the receiving flask, what is the percent of nitrogen in the sample?

44. Describe in your own words what a back titration is.

45. In the Kjeldahl titration
    (a) what acid is used to dissolve/digest the sample?
    (b) what substance (be specific) is added to neutralize this acid?
    (c) what substance (be specific) is distilled into the receiving flask?
    (d) what kind of substance (acid or base) is in the receiving flask before the distillation?
    (e) what kind of substance (acid or base) is the "titrant" in the back titration procedure?
    (f) what kind of substance (acid or base) is the back titrant?

46. Concerning the Kjeldahl method for nitrogen,
    (a) what is concentrated sulfuric acid used for?
    (b) what might boric acid be used for?

47. In the calculation of the percent constituent when using a back titration, the following appears in the numerator:

$$(L_T \times N_T - L_{BT} \times N_{BT})$$

Why is it necessary to do this subtraction?

48. Phenolphthalein indicator changes color in the pH range 8–10, methyl orange changes color in the pH range 3–4.5. Roughly sketch two titration curves as follows:
    (a) one which represents a titration in which phenolphthalein would be useful but methyl orange would not
    (b) one which represents a titration in which a methyl orange would be useful but phenolphthalein would not

49. Roughly sketch the following three titration curves:
    (a) a weak acid titrated with a strong base
    (b) a strong base titrated with a strong acid
    (c) a weak base titrated with a strong acid

50. Bromcresol green was used as the indicator for the Na_2CO_3 titration in Experiment 8. Would phenolphthalein also work? Explain.

51. Would bromcresol green be an appropriate indicator for an acetic acid titration? Explain.

52. The NaH_2PO_4 in Problem #40 is a weak acid. Which indicator would be likely best for the titration, phenolphthalein (pH range 8–10) or methyl orange (pH range 3–5)? Explain.

53. In a titration of 25 mL of 0.10 *N* NaOH with 0.10 *N* HCl calculate the pH in the titration flask at the following points in the titration and plot the pH vs. mL HCl.
    (a) at 0.00 mL of HCl added
    (b) after adding 5.00 mL of HCl
    (c) after adding 10.00 mL of HCl
    (d) after adding 20.00 mL of HCl
    (e) after adding 35.00 mL of HCl
    (f) after adding 50.00 mL of HCl

54. This question pertains to Experiment 40 in Chapter 15. Discuss the calculations for this experiment and explain.
    (a) why the initial mandelic acid solution need not be prepared accurately.
    (b) why the NaOH solution need not be standardized.
    (c) why only the buret readings, and no other data, need appear in the final calculation.

Chapter 7

Complexometric Titrations and Calculations

7.1 INTRODUCTION

Acid-base (neutralization) reactions are only one type of many that are applicable to titrimetric analysis. There are reactions that involve the formation of a precipitate. There are reactions that involve the transfer of electrons. There are reactions, among still others, that involve the formation of a complex ion. This latter type typically involves transition metals and is often used for the qualitative and quantitative colorimetric analysis (Chapter 11) of transition metal ions, since the complex ion that forms can be analyzed according to the depth of a color that it imparts to a solution. This chapter, however, concerns a titrimetric analysis method in which a complex ion-forming reaction is used.

7.2 COMPLEX ION TERMINOLOGY

A complex ion is a polyatomic charged aggregate consisting of a positively charged metal ion combined with either a neutral or negatively charged chemical species. Such a chemical species is called a "ligand". A ligand can consist of a monatomic negatively charged ion such as F^{-1}, Cl^{-1}, etc. or a polyatomic charged or uncharged species, such as H_2O, CN^{-1}, CNS^{-1}, NH_3, CN^{-1}, etc. Some simple examples of the complex ion-forming reaction are

$$Cu^{+2} + 4NH_3 \rightleftarrows Cu(NH_3)_4{}^{+2}$$

$$Fe^{+3} + CNS^{-1} \rightleftarrows Fe(CNS)^{+2}$$

$$Co^{+2} + 4Cl^{-1} \rightleftarrows CoCl_4{}^{-2}$$

In these examples, Cu^{+2}, Fe^{+3}, and Co^{+2} are the metal ions, NH_3, CNS^{-1}, and Cl^{-1} are the ligands, and the products shown are the complex ions. Notice that complex ions can be either positively or negatively charged and also notice that the reactions are equilibrium reactions. We will address this latter observation toward the end of this section.

Ligands can be classified according to the number of bonding sites that are available for forming a coordinate covalent bond to the metal ion. A coordinate covalent bond is one in which the shared electrons are contributed to the bond by only one of the two atoms involved. In the case of complex ion-forming reactions, both electrons are always donated by the ligand. If only one such pair of electrons is available per molecule or ion, we say that the ligand is "monodentate". If two are available per molecule or ion, it is described as "bidentate", etc. Thus, we have the following terminology:

# of Sites	Descriptive Term
1	Monodentate
2	Bidentate
3	Tridentate
6	Hexadentate

Examples of monodentate ligands are all of those given above (Cl^-, CN^-, NH_3, etc.). The pair of electrons from these species available for a coordinate covalent bond is pointed out in the electron dot structures shown in Figure 7.1.

Nitrogen-containing ligands are especially evident in reactions of this kind. This is true because of the pair of electrons occupying the non-bonding orbital found on the nitrogen in many nitrogen-containing compounds. This pair of electrons is present in the Lewis electron dot structure for the nitrogen atom shown in Figure 7.2. It is easy to see that nitrogen would bond in the normal way three times, thus often leaving the non-bonded pair of electrons available for coordinate covalent bonding. An obvious example is the

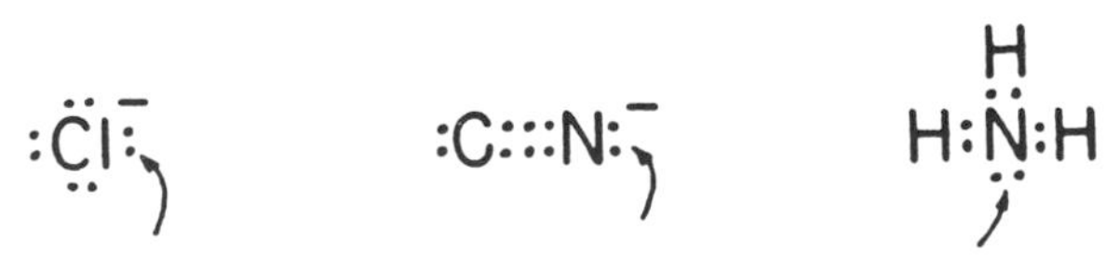

:Cl:⁻ :C:::N:⁻ H:N:H (with H above N)

Figure 7.1 Lewis electron dot structures of some example monodentate ligands.

Figure 7.2. The Lewis electron dot structure of the nitrogen atom, N.

ammonia example above. A good example of a bidentate ligand is the 1,10-phenanthroline molecule which, since it forms a stable complex ion with Fe^{+2} ions that has a deep orange color, is used in the colorimetric analysis of iron(II) ions (see Experiment 22 in Chapter 11). This ligand is shown in Figure 7.3a. The two bonding sites are pointed out with the arrows. Another bidentate ligand is ethylenediamine, shown in Figure 7.3b. These ligands, while bidentate, will form a complex ion with only one metal ion. It is also common for more than one such ligand to combine with a single metal ion, as shown in Figure 7.4.

Complex ions that involve ligands with two or more bonding sites (bidentate, tridentate, etc.) are also called "chelates" and the ligands "chelating agents". Thus, the ligands shown in Figure 7.3 are examples of chelating agents and the complex ions formed are examples of chelates. Another term associated with complex ion chemistry is "masking". Masking refers to the use of ligands and complex ion formation reactions for the purpose of avoiding interferences. When the complex ion formation equilibrium lies far to the right such that the equilibrium constant (more often called "formation constant") is very large, the complex ion formed is very stable. This has the effect of "tying up" or "masking" the metal ion such that the interference does not occur. The ligand used in this application is called the "masking agent". A good example is in the water hardness analysis to be studied in the next section. Dissolved iron ions can interfere with this analysis. This

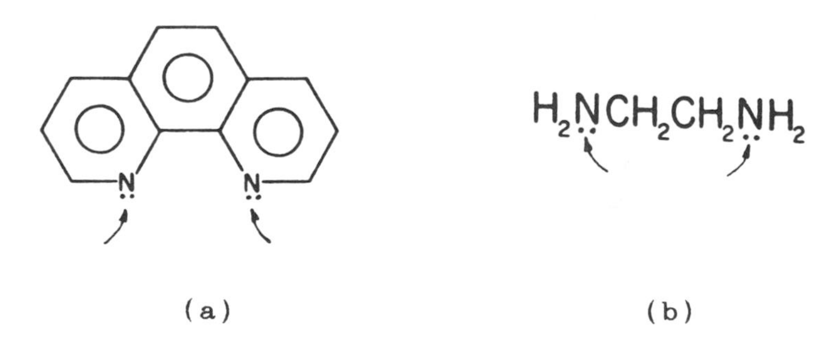

Figure 7.3. The bidentate ligands (a) 1,10 phenanthroline and (b) ethylenediamine. The arrows point out the two bonding sites.

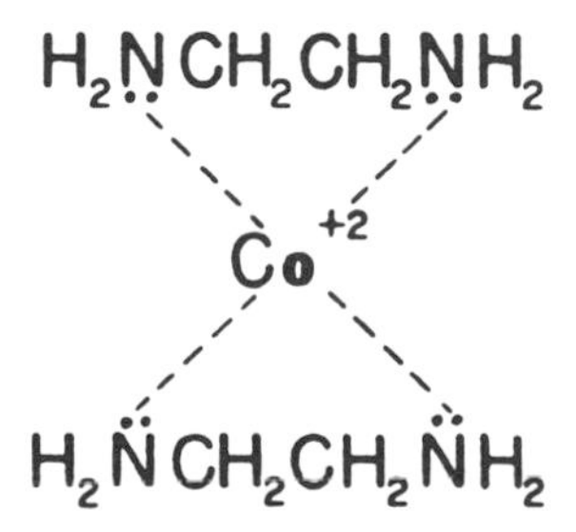

Figure 7.4. The complex ion formed by the combination of a cobalt ion with two ethylenediamine ligands.

interference can be removed if the iron ions are reacted with the cyanide ligand (CN^{-1}). The formation constant of the $Fe(CN)_6^{-3}$ complex ion is large, meaning that the iron ions are effectively removed from the solution, or "masked", cyanide being the "masking agent". The reaction involved and the formation constant, K_f, are as follows:

$$Fe^{+3} + 6CN^{-1} \rightleftarrows Fe(CN)_6^{-3}$$

$$K_f = \frac{[Fe(CN)_6^{-3}]}{[Fe^{+2}]\,[CN^{-1}]^6} = 10^{31}$$

A very important ligand (or chelating agent) for titrimetric analysis is the ethylenediaminetetraacetate ligand (abbreviated EDTA). It is especially useful in reacting with calcium and magnesium ions in hard water such that water hardness can be determined. The next two sections are devoted to this subject.

7.3 THE EDTA LIGAND

EDTA is a hexadentate ligand. Its structure is shown in Figure 7.5. Again, the bonding sites are pointed out by the arrows. In addition to the four charged sites at the carboxyl group oxygens, each of the two nitrogens has an unshared pair of electrons, making six electron pairs available to form coordinate covalent bonds. In forming a complex ion with calcium ions, for example, all six bonding sites bond to calcium, forming a large aggregate consisting of a single EDTA ligand wrapped around the calcium ion as shown in Figure 7.6.

Thus, a one-to-one reaction is involved and, in fact, all reactions of EDTA with metal ions (and most metals do indeed react with EDTA) are one-to-one. We will thus not be concerned with a formal scheme for determining equivalent weights as we were with acid-base reactions.

The usual source of EDTA for use in metal analysis is the disodium dihydrogen salt of ethylenediaminetetraacetic acid. This is the partially neutralized salt of ethylenediaminetetraacetic acid and is shown in Figure 7.7.

Ethylenediaminetetraacetic acid is frequently symbolized H_4Y and the disodium salt, Na_2H_2Y. The hydrogens in these formulas are the acidic hy-

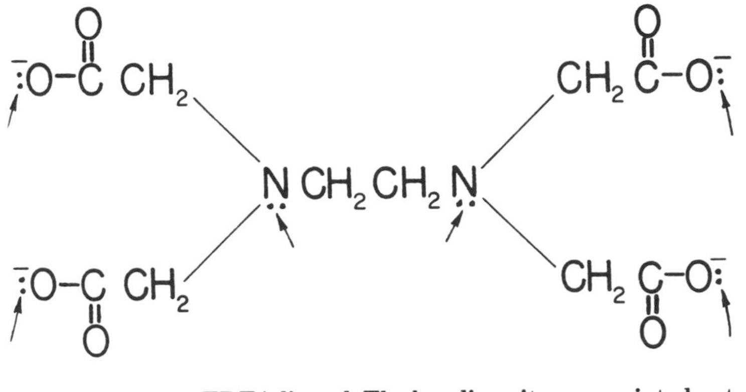

Figure 7.5. The EDTA ligand. The bonding sites are pointed out by the arrows.

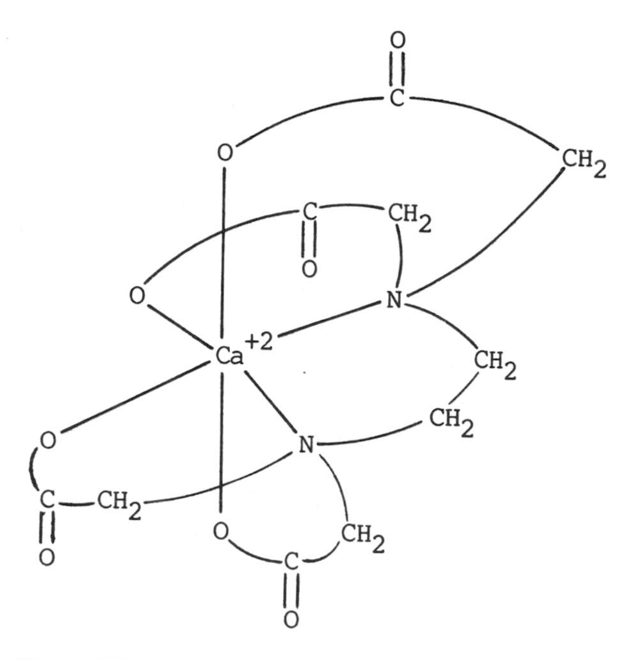

Figure 7.6. The calcium-EDTA complex ion.

drogens associated with the carboxyl groups, as in any weak organic carboxylic acid, and they dissociate from the EDTA ion in a series of equilibrium steps:

Step 1 $H_4Y \rightleftarrows H^{+1} + H_3Y^{-1}$

Step 2 $H_3Y^{-1} \rightleftarrows H^{+1} + H_2Y^{-2}$

Step 3 $H_2Y^{-2} \rightleftarrows H^{+1} + HY^{-3}$

Step 4 $HY^{-3} \rightleftarrows H^{+1} + Y^{-4}$

As discussed briefly in Chapter 6, the exact position of this equilibrium is controlled by controlling the pH of the solution. With extremely basic pH values, Y^{-4} will predominate, while in extremely acidic pH values, H_4Y will

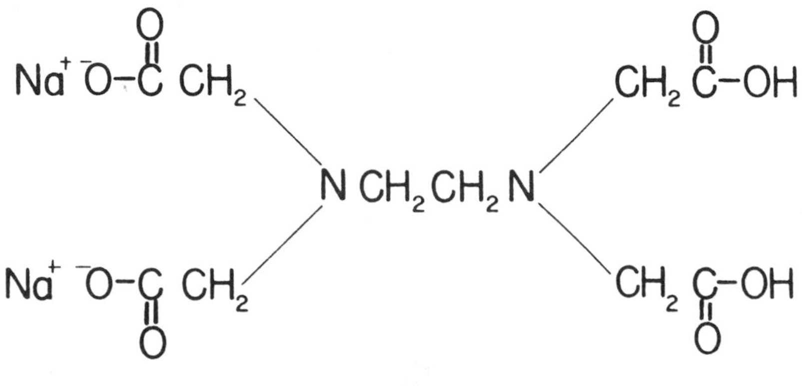

$Na^+ \, ^-O-C(=O)\,CH_2$ $CH_2\,C(=O)-OH$

$N\,CH_2CH_2\,N$

$Na^+ \, ^-O-C(=O)\,CH_2$ $CH_2\,C(=O)-OH$

Figure 7.7. The disodium dihydrogen salt of ethylenediaminetetraacetic acid.

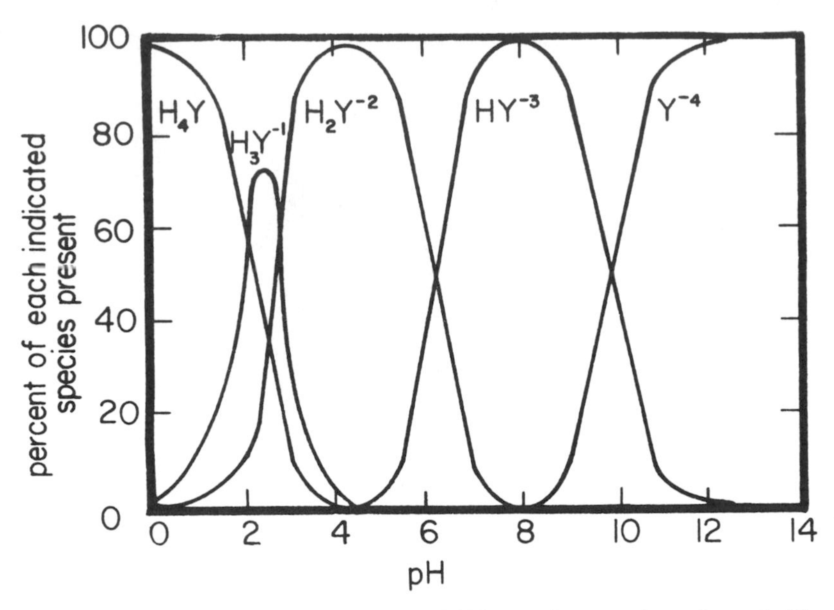

Figure 7.8. The predominance of EDTA species as a function of pH.

predominate. The intermediate species will predominate at intermediate pH values. (See Figure 7.8.)

When Na_2H_2Y is dissolved in water and mixed with a solution of a metal ion, such as calcium, the metal ion will react with the predominant EDTA species such that the following equilibrium is, for example, established:

$$Ca^{+2} + H_2Y^{-2} \rightleftharpoons CaY^{-2} + 2H^{+1}$$

Since it is important for a titrimetric reaction to be quantitative, it is important that this equilibrium be shifted to the right as much as possible. Adjusting the pH to a basic value will cause the H^{+1} ions formed to be consumed and, thus the equilibrium (recall again Le Chatelier's principle) will be shifted to the right. A problem exists with this procedure, however, in that at basic pH values many metal ions precipitate as the hydroxide, e.g., $Mg(OH)_2$, and thus would be lost to the analysis. This occurs with the magnesium in the water hardness procedure alluded to earlier. Luckily, a happy medium exists. At pH = 10, the reaction of the metal ion with the predominant HY^{-3} and Y^{-4} species (Figure 7.8) is shifted sufficiently to the right for the quantitative requirement to be fulfilled, while at the same time the solution is not basic enough for the magnesium ions to precipitate appreciably. Thus, all solutions in the reaction flask in the water hardness determination are buffered at pH = 10. Recall the discussion of buffer solutions and how they work in Chapter 4. In this case the buffer is a solution of the weak base ammonium hydroxide and its salt, ammonium chloride, mentioned briefly in Chapter 4. (See also Problem #35 at the end of Chapter 4.)

7.4 THE DETERMINATION OF WATER HARDNESS

EDTA titrations are routinely used to determine water hardness in a laboratory. Raw well water samples can have a significant quantity of dissolved minerals which contribute to a variety of problems associated with the use of such water. These minerals consist chiefly of calcium and magnesium carbonates, sulfates, etc. The problems that arise are mostly a result of heating or boiling the water over a period of time such that the water is evaporated and the calcium and magnesium salts become concentrated and precipitate in the form of a "scale" on the walls of the container, hence the term "hardness". This kind of problem is thus evident in boilers, domestic and commercial water heaters, humidifiers, tea kettles, and the like. A second problem is that these metals react with soap molecules and form a "scum" to which bathtub rings, etc. are attributed. Hard water is therefore not the best water to use for efficient soapy water cleaning processes, since the metal-soap precipitation reaction competes with the cleaning action. Water from different sources can have very different hardness values. Water samples can vary from simply being "hard" to being "extremely hard". While this description of hardness is anything but quantitative, a quantitative description based on an EDTA titration can be given.

The EDTA determination of water hardness results from the reaction of the EDTA ligand with the metal ions involved, calcium and magnesium. An interesting question is: "How are the results reported?" Hardness is not usually reported precisely as so much calcium plus so much magnesium, etc. There is no distinction made between the metals involved. All species reacting with the EDTA are considered one species and the results reported as an amount of one species, calcium carbonate ($CaCO_3$). That is, in the calculation, when a formula weight is used to convert moles to grams, the formula weight of calcium carbonate is used, and thus a quantity of $CaCO_3$ equivalent to the sum total of all contributors to the hardness is what is reported.

The indicator that is most often used is called Eriochrome Black T or EBT. EBT is actually a ligand that also reacts with the metal ions, like EDTA. In the free uncombined form, it imparts a sky-blue color to the solution, but if it is a part of a complex ion with either calcium or magnesium ions, it is a wine-red color. Thus, before adding any EDTA from a buret, the hard water sample containing the pH = 10 ammonia buffer and several drops of EBT indicator will be wine-red. As the EDTA solution is added, the EDTA ligand reacts with the free metal ions and then actually reacts with the metal-EBT complex ion, complexing the metal and resulting in the free EBT ligand, which, as mentioned earlier, gives a sky-blue color to the solution. The color change, then, is the total conversion of the wine-red color to the sky-blue color, with every trace of red disappearing at the end point.

It is known that this color change is quite sharp when magnesium ions are present. In cases in which magnesium ions are not present in the water samples, the end point will not be sharp. Because of this, a small amount of magnesium chloride is added to the EDTA as it is prepared and thus a sharp end point is assured.

7.5 PARTS PER MILLION

Basic Concepts

As indicated in the last section, the amount of hardness in a water sample is generally reported as parts per million (ppm) $CaCO_3$. "Parts per million" is a unit of concentration similar to molarity, percent, etc. A solution that has a solute content of 1 ppm has one part of solute dissolved for every million (10^6) parts of solution. A "part" can be a weight measurement, it can be a volume measurement, or it can be both. It if is a weight measurement, it might mean, for example, 1 mg/10^6 mg or 1 mg/kg. A conversion of units here indicates that it might also mean 1 μg/10^6 μg or 1 μg/g. If it is a volume measurement, it might mean 1 mL/10^6 mL, 1 mL/1000 L, or 1 μL/L. If the intent is to represent a concentration in a water solution, and this is frequently the case, we can recognize that 1 g of water (and dilute water solutions) is 1 mL of the water and that 1 L of water weighs 1 kg. Thus, 1 ppm might also mean 1 mg/L.

Most commonly, if the sample to be analyzed is a solid material (for example, soil), ppm is expressed as milligrams per kilogram (or micrograms per gram), since the weight (as opposed to volume) of the material is most easily measured. However, if the volume is most easily measured (such as with a liquid sample like water or water solutions), the ppm is expressed as milligrams per liter (or micrograms per milliliter). Thus, if a certain solution is described as being 10 ppm iron, this means 10 mg of iron dissolved per liter of solution. If, however, an analysis of soil for potassium is reported as 250 ppm K, this means 250 mg of potassium per kilogram of soil. (See Figure 7.9.) Some solution preparation procedures a technician may encounter involving the ppm unit are discussed below.

Solution Preparation

If you wanted to prepare 1 L of a 10 ppm solution of zinc (Zn), you would weigh out 10 mg of zinc metal, place it in a 1-L flask, dissolve it, and dilute to the 1-L mark with distilled water. If you need to prepare a volume other than 1 L, the ppm concentration is multiplied by the volume in liters:

$$\text{ppm}_D \times L_D = \text{mg to be weighed} \tag{7.1}$$

$$\left(\frac{\text{mg}}{\text{L}}\right)_D \times L_D = \text{mg to be weighed} \tag{7.2}$$

$$(D = \text{desired})$$

Example 1

Tell how you would prepare 500 mL of a 25 ppm copper solution from pure copper metal.

For Solid Samples

1 ppm = 1 mg/kg
= 1 μg/g

For Liquid Samples and Solution

1 ppm = 1 mg/L
= 1 μg/mL

Figure 7.9. The ppm unit for solid samples and liquid samples and solutions.

Solution

$$\frac{25.0 \text{ mg}}{\text{L}} \times 0.500 \text{ L} = 12.5 \text{ mg}$$

Weigh 12.5 mg of copper into a 500-mL flask, dissolve, and dilute to the mark with water.

Another example is with dilution. It is a very common practice to purchase solutions of metals that are fairly concentrated (1000 ppm) and then dilute them to obtain the desired concentration. As with molarity and normality (see Equations 5.9 and 6.29), the concentration is multiplied by the volume both before and after dilution:

$$\text{ppm}_B \times V_B = \text{ppm}_A \times V_A \tag{7.3}$$

where A and B have their usual significance. This substitution is possible because the conversion factor for converting ppm to molarity (in order to cancel the units properly) is the same on both sides, and thus they cancel, as did the volume conversion factors in Chapter 5 (Equation 5.9). Thus, we write the most general form of this dilution equation, and this is applicable for all concentration and volume units:

$$C_B \times V_B = C_A \times V_A \tag{7.4}$$

in which C represents any concentration unit. This equation is useful for any and all dilution problems, regardless of units.

Example 2

How would you prepare 100 mL of a 20 ppm iron solution from a 1000 ppm solution?

Solution

$$1000 \times V_B = 20 \times 100$$

$$V_B = \frac{20 \times 100}{1000} = 2.0 \text{ mL}$$

Dilute 2.0 mL of the 1000 ppm solution to 100 mL.

There is a third type of solution preparation problem that could be encountered with the ppm unit. This is the case in which the solution of a metal is to be prepared by weighing a metal salt, rather than the pure metal, when only the ppm of the metal in the solution is given. In this case, the weight of the metal must be converted to the weight of the metal salt via a "gravimetric factor" so that the weight of the metal salt is known.

Example 3

How would you prepare 250 mL of a 50 ppm solution of nickel from pure solid nickel chloride, $NiCl_2$, FW = 129.62?

Solution

$$\text{ppm}_D \times L_D \times \text{gravimetric factor} = \text{mg to be weighed}$$

$$\frac{50 \text{ mg}}{\text{L}} \times 0.250 \text{ L} \times \frac{NiCl_2}{Ni} = \text{mg to be weighed}$$

$$50 \times 0.250 \times \frac{129.62}{58.71} = 27.6 \text{ mg of } NiCl_2$$

Weigh out 27.60 mg of $NiCl_2$, place in a 250-mL flask, dissolve, and dilute to the mark.

The substance to be weighed may be a hydrate, as shown in Example 4.

Example 4

How many grams of solid $CuSO_4 \cdot 5H_2O$, FW = 249.68, are needed to prepare 500 mL of a 1000 ppm solution of Cu?

Solution

$$\text{ppm}_D \times L_D \times \text{gravimetric factor} = \text{mg to be weighed}$$

$$\frac{1000 \text{ mg}}{\text{L}} \times 0.500 \text{ L} \times \frac{CuSO_4 \cdot 5H_2O}{Cu} = \text{mg to be weighed}$$

$$1000 \times 0.500 \times \frac{249.68}{63.54} = 1965 \text{ mg} = 1.97 \text{ g}$$

In the next section, we will examine the parts per million calculation involved in a water hardness determination.

7.6 STANDARDIZATION AND CALCULATIONS

As discussed earlier, the hardness of a water sample is reported in ppm $CaCO_3$. The calculation is very similar to the percent constituent calculation discussed in Chapter 6. To calculate percent, or parts per hundred, we calculate the number of grams present in 100 g of sample, which is 100 times what it is per 1 g of sample.

$$\frac{\text{weight of constituent}}{\text{sample weight}} = \text{grams of constituent per gram of sample} \quad (7.5)$$

$$\frac{\text{weight of constituent}}{\text{sample weight}} \times 100 = \text{grams of constituent per 100 g of sample} \quad (7.6)$$

$$= \text{parts per hundred}$$

$$= \text{percent}$$

Obviously, one would use 1 million (10^6) as the multiplication factor, rather than 100 (10^2), for ppm:

$$\text{ppm} = \frac{\text{weight of } CaCO_3}{\text{sample weight}} \times 10^6 \quad (7.7)$$

This approach is somewhat impractical, however, since as previously indicated the volume of water samples, rather than weight, can be measured more conveniently in the laboratory. This measurement, of course, is done with a volumetric pipet. Also, as mentioned earlier, the milligrams per liter unit for ppm is more common than the others for liquid samples. Thus, if we measure the volume of the water sample in liters, rather than the weight, and calculate milligrams of the constituent, rather than grams, we can divide these milligrams by the liters, and our unit would be ppm:

$$\text{ppm} = \frac{\text{weight of CaCO}_3 \text{ (in mg)}}{\text{volume of water (in L)}} \tag{7.8}$$

Milligrams of constituent is computed quite easily starting with computing grams of constituent and tacking on the conversion factor for calculating milligrams (1000 mg/g).

$$L_T \times M_T \times FW_{ST} \times 1000 = \text{mg of constituent} \tag{7.9}$$

Thus the ppm calculation is

$$\frac{L_T \times M_T \times FW_{ST} \times 1000}{\text{liters of sample}} = \text{ppm}_{ST} \tag{7.10}$$

The titrant is EDTA, and the results are reported as ppm $CaCO_3$. Thus we have

$$\frac{L_{EDTA} \times M_{EDTA} \times FW_{CaCO_3} \times 1000}{\text{liters of sample}} = \text{ppm CaCO}_3 \tag{7.11}$$

Example 5

What is the hardness in a water sample, 100 mL of which required 27.95 mL of a 0.01266 *M* EDTA solution for titration?

Solution

$$\text{ppm CaCO}_3 = \frac{0.02795 \times 0.01266 \times 100.09 \times 1000}{0.1000} = 354.2 \text{ ppm CaCO}_3$$

Multiplication by the formula weight of $CaCO_3$ in the numerator has the effect of reporting all metals that reacted with the EDTA as $CaCO_3$. As mentioned earlier, no differentiation is made in this procedure between calcium, magnesium, iron, or whatever else might be reacting with the EDTA. They are all reported as ppm $CaCO_3$ by multiplying by the formula weight of $CaCO_3$.

The molarity of the EDTA solution is determined either by accurately weighing a dry quantity of it when preparing the solution or by standardization with a primary standard – for example, calcium carbonate. The calculation for this latter method is identical to acid-base standardization with a primary standard:

$$\text{moles(EDTA)} = \text{moles(CaCO}_3) \tag{7.12}$$

$$\text{L(EDTA)} \times \text{M(EDTA)} = \frac{\text{wt. (CaCO}_3)}{\text{FW(CaCO}_3)} \tag{7.13}$$

A minor variation will be used here (and this could have been used in previous standardization examples in Chapter 6), however, and that is that a quantity of the primary standard can be weighed, placed in a volumetric flask, dissolved, and diluted to the mark. The concentration of this solution is known, and thus the experiment becomes one of standardization of EDTA with a standard solution of $CaCO_3$ rather than a weighed quantity of the primary standard (as was the case with the KHP in Chapter 6). Thus, the concentration of the $CaCO_3$ is first calculated:

$$M(CaCO_3) = \frac{\text{moles}}{L} \times \frac{\frac{\text{wt.}(CaCO_3)}{FW(CaCO_3)}}{\text{liters of solution}} \tag{7.14}$$

and the standardization calculation becomes

$$L(EDTA) \times M(EDTA) = L(CaCO_3) \times M(CaCO_3) \tag{7.15}$$

in which $L(CaCO_3)$ is the amount of the solution pipetted into the reaction flask. Combining Equations 7.14 and 7.15, we can write

$$L(EDTA) \times M(EDTA) = \frac{L \text{ of } CaCO_3 \text{ taken} \times \frac{\text{wt.}(CaCO_3)}{FW(CaCO_3)}}{L \text{ of } CaCO_3 \text{ prepared}} \tag{7.16}$$

in which "L of $CaCO_3$ taken" is the pipetted volume, while "L of $CaCO_3$ prepared" is the total solution volume.

Example 6

A $CaCO_3$ solution is prepared by weighing 0.5047 g of $CaCO_3$ into a 500-mL volumetric flask, dissolving with HCl, and diluting to the mark. If 25.00 mL of this solution required 28.12 mL of an EDTA solution, what is the molarity of the EDTA?

Solution

$$0.02812 \times M(EDTA) = \frac{0.02500 \times \frac{0.5047}{100.09}}{0.5000}$$

$$M(EDTA) = \frac{0.02500 \times \frac{0.5047}{100.09}}{0.5000 \times 0.02812} = 0.008966\ M$$

EXPERIMENTS

EXPERIMENT 11: PREPARATION AND STANDARDIZATION OF AN EDTA SOLUTION

NOTE: All Erlenmeyer and volumetric flasks used in Experiments 11 and 12 must be rinsed thoroughly with distilled water prior to use. It is obviously important to avoid tap water contamination. The pH = 10 ammonia buffer required for Experiments 11, 12, and 13 can be prepared by dissolving 35 g of NH_4Cl and 285 mL of concentrated ammonium hydroxide in water and diluting to 500 mL. The EBT indicator should be prepared fresh by dissolving 200 mg in a mixture of 15 mL of triethanolamine and 5 mL of ethyl alcohol. Safety glasses are required.

1. Dry a quantity (at least 0.5 g) of primary standard $CaCO_3$ for 1 hr. Your instructor may choose to dispense this to you.
2. Weigh 4.0 g of disodium dihydrogen EDTA dihydrate and 0.10 g of magnesium chloride hexahydrate into a 1-L glass bottle. Add one pellet of NaOH and fill to approximately 1 L with water. Shake well. These ingredients will require some time to dissolve, so it is recommended that this solution be prepared at least 1 hr before its desired use.

NOTE: If the solution is ready, but the $CaCO_3$ is still drying, you can proceed with Experiment 12 and return to the standardization later.

3. Weigh accurately 0.3 to 0.4 g of the dried and cooled $CaCO_3$ by difference into a clean, dry, short-stemmed funnel in the mouth of a 500-mL volumetric flask. Tap the funnel gently to force the $CaCO_3$ into the flask. Wash any remaining $CaCO_3$ into the flask with distilled water using a squeeze bottle. Add a small amount of concentrated HCl (less than 2.0 mL total) to the flask through the funnel such that the funnel is rinsed thoroughly in the process. Rinse the funnel into the flask one more time with distilled water and remove the funnel. Rinse the neck of the flask with distilled water. Swirl the flask until all the $CaCO_3$ is dissolved and effervescence has ceased. Dilute to the mark with distilled water and shake well. If the solution is warm at this point, cool in a cold water bath and then add more distilled water so as to bring the meniscus back to the mark. Shake well again.
4. Pipet a 25.00-mL aliquot (an aliquot is a "portion" of a solution) of this standard calcium solution into each of three 250-mL Erlenmeyer flasks. Add 5.0 mL (graduated cylinder) of the pH = 10 buffer and two drops of EBT indicator to each flask. Give your EDTA solution one final shake to ensure its homogeneity.
5. Clean a 50-mL buret, rinse it several times with your EDTA solution (the titrant), and fill, as usual, with your titrant. Be sure to eliminate the air bubbles from the stopcock and tip. Titrate the solutions in your flasks, one at a time, until the last trace of red disappears in each with a fraction of a drop. This final color change will be from a violet color to a deep sky blue, but should be a sharp change. All three titrations should agree to within 0.05 mL of each other. If they do not, repeat until you have three that do. This assures, as in Experiment 7, that the relative ppt deviation is less than 2.
6. Calculate the molarity of your EDTA for each of the three titrations and calculate the average. Record in your notebook.

EXPERIMENT 12: TITRIMETRIC DETERMINATION OF TOTAL HARDNESS IN A SYNTHETIC OR "REAL" WATER SAMPLE

NOTE: Safety glasses are required.

1. Obtain a water sample. If it is a synthetic sample, it will be contained in a 1-L volumetric flask. Dilute to the mark with distilled water. Shake well. If it is a "real" sample, proceed with Step 2.

2. Pipet 100.0-mL aliquots of the water sample into each of three clean 500-mL Erlenmeyer flasks. Add the buffer and indicator as in Experiment 11. If you are analyzing a "real" water sample in which iron, nickel, or other heavy metals are suspected to be present, add a few small crystals of KCN (*caution* – be sure the pH of the water is neutral or basic before adding cyanide). The CN^{-1} is a ligand ("masking agent" – see Section 7.2) that will form a stable complex ion with metals other than Ca^{+2} or Mg^{+2}. These metals are thus "masked" and will not interfere with the end point detection. Consult with your instructor concerning appropriate safety precautions with the waste solutions in the event cyanide is added.

3. Titrate each, one at a time, with the EDTA solution as before. Again, all three buret readings should agree to within 0.05 mL of each other. If they do not, repeat until you have three that do.

4. Calculate the ppm $CaCO_3$ in the sample and compute the average. Record, as usual, in your notebook.

EXPERIMENT 13: TITRIMETRIC DETERMINATION OF TEMPORARY AND PERMANENT HARDNESS IN A WATER SAMPLE.

NOTE: The fraction of $CaCO_3$ in a water sample that precipitates upon boiling is called "temporary hardness". By boiling a water sample and causing this precipitation, we can determine how much hardness remains after boiling (called permanent hardness) and determine temporary hardness by subtracting the permanent hardness from the total hardness determined in the previous experiment. Safety glasses are required.

1. Measure accurately 250 mL of the "real" water sample from the previous experiment into a 400-mL beaker and gently boil uncovered for 3 min. Since the beaker is uncovered, make certain that the boiling is gentle enough to avoid loss by splashing.

2. Allow to cool. Filter through Whatman #2 filter paper into a clean, dry 250-mL volumetric flask, being careful to catch all the filtrate in the flask. Do not rinse the precipitate, since that would risk partial dissolution of the temporary hardness caught in the filter. Dilute with distilled water to the mark and shake.

3. Titrate 50.00-mL portions of the diluted filtrate using the same procedure as in the previous experiments. The ppm $CaCO_3$ calculated is the permanent hardness. The temporary hardness is then the difference between the total hardness and the temporary hardness.

QUESTIONS AND PROBLEMS

1. Define monodentate, bidentate, hexadentate, ligand, complex ion, chelate, chelating agent, masking, masking agent, formation constant, coordinate covalent bond, water hardness, and aliquot.
2. Define "ligand" and "complex ion". Give an example of each.
3. Given the following reaction, tell which of the three species is the ligand and which is the complex ion:

$$Co^{+2} + 4\ Cl^{-} \rightleftarrows CoCl_4^{-2}$$

4. Give one example each (either structure or name) of a monodentate ligand and a hexadentate ligand. Explain what is meant by "monodentate".
5. Is the ligand ethylenediamine ($H_2NCH_2CH_2NH_2$) monodentate, bidentate, tridentate, or what? Explain your answer.
6. Consider the reaction shown in Figure 7.10.
   (a) Which chemical species is a ligand?
   (b) Which chemical species is a complex ion?
   (c) Is the ligand monodentate, bidentate, or what? Explain your answer.
   (d) Is the complex ion a chelate? Explain.
7. Concerning the EDTA ligand:
   (a) How many bonding sites on this molecule bond to a metal ion when a complex ion is formed?
   (b) How many EDTA molecules will bond to a single metal ion?
   (c) What is the word describing the property pointed out in (a)?
8. Explain why water samples titrated with EDTA need to be buffered at pH = 10 and not at pH = 12 or pH = 8.
9. In the water hardness titration,
   (a) what chemical species is the wine red color at the beginning of the titration due to?
   (b) what chemical species is the sky blue color at the end point due to?
   (c) what does it mean to say that cyanide is a "masking agent"?

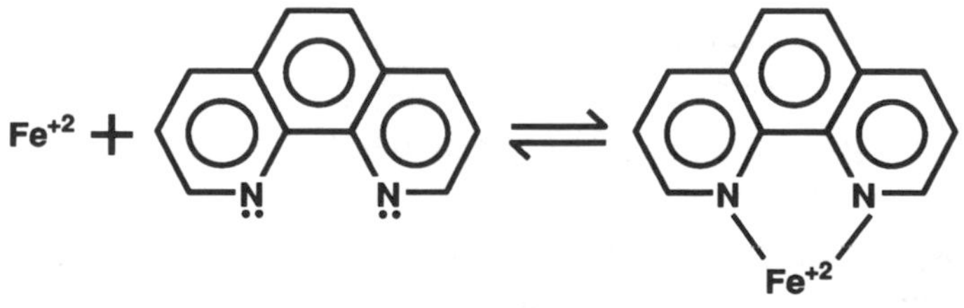

Fe+2 + N N ⇌ N N Fe+2

Figure 7.10. The reaction for Question #6.

10. Explain why the pH = 10 ammonia buffer is required in EDTA titrations for water hardness.
11. How would you prepare 250 mL of a 25 ppm solution of magnesium from pure magnesium metal?
12. How many milligrams of silver metal are needed to prepare 750 mL of a solution that is 30 ppm silver?
13. How many grams of aluminum metal are required to prepare 600 mL of a solution that is 40 ppm aluminum?
14. Tell how you would prepare 500 mL of a 15 ppm Mg solution using pure magnesium for the solute.
15. How many grams of iron are required to prepare 250 mL of a 30 ppm solution of iron?
16. How many milligrams of copper are needed to prepare 100 mL of a 125 ppm solution?
17. How would you prepare 500 mL of a 50 ppm Na solution
    (a) from pure solid NaCl?
    (b) from a solution that is 1000 ppm Na?
18. How many milliliters of 1000 ppm Cu are needed to prepare 100 mL of a 125 ppm solution?
19. How would you prepare 600 mL of a 15 ppm solution of zinc from a solution of zinc that is 1000 ppm?
20. Tell how you would prepare 250 mL of a 25 ppm solution of sodium (Na) from a solution of Na that is 500 ppm.
21. How many grams of $Fe(NO_3)3 \cdot 9H_2O$ (FW = 404.02) are required to prepare 250 mL of a 30 ppm solution of iron?
22. Tell how you would prepare 100 mL of a 50 ppm solution of copper from pure solid anhydrous copper sulfate, $CuSO_4$.
23. How many milligrams of $CuSO_4 \cdot 4H_2O$ are needed to prepare 100 mL of a 125 ppm Cu solution?
24. A technician wishes to prepare 500 mL of a 25.0 ppm solution of barium.
    (a) How many milliliters of 1000 ppm barium would be required if he/she were to prepare this by dilution?
    (b) If he/she were to prepare this from pure barium metal, how many grams would be required?
    (c) If he/she were to prepare this from pure solid $BaCl_2 \cdot 2H_2O$, how many grams would be required?
25. How many milligrams of KBr are needed to prepare 600 mL of a 40 ppm Br solution?
26. How many grams of K_2HPO_4 are needed to prepare 450 mL of a 10 ppm phosphorus solution?
27. How many grams of KNO_3 are needed to prepare 100 mL of a solution that is 50 ppm nitrogen? (Refer to Step 1, Experiment 24, in Chapter 11.)

28. If a laboratory technician needs 900 mL of a solution of potassium that is 50 ppm, how many milliliters of a 1000 ppm solution are needed?
29. Tell how you would prepare 500 mL of a 0.02500 *M* solution of the solid disodium dihydrogen EDTA dihydrate.
30. How many grams of disodium dihydrogen EDTA dihydrate are required to prepare 1000 mL of a 0.01 *M* EDTA solution?
31. An EDTA solution is standardized with pure solid magnesium metal. If 10.0 mg of the Mg required 40.08 mL of the EDTA, what is the molarity of the solution?
32. 0.0236 g of solid $CaCO_3$ were dissolved and exactly consumed by 12.01 mL of an EDTA solution. What is the molarity of the EDTA solution?
33. An EDTA solution is standardized using pure solid $CaCO_3$. If 0.1026 g of $CaCO_3$ required 27.62 mL of the EDTA, what is the molarity of the EDTA?
34. What is the molarity of a solution of EDTA when 30.67 mL of it reacts exactly with 45.33 mg of calcium metal?
35. If 34.29 mL of an EDTA solution is required to react with 0.1879 g of $MgCl_2$, what is the molarity of the EDTA?
36. A 100.0-mL aliquot of a zinc solution required 34.62 mL of an EDTA solution for titration. If the zinc solution was prepared by dissolving 0.0877 g of zinc in 500.0 mL of solution, what is the molarity of the EDTA?
37. An EDTA solution is standardized by preparing a solution of primary standard $CaCO_3$ (0.5622 g dissolved in 1000 mL of solution) and titrating aliquots of it with the EDTA. If a 25.00-mL aliquot required 21.88 mL of the EDTA, what is the concentration of the EDTA?
38. If 25.00 mL of a solution prepared by dissolving 0.4534 g of $CaCO_3$ in 500.0 mL of solution reacts with 34.43 mL of an EDTA solution, what is the molarity of the EDTA?
39. A solution has 0.4970 g of $CaCO_3$ dissolved in 500 mL of solution. If 25.00 mL of it reacts exactly with 29.55 mL of an EDTA solution, what is the molarity of the EDTA?
40. 25.00 mL of a $CaCO_3$ solution reacts with 30.13 mL of an EDTA solution. If there are 0.5652 g of $CaCO_3$ per 500.0 mL of solution, what is the molarity of the calcium carbonate and the EDTA?
41. What is the hardness of a water sample in ppm $CaCO_3$ if a 100.0-mL aliquot required 27.62 mL of 0.01462 *M* EDTA for titration?
42. If 25.00 mL of a hard water sample required 11.68 mL of 0.01147 *M* EDTA for titration, what is the hardness of the water in ppm $CaCO_3$?
43. If 12.42 mL of a 0.01093 *M* EDTA solution were needed to titrate 50.00 mL of a water sample, what is the hardness of the water sample in ppm $CaCO_3$?

44. In an experiment for determining water hardness, 75.00 mL of a water sample required 13.03 mL of an EDTA solution that is 0.009242 *M*. What is the ppm $CaCO_3$ in this sample?

45. A 0.01011 *M* solution of EDTA titrant is used to determine the hardness in a 150.0-mL sample of water. If 16.34 mL of the titrant are needed, what is the hardness of the water in ppm $CaCO_3$?

46. If 14.20 mL of an EDTA solution (4.1198 g of $Na_2H_2EDTA \cdot 2H_2O$ in 500 mL of solution) were needed to titrate 100.0 mL of a water sample, what is the ppm $CaCO_3$ in the water?

47. What is the ppm $CaCO_3$ when 100.0 mL of the water required 13.73 mL of an EDTA solution prepared by dissolving 3.8401 g of $Na_2H_2EDTA \cdot H_2O$ in 500.0 mL of solution?

Chapter 8

Oxidation-Reduction and Other Titrations

8.1 INTRODUCTION

Many chemical species have a tendency to either give up or take on electrons. This tendency is based on the premise that a greater stability, or lower energy state, is achieved as a result of this electron donation or acceptance. A basic example is the reaction between a sodium atom and a chlorine atom:

$$Na + Cl \rightarrow NaCl$$

A sodium atom with only one electron in its outermost energy level would achieve a lower energy state if this electron were released to the chlorine atom, which would also achieve a lower energy state as a result. Both atoms would become ions (Na^+ and Cl^-) and each would have a stable, filled outermost energy level identical with those of the noble gases, neon in the case of sodium and argon in the case of chlorine. Thus, the "electron transfer" does take place and sodium chloride (NaCl) is formed.

Those reactions which this sodium-chlorine case typifies are called oxidation-reduction reactions. The term "oxidation" refers to the loss of electrons, while the term "reduction" refers to the gain of electrons. A number of oxidation-reduction reactions (nicknamed "redox" reactions) are useful in titrimetric analysis, and many are encountered in other analysis methods.

8.2 OXIDATION NUMBER

It may appear strange that the term "reduction" is associated with a gaining process. Actually, the term reduction was coined as a result of what happens to the "oxidation number" of the element when the electron transfer takes place. The oxidation number of an element is a number representing the state of the element with respect to the number of electrons the element has given up, taken on, or contributed to a covalent bond. For example, pure sodium metal has neither given up, taken on, nor shared electrons, and thus its oxidation number is 0. In sodium chloride, however, the sodium has given up an electron and become a +1 charge, and thus its oxidation number is +1. The chlorine in NaCl has taken on an electron and become a -1 charge, and thus its oxidation number is -1, while it, too, had an oxidation number of 0 prior to the reaction. Thus, while it is true that the chlorine gained an electron it is also true that its oxidation number became lower (from 0 to -1), hence the term "reduction".

As we shall see, it is useful to be able to determine the oxidation number of a given element in a compound. Since most elements can exist in a variety of oxidation states, it is necessary to adopt a set of rules or guidelines for this determination. These are listed in Table 8.1.

A discussion of several examples should clarify the general scheme. First, rules 1 through 8 cover many situations in which the sum of the oxidation numbers is not needed. For example, in $BaCl_2$, barium is +2 (rule #3) and Cl is -1 (rule #4). In K_2O, potassium is +1 (rule #2) and O is -2 (rule #7). In CaH_2, calcium is +2 (rule #3) and H is -1 (rule #6).

Next, rule #9 is used for determining all oxidation numbers of all elements that rules 1 through 8 do not cover (this must be all but one element) and then assigning the remaining elements' oxidation numbers, knowing that the total must add up to either zero or the ionic charge.

Example 1

What is the oxidation number of sulfur in H_2SO_4?

Table 8.1. Rules for assigning oxidation numbers.

Element	Oxidation Number
1. Any uncombined element	0
2. Combined alkali metal (Group I)	+1
3. Combined alkaline earth metal (Group II)	+2
4. Combined halogen (Group VIIA) (except in polyatomic ions which include oxygen)	-1
5. Hydrogen combined with nonmetal	+1
6. Hydrogen combined with metal (hydrides)	-1
7. Combined oxygen (except peroxides)	-2
8. Oxygen in peroxides	-1
9. All others—oxidation numbers are determined from the fact that the sum of all oxidation numbers within the molecule or polyatomic ion must either add up to zero (in the case of a neutral molecule) or the charge on the ion (in the case of a polyatomic ion)	

Solution

Oxygen is -2; $4 \times (-2) = -8$
Hydrogen is $+1$; $2 \times (+1) = +2$
Sum $= -8 + 2 = -6$
Sulfur must be $+6$ in order for the total to be zero.

Example 2

What is the oxidation number of manganese in $KMnO_4$?

Solution

Potassium is $+1$; $1 \times (+1) = +1$
Oxygen is -2; $4 \times (-2) = -8$
Sum $= -8 + 1 = -7$
Manganese must be $+7$ in order for the total to be zero.

Example 3

What is the oxidation number of nitrogen in the nitrate ion NO_3^{-1}?

Solution

Oxygen is -2; $3 \text{ X } (-2) = -6$
Sum $= -6 + 5 = -1$
Nitrogen must be $+5$ in order for the net charge to be -1.

The usefulness of determining the oxidation number is twofold. First, it will help determine if there was a change in oxidation number of a given element in a reaction. This always signals the occurrence of an oxidation-reduction reaction. Thus, it helps tell us whether a reaction is a redox reaction or some other reaction. Second, it will lead to the determination of the number of electrons involved, which will aid in balancing the equation and in determining the equivalent weight. These latter points will be discussed in the later sections.

Example 4

Which of the following is a redox reaction?

(a) $Cl_2 + 2NaBr \rightarrow Br_2 + 2NaCl$

(b) $BaCl_2 + K_2SO_4 \rightarrow BaSO_4 + 2KCl$

Solution

The oxidation numbers of Cl and Br in (a) have changed. Cl has changed from 0 to -1, while Br has changed from -1 to 0. (In diatomic molecules, the elements are considered to be in the uncombined state; thus rule #1 applies.) There is no change in oxidation number in (b). Thus, (a) is redox.

8.3 OXIDATION-REDUCTION TERMINOLOGY

The terms oxidation and reduction, as should be obvious from the discussion to this point, can be defined in two different ways – according to the gain or loss of electrons, or according to the increase or decrease in oxidation number:

Oxidation – the loss of electrons – the increase in oxidation number
Reduction – the gain of electrons – the decrease in oxidation number

To say that a substance has been oxidized means that the substance has lost electrons. To say that a substance has been reduced means that the substance has gained electrons. To say that substance A has oxidized substance B means that substance A has caused B to lose electrons. Similarly, to say that substance A has reduced substance B means that substance A has caused B to gain electrons. When a substance (A) causes another substance (B) to be oxidized, substance A is called the "oxidizing agent". When substance A causes substance B to be reduced, substance A is a "reducing agent". To say that substance A "causes" oxidation or reduction means that substance A either removes electrons from or donates electrons to substance B. Thus, oxidation is always accompanied by reduction and vice versa. Also, every redox reaction has an oxidizing agent (which is the substance reduced) and a reducing agent (which is the substance oxidized).

Example 5

Tell what is oxidized and what is reduced and tell what the oxidizing agent and the reducing agent are in the following:

$$Zn + 2HCl \rightarrow ZnCl_2 + H_2$$

Solution

oxidation: $Zn \rightarrow ZnCl_2$
(The oxidation number of Zn has increased from 0 to +2.)

reduction: $HCl \rightarrow H_2$
(The oxidation number of H has decreased from +1 to 0.)

Zn has been oxidized. Therefore, it is the reducing agent. HCl has been reduced. Therefore, it is the oxidizing agent.

8.4 THE ION-ELECTRON METHOD FOR BALANCING EQUATIONS

We have seen how stoichiometry plays a key role in the calculations involved in both gravimetric and titrimetric procedures, and we know that a balanced chemical equation is needed for basic stoichiometry. With redox reactions, balancing equations by inspection can be quite challenging, if not impossible. Thus, several special schemes have been derived for balancing redox equations. The ion-electron method takes into account the electrons that are transferred, since these must also be balanced. That is, the electrons given up must be equal to the electrons taken on. An important byproduct of this method is the information needed to compute equivalent weights of reactants. A review of the ion-electron method of balancing equations will therefore present a simple means of balancing redox equations and also give us the information needed to compute equivalent weights.

The method makes use of only those species, dissolved or otherwise, which actually take part in the reaction. So-called "spectator ions", or ions which are present but play no role in the chemistry, are not included in the balancing procedure. Solubility rules are involved here, since spectator ions result only when an ionic compound dissolves and ionizes. Also, the scheme is slightly different for acid and base conditions. Our purpose, however, is to provide a feel for the basic procedure; thus, spectator ions will be absent from all examples from the start, and acidic conditions will be the only conditions considered. The stepwise procedure we will follow is:

Step 1: Look at the equation to be balanced and determine what is oxidized and what is reduced. This involves checking the oxidation numbers and discovering which have changed.

Step 2: Write a "half-reaction" for both the oxidation process and the reduction process and label as "oxidation" and "reduction". These half-reactions show only the species being oxidized (or the species being reduced) on the left side with only the product of the oxidation (or reduction) on the right side.

Step 3: If oxygen appears in any formula on either side in either equation it is balanced by writing H_2O on the opposite side. This is possible since the reaction mixture is a water solution. The hydrogen in the water is then balanced on the other side by writing H^{+1}, since we are dealing with acid solutions. Now balance both half-reactions for all elements by inspection.

Step 4: Balance the charges on both sides of the equation by adding the appropriate number of electrons (e^-) to whichever side is deficient in negative charges. The charge balancing is accomplished as if the electron is like any chemical species – place the appropriate multiplying coefficient in front of the "e^-".

Step 5: Multiply through both equations by appropriate coefficients so that the number of electrons involved in both half-reactions is the same. This has the effect of making the total charge loss equal to the total charge gain and thus eliminates electrons from the final balanced equation, as you will see in Step 6.

Step 6: Add the two equations together. The number of electrons, being the same on both sides, cancels out and thus does not appear in the final result. One can also cancel out some H^{+1} and H_2O if they appear on both sides at this point.

Step 7: Make a final check to see that the equation is balanced.

Example 6

Balance the following equation by the ion-electron method:

$$MnO_4^- + Fe^{+2} \rightarrow Fe^{+3} + Mn^{+2}$$

Solution

Step 1: Fe^{+2} oxidized (Fe: +2 → +3, loss of electrons)
MnO_4^- reduced (Mn: +7 → +2, gain of electrons)

Step 2: Oxidation: $Fe^{+2} \rightarrow Fe^{+3}$
Reduction: $MnO_4^- \rightarrow Mn^{+2}$

Step 3: Oxidation: $Fe^{+2} \rightarrow Fe^{+3}$
Reduction: $8H^+ + MnO_4^- \rightarrow Mn^{+2} + 4H_2O$

Step 4: Oxidation: $Fe^{+2} \rightarrow Fe^{+3} + 1e^-$
Reduction: $5e^- + 8H^+ + MnO_4^- \rightarrow Mn^{+2} + 4H_2O$

Step 5: Oxidation: $5(Fe^{+2} \rightarrow Fe^{+3} + 1e^-)$
Reduction: $1(5e^- + 8H^+ + MnO_4^- \rightarrow Mn^{+2} + 4H_2O)$

Step 6: Oxidation: $5(Fe^{+2} \rightarrow Fe^{+3} + 1e^-)$
Reduction: $1(5e^- + 8H^+ MnO_4^- \rightarrow Mn^{+2} + 4H_2O)$

$$5Fe^{+2} + 8H^+ + MnO_4^- \rightarrow 5Fe^{+3} + Mn^{+2} + 4H_2O$$

Step 7: Fe on each side
8 H on each side
1 Mn on each side
4 O on each side

The equation is balanced.

8.5 EQUIVALENT WEIGHTS IN REDOX REACTIONS

With acid-base reactions, equivalent weights (EW) are calculated by dividing the formula weights by the number of hydrogen ions donated or accepted by the acid or base. With redox reactions, the equivalent weights are calculated by dividing the formula weights by the number of electrons given up or taken on in the balanced (both charge-balanced and element-balanced) half-reaction equations (on a "per mole" basis). A couple of examples should clarify the concept.

Example 7

In the reaction of potassium permanganate (FW = 158.04) with Mohr's salt, $Fe(NH_4)2(SO_4)_2{\cdot}6H_2O$ (FW = 392.2), the following is the (unbalanced) ionic reaction:

$$Fe^{+2} + MnO_4^- \rightarrow Mn^{+2} + Fe^{+3}$$

What is the equivalent weight of the Mohr's salt and the potassium permanganate?

Solution

$$Fe^{+2} \rightarrow Fe^{+3} + 1e^-$$

$$\text{EW (Mohr's salt)} = \frac{392.2}{1} = 392.2$$

$$5e^- + 8H^+ + MnO_4^- \rightarrow Mn^{+2} + 4H_2O$$

$$\text{EW } (KMnO_4) = \frac{158.04}{5} = 31.608$$

Example 8

In the reaction of potassium dichromate (FW = 294.19) with potassium iodide (FW = 166.01)

$$I^- + Cr_2O_7^{-2} \rightarrow I_2 + Cr^{+3}$$

What are the equivalent weights of the two reactants?

Solution

$$2I^- \rightarrow I_2 + 2e^- \text{ (1e}^- \text{ on a "per mole" basis)}$$

$$\text{EW (KI)} = \frac{166.01}{1} = 166.01$$

$$6e^- + 14H^+ + Cr_2O_7^{-2} \rightarrow 2Cr^{+3} + 7H_2O$$

$$\text{EW } (K_2Cr_2O_7) = \frac{294.19}{6} = 49.033$$

Calculations involving standardization solution preparation and percent constituent, once the equivalent weights have been determined, if necessary, are identical with those previously presented for the acid-base case (Chapter 6).

8.6 POTASSIUM PERMANGANATE

An oxidizing agent that has significant application in redox titrimetry is potassium permanganate, $KMnO_4$. As discovered earlier, the manganese in $KMnO_4$ has a +7 oxidation number. It is as if the manganese has contributed all seven of its outermost electrons ($4s^2$, $3d^5$) to a bonding situation, an observation which implies significant instability. Manganese atoms are in lower energy states if they are found with +4 or +2 oxidation numbers. Thus, like sodium and chlorine in the zero state, manganese in the +7 state is very unstable and will grab onto electrons, given the opportunity, and be reduced. Hence, it is a strong (relatively speaking) oxidizing agent. Some notable examples of its oxidizing powers include Fe^{+2} to Fe^{+3}, $H_2C_2O_4$ to CO_2, As (III) to As (V), and H_2O_2 to O_2. Organic compounds could also be included in this list and, in fact, potassium permanganate solutions, which are deep purple, are used to test qualitatively for alkenes, the reaction being the reduction of MnO_4^- to MnO_2. The purple color disappears in this test, and the brown precipitate, MnO_2, forms indicating a positive test.

Working with $KMnO_4$ presents some special problems because of its significant oxidizing properties. The keys to successful redox titrimetry using $KMnO_4$ are (1) to prepare solutions well in advance so that any oxidizable impurities (usually organic in nature) in the distilled water used are completely oxidized and (2) to protect the standardized solution from additional oxidizable materials (such as lint, fingerprints, rubber, etc.) so that its concentration remains constant until the solution is no longer needed. One additional problem is that once some MnO_2 has formed through the oxidation of such organic substances as those listed, it can catalyze further decomposition and thereby cause further concentration changes. It is obvious that if the solutions are not carefully protected, the concentration of the standardized solution cannot be trusted. Even ordinary light can catalyze the reaction, and for this reason solutions must be protected from light. The MnO_2 present from the initial reactions (prior to standardization) is filtered out so that the catalytic reactions are minimized.

There are three major points to be made concerning the actual titrations using $KMnO_4$. First, since the solutions are unstable when first prepared due to the presence of oxidizable materials in the distilled water, it cannot be used as a primary standard. Thus, $KMnO_4$ solutions must always be standardized prior to use. Second, all titrations are carried out in acid solution. Thus, Mn^{+2} is the product, rather than MnO_2, and no brown precipitate forms in the reaction flask. Third, the fact that the color of the solution is a deep purple means that an indicator is not required. At the point when the substance titrated is exactly consumed, the $KMnO_4$ will no longer react, and since it is a highly colored chemical species, the slightest amount of it in the reaction flask is easily seen. Thus, the end point is the first detectable pink color due to unreacted permanganate. $KMnO_4$ is its own indicator.

$KMnO_4$ solutions are standardized using primary standard grade reducing agents. Typical reducing agents include sodium oxalate, $Na_2C_2O_4$, and ferrous ammonium sulfate hexahydrate, $Fe(NH_4)_2(SO_4)_2 \cdot 6H_2O$, also known as "Mohr's salt" (see Example 7 and also Problems #21 and #31). Experiment 14 includes the standardization of a $KMnO_4$ solution using Mohr's salt.

8.7 IODOMETRY–AN INDIRECT METHOD

Another important reactant in redox titrimetry is potassium iodide (KI). KI is a reducing agent ($2I^- \rightarrow I_2 + 2e^-$) that is useful in analyzing for oxidizing agents. The interesting aspect of the iodide-iodine chemistry is that it is most often used as an *indirect* method (recall the indirect Kjeldahl titration involving boric acid). This means that the oxidizing agent analyzed for is not measured directly by a titration with KI, but is measured indirectly by the titration of the iodine that forms in the reaction. The KI is actually added in excess, since it need not be measured at all. The experiment is called iodometry. Figure 8.1 shows the sequence of events. Thus, the percent of the oxidizing agent ("O" in Figure 8.1) is calculated indirectly

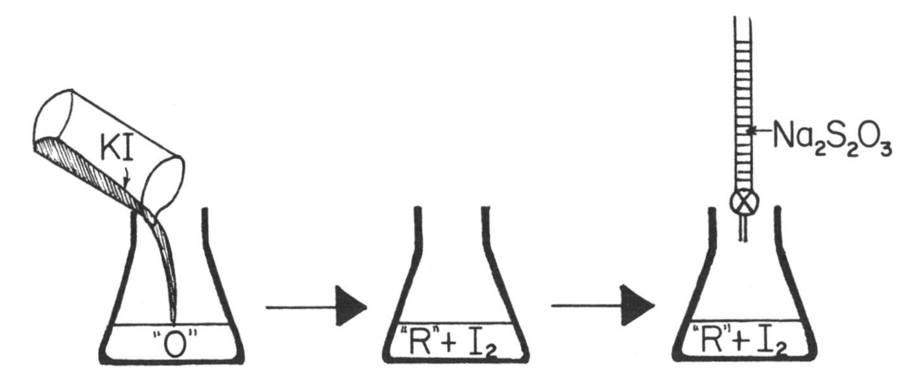

Figure 8.1. In iodometry, a solution of KI is added to a solution of the analyte ("O"). The products are "R" and iodine (I_2). The amount of I_2, which is proportional to the amount of "O", is titrated with $Na_2S_2O_3$.

from the amount of titrant since the titrant actually reacts with I_2 and not "O". This titrant is normally sodium thiosulfate ($Na_2S_2O_3$).

The sodium thiosulfate solution must be standardized. Several primary standard oxidizing agents are useful for this. Probably the most common one is potassium dichromate, $K_2Cr_2O_7$. (See Example 8 for the half-reaction of the dichromate reduction.) Primary standard potassium bromate, $KBrO_3$, or potassium iodate, KIO_3, can also be used. Even primary standard iodine, I_2, can be used. Usually in the standardization procedures, KI is again added to the substance to be titrated ($Cr_2O_7^{-2}$, etc.) and the liberated iodine titrated with thiosulfate. If I_2 is the primary standard, it is titrated directly. The end point is usually detected with the use of a starch solution as the indicator. Starch, in the presence of iodine, is a deep blue color. When all of the iodine is consumed by the thiosulfate, the color changes sharply from blue to violet, providing a very satisfactory end point. Some important precautions concerning the starch, however, are to be considered. The starch solution should be fresh, should not be added until the end point is near, cannot be used in strong acid solutions, and cannot be used with solution temperatures above about 40°C.

An important application of iodometry can be found in many wastewater treatment plant laboratories. In many wastewater treatment plants, chlorine (Cl_2) is used in a final treatment process prior to allowing the wastewater effluent to flow into a nearby river. Of course, the chlorine in both the free and combined forms can be just as harmful environmentally as many components in the raw wastewater. Thus, an important measurement for the laboratory to make is the amount of residual chlorine remaining unreacted in the effluent. Such chlorine, which is an oxidizing agent, can be determined by iodometry. It is the "O" in Figure 8.1. (See Experiments 16 and 17.)

8.8 SUMMARY OF REDOX APPLICATIONS

Permanganate

Potassium permanganate, a strong oxidizing agent, probably has the widest application possibilities of all redox chemicals. It will accept electrons from a wide range of substances having even the slightest oxidation tendency. If you have performed Experiments 14 and 15, you have already experienced first-hand the use of permanganate for the Fe^{+2} and $C_2O_4^{-2}$ oxidations. Application to arsenic and peroxide analysis has also been briefly mentioned. In addition, tin, ferrocyanide [$Fe(CN_6^{-4})$], vanadium, molybdenum, tungsten, uranium, titanium, and nitrite analyses have been accomplished. Finally, metals whose oxalate salts are insoluble (magnesium, calcium, zinc, cobalt, lead, and silver) can be determined by permanganate solutions following the acid dissolution of the oxalate precipitates. The titration reaction is identical to the oxalate determination discussed earlier. The analysis of limestone for calcium (Experiment 18) is an example of this.

Iodine

Iodometric methods have also been studied. Dichromate, bromate, iodate, and chlorine have been previously mentioned as oxidizing agents that can be titrated by this indirect method. In addition, standard iodine can be used as the titrant (known as iodimetry rather than iodometry). Determination by this method include arsenic, antimony (especially in stibnite ores), tin, hydrogen sulfide, sulfur dioxide, thiosulfate, and N_2H_4.

Cerium and Dichromate

Cerium in the +4 state is as potent as permanganate in oxidizing power. The reaction is

$$Ce^{+4} + 1e^- \rightarrow Ce^{+3}$$

Typical uses include all of those previously mentioned for permanganate. The most significant disadvantage is that an indicator is needed. The most widely used indicator is an iron/*o*-phenanthroline (complex ion) solution (Chapter 7) known as "ferroin". The end point may also be detected potentiometrically (Chapter 19). Sources of cerium include cerium (IV) ammonium nitrate and cerium (IV) ammonium sulfate.

Potassium dichromate is also a fairly common titrant. This chemical and its solutions are bright orange, which generally is not intense enough to serve as the indicator. The determination of iron as Fe^{+2} is the major application.

Reducing Agents

Solutions of reducing agents are not nearly as useful for titrants as oxidizing agents. The principal reason for this is the fact that oxygen is present in all solutions exposed to air and oxygen is easily reduced by these compounds, thus altering their concentration. The must notable examples of useful reducing agents are solutions of iron II from Mohr's Salt (ferrous ammonium sulfate) and solutions of sodium thiosulfate (previously mentioned). The decomposition of Mohr's salt solutions by oxygen is inhibited by the addition of an acid, but such solutions should be re-standardized frequently.

Thiosulfate solutions are more stable if they are somewhat basic (pH = 9–10 is best), although neutral solutions are acceptable if used soon after they are standardized. Decomposition can also be caused by bacteria. The primary use for thiosulfate as a titrant is in iodometry, as discussed earlier.

Pre-Reduction and Pre-Oxidation

Perhaps the most important application of redox chemicals in the modern laboratory is in oxidation or reduction reactions that are required as part of a preparation scheme. Such application has been used for years in titrimetric analysis procedures in order to get the substance titrated in an oxidation state that will react with a redox titrant. A common procedure has been to pack a glass tube with an oxidizable metal amalgam (typically zinc or cadmium), to pour the solution to be titrated through this tube, and then to collect it in the titration flask, so that the analyte is reduced to the correct oxidation state. The analyst then immediately titrates this solution, thus minimizing error due to rapid redox reactions, such as reaction with dissolved oxygen from the air. Such glass tubes are usually extra burets, since their use is convenient. They are called, for example, the "Jones reductor" (zinc) or the "Walden reductor" (silver).

Pre-oxidation or pre-reduction is also frequently required for certain instrumental procedures for which a specific oxidation state is required in order to measure whatever property is measured by the instrument. Examples in this textbook can be found in Experiment 22 (the hydroxylamine hydrochloride keeps the iron in the +2 state) and 26 (the potassium periodate oxidizes Mn^{+2} to MnO_4^{-1}).

Also, in wastewater treatment plants it is important to measure dissolved oxygen ("DO"). In this procedure, $Mn(OH)_2$ reacts with the oxygen in basic solution to form $Mn(OH)_3$. When acidified and in the presence of KI, iodine is liberated and titrated. This method is called the "Winkler method".

8.9 NON-REDOX PRECIPITATION TITRATIONS

Precipitation reactions are used for some determinations. These involve principally reactions utilizing the highly insoluble nature of silver compounds. Two example reactions are

Volhard method for silver

$$Ag^+ + CNS^- \rightarrow AgCNS$$

Mohr method for chloride and also
the "chloride method" for silver

$$Ag^+ + Cl^- \rightarrow AgCl$$

End points in these are detected in any one of several ways. Most utilize a color change resulting from the first excess of the titrant reacting with another added component, the product being colored. With the precipitate also present in the solution, this color is often very difficult to see. In these cases, a blank correction is needed. (See Experiment 19.)

EXPERIMENTS

EXPERIMENT 14: PREPARATION AND STANDARDIZATION OF A $KMnO_4$ SOLUTION

NOTE: Safety glasses are required.

1. Place about 200 mL of water in a beaker on a hot plate and heat until warm to the touch. Weigh 1.6 g of $KMnO_4$ and dissolve in the warm water in the beaker. Stir with a stirring rod and then pour into a clean 500-mL amber bottle. Add water up to approximately 500 mL total. Shake. Store until the next laboratory period.
2. Do not shake the solution after it has been stored. The MnO_2 precipitate has settled to the bottom and should not be disturbed. This precaution will result in a shorter filtration time.
3. Set up a suction filtration apparatus to include a vacuum pump, a clean filter flask, a rubber ring, and a clean sintered glass filtering crucible (instructor may want to assist). Proceed to filter the solution with the use of suction, being careful not to allow the filtered permanganate to contact any oxidizable surface, such as the rubber ring. Pour the solution into the crucible using a stirring rod as a guide. Discard the last half inch of solution in the bottle.

4. Rinse the amber bottle thoroughly with water and then rinse (with the lid on) with a fairly concentrated HCl solution. (Use caution.) The purpose of this step is to dissolve any MnO_2 adhering to the walls of the bottle. Discard the acid, rinse thoroughly with distilled water, and pour the $KMnO_4$ from the filter flask back into the amber bottle. Shake. It is now ready for standardization.
5. Place a quantity of Mohr's salt (ferrous ammonium sulfate) into a dry weighing bottle. Your instructor may choose to dispense this to you. Weigh accurately, by difference, three samples each weighing between 0.8 and 1.2 g into 250-mL Erlenmeyer flasks. Rinse and fill a buret with your $KMnO_4$ solution as usual. Due to the dark color of the permanganate solution you may elect to read the top of the meniscus rather than the bottom.
6. Dissolve the lightest of the three samples in 100 mL (graduated cylinder) of water. Add 5 mL of concentrated H_2SO_4 (use extreme caution) and mix thoroughly by swirling. Titrate with the permanganate to the faint pink permanganate end point. Record the volume as usual.
7. Repeat Step 6 for the remaining samples and calculate the three normalities. The formula weight of Mohr's salt is 392.2. Calculate the relative parts per thousand (ppt) deviation from the average, as before, and perform additional titrations if the values are not within 2 ppt of the average or until you have a precision agreeable to your instructor.

EXPERIMENT 15: TITRIMETRIC DETERMINATION OF SODIUM OXALATE IN A COMMERCIAL UNKNOWN

NOTE: Safety glasses are required.

1. In advance of this lab, perhaps while performing Experiment 14, obtain an unknown oxalate sample from your instructor, record the number, place in a weighing bottle, and dry for 1 hr in a drying oven.
2. Weigh, by difference, samples weighing between 0.7 and 1.0 g into each of three 500-mL Erlenmeyer flasks. Add 200 mL of water and 6 mL of concentrated H_2SO_4 (caution) to each.
3. Place your lightest sample on a hot plate and bring to boiling. Titrate while hot at a rate not to exceed 10 mL/min with your standard $KMnO_4$.
4. Repeat Step 3 for your remaining samples and calculate the percent of $Na_2C_2O_4$ for each. Calculate the relative ppt deviations and repeat until you have three that agree to within 10 ppt of the average or until you have a precision agreeable to your instructor.

EXPERIMENT 16: THE PREPARATION AND STANDARDIZATION OF A SODIUM THIOSULFATE SOLUTION

NOTE: Safety glasses are required. The starch indicator for this experiment and for Experiment 17 is prepared by mixing 0.5 g of soluble starch and 3–5 mg of HgI_2 with 2–3 mL of water. This mixture is then diluted to about 50 mL with boiling water and heated for an additional 5 min. After cooling it is ready for use.

1. Place a small quantity of primary standard potassium dichromate ($K_2Cr_2O_7$) in a weighing bottle and dry in the oven for 1 hr. After drying, accurately weigh about 5 g into a 100-mL volumetric flask, add water to dissolve, and then dilute to the mark and shake. Calculate and record the normality, assuming the following half-reaction occurs:

$$Cr_2O_7^{-2} \rightarrow Cr^{+3}$$

2. Place 25 g of $Na_2S_2O_3 \cdot 5H_2O$ into a 1-L bottle, dissolve, and then fill with freshly distilled or boiled water.
3. With graduated cylinders, add 80 mL distilled water and 1 mL concentrated H_2SO_4 to a 250-mL Erlenmeyer flask. Pipet 10.00 mL of the $K_2Cr_2O_7$ solution to this flask and add 1 g of potassium iodide. Swirl, let stand for 6 min, and then titrate with the thiosulfate solution until the yellow color of the liberated iodine is almost discharged. Add 1 mL of starch indicator solution and continue titrating until the blue color disappears. This is the end point. Record the buret reading and calculate the normality of the thiosulfate.

$$S_2O_3^{-2} \rightarrow S_4O_6^{-2}$$

4. Repeat Step 3 until you have three normalities that agree to within 2 ppt relative deviation or until you have a precision agreeable to your instructor.

EXPERIMENT 17: THE DETERMINATION OF RESIDUAL CHLORINE IN WASTEWATER PLANT EFFLUENT BY IODOMETRY

NOTE: Safety glasses are required.

1. Prepare a standard 0.01 *N* $Na_2S_2O_3$ solution from the 0.10 *N* solution prepared in Experiment 16. Pipet accurately and use a volumetric flask for the new solution.

2. Obtain at least 500 mL of a wastewater plant effluent or water from a creek downstream from a wastewater plant as directed by your instructor.

3. Measure exactly 500 mL of the sample in a large graduated cylinder and then pour into a 1-L Erlenmeyer flask. Use a stirring bar and a magnetic stirrer for stirring the solution. Add 5 mL acetic acid and 1 g potassium iodide. Titrate with the sodium thiosulfate prepared in Step 1 until the yellow color of the iodine is almost discharged. Add 1 mL starch and continue titrating until the blue color disappears.

4. Calculate the chlorine as ppm Cl.

EXPERIMENT 18: TITRIMETRIC DETERMINATION OF CALCIUM IN A COMMERCIAL LIMESTONE UNKNOWN

NOTE: Limestone is mostly calcium carbonate, although some limestones can contain significant amounts of magnesium carbonate as well. Metal silicates may also be present and do not dissolve in the HCl solution used here. This procedure assumes that silicates are absent. Safety glasses are required.

1. Obtain an unknown and dry as usual for 2 hr.
2. Weigh accurately, by difference, three samples into 250-mL beakers, each sample weighing between 0.20 and 0.30 g. Place a clean watch glass on each beaker. Add 20 mL of a 50:50 mixture of distilled water and concentrated HCl to each beaker. Swirl to dissolve.
3. Add five drops of saturated bromine water to oxidize any iron to Fe^{+3}, swirl, and then boil for 5 min to remove the excess Br_2.
4. If not already prepared for you by your instructor, prepare a 5% ammonium oxalate solution. This solution should be filtered to avoid introducing any insoluble materials on the beakers. You will need 100 mL for each beaker. Add approximately 30 mL of distilled water to each beaker.
5. Bring all four solutions to near boiling on a hot plate and add 100 mL (graduated cylinder) of the oxalate solution to each. Add three drops of methyl red indicator, and add to each beaker (dropwise, with stirring) a 50:50 solution of distilled water and concentrated ammonia until the solution turns an orange-yellow color. Allow the solution to stand for 30 min. The precipitate is CaC_2O_4.
6. Set up a filtering apparatus similar to that in Experiment 14, using a medium-porosity filtering crucible, and filter each solution. Rinse the beaker with several 10-mL portions of distilled water and add the rinsings to the crucible. The transfer of the precipitate need not be quantitative, however. Return each crucible to the original beakers, and add 150 mL of distilled water and 50 mL of 3 *M* H_2SO_4.
7. Heat each solution to near boiling and titrate while hot with the standardized $KMnO_4$ solution from Experiment 14. Calculate and report the percent of CaO for each.

EXPERIMENT 19: THE DETERMINATION OF CHLORIDE IN A COMMERCIAL UNKNOWN BY THE MOHR METHOD

NOTE: Safety glasses are required.

1. Clean two weighing bottles and obtain an unknown sample of chloride (record unknown number) and 6.0–6.5 g of primary standard silver nitrate from your instructor. Dry both for 1 hr. (Step 1 can be done in advance of the lab. Do not leave the silver nitrate in the oven more than 1 hr.)
2. Weigh the weighing bottle containing the silver nitrate and transfer all of the contents into a 250-mL volumetric flask. Reweigh the weighing bottle, and subtract the two weights. You can use a funnel to avoid

spillage, but remember to rinse the funnel into the flask with distilled water. Dissolve the $AgNO_3$ and dilute to the mark with water. Shake well. Calculate the concentration from the weight and write it in your notebook.

3. Weigh by difference 0.25–0.35 g samples in each of three 250-mL Erlenmeyer flasks. Dissolve each in about 75 mL of distilled water. Add 1 mL (20 drops) of 5% K_2CrO_4 indicator solution. These samples are now ready for titration.
4. In order to determine a good end point, prepare a "blank" solution by placing 1.0 g of $CaCO_3$ (weighed on the triple-beam balance or other less accurate balance) in a 250-mL Erlenmeyer flask. Also add 75 mL of water and the indicator to this flask.
5. Place the silver nitrate solution in your buret (rinse!) and titrate the blank first. The purpose of the blank is to give you an idea what the end point looks like and also to provide a blank "correction" for your samples. Add the silver nitrate dropwise to your blank until a permanent faint red-orange tinge is produced, just noticeably different from the yellow color of chromate ion. Only a few drops should be required for the blank. Do not discard this blank solution. You will use it to match the color to your three samples as a means of detecting the end point. Record the volume used for the blank.
6. Now titrate your three samples to this same end point. Subtract your blank volume readings from that of each sample and use these volumes to calculate the percent of Cl in each sample. Use the silver concentration calculated in Step 2 for the normality of the titrant. Use the usual guidelines (see Experiment 15, Step 4) for deciding whether to titrate additional samples.

QUESTIONS AND PROBLEMS

1. Define oxidation, reduction, oxidation number, oxidizing agent, reducing agent.
2. What is the oxidation number of each of the following?

   (a) P in H_3PO_4 (b) Cl in $NaClO_2$
   (c) Cr in CrO_4^{-2} (d) Br in $KBrO_3$
   (e) I in IO_4^{-1} (f) N in N_2O
   (g) S in H_2SO_3 (h) S in H_2SO_4
   (i) N in NO_2^{-1} (j) P in PO_3^{-3}

3. What is the oxidation number of bromine (Br) in each of the following?

   (a) HBrO (b) NaBr (c) BrO_3^{-1} (d) Br_2 (e) $Mg(BrO_2)_2$
   (f) BrO_4^{-1}

4. What is the oxidation number of chromium (Cr) in each of the following?

   (a) $CrBr_3$ (b) Cr (c) CrO_3 (d) CrO_4^{-2} (e) $K_2Cr_2O_7$

5. What is the oxidation number of iodine (I) in each of the following?

   (a) HIO_4 (b) CaI_2 (c) I_2 (d) IO_2^- (e) $Mg(IO)_2$

6. What is the oxidation number of sulfur (S) in each of the following?

   (a) SO_2 (b) H_2S (c) S (d) SO_4^{-2} (e) K_2SO_3

   (f) K_2S (g) H_2SO_4 (h) SO_3^{-2} (i) SO_3 (j) SF_6

7. What is the oxidation number of phosphorus (P) in each of the following?

   (a) P_2O_5 (b) Na_3PO_3 (c) H_3PO_4 (d) PCl_3 (e) HPO_3^{-2}

   (f) PO_4^{-3} (g) P^{-3} (h) $Mg_2P_2O_7$ (i) P (j) NaH_2PO_4

8. In the following redox reactions tell what has been oxidized and what has been reduced and explain your answers.

   (a) $3CuO + 2NH_3 \rightarrow 3Cu + N_2 + 3H_2O$

   (b) $Cl_2 + 2KBr \rightarrow Br_2 + 2KCl$

9. In each of the following reactions tell what is the oxidizing agent and what is the reducing agent and explain your answers.

   (a) $Mg + 2HBr \rightarrow MgBr_2 + H_2$

   (b) $4Fe + 3O_2 \rightarrow 2Fe_2O_3$

10. Which of the following are redox reactions? Explain your answers.

    (a) $H_2SO_4 + NaOH \rightarrow Na_2SO_4 + H_2O$

    (b) $H_2S + HNO_3 \rightarrow S + NO + H_2O$

    (c) $Na + H_2O \rightarrow NaOH + H_2$

    (d) $H_2SO_4 + Ba(OH)_2 \rightarrow BaSO_4 + H_2O$

    (e) $K_2CrO_4 + Pb(NO_3)_2 \rightarrow KNO_3 + PbCrO_4$

    (f) $K + Br_2 \rightarrow KBr$

    (g) $KClO_3 \rightarrow KCl + O_2$

    (h) $KOH + HCl \rightarrow KCl + H_2O$

    (i) $BaCl_2 + Na_3PO_4 \rightarrow Ba_3(PO_4)_2 + NaCl$

    (j) $Mg + HCl \rightarrow MgCl_2 + H_2$

11. Consider the following two reactions:

    (a) $KOH + HCl \rightarrow KCl + H_2O$

    (b) $Cu + HNO_3 \rightarrow Cu(NO_3)_2 + NO + H_2O$

    Which is redox, (a) or (b)? In the redox reaction, what has been oxidized and what has been reduced? In the redox reaction, what is the oxidizing agent and what is the reducing agent?

12. One of the following is a redox reaction and one is not. Select the one that is a redox reaction and answer the questions that follow.

    (1) $Pb(NO_3)_2 + K_2CrO_4 \rightarrow PbCrO_4 + 2KNO_3$

    (2) $Zn + HCl \rightarrow ZnCl_2 + H_2$

    (a) Which one is redox, (1) or (2)?
    (b) What is the oxidizing agent?
    (c) What has been oxidized?
    (d) Did the reducing agent lose or gain electrons?

13. Balance the following equations by the ion-electron method:
    - (a) $Cl^- + NO_3^- \rightarrow ClO_3^- + N^{-3}$
    - (b) $Cl^- + NO_3^- \rightarrow ClO_2^- + N_2O$
    - (c) $ClO^- + NO_3^- \rightarrow ClO_3^- + NO_2^-$
    - (d) $ClO^- + NO_3^- \rightarrow ClO_2^- + NO$
    - (e) $ClO_3^- + SO_4^{-2} \rightarrow ClO_4^- + S^{-2}$
    - (f) $BrO_3^- + SO_4^{-2} \rightarrow BrO_4^- + SO_3^{-2}$
    - (g) $IO_3^- + SO_3 \rightarrow IO_4^- + S^{-2}$
    - (h) $Cl^- + SO_4^{-2} \rightarrow ClO^- + SO_2$
    - (i) $Cl^- + SO_4^{-2} \rightarrow ClO_4^- + S$
    - (j) $Br^- + SO_3 \rightarrow Br_2 + S$
    - (k) $I^- + NO_3^- \rightarrow IO_2^- + N_2$
    - (l) $P + IO_4^- \rightarrow PO_4^{-3} + I^-$
    - (m) $SO_2 + BrO_3^- \rightarrow SO_4^{-2} + Br^-$
    - (n) $SO_3 + BrO_3^- \rightarrow SO_4^{-2} + Br_2$
    - (o) $P + IO_4^- \rightarrow PO_3^{-3} + I_2$
    - (p) $Fe + P_2O_5 \rightarrow Fe^{+3} + P$
    - (q) $Cr + PO_4^{-3} \rightarrow Cr^{+3} + PO_3^{-3}$
    - (r) $Ni + PO_4^{-3} \rightarrow Ni^{+2} + P$
    - (s) $MnO_4^{-1} + H_2C_2O_4 \rightarrow Mn^{+2} + CO_2$
    - (t) $I^{-1} + Cr_2O_7^{-2} \rightarrow I_2 + Cr^{+3}$
    - (u) $Cl_2 + NO_2^{-1} \rightarrow Cl^{-1} + NO_3^{-1}$
    - (v) $S^{-2} + NO_3^{-1} \rightarrow S + NO$
    - (w) $SO_3^{-2} + NO_3^{-1} \rightarrow SO_4^{-2} + NO_2^{-1}$

14. (a) Indicate how to calculate the equivalent weight of $K_2Cr_2O_7$ if the product of its reduction is Cr^{+3}.
    (b) Indicate how to calculate the equivalent weight of Cl_2 as used in #13u above.

15. What is the equivalent weight of
    (a) Na_2SO_3 as used in $SO_3^{-2} \rightarrow S$?
    (b) KI as used in $I^- \rightarrow I_2$?
    (c) Cu and HNO_3 as in #11b above?

16. What is the equivalent weight of $KClO_4$ if it is to be used as in the following equation?

$$ClO_4^{-1} + Mn^{+2} \rightarrow Cl^{-1} + MnO_4^{-1}$$

17. How many grams of P_2O_5 are required to prepare 300.0 mL of a 0.450 *N* solution if it is to be used as in Equation p of Question 13?

18. How many grams of phosphorus are required to prepare 500.0 mL of a 0.200 *N* solution if it is to be used as in Equation o of Question 13?

19. How many grams of NaClO are needed to prepare 500 mL of a 0.20 *N* solution if it is to be used as in the following equation?

$$ClO^{-1} + I^{-1} \rightarrow Cl^{-1} + I_2$$

20. How many grams of KNO_3 are needed to prepare 500 mL of a 0.250 *N* solution of KNO_3 if it is to be used as in the following half-reaction?

$$NO_3^{-1} \rightarrow NO_2^{-1}$$

21. How would you prepare 500 mL of 0.10 *N* $KMnO_4$ for use as in Example 7 in the text?
22. How would you prepare 500 mL of 0.10 *N* KNO_3 for use in the following? (Assume KNO_3 is a pure solid chemical.)

$$S^{-2} + NO_3^{-1} \rightarrow S + NO$$

23. Consider the standardization of a solution of Fe^{+2} with $K_2Cr_2O_7$ according to the following:

$$Fe^{+2} + Cr_2O_7^{-2} \rightarrow Fe^{+3} + Cr^{+3}$$

(a) How many grams of $K_2Cr_2O_7$ are required to prepare 500 mL of a 0.15 *N* solution?
(b) What is the exact normality of the solution prepared in part (a) if 1.8976 g of Mohr's salt were exactly reacted with 24.22 mL of the solution?

24. (a) Balance the following equation

$$S_2O_3^{-2} + Cr_2O_7^{-2} \rightarrow S_4O_6^{-2} + Cr^{+3}$$

(b) How would you prepare 500 mL of 0.100 *N* $Na_2S_2O_3$ from pure solid $Na_2S_2O_3 \cdot 5H_2O$ for use in this reaction?
(c) If 0.5334 g of $K_2Cr_2O_7$ were titrated with 24.31 mL of the $Na_2S_2O_3$ solution what is the normality of the $Na_2S_2O_3$?

25. Consider the reaction of Fe^{+2} with $K_2Cr_2O_7$ according to the following:

$$Fe^{+2} + Cr_2O_7^{-2} \rightarrow Fe^{+3} + Cr^{+3}$$

(a) How many grams of $K_2Cr_2O_7$ are required to prepare 500 mL of a 0.25 *N* solution?
(b) What is the exact normality of the solution prepared in part (a) if 1.7976 g of Mohr's salt [an Fe^{+2} compound, $Fe(NH_4)_2(SO_4)_2 \cdot 6H_2O$] were exactly reacted with 22.22 mL of the solution?

26. Consider the preparation and standardization of a solution of KIO_4 with Mohr's salt according to the following:

$$Fe^{+2} + IO_4^{-1} \rightarrow Fe^{+3} + I^{-1}$$

(a) How many grams of KIO_4 are required to prepare 500 mL of a 0.15 *N* solution?

(b) What is the exact normality of the solution prepared in part (a) if 1.8976 g of Mohr's salt were exactly reacted with 24.22 mL of the solution?

27. Consider the standardization of a solution of $K_2Cr_2O_7$ with iron metal according to the following:

$$Fe + Cr_2O_7^{-2} \rightarrow Fe^{+3} + Cr^{+3}$$

(a) How many grams of $K_2Cr_2O_7$ are required to prepare 500 mL of a 0.13 *N* solution?
(b) What is the exact normality of the solution prepared in part (a) if 0.1276 g of iron metal were exactly reacted with 48.56 mL of the solution?

28. What is the normality of a solution of $KBrO_3$ if 34.23 mL of it exactly reacted with 0.6012 g of SO_2 as in Equation m of Question 13?

29. What is the normality of a solution of KIO_4 if 28.90 mL of it exactly reacted with 0.5293 g of phosphorus as in Equation l of Question 13?

30. What is the normality of a $KBrO_3$ solution if 23.59 mL of it reacted with 0.3291 g of Na_2SO_3 as in the following reaction?

$$BrO_3^{-1} + SO_3^{-2} \rightarrow Br^{-1} + SO_4^{-2}$$

31. What is the normality of a $KMnO_4$ solution that has been standardized against primary standard $Na_2C_2O_4$ (see Question 13s), 0.9324 g requiring 32.06 mL?

32. What is the normality of a solution of KIO_3 if 0.6729 g of $K_2Cr_2O_7$ are exactly consumed by 24.92 mL of it as in the following equation?

$$IO_3^{-1} + Cr_2O_7^{-2} \rightarrow IO_4^{-1} + Cr^{+3}$$

33. What is the percent SO_3 in a sample if 45.69 mL of a 0.2011 *N* solution of KIO_3 are needed to consume 0.9308 g of sample according to Equation g of Question 13?

34. What is the percent of K_2SO_4 in a sample if 35.01 mL of 0.09123 *N* $KBrO_3$ solution are needed to consume 0.7910 g of sample according to Equation f of Question 13?

35. What is the percent of Fe in a sample titrated with $K_2Cr_2O_7$ according to the following equation if 0.6426 g of the sample required 40.12 mL of 0.1096 *N* $K_2Cr_2O_7$?

$$Fe^{+2} + Cr_2O_7^{-2} \rightarrow Fe^{+3} + Cr^{+3}$$

36. What is the percent of Sn in a sample of ore containing Sn that was dissolved and titrated with $KMnO_4$ if 4.2099 g of the ore were dissolved and titrated with 36.12 mL of 0.1653 *N* $KMnO_4$?

$$MnO_4^{-} + Sn^{+2} \rightarrow Sn^{+4} + Mn^{+2}$$

37. What is the percent of KIO in a sample that was titrated with $KMnO_4$ if 4.0099 g of the sample were dissolved and titrated with 32.12 mL of 0.1603 *N* $KMnO_4$?

$$MnO_4^- + IO^- \rightarrow IO_3^- + Mn^{+2}$$

38. What is the percent of NaBr in a sample if 0.9402 g of it needed 34.29 mL of a 0.2103 *N* solution of $KMnO_4$ for titration according to the following equation?

$$Br^{-1} + MnO_4^{-1} \rightarrow BrO_4^{-1} + Mn^{+2}$$

39. What does it mean to say that potassium permanganate is its own indicator?

40. (a) Why is potassium permanganate, $KMnO_4$, termed an oxidizing agent?
    (b) Why must standardized potassium permanganate solutions be carefully protected from oxidizable substances if you expect them to remain standardized?

41. Explain what the difference is between an indirect titration and a back titration.

42. One redox method we have discussed is called iodometry. What is iodometry and why is it called an "indirect" method?

43. Briefly explain the use of the following substances in iodometry:
    (a) KI
    (b) $Na_2S_2O_3$
    (c) $K_2Cr_2O_7$

44. What is the difference between iodimetry and iodometry? Give an example of an iodimetry experiment.

45. What is "ferroin" and what is used for?

46. What is a "Jones reductor"?

47. Explain the use of redox chemistry in Experiment 22 in Chapter 11.

Chapter 9

Analysis with Instruments and Computers

9.1 INTRODUCTION

The remainder of this text is devoted to techniques and methods of chemical analysis involving electronic instrumentation. In Chapter 1 we made a distinction between wet methods and instrumental methods, saying that the wet methods known as gravimetric and titrimetric analysis are more classical as compared to instrumental techniques. They are classical techniques in the sense that they have adequately served the chemical analysis laboratory for many years but have now largely been replaced by the streamlined and computerized instrumental methods, especially where there is a need to detect trace amounts of analytes and to analyze more complicated samples. There are certain aspects of instrumental analysis that are common to virtually all instrumental techniques. These aspects will be dealt with in detail in this chapter. Characteristics of the individual instruments and associated procedures will be covered in the chapters to follow.

9.2 INSTRUMENTAL DATA AND READOUT

Recorders

Most laboratory instruments measure some physical or chemical parameter of a solution of a sample and display the measurement as a proportional voltage on some type of readout device. This device can be a simple meter (like the display on a pH meter), digital or otherwise; a recorder, which may either be a strip-chart recorder or x-y recorder; a computer monitoring screen and/or a printer-plotter; or any combination of these. Let us first discuss the details of the two types of recorders and how they are used.

A recorder has one or more writing pens driven by a servomotor and records information on paper. The position of the pen on the paper represents the voltage or voltages fed into the recorder from the instrument, just as the position of the pointer on a meter represents such a voltage. The advantage of a recorder is that the voltage can be permanently recorded over a period of time, a capability which is of significant importance when the voltage output of an instrument changes with time, such as with gas chromatography (GC; Chapter 16) or high-performance liquid chromatography (HPLC; Chapter 17), or when *two* parameters change with time, one as a function of the other, such as when recording a molecular absorption spectrum with a spectrophotometer (Chapter 11). This characteristic of voltage signals changing with time is quite common, and thus recorders are very commonly used in instrumental laboratories as readout devices.

The differences between the strip-chart recorder and the x-y recorder are noteworthy. The strip-chart recorder records the voltage output of an instrument strictly as a function of time. In other words, the pen records the voltage on the paper as the paper moves at a certain rate through the recorder. Thus, one axis of the resulting graph is time and the other is the voltage level. Multiple pens in a strip-chart recorder can be used to record multiple voltage levels and how they change with time. The x-y recorder is a recording device which accepts two voltages from the instrument for recording with only one pen. One of the voltage levels determines the x-axis position of the pen while the other determines the y-axis position. The paper doesn't move, but rather the pen moves over the paper as the two voltage levels change. Figure 9.1 shows representations of (a) a strip-chart recorder and (b) an x-y recorder, while Figure 9.2 shows some sample recordings.

Instrument Readout and Concentration

The goal of instrumental methods of analysis most of the time is determination of the concentration of a constituent in an unknown sample. Thus, the concentration of the desired constituent in the solution measured is to be determined from the readout resulting from that solution. As we have said, the readout (let's call it "R"), whatever it represents, is proportional to this concentration:

$$R = KC \tag{9.1}$$

If the proportionality constant, K, is known, the concentration can be calculated:

$$C = R/K \tag{9.2}$$

Often, however, the proportionality constant is not known, and the experiment is not as simple as implied above. One alternate method would be to prepare a solution of the constituent so that the concentration is known (a standard solution), measure R, and then calculate K:

$$K = R_s/C_s \tag{9.3}$$

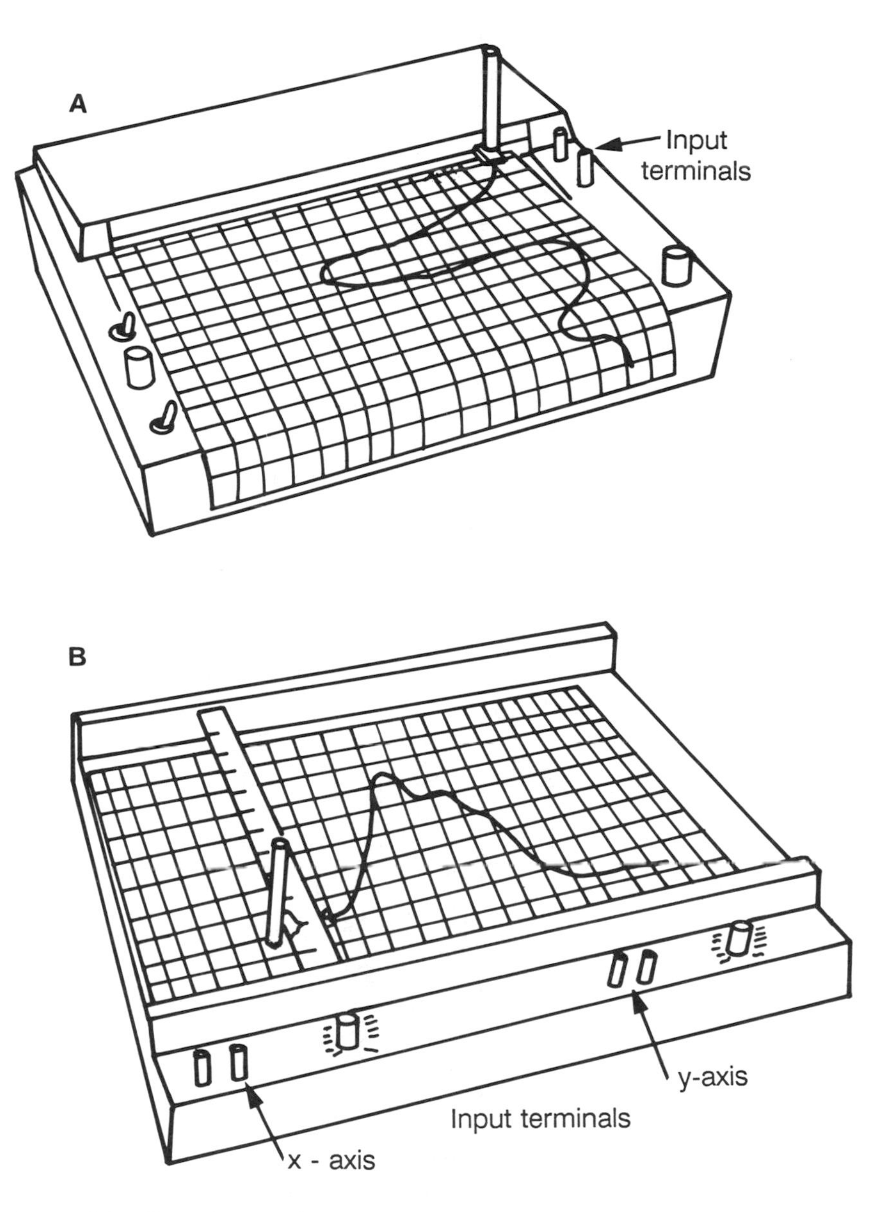

Figure 9.1. Drawings of (a) a strip-chart recorder (note only one set of input terminals for only one incoming signal) and (b) an x-y recorder (note two sets of input terminals for two incoming signals, one for the x-axis and one for the y-axis).

In this equation, R_s is the readout for the standard and C_s is the concentration of the standard. The value of K is then used to calculate C for the unknown, assuming, of course, that the parameters which contribute to the value of K for both solutions are identical at both concentrations.

$$C_u = R_u/K \tag{9.4}$$

Here, R_u is the readout for the unknown and C_u is the concentration of the unknown. Actually, the value of K need not be calculated at all, as is obvious from the following:

$$\frac{R_s = KC_s}{R_u = KC_u} \tag{9.5}$$

or

$$\frac{R_s}{R_u} = \frac{C_s}{C_u} \tag{9.6}$$

and

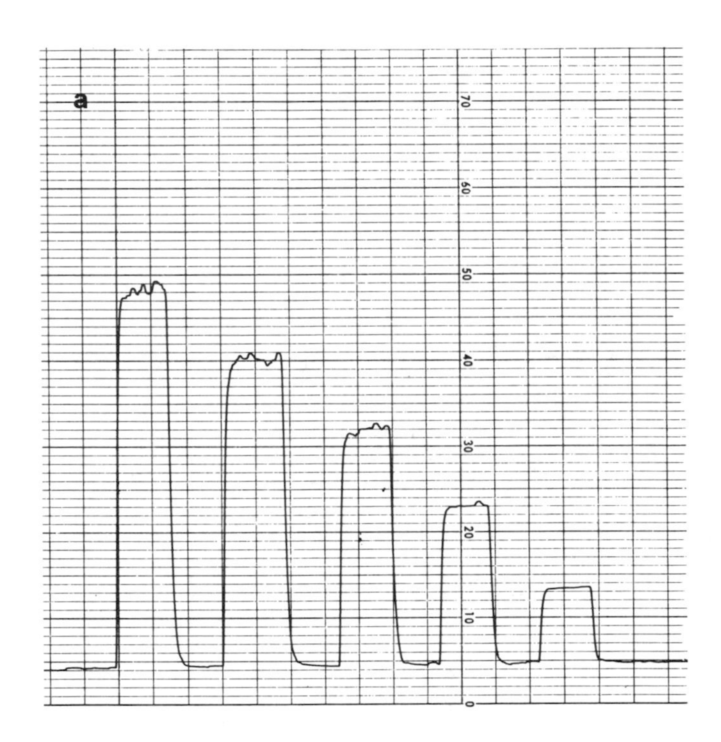

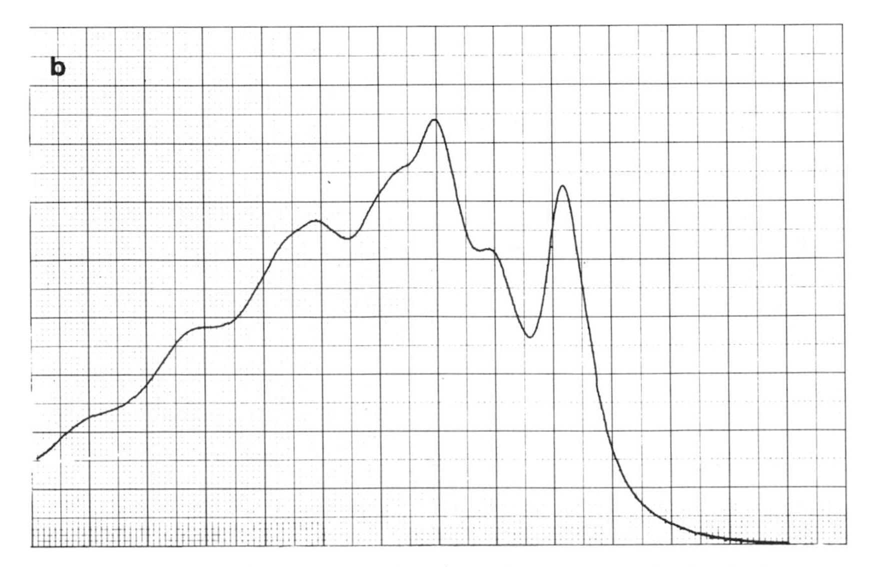

Figure 9.2. (a) The absorbances of a series of standard solutions recorded on a strip-chart recorder; (b) a molecular absorption spectrum recorded with an x-y recorder.

$$C_u = \frac{C_s R_u}{R_s} \tag{9.7}$$

In other words, the value of C_u is related to that of C_s by the ratio of their instrument reading, or R_s is to R_u as C_s is to C_u.

This treatment requires that the parameters that contribute to the value of K not change between the two concentration levels, as indicated above. If the concentrations are very nearly the same, or if the relationship between the readout and concentration is of the same proportion at the different concentration levels, then this treatment is entirely valid. However, this may not be the case, and it is therefore useful to determine the linearity of R as a function of C over the concentration range in question. Such a linearity would verify the constancy of K and provide confidence in the calculated result. The procedure involves preparing a *series* of standard solutions and making a graph of R vs. C to confirm this linearity. (See Figure 9.3a.) This graph is sometimes called the "standard curve". If the results show linearity over the concentration range in question (the range in which the unknown's concentration falls), then the unknown concentration may be determined from the graph as shown in Figure 9.3b. Because of the uncertainty in knowing whether the proportionality constant is indeed constant over the concentration range studied, this series of standard solution method is commonplace in an instrumental analysis laboratory for virtually all quantitative instrumental procedures. Examples of this abound in this text for many spectrophotometric, chromatographic, and other techniques.

Mathematics of Linear Relationships

The mathematics of a straight-line relationship are described by the following equation:

$$y = mx + b \tag{9.8}$$

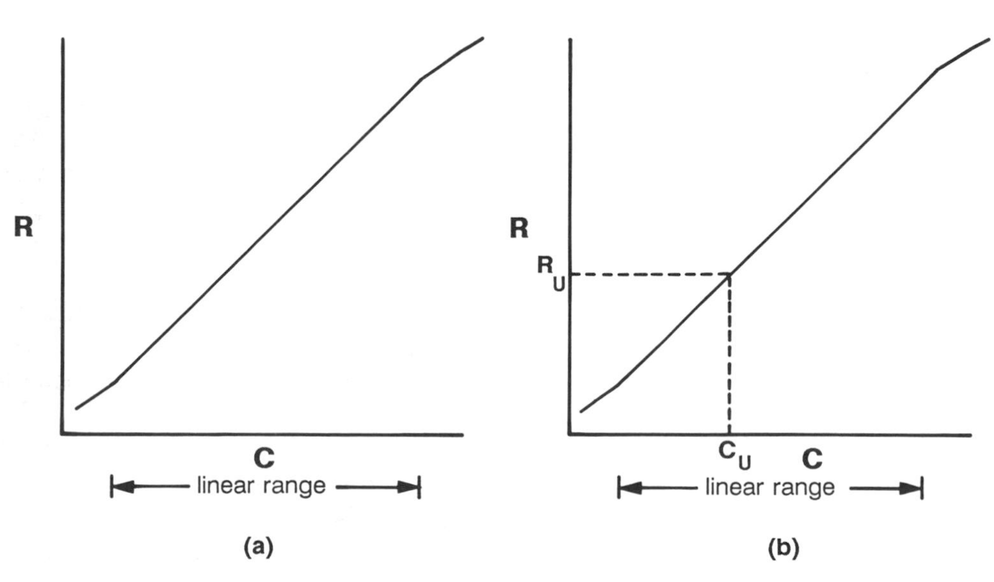

Figure 9.3. (a) A plot of an instrument readout (R) vs. the concentration (C) of an analyte, showing a linear relationship in the concentration range chosen; (b) the determination of an unknown's concentration from the graph in (a).

Thus, if the variables x and y bear a linear relationship to one another, then "m" and "b" are constants defining the "slope" (m) of the line (how "steep" it is) and the "y-intercept" (b), the value of y when x is zero. The slope of a line can be determined very simply if two or more points, (x_1,y_1) and (x_2,y_2), that make up the line are known.

$$m = \frac{\Delta y}{\Delta x} = \frac{(y_2 - y_1)}{(x_2 - x_1)} \tag{9.9}$$

The y-intercept can be determined by observing what the y value is on the graph where the line crosses the y-axis or, if the slope and one point are known, may be calculated using Equation 9.8.

The important observation for instrumental analysis is that the "K" in Equation 9.1 is the "m" in Equation 9.8. Thus, in a graph of R vs. C, the slope of the line is the "K" value, the proportionality constant.

It would seem that the y-intercept would always be zero, since if the concentration is zero the instrument readout would logically be zero, especially since a "blank" is often used. The blank, a solution prepared so as to have all sample components in it except the analyte (see Section 9.4), is a solution of zero concentration. Such a solution is often used to "zero" the readout, meaning that the instrument is manually set to read zero when the blank is being read by the instrument. This represents a "calibration" for the instrument. However, "real data" sometimes do not fit a straight line graph all the way to the y-intercept, and thus the y-intercept is *not* always zero and the (0,0) point is usually not included in the graph.

Another problem with "real" data is that due to random indeterminate errors (Chapter 3) the analyst cannot expect the measured points to fit a straight line graph exactly. Thus, it is often true that the *best* straight line that can be drawn through a set of data points is what is drawn and the unknown is determined from *this* line. A "linear regression", or "least squares", procedure is often performed to maximize accuracy.

Method of Least Squares

The linearity of instrument readout vs. concentration data must be established for best results. As stated in the previous subsection, random indeterminate errors during the solution preparation and during the measurement of R will likely cause the results to appear to deviate from linearity to some extent. In that case a method must be adopted which will fit a straight line to the data as well as possible. It may happen that some (or even all) of the points may not fall exactly on the line because of these random errors, but a straight line must still be drawn since the random errors are indeed random and cannot be compensated for directly. Thus, the best straight line possible is drawn through the points. (See Figure 9.4.)

"Eyeballing" the line through the points with a straight edge on the graph paper may easily result in significant error. Therefore, a procedure called the method of least squares (also called linear regression analysis) is best applied to the data; this method results mathematically in the best straight line possible through a given set of data points. By this method, the best straight line fit is obtained when the sum of the squares of the individual y-axis value deviations (deviations between the plotted y values and the val-

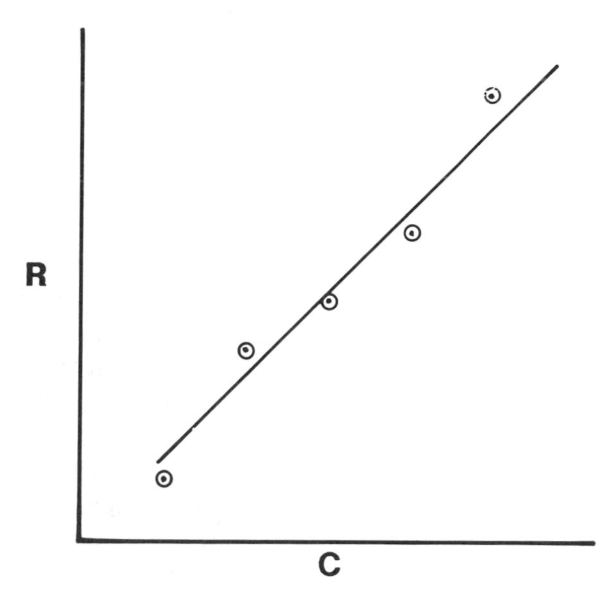

Figure 9.4. An example of a straight line fitted to a set of data.

ues on the proposed line) are at a minimum. This "proposed line" is actually calculated from the given data, a slope and y-intercept ("m" and "b", respectively, in Equation 9.8) are then obtained, and the deviations ($y_{point} - y_{line}$) for each given x are calculated. Finding the values of the slope and the y-intercept that minimize the sum of the squares of the deviations involves some complicated mathematics that is beyond the scope of this text. Computers and programmable calculators, however, handle this routinely in the modern laboratory, and the results are very important to a given analysis since the line that is determined is statistically the most correct line that can be drawn with the data obtained. The concentrations of unknown samples are also readily obtainable on the calculator or computer since the equation of the straight line, including the slope and y-intercept, is known as a result of the least squares procedure. Other statistically important parameters are readily obtainable as well, including the "correlation coefficient".

The Correlation Coefficient

The correlation coefficient is one measure of how well the straight line fits the analyst's data—how well a change in one variable correlates with a change in another. Laboratories can establish their own criteria as to what numerical value for such a coefficient is required for the accuracy desired. A correlation coefficient of exactly "1" indicates perfectly linear data. This, however, rarely occurs in practice. It occurs if all instrument readings increase by exactly the same factor as the concentration level increases, as in the following sample data:

R	C
4	2
8	4
12	6
24	12

Due to random errors, these data are more ideal rather than real. Data that approach such linearity will show a correlation coefficient less than 1 but very nearly equal to 1. Numbers such as 0.9997 or 0.9996 are considered excellent and attainable correlation coefficients for many instrumental techniques. Good pipetting and weighing technique when preparing standards and well-maintained and calibrated instruments can minimize random errors and can produce excellent correlation coefficients and therefore accurate results. The analyst usually strives for *at least* two nines, or possibly three, in their correlation coefficients, depending on the particular instrumental method used. Again, these coefficients can be determined on programmable calculators and laboratory computers.

9.3 METHODS FOR QUANTITATIVE ANALYSIS

As we determined in the last section, the usual procedure for instrumental analysis is to plot the instrument readout for a series of standard solutions vs. the concentrations of those solutions, establish the linear range, and then, after determining an unknown readout, find its concentration by interpolation within this linear range. We now describe further details and several options that are useful when specific instrument and experiment designs dictate.

Series of Standard Solutions (or Serial Dilution) Method

When the instrument readout is totally free of matrix effects, sample loss effects, or any other such potential error-causing effects, the method described thus far works well. The usual procedure is to accurately prepare a stock standard solution of the analyte and then to prepare the series of standards by diluting this stock. Various solution preparation schemes may be employed for both the stock and the series of standards, including preparing solutions by weighing a pure solid chemical and also by dilution. A possibility is to prepare the stock, dilute it to make the first standard, dilute the first to make the second, the second to make the third, etc. This is called "serial dilution". In either case, the solution to be diluted is pipetted into a volumetric flask and diluted to the mark with either the pure solvent, often distilled water, or a solution determined to be an approximation of the sample matrix, if that is important, or having other additives, such as for pH adjustment. It should be mentioned that stock solutions of many analytes, especially metals ("atomic absorption reference standards"), are available commercially, and this eliminates the need to devote lab time to such preparation. This not only saves time but also minimizes the possibility of error in this part of the procedure.

Internal Standard Method

In some instrumental procedures, the instrument readout can vary from standard to standard not only because of the concentration differences, but also because of uncontrollable experiment parameters. Such parameters may involve, for example, the problem of irreproducibility of sample injection volume in GC (Chapter 16) or nonlinear changes in solution viscosity as analyte concentration increases in the flame atomic absorption technique (Chapter 14). The solution to these problems is the internal standard method.

In this method, all standards and the sample are spiked with a constant known amount of a substance to act as what is called an internal standard. The purpose of the internal standard is to serve as a reference point for instrument readout values for the analyte such that in a ratio of the readout for the analyte to the readout for the internal standard the effect cancels out. Thus, if one were to plot this ratio vs. the analyte concentration, one expects a linear relationship and the problem goes away. Slight variations in GC injection volume, for example, are compensated for by the fact that the internal standard peak size and the analyte peak size (refer to Chapter 16) are both affected proportionally by such variations. A plot of the peak size ratio vs. analyte concentration would be free of peak size variations due to irreproducible sample volume and thus would give accurate results.

Solution preparation here is identical with the series of standard solutions method, except that the constant known amount of internal standard is added (usually pipetted) to all standards and also to the unknown.

Method of Standard Additions

It is possible for the analyte readout in an instrumental procedure to be either suppressed or enhanced by some components of the sample. There is therefore a potential for error when these components are present in the sample but not in the standards. One solution to this problem would be to add the interfering substances to the standards as well so that the effect would be measured in both the standards and the sample. This procedure is called "matrix matching". The requirements of matrix matching are (1) that the interfering substances be identifiable and (2) that the concentrations of the interfering chemicals in the sample be known so that they can be matched in the preparation of the standards. The absence of either one or both of these would make matrix matching impossible. The answer is the method of standard additions.

In this method, small amounts of a standard solution of (or pure) analyte are added to the sample and the readout measured after each addition. In this way, the interfering components need not be identified and the sample matrix is always present at the same component concentrations as in the sample, with only perhaps a minor dilution.

The plotting procedure and the determination of the unknown concentration are altered somewhat, however. The plot is a graph of instrument readout vs. concentration *added*. The first point to be plotted would be for zero added (the sample readout), and the readout would increase (presumably linearly) for each addition of analyte. Extrapolation of the resulting line to zero readout (the x-intercept), as shown in Figure 9.5, results in a length of x-axis on the negative side of zero added, which represents the concentration in the unknown as shown. In this method, one must presume that the

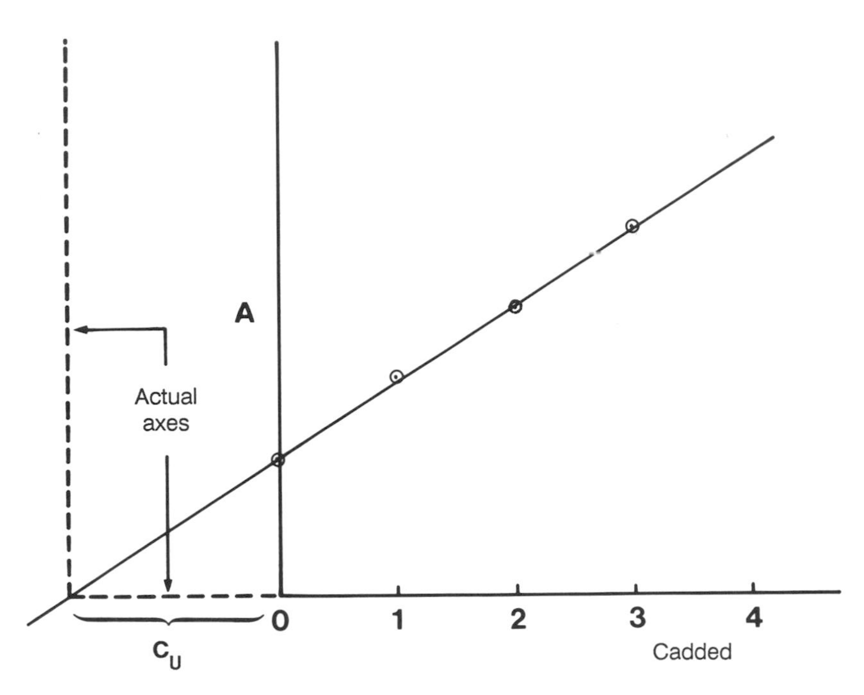

Figure 9.5. An example of a graph for a standard additions absorbance experiment. (From Kenkel, J., *Analytical Chemistry Refresher Manual*, Lewis Publishers, Chelsea, MI, 1992. With permission.)

plot is linear between the real zero and zero added since the standards will not encompass that concentration region.

This method can be used in cases in which there is some sample preparation as well – for example, in cases in which lanthanum must be added in an atomic absorbance (AA) analysis for calcium (Chapter 14). Once the pretreatment establishes the sample matrix, the standard additions can be performed and data graphed.

Since some sample may be consumed, such as in sample aspiration in AA or sample injection in GC, prior to making the second and subsequent additions of the analyte, the standard additions method could result in an error due to concentration changes that result. One way to partially compensate is to prepare a series of standard solutions using the sample matrix as the diluent. With either method, volumes of highly concentrated (or pure) analyte can be quite small (on the order of microliters) so that the dilution effect is negligible. A correction factor for the dilution can also be calculated. Figure 9.6 shows the data from a GC experiment which involved standard additions.

9.4 BLANKS AND CONTROLS

Reagent Blanks

In addition to the series of standard solutions needed for an instrumental analysis, there are often other solutions needed for the procedure. We have already briefly mentioned the need for and use of a "blank" (Section 9.2). As stated previously, the blank is a solution that contains all the ingredients present in the standards and the unknown (if possible) except for the ana-

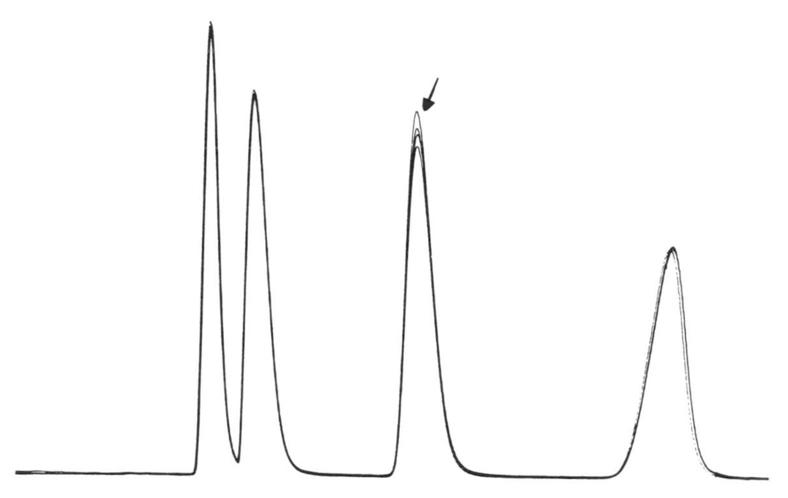

Figure 9.6. A series of chromatograms superimposed to illustrate standard additions in gas chromatography. The peak size of the analyte grows (arrow) while the others remain unchanged.

lyte. The readout for such a solution should be zero and, as we've indicated, the readout is often manually adjusted to read zero when this blank is being measured. Thus, the blank is useful as a sort of calibration step for the instrument.

Such a blank is appropriately called a "reagent blank". While an analyst may be tempted to use distilled water or some other pure sample solvent as a blank, this may not be desirable because other chemicals (reagents) may have been added to enhance the readout for the analyte in the standards and unknowns and these chemicals may independently affect the readout in some small way. The blank must take into account the effects of all the reagents (and any contaminants in the reagents) used in the analysis. Thus, if the reagents or their contaminants contribute to an instrument's readout when the analyte is being measured, such effects would be cancelled out as a result of the zeroing step. The value of a reagent blank is thus obvious.

Sample Blanks

While a reagent blank is frequently prepared and used as described above, a "sample blank" is sometimes appropriate instead. A sample blank takes into account any chemical changes that may take place as the sample is taken and/or prepared. For example, the analysis of an air sample for a component in the particulates in the air involves drawing the air sample through a filter in order to capture the particulates so that they can be dissolved. This dissolution step involves not just the particulates but also the filter itself. Thus, the dissolved filter changes the chemistry of the sample and may itself contribute to the instrument readout, and this would not be taken into account by a simple reagent blank. The answer to this problem is to take a clean filter, run it through the dissolution procedure alongside the sample, and use the resulting solution as the blank. Such a blank is called a sample blank.

Of course, anytime the dissolving of a sample includes a heating step, or any other step that may produce chemical changes in the solution, the blank should undergo the same steps. The reagent blank then becomes a sample blank and the resulting solution would represent a matrix as close to the composition of the sample solution as possible, thus enhancing accuracy when reading the samples.

Controls

A "control" is a standard solution of the analyte prepared independently, often by other laboratory personnel, for the purpose of cross-checking the analyst's work. It is a solution that the analyst measures alongside all the other standards and samples. If the concentration found for such a solution agrees with the concentration it is known to have (within an acceptable standard deviation), then this increases the confidence a laboratory has in the answers found for the real samples. If, however, the answer found differs significantly from the concentration it is known to have, then this signals a problem that would not have otherwise been detected. The analyst then knows to scrutinize his or her work for the purpose of discovering an error. The control chart mentioned in Chapter 3 is often used here in order to visualize the history of the analysis in the laboratory so that a date and time can be identified as to when the problem was first detected. Thus, the problem can be traced to a bad reagent, instrument, or other component of the procedure if such a component was first put into use the day the problem was first detected. Your instructor may want you to use controls in various experiments in this text.

9.5 EFFECTS OF SAMPLE PRETREATMENT ON CALCULATIONS

The concentration obtained from the various graphs of the instrument readout (or ratio, etc.) vs. concentration is rarely the final answer in a real-world instrumental analysis. In most procedures, the sample has undergone some form of pre-analysis treatment prior to the actual measurement. In some cases, the sample must be diluted prior to the measurement. In other cases, a chemical must be added prior to the measurement, possibly changing the analyte's concentration. In still other cases, the sample is a solid and must be dissolved or extracted prior to the measurement.

The instrument measurement is the measurement of the solution tested, and the concentration found is the concentration in that solution. What the concentration is in the original, untreated solution or sample must then be calculated based on what the pretreatment involved. Often this is merely a dilution factor. It may also be a calculation of the grams of the constituent from the molar concentration of the solution, or the calculation of the ppm in a solid material based on the weight of the solid taken and the milliliters of extraction solution used and whether or not the extract was diluted to the mark of a volumetric flask. Some examples of this follow. Remember that ppm for liquid samples is milliliters per liter while for solid samples it is milligrams per kilogram. Review Chapter 7 for more information about the ppm unit.

In order to set up a calculation for a given real-world analysis, it may be useful to first decide how the answer is to be reported and then work backwards from this to the nature of the data obtained. For example, if the ppm of a constituent in a solid material sample is to be reported, the final step will involve dividing the milligrams of constituent by the kilograms of sample. Thus the sample weight in kilograms will be in the denominator. The milligrams of constituent in the numerator will likely need to be calculated from the concentration (such as ppm in mg/L) of the solution tested as discovered from the plot of the standard curve. The first step then is frequently the conversion of mg/L to milligrams, which is done by multiplying the ppm found by the volume of the solution tested (in liters). If this is the total milligrams in the sample (if there is no additional dilution or pretreatment), then dividing by the kilograms of sample weighed out would give the answer.

Example 1

A water sample was tested for iron content, but was diluted prior to obtaining the instrument reading. This dilution involved taking 10 mL of the sample and diluting it to 100 mL. If the instrument reading gave a concentration of 0.891 ppm for this diluted sample, what is the concentration in the undiluted sample?

Solution

$$\frac{0.891 \text{ mg}}{\text{L}} \times 0.100 \text{ L} = 0.0891 \text{ mg Fe in flask}$$

This is also the milligrams of Fe in the 10 mL of undiluted sample. Thus,

$$\frac{0.0891 \text{ mg Fe}}{0.010 \text{ L water}} = 8.91 \text{ ppm Fe in original water}$$

Alternate Solution:

The "dilution factor" is:

$$\frac{0.100 \text{ L}}{0.010 \text{ L}} \text{ or } \frac{100 \text{ mL}}{10 \text{ mL}} \text{ or } 10$$

Therefore, $0.891 \times 10 = 8.91$ ppm Fe.

Example 2

A 0.5693-g sample of insulation was analyzed for formaldehyde residue by extraction with 25.00 mL of extracting solution. After extraction, the sample was filtered and the filtrate analyzed without further dilution. The concentration of formaldehyde in this extract was determined to be 4.20 ppm. What is the concentration of formaldehyde in the insulation?

Solution

$$\frac{\frac{4.20 \text{ mg}}{\text{L}} \times 0.025 \text{ L}}{0.0005693 \text{ kg}} = 184 \text{ ppm}$$

Example 3

A 2.000-g soil sample was analyzed for potassium content by extracting the potassium using 10.00 mL of aqueous ammonium acetate solution. The soil was rinsed and the solution diluted to exactly 50.00 mL. Then, 1.00 mL of this solution was diluted to 25 mL and measured with an instrument. The concentration in this 25 mL was found to be 3.18 ppm. What is the concentration of the potassium in the soil in ppm?

Solution

The dilution factor for the final dilution is 50:1. Thus, there is 50 times more K in the extract than there is in the solution tested and it all originated from the 2 g of soil. Thus,

$$\text{ppm K} = \frac{\text{mg K}}{\text{kg soil}} = \frac{3.18 \text{ mg/L} \times 0.02500 \text{ L} \times 50}{0.002000 \text{ kg}}$$

$$\frac{3.18 \text{ mg}}{\text{L}} \times 625.0 = 1.99 \times 10^{3} \text{ ppm K in soil}$$

9.6 USE OF COMPUTERS

The role that computers play in the analytical laboratory has become extremely important. We will describe here four important functions that microprocessors, microcomputers, and computers in general perform in the analytical laboratory. These are (1) data acquisition, (2) data manipulation and storage, (3) graph plotting, and (4) experiment control.

1. *Data Acquisition.* It is very commonplace today for the analyst to use a computer for acquiring data. The term "data acquisition" refers to the fact that the data that are normally fed into a recorder (like so many of the instrumental techniques described in this text) can also simultaneously be fed into and acquired by a computer, perhaps even alongside the recorder and sharing the same output line from the instrument. Such data are thereby stored in short term in the computer's memory and in the long term on disk or tape. The actual experiment, however, requires the use of another piece of hardware most of the time because most computers will accept only "digital" signals, while the output signal from an instrument is most often of the continuous or "analog" variety. Thus, one needs what is called an "analog to digital" ("A to D") converter, sometimes referred to as an "interface".* The continuously changing nature of an analog signal is evident on the chart recording of a chromatography peak or molecular absorption spectrum, for example. The A to D converter samples these data at predetermined time intervals and feeds these "digital" values into the computer, where they are stored and/or displayed on the CRT screen. Individual discreet "digital" values are what the computer then sees.

The A to D converter, while described here as a separate piece of hardware, is most often incorporated either into a circuit board designed to plug into a slot inside the microcomputer or within the instrument itself such

*The term "interface" is also used to describe the connection of any piece of hardware to a computer, and not just an A to D converter.

that an output for computer data acquisition is provided. In either configuration it appears externally that the computer is actually accepting an analog signal, which in a sense it is.

When in actual use, a program is often run on the computer which activates the A to D converter, establishes the sampling time interval and other parameters at the discretion of the operator, and begins the data acquisition at the touch of a key on the computer keyboard.

2. *Data Manipulation and Storage.* Any voltage signal generated by laboratory instruments and output to a recorder can be fed into a computer in the manner just described. This includes pH meters, spectrophotometers, chromatographs, polarographs, etc. Even automated analysis recorder signals can be fed into a computer. The real value of this, however, lies in the fact that the data can be permanently stored on a computer disk or tape and recalled later for the purpose of manipulation, calculation, or whatever form of "data reduction" is needed for a given type of data.

The value of being able to permanently store data on disk or tape is in the achievement of a "paperless" laboratory. Before the advent of the computer, huge, cumbersome volumes of chart recordings had to be stored in the laboratory. Today, if the analyst wants to review data acquired even years earlier, he/she can call the data from disk or tape and view it in a matter of seconds.

3. *Graph Plotting.* Most instrumental analytical procedures require the plotting of a linear graph discussed earlier in this chapter for the purpose of establishing the exact relationship between the instrument reading and the concentration of the analyte. The least squares fit of the data points to the line is important in order to be as accurate as possible with the location of the line. Also, the calculation of the correlation coefficient is important. We have already indicated that a computer can be used to perform these functions. Indeed, the use of a computer to plot data, perform a least squares fit, print out a copy of the fitted curve, and also calculate the correlation coefficient is a very important computer function in the modern lab. The improved accuracy of the determination can be demonstrated by manual plotting and curve fitting and comparing the answer to what a computer found. You will find that the speed of the determination also improves dramatically.

4. *Experiment Control.* Finally, entire experiments can be controlled with the use of microcomputers and microprocessors. Pressing a key on a computer keyboard can cause an experiment to begin, be altered, or end. Indeed the more sophisticated modern instruments have the microprocessor for this built right into the instrument (essentially a microcomputer complete with disk drive and CRT, and the instrument all in one unit) such that experiment control is quite easily accomplished.

Essentially any instrument can be told when to begin executing its function in this manner, even GC chromatographs in which the experiment begins when the operator pushes in the syringe plunger. An electrical contact closure can be incorporated into the syringe or injection port such that the syringe injection actually activates the computer for data acquisition.

9.7 MODERN DATA HANDLING

Introduction

The molecular spectroscopic techniques of ultraviolet, visible, and infrared spectrometry; the atomic spectroscopic techniques of flame atomic absorption, graphite furnace atomic absorption, and inductively coupled plasma; the instrumental chromatography techniques of gas chromatography and high performance liquid chromatography – all of these and the myriad of others discussed in the chapters to follow have become very popular analysis techniques of the modern analytical chemistry laboratory. The ability to handle the quantity of data produced and the ability to assure the quality of the decisions made as a result of the data produced are uppermost in the minds of laboratory managers everywhere. It is obviously important for these managers to be keenly aware of modern data-handling procedures and to be able to execute them.

At the heart of modern laboratory data handling is the laboratory computer, often termed the laboratory "workstation" or "data station". The use of the computer in the laboratory was briefly discussed above. The modern laboratory utilizes computer workstations to (1) acquire data from instruments, (2) store the acquired data, (3) display and manipulate the acquired data, (4) control instruments, and finally (5) gather and print the results in a way that is acceptable to clients. Professional analytical chemists must apply their knowledge of chemistry to the laboratory methods and procedures, but they must also be knowledgeable in computer usage for all the categories listed above.

Data Display and Analysis

The conversion of the analog signal typical of the output of many types of modern laboratory instruments to the digital signal required by a computer was discussed previously in this chapter. The display of the digitized data on a monitor screen or a printer/plotter for subsequent interpretation and analysis is a common activity in these laboratories. Such a display amounts to a reconstruction of the analog data which allows the analyst to quickly determine maxima or minima in the data (such as may be required for chromatography peak heights or the absorbance at the wavelength of maximum absorption), perform simple mathematical operations on all or part of the data (such as multiplying by a dilution factor), average all or a portion of the data (which may be required for statistical analysis – see the next section), and many other similar manipulations. The computer especially enhances chromatographic data analysis, since it is capable of quickly determining retention times (and thus quickly identify mixture components) as well as quantitation parameters, such as where to begin and end a peak integration, whether the peak threshold limit defined by the analyst has been reached, and the actual determination of the peak sizes by the integration (see Chapter 16).

Various companies market the hardware and software required for computer workstations. Such systems are readily available in various price ranges and are commonplace in modern laboratories.

Reporting and Managing Results

Modern computerized data stations are capable of automatically accumulating and assembling data files for the purpose of preparing and storing personal reports for the analyst as well as customized reports for clients. Such reports can be assembled for presentation before a group, or they can also be formatted for importing to standard microcomputer software, most notably word processing, data base, and spreadsheet software, which is useful if a formal written report of the results of an analysis is required.

In laboratories in which the volume of laboratory data generated is large, the computer is a godsend. The analyst doesn't even need to view individual routine results. Often the computer takes over, accumulates and analyzes the data, and generates reports which it automatically imports to data base or spreadsheet files for a formal printed laboratory report. A visual check of the resulting report can pinpoint problems and suggest remedies.

The State of the Art

The strip-chart recorders and x-y recorders discussed previously are fast becoming obsolete, if they aren't already. They are useful for immediate visual inspection of the analog data and are inexpensive, but they respond slowly and are not capable of automatic range switching and attenuation. Instrument output signals that "go off scale" or are too small to see need to be generated a second time.

Computing integrators (Chapter 16) are in common use in GC and HPLC laboratories. These perform retention time and peak integration determinations quickly and accurately, but usually have limited report-generating capability and cannot store and reconstruct chromatograms, items we discussed above as advantages of modern computer workstations. They are, however, relatively inexpensive and portable.

Thus, personal computers and computer workstations, as well as minicomputer and mainframe systems, have transformed laboratory data handling into job functions that are often as easy as pressing a key on a keyboard, especially given the increasing speed of the computers' central processors. Virtually all the functions listed in the preceding discussions can be performed quickly and accurately, thus making the handling of analytical data electronically automated. Modern chemical literature* addresses these issues and should be scanned regularly for modern developments.

EXPERIMENTS

EXPERIMENT 20: ELECTRICAL CONNECTIONS BETWEEN INSTRUMENTS, RECORDERS, AND COMPUTERS

Introduction

The objective of this experiment is to familiarize you with how to make the electrical connections between instruments and the various recorders,

*See, for example, a column entitled "The Data File" appearing monthly in the journal *GC·LC*, which is a journal dedicated to modern instrumental chromatography.

plotters, and computers. In each case, you will need to examine the output ports of the instruments and the input ports of the recorders or computers in order to know what type of connector (ie., banana plug or screw-type, etc.) is needed and also to observe the color codes, and then to actually install the connectors. Output and input terminals always consist of at least two posts, one for the signal and one for ground. A third post may also be present for a connection to "true ground" and to interconnect the true ground circuit of all instruments and recorders. This interconnection may reduce electrical interference that may be picked up by these devices. The signal post is usually red, while the ground post is usually black. The true ground post is usually some other color. Thus, the red post on an instrument is connected to the red post on the recorder and the black one is connected to the black post, etc. Your instructor may want to assist with or demonstrate parts of the experiment.

UV/Vis Spectrophotometer + Recorder

1. Examine the UV/vis spectrophotometer designated by your instructor and find the output ports for the electrical signal corresponding to absorbance and wavelength. Using the wire leads supplied, connect one end to the instrument's output for the absorbance signal (one lead each for the signal, ground and true ground, if applicable) and the other end to the recorder. (In lieu of a third lead, you may short the ground post to the true ground on both the instrument and recorder.) If it is a strip-chart recorder, there is only one set of input terminals. If it is an x-y recorder, connect these leads to the y-axis input. If the recorder is an x-y recorder, connect the instrument's output terminal for wavelength to the x-axis of the recorder. Ask your instructor to check your work.

Atomic Absorption Spectrophotometer + Strip-Chart Recorder

2. Examine the atomic absorption spectrophotometer designated by your instructor and find the output ports for the electrical signal corresponding to absorbance. Make the electrical connections from this instrument to a strip-chart recorder in the same manner as described in Step 1. Ask your instructor to check your work.

Gas Chromatograph + Strip-Chart Recorder/Computing Integrator

3. Examine the gas chromatograph designated by your instructor and find the output ports for the detector's electrical signal for a strip-chart recorder. Make the connection to a strip-chart recorder as described in Step 1. Ask your instructor to check your work.
4. Disconnect the strip-chart recorder. Find the output ports for the detector's electrical signal for a computing integrator and the input terminals on the computing integrator designated by your instructor and make the connections. Ask your instructor to check your work.

Liquid Chromatograph + Strip-Chart Recorder/ Computing Integrator

5. Repeat Steps 3 and 4 above for the liquid chromatograph designated by your instructor.

Connections to a Microcomputer

6. Open the case of the microcomputer designated by your instructor, examine the area where external interface cards are installed (instructor may demonstrate), and identify the card that is used for instrument interfacing and data acquisition. Trace the electrical connection from the card and identify the terminal for external connection to the instrument. Once you've identified the terminal, close the case. Now examine the nature of the external connection from the computer to an instrument and make appropriate preparations for this connection. Such a connection may involve a third device, such as an A to D converter or a switchbox.
7. Connect the instrument designated by your instructor to the computer so that it is readied for a data acquisition experiment. Ask your instructor to check your work.

EXPERIMENT 21: THE MEASUREMENT OF A STRONG ACID-STRONG BASE TITRATION CURVE BY DATA ACQUISITION WITH A MICROCOMPUTER

NOTE: This experiment is an acid-base titration similar to those performed in Chapter 6. It is the titration of 0.10 *M* HCl with 0.10 *M* NaOH, as in Experiment 7, but a combination pH probe will be used to monitor the pH during the titration, as in Experiment 9. The pH meter used has output terminals that may be used to output the measured pH to a recorder or computer. We will use this special feature of the pH meter to feed the pH data directly to a microcomputer in an example of data acquisition by computer. Safety glasses are required.

1. Prepare 100 mL of a 0.10 *M* HCl solution and 100 mL of a 0.10 *M* NaOH solution.
2. Measure 25 mL of the HCl into a 250-mL beaker. Fill a 50-mL buret with the NaOH.
3. Place a magnetic stirring bar in the beaker and place the beaker on a magnetic stirrer.
4. Prepare a pH meter and a combination pH electrode. The pH meter must have a recorder output. Place the electrode in the beaker such that the tip of the electrode is immersed but suspended with a clamp so as to avoid contact with the stirring bar.
5. Turn on the stirrer and measure the pH. The pH should be between 1 and 2. Check to be sure that the recorder output is fed into an A to D converter or a microcomputer with an internal A to D converter. Your instructor will likely have made this connection for you. Load and run

the data acquisition program—your instructor will guide you in getting this ready.

6. Begin the titration at the same time that the data acquisition begins. Your instructor will suggest an appropriate titration rate. Titrate until all 50 mL have been added. At that point, stop the data acquisition and store the data on disk.
7. Using the computer, plot the titration curve and display it on the computer screen. If possible, obtain a hard copy using a printer.

QUESTIONS AND PROBLEMS

1. Distinguish between "wet" methods of analysis and "instrumental" methods of analysis.
2. In what ways does electronic instrumentation improve upon the classical wet chemical methods of analysis?
3. Distinguish between a strip-chart recorder and an x-y recorder.
4. The proportionality constant between an instrument readout and concentration is 54.2. Assuming a linear relationship between the readout and concentration, what is the numerical value of the concentration of a solution when the instrument readout is 0.922?
5. What is the numerical value of the concentration in a solution that gave an instrument readout of 53.9 when the proportionality constant is 104.8? (Assume a linear relationship.)
6. An instrument reading for a standard solution whose concentration is 8.0 ppm is 0.651. This same instrument gave a reading of 0.597 for an unknown. What is the concentration of the unknown?
7. Calculate the concentration for the unknown given the following data:

R	C (ppm)
72.0	0.693
68.1	C_u

8. Plot the following data and give the concentration of the unknown solution:

R	C (ppm)
8.2	2.00
17.0	4.00
24.9	6.00
31.9	8.00
40.5	10.00
26.7	C_u

9. Why must the instrument readout for an unknown fall within the range of the series of readouts for a series of standard solutions in order to be accurate?

10. Calculate the slope of the straight line defined by the following two points: (0.20,0.439) and (0.50,0.993).

11. Given the following data, calculate the slope of the line drawn between the two points:

y	x
0.542	4.00
0.819	8.00

12. Given the following data, calculate the proportionality constant, K, between A and C. (Assume that the y-intercept is zero.)

A	C
0.419	3.00
0.837	6.00

13. Given the data from Problem #12, what is the concentration in a solution that gave an "A" value of 0.677?

14. When plotting the results of the measurement of a series of standard solutions, why do we draw "the best straight line" possible through the points rather than just connect the points?

15. What is the method of least squares and why is it useful in instrumental analysis?

16. What is it about the calculations involved in the method of least squares that gives this method its name?

17. What is meant by "linear regression" analysis?

18. Give four parameters that are readily obtainable as a result of the method of least squares treatment of a set of data.

19. What is meant by perfectly linear data? What value of what parameter would indicate that a data set is perfectly linear?

20. What are some realistic values of a correlation coefficient that would indicate to a laboratory worker that the error associated with his or her data is probably minimal?

21. What is meant by "serial dilution"?

22. The "series of standards solution" method works satisfactorily most of the time. When does it *not* work well?

23. A stock standard solution of copper (1000 ppm) is available. Tell how you would prepare 50 mL each of five standards solutions, giving the amount of the 1000 ppm solution required for each one and how you would proceed with the preparation, including the kind of pipet needed. The concentrations of the standards should be 1, 2, 3, 4, and 5 ppm.

24. How many milliliters of a 100 ppm stock are needed to prepare 25 mL of each of four standards with concentrations of 2, 4, 6, and 8 ppm?

25. What pipet is needed for the preparations in #24?

26. What is meant by the "internal standard" method of analysis and why is it important?
27. What is an "internal standard"?
28. A series of standard solutions of benzene in toluene solvent is to be prepared for an internal standard procedure with ethylbenzene serving as the internal standard. Tell how you would prepare the solutions, including the milliliters of benzene and ethylbenzene measured out for each and what pipet to use. The benzene concentrations are to be 0.5%, 1.0%, 2.0%, and 3.0% and the internal standard concentration is to be 1.0%.
29. What is the method of standard additions and under what conditions is it most useful?
30. What is meant by "matrix matching"?
31. The following data were obtained through the method of standard additions. Plot the data and report the ppm for the unknown.

R	C added (ppm)
0.138	0.00
0.241	1.00
0.356	2.00
0.460	3.00

32. What is a "blank"? What is the difference between a "reagent blank" and a "sample blank"?
33. What is a "control" sample and how is it useful as an accuracy check?
34. Explain why the concentration of an unknown obtained from a graph of a series of standard solutions may not be the final answer in an instrumental analysis.
35. An unknown solution of riboflavin, contained in a 25-mL volumetric flask, was determined to have a concentration of 0.525 ppm. How many milligrams of riboflavin are in the flask?
36. After analysis for iron content, a water sample was found to have a concentration of 4.62 ppm iron. How many milligrams of iron are contained in 100 mL of the water?
37. A 2.000-g sample of a soil is found to yield 3.73 mg of phosphorus after extraction. What is the concentration of phosphorus in the soil in ppm?
38. After dissolving a 5.000-g sample of concrete, the resulting solution was found to contain 0.229 g of manganese. What is the concentration of manganese in the concrete in ppm?
39. A certain water sample was diluted from 1 mL to 50 mL with distilled water. After an analysis for zinc was performed, this diluted sample was found to contain 10.7 ppm zinc. What is the zinc concentration in ppm in the undiluted sample?
40. A rag that a farmer was using was analyzed for pesticide residue. The rag weighed 49.22 g and yielded 25.00 mL of an extract solution that was determined to have a pesticide concentration of 102.5 ppm. How

many grams of pesticide were in the rag and what is the concentration of pesticide in the rag in ppm?

41. The soil around an old gasoline tank buried in the ground is analyzed for benzene. If 10.00 g of soil shows the concentration of benzene in 100 mL of soil extract to be 75.0 ppm, what is the concentration of benzene in the soil in ppm?

42. Suppose a 4.272-g soil sample undergoes an extraction with 50 mL of extracting solvent to remove the potassium. This 50 mL was then diluted to 250 mL and tested with an instrument. The concentration of potassium in this diluted extract was found to be 35.7 ppm. What is the potassium concentration, in ppm, in the untreated soil sample?

43. A city's water supply is found to be contaminated with carbon tetrachloride. The chemical analysis procedure involves the extraction of the carbon tetrachloride from the water with hexane. A 4.00-L sample of water is extracted with 10 mL of hexane. If this hexane solution is diluted, 1 mL to 25 mL, and the concentration of carbon tetrachloride in the diluted solution is found to be 10.4 ppm, what is the concentration in the original water?

44. What is meant by "data acquisition" by computer?

45. What is an "A to D converter" and why is it an important piece of hardware for data acquisition?

46. Why has computerized data storage developed into an important function of computers in the modern laboratory?

47. What is meant by a laboratory "workstation" or "data station" and what is their function in the modern laboratory?

48. How has spreadsheet computer software been useful in the modern laboratory?

49. Why are strip-chart recorders and x-y recorders becoming obsolete?

Chapter 10

Fundamentals of Light and Light Absorption

10.1 INTRODUCTION

More than half of all instrumental methods of analysis involve the measurement of either light absorption or light emission by a sample. For this reason, the analyst needs to have a basic understanding of the modern theory of light and parameters of light, namely wavelength, frequency, wave number, and energy, as well as the concept of how light interacts with matter, the theory behind the emission and absorption phenomena. This chapter introduces these concepts.

10.2 NATURE AND PARAMETERS OF LIGHT

The modern theory of light says that light has a "dual nature". This means that light exhibits both the properties of waves (the wave theory) and the properties of particles (the particle theory). The wave theory says that light travels from its source via a series of repeating waves much like waves of water move across the surface of a body of water. The particle theory says that light consists of a stream of particles called "photons" emanating from the source. For our purposes, the wave theory seems to have the most applicability.

The one important difference between waves of light and waves of water moving across a body of water is that the light waves are not mechanical waves like water waves. Light waves do not require matter to move or to exist. Rather than being mechanical disturbances, they are electromagnetic disturbances, and, as such, they can travel through a vacuum, from the sun to the earth, for example. Since light waves are electromagnetic disturbances, light is often referred to as electromagnetic radiation. It has an electrical component and a magnetic component. Of particular importance in analytical chemistry are the wavelength, frequency, wave number, and energy of light as described by the wave theory.

Since light consists of a series of repeating waves, the physical distance from a point on one wave to the same point on the next wave is an important parameter. This distance is termed the wavelength. It is given the Greek symbol lower-case lambda (λ). (See Figure 10.1.) Wavelength can vary from distances as little as fractions of atomic diameters to several miles. This suggests the existence of a very broad spectrum of wavelengths. Indeed, the distinction between visible (vis) light, ultraviolet (UV) light, infrared (IR) light, etc. is the magnitude of the wavelength. Each of these, along with the others mentioned previously, encompass a particular "region" of the total electromagnetic spectrum. Thus, we have the visible region, the ultraviolet region, the infrared region, etc. Figure 10.2 shows a representation of the entire electromagnetic spectrum, with wavelength increasing from left to right, and indicates the approximate wavelength borders of the various regions in nanometers. In the ultraviolet, visible, and infrared regions, which are the regions that we will be emphasizing, the nanometer (nm) and the micrometer (μm; also called the micron [μ]) are the most commonly used units of wavelength.

Another parameter of light derived from the wave theory is frequency. Frequency is defined as the number of waves that pass a fixed point in 1 sec (cycles per second). It is given the Greek symbol nu (ν). Frequency would obviously depend on how fast the wave travels. However, all light travels at

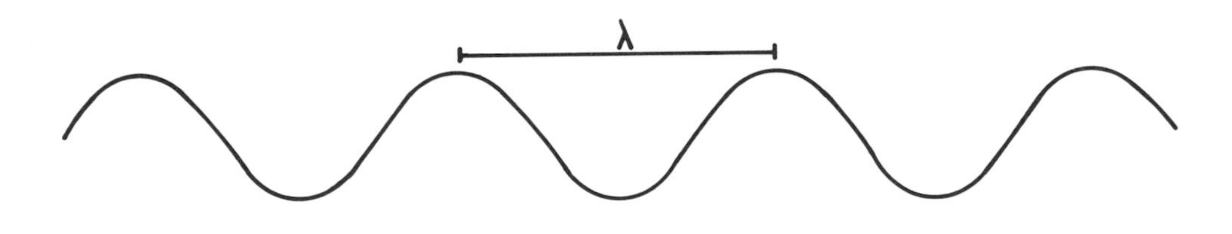

Figure 10.1. The definition of wavelength (λ).

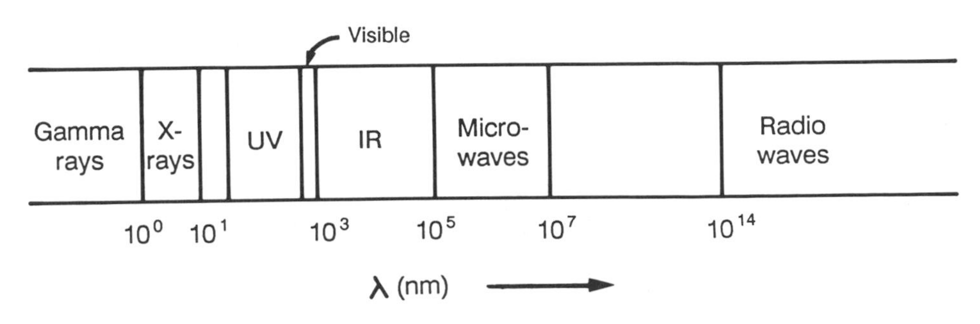

Figure 10.2. The electromagnetic spectrum.

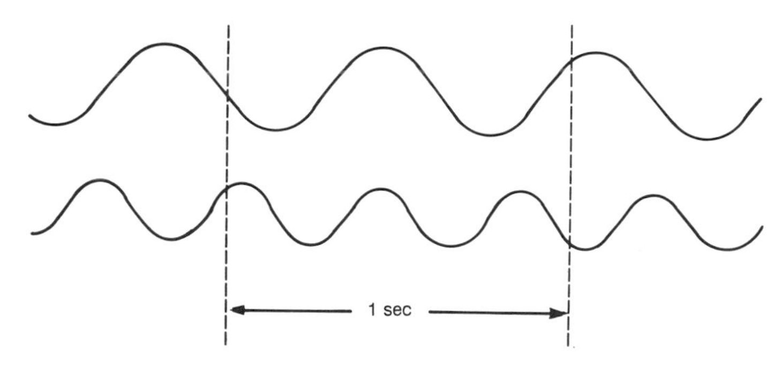

Figure 10.3. Light with a shorter wavelength has more waves (cycles) per second. (From Kenkel, J., *Analytical Chemistry Refresher Manual*, Lewis Publishers, Chelsea, MI, 1992. With permission.)

the same speed in a vacuum, 3.00×10^{10} cm/sec, the "speed of light".* Thus, a change in wavelength is the only change that would produce a different frequency. In other words, a particular wavelength corresponds to a particular frequency. Figure 10.3 should help clarify this concept. The most common unit of frequency is the hertz (Hz) or sec^{-1}. Speed, wavelength, and frequency are mathematically related as follows:

$$c = \lambda \nu \tag{10.1}$$

$$\lambda = \frac{c}{\nu} \tag{10.2}$$

or

$$\nu = \frac{c}{\lambda} \tag{10.3}$$

in which c is the speed of light, 3.00×10^{10} cm/sec.

A parameter closely related to frequency is the wave number. Wave number is defined as the reciprocal of the wavelength when the wavelength is expressed in centimeters, and it is given the symbol $\bar{\nu}$ (read "nu bar").

$$\bar{\nu} = \frac{1}{\lambda \text{ (cm)}} \tag{10.4}$$

Wave number is an important parameter for the infrared region, as we will see.

Last, but certainly not least, is energy. Different energies are associated with the different wavelengths and frequencies of light. Mathematically, we define energy as follows,

$$E = h\nu \tag{10.5}$$

and in terms of wavelength,

*In a medium other than a vacuum, the speed of light is different from that in a vacuum. This creates a phenomenon known as refraction, a parameter known as refractive index, and a technique known as refractometry. These will be discussed in Chapter 17.

$$E = h\left(\frac{c}{\lambda}\right) \tag{10.6}$$

The symbol "h" is a proportionality constant known as "Planck's Constant" which has the value 6.62×10^{-27} erg sec/photon. The photon, as indicated previously, is the name given to a "particle" of light in the particle theory. The erg is a unit of energy in the metric system.

As we consider the theories of the interaction of ultraviolet, visible, and infrared light with matter and the theories of molecular and atomic absorption in this and succeeding chapters, it is important to understand what the above equations mean in terms of how one parameter relates to any of the others. (See Table 10.1.) In addition, we can indicate which regions of the electromagnetic spectrum are high energy regions, which are low-energy regions, which are high-frequency regions, which are low-frequency regions, etc. See Figure 10.4 for this. Concerning the three regions to be emphasized in this chapter, then, we can say the following:

energy: UV > vis > IR
wavelength: UV < vis < IR
frequency: UV > vis > IR

Table 10.1. Statements of interpretation of equations presented in Section 10.2.

Equations	Interpretation
10.2, 10.3	Frequency is inversely proportional to wavelength and vice versa. As frequency increases, wavelength decreases and vice versa. Long wavelength corresponds to low frequency, short wavelength to high frequency.
10.4	Wave number is inversely proportional to wavelength. As wavelength increases, wave number decreases and vice versa. Long wavelength corresponds to low wave number, short wavelength to high wave number.
10.5	Energy is directly proportional to frequency. As frequency increases, energy increases and vice versa. High frequency corresponds to high energy, low frequency to low energy.
10.6	Energy is inversely proportional to wavelength. As wavelength increases, energy decreases and vice versa. Long wavelength corresponds to low energy, short wavelength to high energy.

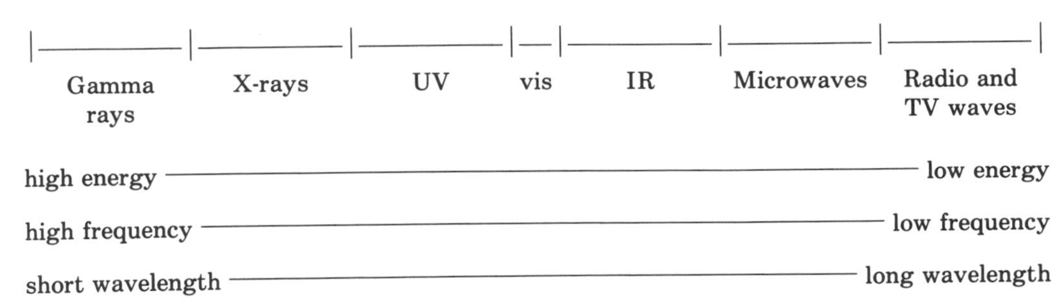

Figure 10.4. Comparisons of the various regions of the electromagnetic spectrum in terms of energy, frequency, and wavelength.

The visible region of the spectrum (vis), as shown in Figures 10.2 and 10.4, is a very "narrow" region compared to the others. Visible of course means that to which our human eye is sensitive. Thus, it is apparent that we can see only a very small fraction of all the wavelengths of light that continuously criss-cross our atmosphere and universe. In addition, those individual wavelengths of visible light that enter our eyes are characteristically interpreted; i.e., different wavelengths correspond to different colors. Thus, we have red light, yellow light, green light, etc. Since the different colors correspond to different wavelengths, they also correspond to different energies and frequencies. (See Figure 10.5.)

10.3 CONVERTING AND CALCULATING LIGHT PARAMETERS

In order to give you a feel for the magnitude of the parameters of light, especially in the visible region, we now present a series of example calculations and conversions.

The interconversion of the units of wavelength is sometimes necessary when performing calculations with the relationships in the previous section. If 3.00×10^{10} cm/sec is used as the speed of light, c, then the wavelength units must be in centimeters in order to allow appropriate cancellation. An example is the calculation of the frequency given the wavelength (Equation 10.3). Since the units of frequency are cycles per second, or sec^{-1}, as they are most often written, the wavelength unit must be centimeters since the centimeters from the units of c must cancel:

$$\frac{3.00 \times 10^{10}\ \cancel{cm}/sec}{\cancel{cm}} = sec^{-1} \tag{10.7}$$

Wavelengths units are not always given in centimeters, thus the need for a conversion to centimeters if one is to calculate frequency. Table 10.2 lists some useful conversion factors for converting metric system units of length.

Example 1

How many centimeters does 452 nm represent?

Solution

$$452\ \cancel{nm} \times \frac{10^{-7}\ cm}{\cancel{nm}} = 452 \times 10^{-7}\ cm \text{ or } 4.52 \times 10^{-5}\ cm$$

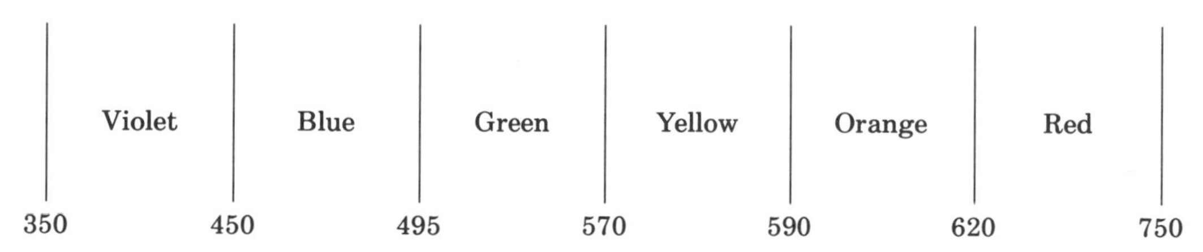

Figure 10.5. The visible region of the electromagnetic spectrum and the approximate wavelength boundaries of the colors.

Table 10.2. Some useful conversion factors for interconverting units of wavelength of light.

Unit to Be Converted	Conversion Factor
To centimeters (cm):	
Meters (m)	10^2 cm/m
Millimeters (mm)	10^{-1} cm/mm
Nanometers (nm)	10^{-7} cm/nm
Angstroms (Å)	10^{-8} cm/Å
To meters (m):	
Nanometers (nm)	10^{-9} m/nm
Millimeters (mm)	10^{-3} m/mm
Angstroms (Å)	10^{-10} m/Å

Example 2

What is the frequency of light with a wavelength of 537 nm?

Solution

$$\nu = \frac{c}{\lambda} = \frac{3.00 \times 10^{10}\ \text{cm/sec}}{537\ \text{nm} \times \frac{10^{-7}\ \text{cm}}{\text{nm}}} = 5.59 \times 10^{14}\ \text{sec}^{-1}$$

Example 3

What is the wavelength of light having a frequency of 7.89×10^{16} sec^{-1}? Express your answer in both nanometers and centimeters.

Solution

$$\lambda = \frac{c}{\nu} = \frac{3.00 \times 10^{10}\ \text{cm/sec}}{7.89 \times 10^{16}\ \text{sec}^{-1}} = 3.80 \times 10^{-7}\ \text{cm}$$

$$3.80 \times 10^{-7}\ \text{cm} \times \frac{\text{nm}}{10^{-7}\ \text{cm}} = 3.80\ \text{nm}$$

This same unit interconversion may be necessary when calculating $\bar{\nu}$ from λ or ν (or vice versa) and when calculating E from λ (or vice versa). The usual units of $\bar{\nu}$ are cm^{-1}, and thus λ must be in centimeters. The usual units of E are ergs, and thus again the units of λ must be centimeters in order to cancel with the centimeter units in c.

Example 4

What is the wave number, $\bar{\nu}$, of light having a wavelength of 24.8 nm?

Solution

$$\bar{\nu} = \frac{1}{\lambda} = \frac{1}{24.8\ \text{nm} \times \frac{10^{-7}\ \text{cm}}{\text{nm}}} = 4.03 \times 10^{5}\ \text{cm}^{-1}$$

Example 5

What is the energy of light that has a wavelength of 4966 Å?

Solution

$$E = h\left(\frac{c}{\lambda}\right) = \frac{6.62 \times 10^{-27}\ \text{erg}\ \cancel{\text{sec}} \times 3.00 \times 10^{10}\ \cancel{\text{cm}}/\cancel{\text{sec}}}{4966\ \cancel{\text{Å}} \times \dfrac{10^{-8}\ \cancel{\text{cm}}}{\cancel{\text{Å}}}}$$

$$= 4.00 \times 10^{-12}\ \text{erg}$$

10.4 LIGHT ABSORPTION AND EMISSION

Evidence of light absorption abounds in our world. All colored objects constitute such evidence. Visible white light that comes from the sun, a light bulb, or any visible light source actually consists of all the visible wavelengths represented in Figure 10.5. Consider a rainbow, for example. White light coming to earth from the sun can, under the right conditions, be dispersed into the rainbow colors. The right conditions usually means a large concentration of water droplets in the atmosphere. This is evidence that white light consists of all the colors and that these colors are separable. Why does a red sheet of paper appear red? All the wavelengths of visible light from the visible light source in the room are absorbed except for the red wavelengths, and these are reflected to our eye. Why does a solution of potassium permanganate appear to be a deep purple color? The wavelengths of visible light which are incident on the solution from the light in the room are all absorbed except for those in the violet and red ends of the visible region. The result is an intense purple color. Thus, whatever color of light is not absorbed by a sample of matter is the color that is reflected, or transmitted, to our eye, and this is the reason for our perception of the colors that forms of matter appear to have. The purpose of the present discussion is to explain the phenomenon of light absorption in terms of the exact nature of the interaction of light with the atoms, molecules, and electrons found in samples of matter.

The modern theory of the atom states that electrons exist in energy levels outside the nucleus of an atom. The modern theory also holds that electrons can be moved from one energy level to a higher one if conditions are right. These conditions consist of (1) the addition of sufficient energy to the electron and (2) a vacancy for the electron with this greater energy in a higher energy level. In other words, if an electron absorbs the energy required to be promoted to a higher vacant energy level then it will be promoted to that level.

Where can an electron obtain the required energy? One way is for it to come into contact with light of the required energy. Figure 10.6 represents a picture of this idea. If the light is indeed absorbed it will not be observed, transmitted, or reflected by the sample as Figure 10.6 depicts. Thus we have a mechanism for light absorption by atoms. When the light "strikes" an electron, causing it to be promoted to a higher energy level, the energy that once was light is now possessed by the electron. This light is thus "absorbed" and is not observed emerging from the sample. An important point concerning this process, however, is that the light coming in must be ex-

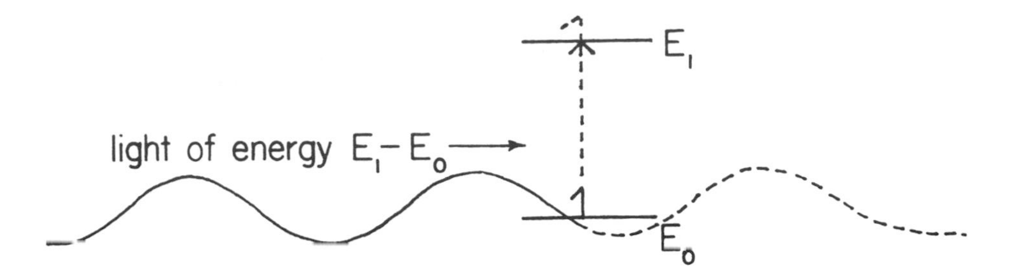

Figure 10.6. The absorption of light by an electron, causing the promotion from an energy level E_0 (ground state) to an energy level E_1 (excited state). The energy that was once light now belongs to the electron.

actly the same energy as the energy difference between the two electronic energy levels; otherwise it will not be absorbed at all. This latter point will have a bearing on the discussion in the next section and in succeeding chapters.

There are other mechanisms by which light interacts with matter. While the electronic transitions just described are possible and common, so-called "vibrational transitions" and "rotational transitions" are also possible and common. In these cases, molecules are thought of as existing in vibrational and rotational energy states which can change depending on the energy available to them. Vibrational states correspond to different modes of vibration of molecules. Rotational states correspond to different rotational modes of molecules. Figure 10.7 shows a molecule consisting of three atoms bonded in a straight line, such as CO_2, existing in three vibrational modes. Each of these vibration modes can be initiated by the absorption of a certain amount of energy, just as with the electronic "modes". Thus, we can picture vibrational energy transitions via levels as in Figure 10.6, whereby light of a particular energy is required to cause the jump from one level to the next. As with electronic transitions, the light causing such a jump is absorbed and cannot be observed reflecting from or transmitting through the sample. (See Figure 10.8.) We can make similar statements about rotational transitions.

The magnitude of the electronic, vibrational, and rotational energy jumps are of importance, at least from a comparison point of view. The energy required to cause electronic transitions is greater than that required for vibrational transitions, which is greater than that required for rotational transitions. Thus, the region of the electromagnetic spectrum required for each transition is different. Ultraviolet and visible light cause electronic transitions when they are absorbed, while infrared light, being lower in energy, causes vibrational and rotational transitions. Short-wavelength IR, or "near" IR, causes vibrational transitions, while long-wavelength IR, or "far" IR, causes rotational transitions. (See Table 10.3.)

Figure 10.7. Three vibrational modes of a three-atom molecule.

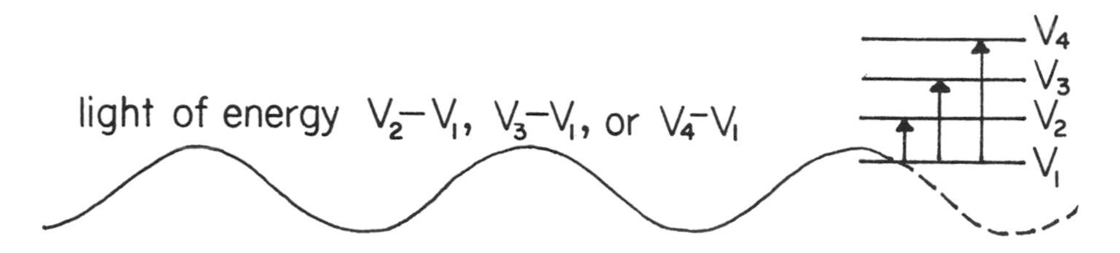

Figure 10.8. The absorption of light by molecules can cause elevation from V_1 to higher vibrational states V_2, V_3, V_4.

Table 10.3. Comparison of UV, visible, IR and far IR in terms of the type of atomic or molecular activity generated when absorbed by matter.

Light	Activity
Ultraviolet	Electronic
Visible	Electronic
Near infrared	Vibrational
Far infrared	Rotational

Finally, as we'll see in the next section, it is important to point out that vibrational and rotational levels are superimposed on all electronic levels. This means that, regardless of the state of a particular electron in a molecule, the molecule will still have its vibrational and rotational modes. Thus, there is an entire set of vibrational modes in each electronic state and an entire set of rotational modes in the vibrational states. Figure 10.9 should clarify this.

10.5 ATOMIC VS. MOLECULAR

The subject of the absorption of light by individual non-bonded atoms must be considered separately from the subject of molecule absorption. The reason for this is that, since elemental atoms are not bonded to each other as

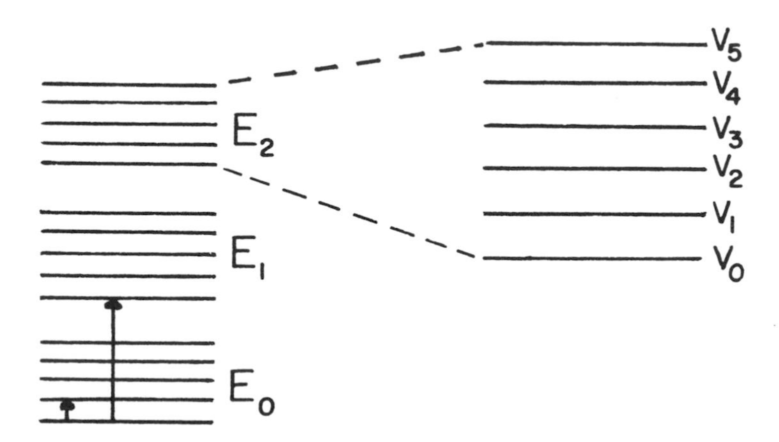

Figure 10.9. Vibrational levels superimposed on electronic levels. The level E_2 is expanded and on the right for clarity. The relative magnitude of the two types of transitions is indicated.

in molecules, they do not have vibrational modes and, therefore, cannot have vibrational energy transitions. Thus, all energy transitions that take place are purely electronic. A consequence of this is that only individual, discrete, electronic energy transitions are possible. These transitions elevate an electron from one electronic level to the next, as depicted in Figure 10.10. Since no vibrational transitions take place, only a limited number of wavelengths, those corresponding to the electronic transitions such as in Figure 10.10, have a chance of being absorbed. We can show this by plotting the amount of absorbed light (lets call this "A") vs. wavelength. (See Figure 10.11.) Each vertical line in this graph represents an absorption corresponding to a particular electronic transition which in turn corresponds to a particular wavelength of light [$E = h(c/\lambda)$]. Thus, we could say that the transition $E_0 \rightarrow E_1$ is due to λ_1, $E_0 \rightarrow E_2$ is due to λ_2, etc. Each discrete energy increase is due to the absorption of the wavelength corresponding to that energy, and therefore only those wavelengths are absorbed and only those wavelengths show up in the atomic spectrum exemplified by Figure 10.11. The spectrum in Figure 10.11 is sometimes referred to as a "line spectrum" because of the existence of the vertical lines.

You have probably noticed that the *amount* of light absorbed at each of the wavelengths in Figure 10.11 is different. This is because the different energy transitions have different likelihoods of occurring. There are a variety of possible reasons for this, such as the number of electrons in the sample which are able to make a given transition. When performing an

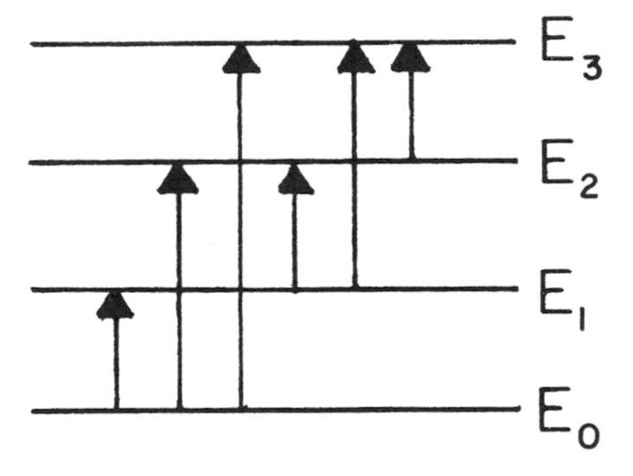

Figure 10.10. An energy level diagram showing four electronic levels and the transitions that are possible with these levels.

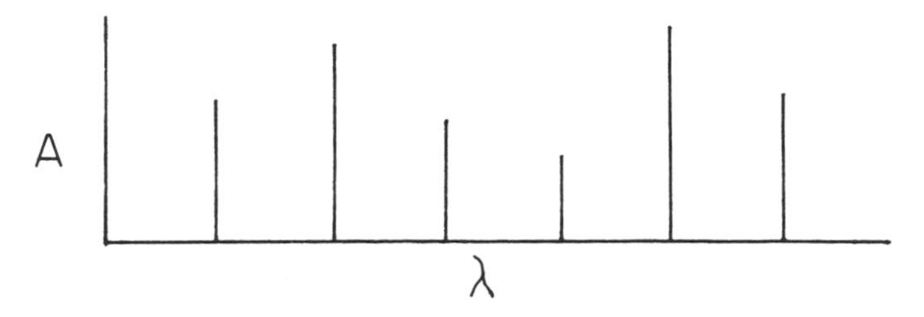

Figure 10.11. An atomic absorption spectrum, also referred to as a line spectrum.

analytical experiment with atoms (see Chapter 14) the most intense line is usually chosen.

With molecular absorption, both electronic and vibrational transitions are possible and, in fact, the picture becomes much more scrambled. Not only can transitions occur between vibrational levels within a given electronic level, but transitions can also occur between any vibrational level in one electronic state and any vibrational level in any other electronic state. Figure 10.12 shows the transitions possible between hypothetical electronic levels E_0 and E_1. One can imagine that the number of allowed transitions mushrooms when the additional electronic levels are considered. The result of this is that rather than obtaining a line spectrum for the absorption we observe a "continuous" or "band" spectrum (see Figure 10.13), since virtually all wavelengths have a chance of being absorbed to some degree because transitions are not limited to those that are purely electronic.

In summary, a given wavelength of light is absorbed if an energy transition exists which corresponds to it. In the case of atoms, only a few wavelengths are absorbed since only a few energy transitions are possible. The result is a line absorption spectrum. In the case of molecules, many transitions are possible and thus a large number of wavelengths are absorbed. The result is a continuous absorption spectrum.

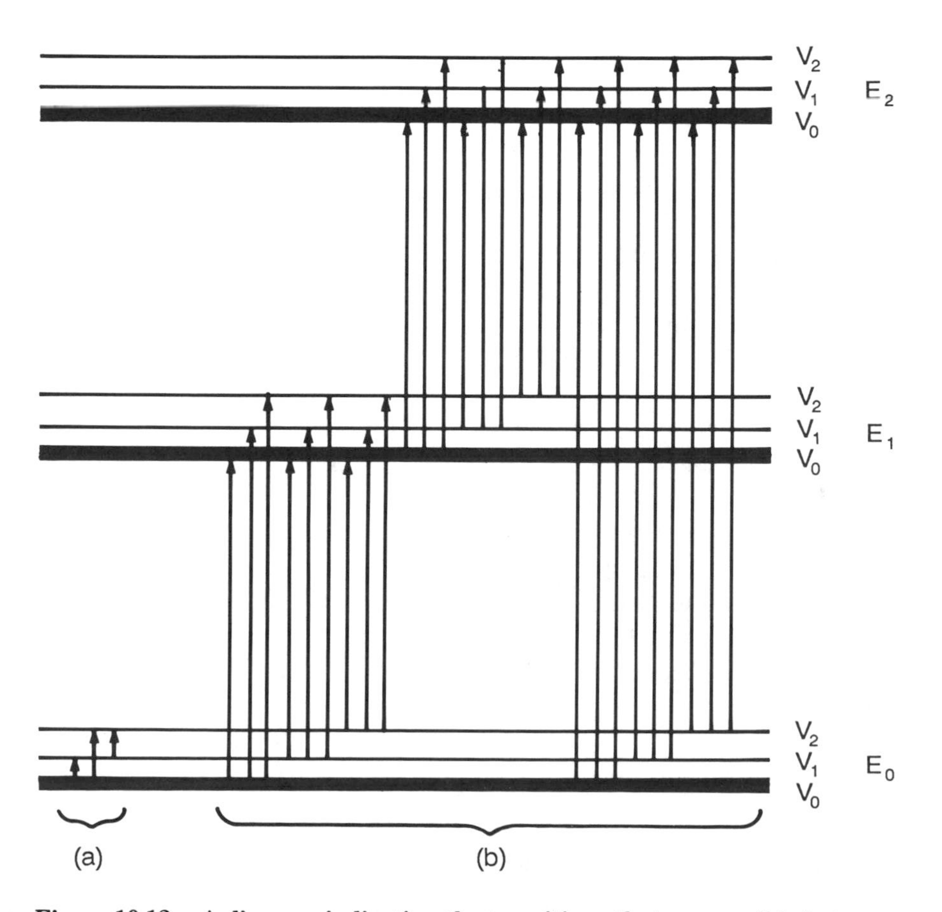

Figure 10.12. A diagram indicating the transitions that are possible between energy levels of a hypothetical molecule (compare with Figure 10.10 for atoms).

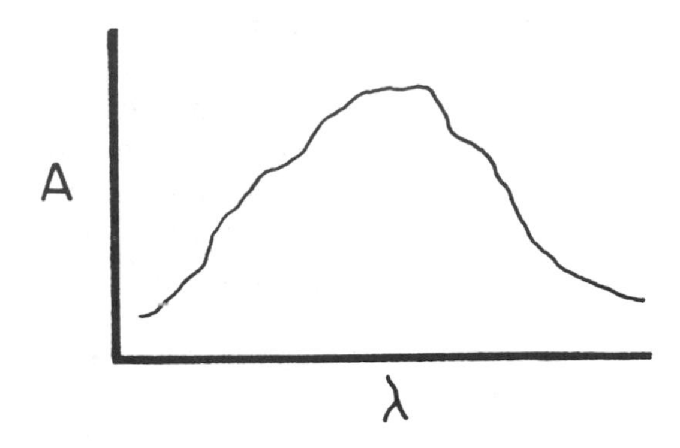

Figure 10.13. A hypothetical example of a molecular absorption spectrum. This would be described as a continuous or band spectrum as opposed to the line spectrum in Figure 10.11.

10.6 NAMES OF TECHNIQUES AND INSTRUMENTS

The absorption of light at particular wavelengths is obviously an important measurement. It is therefore important for all instruments designed to measure such absorption to have, as part of the absorption measurement process, the capability of resolving an entire spectral region ("polychromatic" light) into these particular wavelengths ("monochromatic" light). Two general terms which describe all techniques involving light, including those requiring monochromatic light as just described, are "spectroscopy" and "spectrometry". The general term for the instrument used is "spectrometer". Any spectrometer that resolves polychromatic light into monochromatic light and utilizes a "photomultiplier" tube (described in Chapter 11) as a detector of light is properly called a "spectrophotometer" and the corresponding technique "spectrophotometry". This latter term correctly applies to UV and visible instruments, but technically not to infrared instruments since in that case, as we will see, the detector is not a photomultiplier tube. Nonetheless, infrared spectrometers are often referred to as spectrophotometers. It is most correct to refer to an infrared instrument as a spectrometer and to the technique as infrared spectrometry. In addition, an instrument constructed only to be used in the visible region is often called a colorimeter, since the monochromatic light and the solutions measured would display a color. The technique which utilizes a colorimeter is called colorimetry. This term is also often used to describe a technique in which the color of an unknown is matched visually (rather than with an instrument) to a set of standards in order to determine concentration.

QUESTIONS AND PROBLEMS

1. What is meant by the "dual nature" of light?
2. Light waves are "electromagnetic" and not "mechanical". What does this have to do with the fact that light can travel through outer space?

3. Define the following in words: wavelength, frequency, speed of light, energy of light, wave number, refraction.

4. Compare IR light with UV light in terms of wavelength, frequency, and wave number.

5. List the following in order of increasing energy, frequency, and wavelength: X-rays, infrared light, visible light, radio waves, ultraviolet light.

6. What are the upper and lower wavelength limits of visible light?

7. Express each of the following in centimeters:
   (a) 831 nm
   (b) 749 nm
   (c) 4927 Å
   (d) 3826 Å

8. Express each of the following in nanometers:
   (a) 0.0000000317 cm
   (b) 5.11×10^{-4} cm

9. What is the wavelength of light that has a frequency of
   (a) 6.26×10^{11} sec^{-1}?
   (b) 7.82×10^{12} sec^{-1}?
   (c) 3.94×10^{13} sec^{-1}?

10. What is the frequency of light that has a wavelength of
   (a) 4.26×10^{-4} cm?
   (b) 7.27×10^{2} nm?
   (c) 654 nm?

11. What is the energy of a light that has a frequency of
   (a) 3.02×10^{13} sec^{-1}?
   (b) 3.72×10^{11} sec^{-1}?
   (c) 7.65×10^{10} sec^{-1}?

12. What is the frequency of light that has an energy of
   (a) 6.88×10^{-23} erg?
   (b) 2.72×10^{-25} erg?

13. What is the energy of light with a wavelength of
   (a) 462 cm?
   (b) 46.9 nm?

14. What is the wavelength of light with energy of
   (a) 4.29×10^{-24} erg?
   (b) 9.03×10^{-22} erg?

15. What is the wave number of light with a frequency of
   (a) 5.28×10^{12}?
   (b) 3.17×10^{13}?

16. What is the frequency of light with a wave number of
   (a) 5.67×10^{4} cm^{-1}?
   (b) 3.15×10^{5} cm^{-1}?

17. What is the wave number of light that has a wavelength of
    (a) 792 nm?
    (b) 335 nm?
18. What is the wavelength of light that has a wave number of
    (a) 4.99×10^7 cm^{-1}?
    (b) 6.18×10^5 cm^{-1}?
19. A wavelength of 254 nm is a frequently used wavelength when analyzing for organic compounds.
    (a) What is the wavelength in centimeters?
    (b) What is the frequency corresponding to this wavelength?
    (c) What is the energy of light of this wavelength?
    (d) What is the wave number corresponding to this wavelength?
20. Which has a greater frequency, light of wavelength 627 Å or light of wavelength 462 nm?
21. Which has the longer wavelength, light with a frequency of 7.84×10^{13} sec^{-1} or light with an energy of 5.13×10^{-13} erg?
22. Which has the greater energy, light of wavelength 591 nm or light with a frequency of 5.42×10^{12} sec^{-1}?
23. Which has the greater wave number, light with a frequency of 7.34×10^{13} sec^{-1} or light with an energy of 5.23×10^{-14} erg?
24. Which has the lower energy, light with a frequency of 7.14×10^{13} sec^{-1} or light with a wave number of 1.91×10^4 cm^{-1}?
25. A certain light, A, has a greater frequency than a second light, B.
    (a) Which light has the greater energy, A or B?
    (b) Which light has the shorter wavelength, A or B?
    (c) Which light has the higher wave number, A or B?
26. If the wavelength used in an instrument is changed from 460 nm to 560 nm
    (a) has the energy been increased or decreased?
    (b) has the frequency been increased or decreased?
    (c) has the wave number been increased or decreased?
27. Compare infrared and ultraviolet radiation
    (a) in terms of energy.
    (b) in terms of the type of disturbance they cause within molecules.
28. Why does a yellow sweatshirt appear yellow and not some other color?
29. Explain briefly the phenomenon of light absorption in terms of the energy associated with light and in terms of the energy levels in atoms and molecules.
30. What is meant by each of the following: electronic transition, vibrational transition, rotational transition.
31. Which of the transitions in #30 requires the most energy? Which requires the least energy?
32. In terms of electrons and energy levels, explain what happens when light is absorbed by a sample.

33. Why is it that so many more wavelengths can be absorbed in the case of molecules than with atoms?
34. What is the difference between a molecular absorption spectrum and an atomic absorption spectrum and why does this difference exist?
35. What is a line spectrum?
36. Why is an atomic absorption spectrum a "line" spectrum?
37. Why is it that with atoms only certain specific wavelengths get absorbed (resulting in line spectra), while with molecules broad bands of wavelengths get absorbed (resulting in band spectra)?
38. What is the difference between polychromatic light and monochromatic light?
39. Define spectroscopy, spectrometry, spectrometer, spectrophotometer, spectrophotometry, colorimetry, colorimeter.

Chapter 11

UV-Vis Spectrophotometry

11.1 GENERAL DESCRIPTION

We will first study the analytical technique involving the absorption of UV and visible light. This technique, like the infrared technique to be discussed in the next chapter, involves the use of a laboratory instrument that measures the degree of light absorption by a sample. In addition to the detector (Section 11.2), there are other features that distinguish the UV/vis instruments from the IR instruments, as we will see. The design of the UV/vis instrument is such that a monochromatic wavelength of light from a light source (inside the instrument) is allowed to pass through a sample solution. The amount of light absorbed by this solution is electronically measured by a photomultiplier tube and displayed on a readout device. The analyst is able to quantitate a constituent in the sample by relating this degree of absorption displayed by the instrument to the constituent's concentration. In addition to this quantitative analysis, a qualitative analysis can also be performed by observing the pattern of absorption that a sample exhibits over a range of wavelengths, the so-called "molecular absorption spectrum". No two such patterns from any two chemical species are exactly alike. We therefore have what can be called a molecular "fingerprint", and this is what makes identification, or qualitative analysis, possible. The modern instrument is capable of both qualitative and quantitative measurements. The specific schemes by which qualitative and quantitative analysis is accomplished will be described, but let us begin by detailing the inner workings of the instrument.

11.2 INSTRUMENT DESIGN

There are two types of UV/visible instruments in common use, the single-beam instrument and the double-beam instrument. These instruments have many common features and utilize the same components. First, the single-beam instrument and its components will be described, and this will be followed by a description of the double-beam instrument. A diagram of the basic single-beam spectrophotometer is shown in Figure 11.1. The light source (polychromatic) provides the light to be directed at the sample. The wavelength selector, or monochromator, isolates the wavelength to be used. The sample holder/compartment is a light-tight "box" in which the sample solution is held, and the detector/readout components are the electronic modules which measure and display the degree of absorption. Let us describe each of these components in more detail.

The Light Source. The light source for the visible region is different from the light source for the ultraviolet region. Instruments which have the capability of measuring the absorption in both regions must have two independently selectable light sources. For the visible region, a light bulb with a tungsten filament is used. Such a source is very bright and emits light over the entire visible region and into the near infrared region. The intensity of the light varies dramatically across this wavelength range (Figure 11.2). This creates a bit of a problem for the analyst; we will discuss this later.

For the UV region, the light source is usually the deuterium discharge lamp. Its wavelength output ranges from 185 nm to about 375 nm, satisfactory for most UV analyses. Here again the intensity varies with wavelength.

The Monochromator. The monochromator, or wavelength selector, consists of three main parts: an entrance slit, a dispersing element, and an exit slit. In addition to these, there is often a network of mirrors situated for the purpose of aligning or collimating a beam of light before and after it contacts the dispersing element. A slit is a small circular or rectangular opening cut into an otherwise opaque plate, such as a painted metal plate. The size of the opening is often variable – a "variable slit width". The entrance slit is

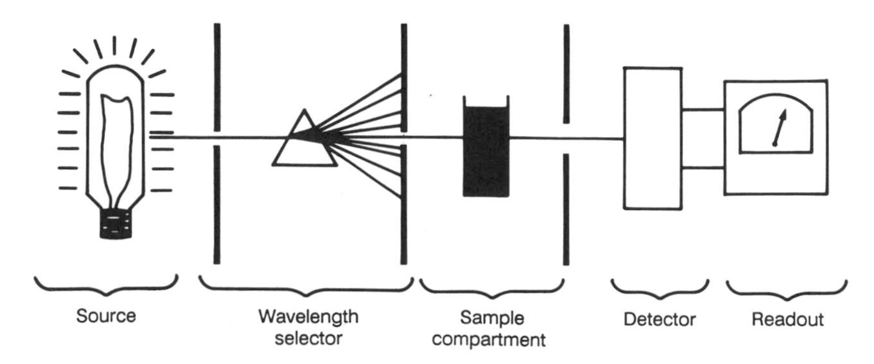

Figure 11.1. The essential components of a spectrophotometer. (From Kenkel, J., *Analytical Chemistry Refresher Manual*, Lewis Publishers, Chelsea, MI, 1992. With permission.)

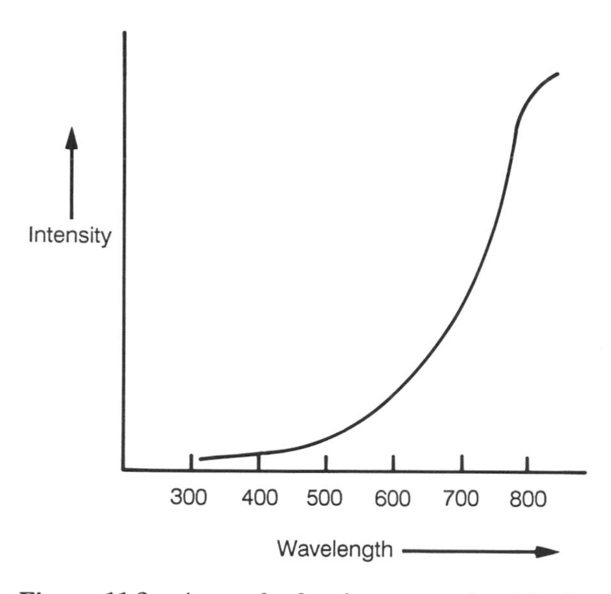

Figure 11.2. A graph showing approximately how the intensity of light from a tungsten filament source varies with wavelength. (From Kenkel, J., *Analytical Chemistry Refresher Manual*, Lewis Publishers, Chelsea, MI, 1992. With permission.)

where light enters the monochromator from the source. Its purpose is to create a unidirectional beam of light of appropriate intensity from the multidirectional light emanating from the source. Its slit width is usually variable so that the intensity of the beam can be varied – the wider the opening the more intense the beam.

After passing through the entrance slit, the beam strikes a dispersing element. The dispersing element disperses the light into its component wavelengths. For visible light, for example, this would mean that a beam of white light is dispersed into a spray of rainbow colors, the violet/blue wavelengths on one end to the red wavelengths on the other, with the green and yellow in between. The monochromatic light is then selected by the exit slit. As the dispersing element is rotated, the spray of colors moves across the exit slit such that a particular wavelength emerges at each position of rotation. The exit slit width can be variable, too, but making it wider would result in a wider wavelength band (the "band pass") passing through, which is usually undesirable. The light emerging from this exit slit is therefore monochromatic and is passed on to the sample compartment to strike the sample. The concept is the same for UV light. (See Figure 11.3.) The rotation of the dispersing element is accomplished either by manually turning a knob on the face of the instrument or internally by programmed scanning controls. The position of the knob is coordinated with the wavelength emerging from the exit slit such that this wavelength is read from a scale of wavelengths on the face of the instrument or on a readout meter. For manual control, the operator thus simply dials in the desired wavelength.

The dispersing element is either a diffraction grating or a prism. A prism is a three-dimensional triangularly shaped glass or quartz block as indicated

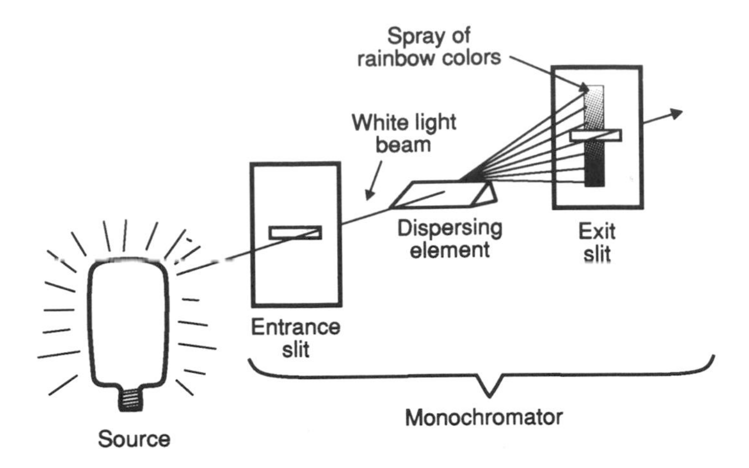

Figure 11.3. The monochromator component. See text for full description. (From Kenkel, J., *Analytical Chemistry Refresher Manual*, Lewis Publishers, Chelsea, MI, 1992. With permission.)

in Figures 11.1 and 11.3. When the light beam strikes one of the three faces of the prism, the light emerging through another face is dispersed.

A diffraction grating is used more often than a prism. A diffraction grating is like a highly polished mirror that has a large number of precisely parallel lines or grooves scribed onto its surface. Light striking this surface is reflected, diffracted, and dispersed into the component wavelengths. See Figure 11.4 for more explicit diagrams of a prism and a diffraction grating.

Sample Compartment. Next, the light from the monochromator passes through the sample in the sample compartment. For the UV/visible instrument, this is a light-tight box in which the container holding the sample solution is placed. The container is called a "cuvette". The material making up the cuvette walls is of interest.

For optimum performance, this material must be transparent to all wavelengths of light that may pass through it. For visible light, this means that the material must be completely clear and colorless. Inexpensive materials, such as plastic and ordinary glass, are perfectly suitable in that case. How-

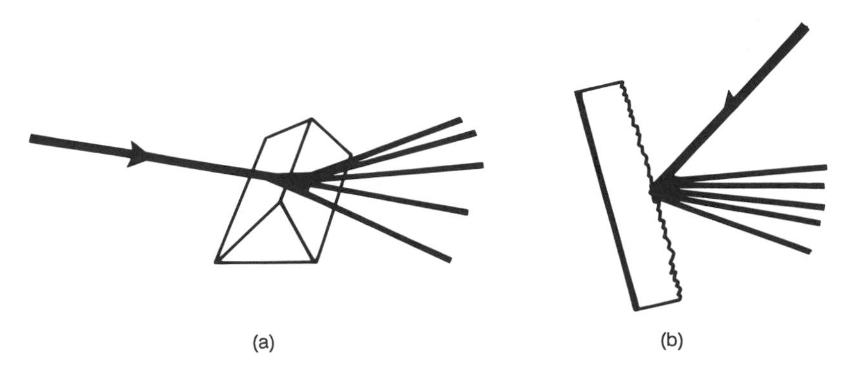

Figure 11.4. (a) A prism; (b) a diffraction grating. (From Kenkel, J., *Analytical Chemistry Refresher Manual*, Lewis Publishers, Chelsea, MI, 1992. With permission.)

ever, such materials may not be transparent to light in the ultraviolet and infrared regions. For the ultraviolet region the more expensive quartz cuvettes must be used, while in the infrared region the walls of the cuvette are often made of inorganic salt crystals (Chapter 12).

Detector/Readout. The most common detector for UV/vis spectrophotometry is the photomultiplier tube, which is comparable to a solar cell. When light of a particular intensity strikes it, a current, or electrical signal, of proportional magnitude is generated, amplified, and sent on to the readout. The process is not energy efficient like a solar cell needs to be, however, since it requires more electricity to run it than it generates.

The photomultiplier tube consists of a photocathode, an anode, and a series of dynodes for multiplying the signal, hence its name. The dynodes are situated between the photocathode and the anode. A high voltage is applied between the photocathode (an electrode which emits electrons when light strikes it) and the anode. When the light beam from the sample compartment strikes the photocathode, electrons are emitted and accelerated, because of the high voltage, to the first dynode, where more electrons are emitted. These electrons pass on to the second dynode, where even more electrons are emitted, etc. When the electrons finally reach the anode, the signal has been sufficiently multiplied as to be treated as any ordinary electrical signal capable of being amplified by a conventional amplifier. This amplified signal is then sent on to the readout in one form or another. Figure 11.5 illustrates this process.

Most instruments offer a choice as to what is displayed on the readout, since there are several parameters that can have analytical importance or convenience. These parameters, notably "transmittance", "percent transmittance", "absorbance", and analytical concentration, will be defined and discussed. The display of each of these requires associated mathematical operations so as to obtain it from the intensity of the light to which the detector's output signal is proportional. Thus, electronic circuits which compute a

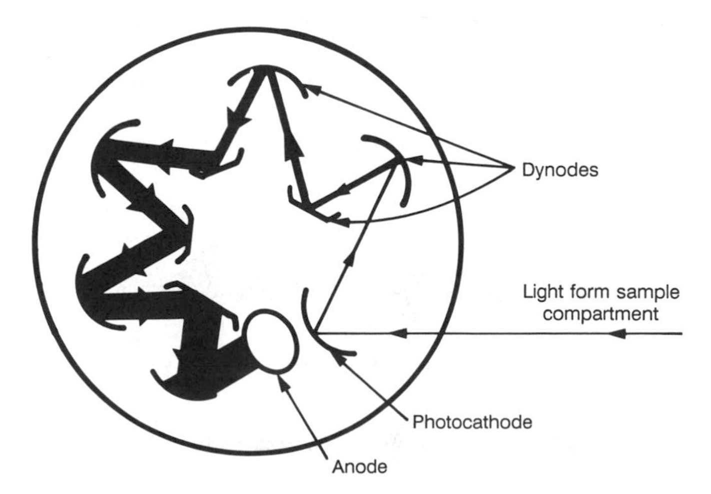

Figure 11.5. The process occurring in a photomultiplier tube. (From Kenkel, J., *Analytical Chemistry Refresher Manual*, Lewis Publishers, Chelsea, MI, 1992. With permission.)

ratio of intensities (transmittance), the percent transmittance, the negative logarithm of this ratio (absorbance), etc. are also often included as part of the total circuitry of these instrument components. In addition, the signal may be digitized using an analog to digital converter (Chapter 9) for display on a digital readout or for output to a computer.

Calibration. The calibration of the single-beam instrument that has just been described is important to consider. Such calibration involves adjustment of the "dark current" control when all the light is physically blocked from striking the detector and adjustment of the entrance slit opening to the optimum value when no analyte species is present in the path of the light. In order to understand these two steps, we need to define some parameters involved.

The intensity of light striking the detector when a "blank" solution (no analyte species) is held in the cuvette is given the symbol "I_o". The blank, as discussed in Section 9.4, is a solution that contains all chemical species that will be present in the standards and samples to be measured (at equal concentration levels) except for the analyte species. Such a solution should not display any absorption due to the analyte, and thus I_o represents the maximum intensity that can strike the detector at any time. When the blank is replaced with a solution of the analyte, a less intense light beam will be detected. The intensity of the light for this solution is given the symbol "I". (See Figure 11.6.) The fraction of light transmitted is thus I/I_o. This fraction is defined as the transmittance, "T".

$$T = \frac{I}{I_o} \tag{11.1}$$

The percent transmittance is similarly defined:

$$\%T = T \times 100 \tag{11.2}$$

Transmittance and percent transmittance are two parameters that can be displayed on the readout. Calibration thus involves adjustment to 0% transmittance using the dark current control when the light is physically blocked from the detector and adjustment to 100%T using the entrance slit opening control when the blank solution is in the path of the light. Opening or closing the entrance slit increases or decreases the light intensity eventually

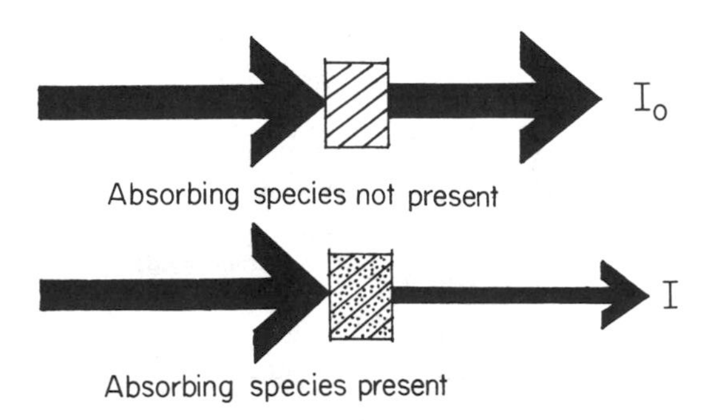

Figure 11.6. Illustration of the definitions of I and I_o.

striking the detector and thus is useful for such calibration. Following calibration, the samples and standards, in the same or identical cuvettes (in order to measure only the effect of the solution—see Section 11.4), are consecutively placed in the path of the light and the readouts recorded.

Double-Beam Instruments. The discussion in this section thus far has been concerned with instruments in which a single beam of monochromatic light passes through the sample compartment. While such instruments are in very common use (they are comparatively inexpensive), an instrument design which utilizes two beams of light passing through the sample compartment is also quite common and offers some important advantages.

In any spectrophotometer, if the intensity of the light passing through the sample changes following calibration and before a sample or standard is read, an error will result. Such a change can occur due to fluctuations in the power supply to the source and/or detector (the line voltage), due to an unstable source, or when the operator selects a different wavelength via the monochromator control. The intensity of the light changes dramatically as the wavelength through the monochromator is changed. Refer back to Figure 11.2. If a given experiment involves a change in the wavelength, such as when determining the pattern of absorption over a range of wavelengths (the "molecular absorption spectrum"), or if one suspects that the line voltage fluctuates, recalibration with the blank is necessary periodically or each time the wavelength is changed. This can result in an extraordinarily long and tedious procedure and/or a loss of accuracy since the procedure of recalibration with the blank can take up to 5 to 10 sec and a power fluctuation can occur in this length of time. The double-beam design allows the blank to be checked and calibration to take place only a split second before the sample is read. Not only does this take much less time, eliminating the need for continuous manual monitoring of the blank, but it also increases accuracy since the time between blank calibration and sample measurement is dramatically decreased. It also allows rapid wavelength scanning in order to conveniently obtain the molecular absorption spectrum.

A schematic diagram of a typical double-beam design for UV/vis spectrophotometry is shown in Figure 11.7. The light coming from the monochro-

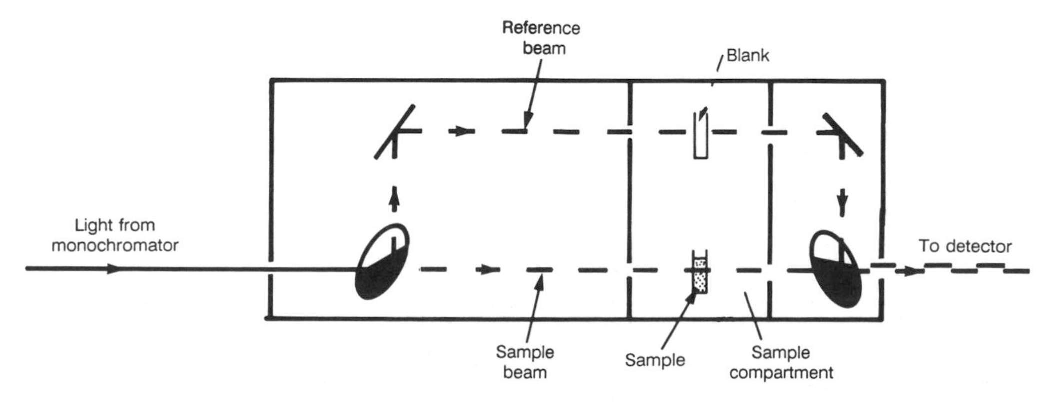

Figure 11.7. A schematic diagram of a typical double-beam design. (From Kenkel, J., *Analytical Chemistry Refresher Manual*, Lewis Publishers, Chelsea, MI, 1992. With permission.)

mator is directed along either one of two paths with the use of a "chopper". The chopper in this case is a rotating circular half-mirror used for splitting a light beam into two beams. At one moment the light passes through the sample, while at the next moment it passes through the blank. Both beams are joined again with a beam combiner, such as another rotating half mirror, prior to entering the detector. The detector sees alternating light intensities, I and I_0, and thus immediately and automatically compensates for fluctuations and wavelength changes, usually by automatically widening or narrowing the entrance slit to the monochromator. If the beam becomes less intense, the slit is opened; if the beam becomes more intense, the slit is narrowed. Thus the signal relayed to the readout is free of effects of intensity fluctuations that cause errors.

Sample compartments in such instruments have two cuvette holders, one for a cuvette containing the sample or standard and one for a cuvette containing the blank. The two beams of light pass through the sample compartment, one through the blank (the reference beam) and one through the sample or standard (the sample beam). The two cuvettes must be matched in terms of path length and reflective and refractive properties.

Scanning double-beam UV/visible spectrophotometers have become very commonplace in analytical laboratories. Modern instruments can include substantial microprocessor control, including control of scan time, output functions, and data storage as well as internal storage of Beer's law data (see next section) such that sample concentrations can be displayed on the readout. These instruments also have output terminals capable of transferring concentration, absorption/transmittance, and wavelength data to an external recorder or computer.

11.3 QUALITATIVE AND QUANTITATIVE ANALYSIS

The transmittance and percent transmittance of a sample have been defined (Equations 11.1 and 11.2). The unfortunate aspect of transmittance is that it is not linear with concentration. Low concentrations give a high transmittance and high concentrations give a low transmittance. However, the relationship is not linear but logarithmic. (See Figure 11.8a.) Without a

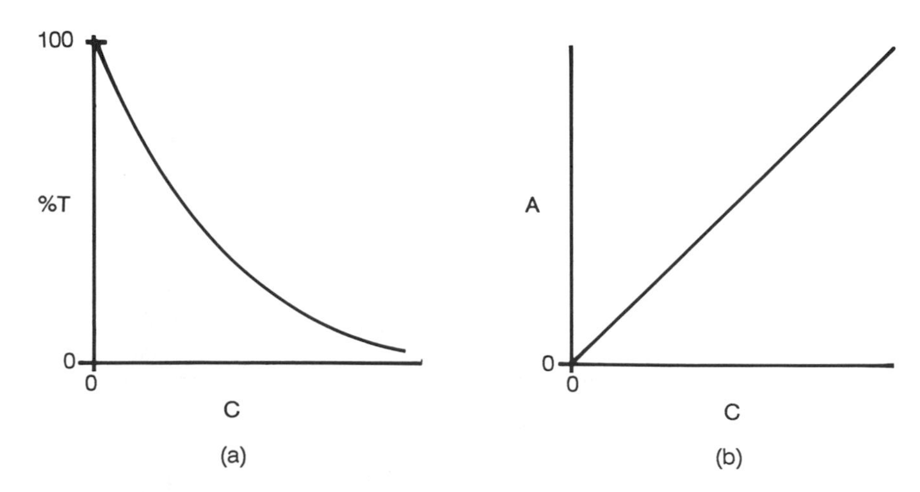

Figure 11.8. (a) A plot of percent transmittance vs. concentration. (b) A plot of absorbance vs. concentration.

linear relationship, the concentration of an unknown sample cannot be determined by the usual procedures outlined in Chapter 9. However, since the relationship is logarithmic, the logarithm of the transmittance can be expected to be linear. Thus, we define the parameter of "absorbance" as being the negative logarithm of the transmittance and give it the symbol "A":

$$A = -\log T \tag{11.3}$$

Absorbance is a parameter then that increases linearly with concentration. (See Figure 11.8b.) Absorbance is thus the parameter that is important for quantitative analysis. If transmittance is measured by an instrument, it must be converted to absorbance via Equation 11.3 (or by using semi-log graph paper) before plotting vs. concentration and obtaining the unknown concentration according to procedures outlined in Chapter 9.

Example 1

What is the absorbance of a sample if the transmittance is 0.347?

Solution

$$A = -\log T$$

$$A = -\log (0.347)$$

$$A = 0.460$$

Example 2

What is the absorbance of a sample that has a %T of 49.6%?

Solution

$$\%T = T \times 100$$

$$T = \frac{\%T}{100} = \frac{49.6}{100} = 0.496$$

$$A = -\log T$$

$$A = -\log (0.496) = 0.305$$

The conversion of A to T or %T, the reverse of the above, may also be important.

Example 3

What is the %T if A = 0.774?

Solution

$$A = -\log T$$

$$0.774 = -\log T$$

$$T = 0.168$$

$$\%T - T \times 100$$

$$\%T = 16.8\%$$

As indicated previously, most instruments have the electronic circuitry for calculating absorbance built into the detector/readout system and thus are capable of displaying absorbance on the readout, including on a recorder, such as for a display of the molecular absorption spectrum. Such a display is useful for a qualitative analysis since, as indicated previously, it is a molecular fingerprint of the system studied. Let us study quantitative and qualitative analysis in UV/visible spectrophotometry beginning with this latter concept.

Qualitative Analysis. The molecular absorption spectrum, the pattern of absorption over a range of wavelengths, which is the molecular "fingerprint" of a particular chemical species, has been referred to at several junctures in this chapter and in Chapters 9 and 10. We are now ready to expand on this concept and look at some examples. The absorption pattern can be displayed as either a plot of absorbance vs. wavelength or transmittance vs. wavelength. Since absorbance and transmittance are opposites, the transmission spectrum appears as an inversion of the absorption spectrum. Some examples of absorption spectra are given in Figure 11.9, and some examples of transmission spectra are shown in Figure 11.10. All represent molecular fingerprints — no two chemical species will display identical absorption and transmission spectra. Qualitative analysis can simply involve a matching-up of spectra, known with unknown.

Absorption in the ultraviolet region warrants additional comment. Certain characteristics of the structure of organic compounds will display certain unique characteristics in ultraviolet absorption spectra. In the first place, organic structures with nothing but single bonds, such as alkanes and ordinary alcohols, do not absorb at all in the ultraviolet region. In fact, they are often used as solvents for compounds that do absorb in this region. However, structures with double bonds, triple bonds, or benzene rings have very strong absorption patterns. All of these types of structures contain pi electrons, and pi electrons are especially susceptible to ultraviolet absorption. Any such group present in a structure, causing the appearance of absorption bands in the ultraviolet (or visible) region, is called a "chromophore". A -C=C- bond, a -C=O bond, a -N=O bond, and a benzene ring are thus all classified as chromophores.

If two or more chromophores appear in the same molecule, the location of one relative to the other dictates a particular pattern. If they are separated by more than one carbon, the absorption pattern represents a simple summation of the two individual patterns. If the two chromophores are on adjacent carbons, an apparent "shifting" (called a "bathochromic" shift) of

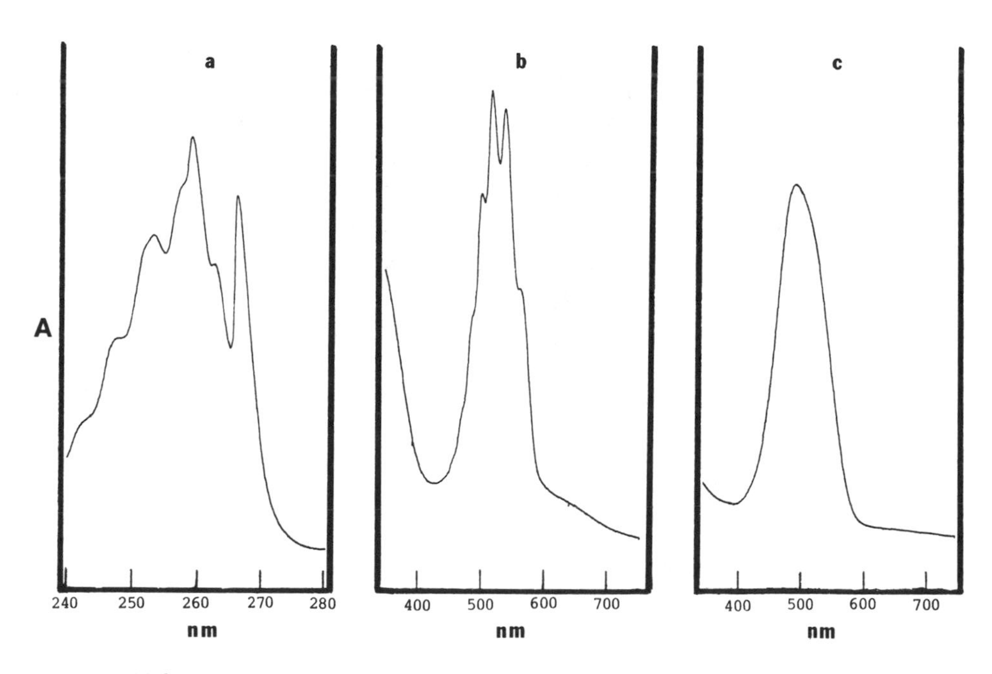

Figure 11.9. Absorption spectra of (a) toluene in cyclohexane, (b) potassium permanganate in water, (c) methyl red in water. (From Kenkel, J., *Analytical Chemistry Refresher Manual*, Lewis Publishers, Chelsea, MI, 1992. With permission.)

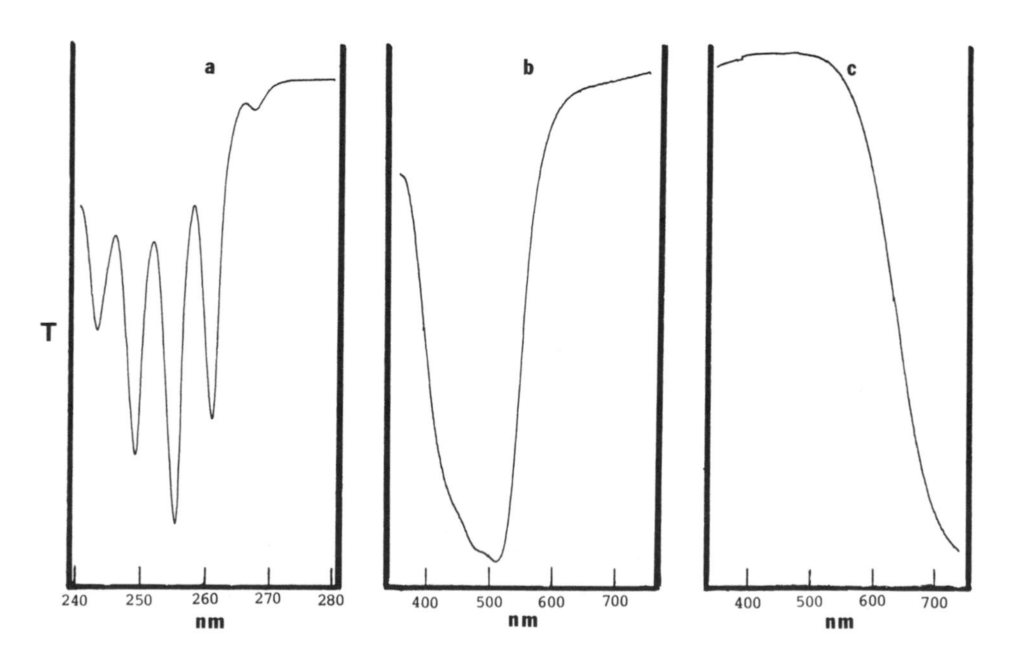

Figure 11.10. Transmission spectra of (a) benzene in cyclohexane, (b) iron/*o*-phenanthroline in water, and (c) copper sulfate in water. (From Kenkel, J., *Analytical Chemistry Refresher Manual*, Lewis Publishers, Chelsea, MI, 1992. With permission.)

the absorption maximum to a longer wavelength is usually observed and the absorbance is increased (a "hyperchromic" effect). If the opposite is observed (a shift to a shorter wavelength), it is called a "hypochromic" effect. When the two chromophores are attached to the same atom, the two observations just noted also occur, but to a lesser extent. Additional shifting of the absorption maximum can occur when an ordinarily non-absorbing group is attached near a chromophore. Such a group is called an "auxochrome". An example would be an –OH group attached near a double bond. Additionally, effects of solvent and pH are important to observe. Thus, in addition to fingerprinting (spectra matching), observations of shifting of absorption maxima and absorbance intensity can be very important in a qualitative analysis.

Quantitative Analysis. The exact relationship between absorbance and concentration is a famous one. It is known as the Beer-Lambert law, or simply as Beer's law. A statement of Beer's law is

$$A = abc \tag{11.4}$$

in which A is absorbance, a is absorptivity or the extinction coefficient, b the path length, and c the concentration. Absorbance was defined previously in this section. Absorptivity is the inherent ability of a chemical species to absorb light and is constant at a given wavelength. Path length is the distance the light travels through the measured solution. It is the inside diameter of the cuvette. Path length is measured in units of length, usually centimeters or millimeters. Concentration can be expressed in a variety of units, but usually is in molarity, ppm, or grams per 100 mL. The unit of absorptivity depends on the units of these other parameters, since absorbance is a dimensionless quantity. When the concentration is in molarity and the path length is in centimeters, the units of absorptivity must be L/(mol cm). Under these specific conditions, the absorptivity is called the "molar absorptivity", or the "molar extinction coefficient", and is given a special symbol, the Greek letter epsilon (ϵ). Beer's law is therefore sometimes given as

$$A = \epsilon bc \tag{11.5}$$

Example 4

The measured absorbance of a sample in a 1.00-cm cuvette is 0.544. If the concentration is 1.4×10^{-3} *M*, what is the molar absorptivity for this species?

Solution

$$A = \epsilon bc \qquad \text{(b in centimeters, c in molarity)}$$

$$\epsilon = \frac{A}{bc}$$

$$\epsilon = \frac{0.544}{1.00 \times 1.4 \times 10^{-3}}$$

$$\epsilon = 389 \text{ L/(mol cm)}$$

Defining the molar absorptivity parameter presents analytical chemists with a standardized method of comparing one spectrophotometric method with another. The larger the molar absorptivity, the more sensitive the method. [It is not unusual for molar absorptivity values to be as large as 10,000 L/(mol cm) and higher]. In addition, it was stated above that the absorptivity is constant "at a given wavelength", implying that it changes with wavelength. The greatest analytical sensitivity occurs at the wavelength at which the absorptivity is at a maximum. This is the same wavelength that displays the maximum absorbance in the molecular absorption spectrum for that species. In fact, the molecular absorption spectrum is sometimes shown as a plot of absorptivity vs. wavelength rather than absorbance vs. wavelength. For a given absorbing species such a plot would display the same characteristic shape and the same wavelength of maximum absorbance.

Beer's law constitutes the spectrophotometry application of the discussion in Chapter 9 which states that in most instrumental quantitative analyses an instrument readout is proportional to concentration. In this case, absorbance is the readout. Thus, most quantitative analyses by Beer's law involve preparing a series of standard solutions, measuring the absorbance of each in identical cuvettes, and plotting the measured absorbance vs. concentration, creating the so-called "standard curve". The absorbance of an unknown solution is then measured and its concentration determined from the graph. Such a graph is often called a Beer's law plot. (See Figure 11.11.)

Of course, an unknown's concentration can be determined by comparing its absorbance with just one standard:

$$\frac{c_u}{c_s} = \frac{A_u}{A_s} \tag{11.6}$$

in which c_u is the concentration of the unknown, c_s is the concentration of the standard, A_u is the absorbance of the unknown, and A_s is the absorbance of the standard. It may also be determined by direct calculation if the absorptivity value and path length are precisely known:

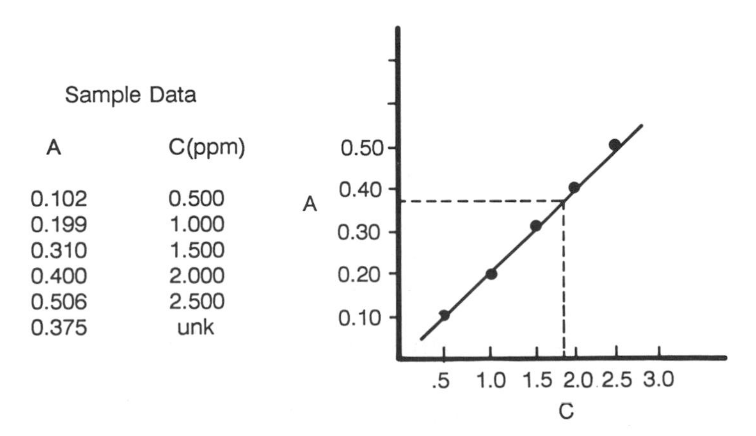

Figure 11.11. Some sample data and a Beer's law plot of the data, showing the determination of the unknown concentration.

$$c = \frac{A}{ab} \tag{11.7}$$

See Chapter 9 for a more thorough discussion of each of these methods, including limitations, and also the calculations that are often involved after the unknown solution concentration is determined.

Example 5

What is the concentration of an absorbing species if its molar absorptivity is 1500 L/(mol cm) and the measured absorbance in a 1.00-cm cuvette is 0.742?

Solution

$$A = \epsilon bc$$

$$c = \frac{A}{\epsilon b}$$

$$c = \frac{0.742}{1500 \times 1.00} = 4.95 \times 10^{-4}\ M$$

Example 6

The absorbance of a standard 2.0 ppm sample in a 1.00-cm cuvette is 0.246. What is the concentration of an unknown solution (same absorbing species) if the absorbance in the same cuvette is 0.529? (Assume a linear relationship in the concentration range in question.)

Solution

$$A_s = abc_s \qquad A_u = abc_u$$

$$ab = \frac{A_s}{c_s} \qquad A_u = \frac{A_s}{c_s} \times c_u$$

$$c_u = \frac{A_u \times c_s}{A_s}$$

$$c_u = \frac{0.529 \times 2.0}{0.246}$$

$$c_u = 4.3 \text{ ppm}$$

Alternate Solution

$$\frac{A_s}{A_u} = \frac{abc_s}{abc_u} \qquad \frac{A_s}{A_u} = \frac{c_s}{c_u}$$

$$C_u = \frac{A_u \times C_s}{A_s} = \frac{0.529 \times 2.0}{0.246} = 4.3 \text{ ppm}$$

Interferences and Deviations. Interferences are quite common in qualitative and quantitative analysis by UV/vis spectrophotometry. An interference is a contaminating substance that gives an absorbance signal at the same wavelength or wavelength range selected for the analyte. For qualitative analysis this would show up as an incorrect absorption spectrum, thus possibly leading to erroneous conclusions if the contaminant was not known to be present. For quantitative analysis this would result in a higher absorbance than one would measure otherwise. Absorbances are additive. This means that the total absorbance measured at a particular wavelength is the sum of absorbances of all absorbing species present. Thus, if an interference is present, the correct absorbance can be determined by subtracting the absorbance of the interference at the wavelength used, if it is known. The modern solution to these problems is to utilize separation procedures such as extraction or liquid chromatography to separate the interfering substance from the analyte prior to the spectrophotometric measurement. These techniques will be discussed in Chapters 15 and 17.

Deviations from Beer's law are in evidence when the Beer's law plot is not linear. This is probably most often observed at the higher concentrations of the analyte. (See Figure 11.12.) Such deviations can be either chemical or instrumental. Instrumental deviations occur because it is not possible for an instrument to be accurate at extremely high or extremely low transmittance values – values that are approaching either 0% T or 100% T. The normal working range is between 15% and 80%, corresponding to absorbance values between 0.10 and 0.82. It is recommended that standards be prepared so as to measure in this range and that unknown samples be diluted if necessary. Chemical interferences occur when a high or low concentration of the analyte causes chemical equilibrium shifts in the solution which directly or indirectly affect its absorbance. It may be necessary in these instances to work in a narrower concentration range than expected. This means that unknown samples may also need to be further diluted as in the instrumental deviation case.

11.4 THE QUESTION OF MATCHED CUVETTES

If two or more different cuvettes are used in an analysis, one should be sure that they are "matched". Matched cuvettes are identical with respect to path length and reflective and refractive properties in the area where the light beam passes. If the path lengths were different, or if the wall of one cuvette reflects more or less light than another cuvette, then the absorbance measurement could be different for that reason, and not because the

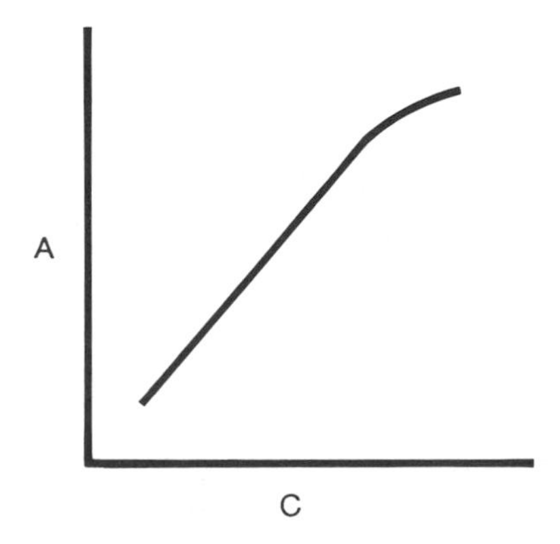

Figure 11.12. A Beer's law plot that shows a deviation from linearity at higher concentrations.

solution concentration is different. Thus there would be an error. A general guideline with respect to matching cuvettes is that a solution which transmits 50% of the incident light should not have a reading differing by more than 1% in any cuvette. Also, such cuvettes must be placed in the instrument in exactly the same way each time, since the path length and reflective/refractive properties can change by rotating the cuvette. Obviously, cuvettes that have scratches, or cleaning procedures that may cause scratching, should be avoided. Any liquid or fingerprints adhering to the outside wall must be removed with a soft cloth or towel.

11.5 THE ABSORPTION SPECTRUM AND λ_{MAX}

Previously we described the use of the monochromator as the device for selecting the wavelength to be used for a spectrophotometric analysis. We indicated that the choice of wavelength can be accomplished by turning a knob on the face of the instrument so as to obtain the correct position of the dispersing element relative to the slit. The question remains as to how one knows what wavelength is to be used in an analysis.

Just as one chooses the *method* which gives the greatest absorptivity, so also one chooses the *wavelength* that also gives the greatest absorptivity. This wavelength is ascertained by examining the absorbance (or absorptivity) vs. λ behavior, the molecular absorption spectrum, for a given species. The wavelength that corresponds to the greatest absorption, the highest "peak" in this spectrum, is the wavelength to be used.

It should be mentioned here that "transmittance spectra" are also commonly recorded in the laboratory. Since transmittance and absorbance are inverse terms (when transmittance is high, absorbance is low and vice versa), the transmittance spectrum of a compound would appear to be an inversion of the absorption spectrum. Thus the proper wavelength to use in an analysis, the wavelength of maximum absorbance, would be the wavelength of minimum transmittance. Infrared absorption spectra (Chapter 12) are typically recorded as transmittance spectra.

11.6 FLUOROMETRY

Fluorometry is an analytical technique which utilizes the ability of some substances to exhibit fluorescence. Fluorescence is a phenomenon in which the substance appears to glow when a light shines on it. In other words, light of a wavelength different from the irradiating light is released or "emitted" following the absorption process. Most often the irradiating light is ultraviolet light and the emitted light is visible light. The phenomenon is explained based on light absorption theory and what can happen to a chemical species in order to revert back to the ground state once the absorption—the elevation to an excited state—has taken place. Fluorescence can occur with both molecules and atoms. The present discussion will focus on molecules and complex ions. Atomic fluorescence will be discussed in Chapter 14.

All atoms and molecules seek to exist in their lowest possible energy state at all times. When a molecule is raised to an excited electronic energy state through the absorption of light, it is no longer in its lowest possible energy state and will seek to lose the energy it gained any way it can. Most often, the energy is lost through mechanical means, such as through collisions with other chemical species in the solution. However, there can be a direct jump back to the ground state with only some intermediate stops at some lower vibrational states in between. With such a jump back to the ground state, the energy the molecule gained as a result of the absorption process is lost in the form of light, and since it is light of less energy due to the accompanying small energy losses in the form of vibrational loss, the wavelength is longer. See Figure 11.13 for a graphical picture of this process.

The instrument for measuring fluorescence intensity for quantitative analysis is constructed with two monochromators, one to select the wavelength to be absorbed and one to select the fluorescence wavelength to be measured. In addition, the instrument components are configured so that the fluorescence measurement is optimized to be free of interference from transmitted light from the source.

This latter point means that the fluorescence monochromator and detector are not placed in a straight line with the source, absorption monochromator, and sample (such as in an absorption spectrophotometer), but are rather placed at a right angle as shown in Figure 11.14. Thus, there is a "right angle configuration" in a fluorometer to avoid any interference from the transmitted light from the light source.

Colored glass light filters are often used for the monochromators in fluorometers. Instruments with such filters are called filter fluorometers and are considerably less expensive than spectrophotofluorometers, which have standard slit/dispersing element/slit monochromators. Although these latter instruments are excellent for determining the proper wavelengths to be used and for precise work, filter fluorometers have proved quite satisfactory for most routine work and are much less expensive.

We have indicated that the intensity of the fluorescence emitted by the fluorescing species is proportional to the concentration of this species in solution. Fluorescence intensity is therefore the parameter to be measured and related to concentration. This means that the fluorescence intensity of a single standard solution, or a series of standards, is measured and related to concentration. A graph of fluorescence intensity vs. concentration is expected to be linear in the concentration range studied. Procedures outlined in Chapter 9, and also Section 11.3 for Beer's law, are also applicable here.

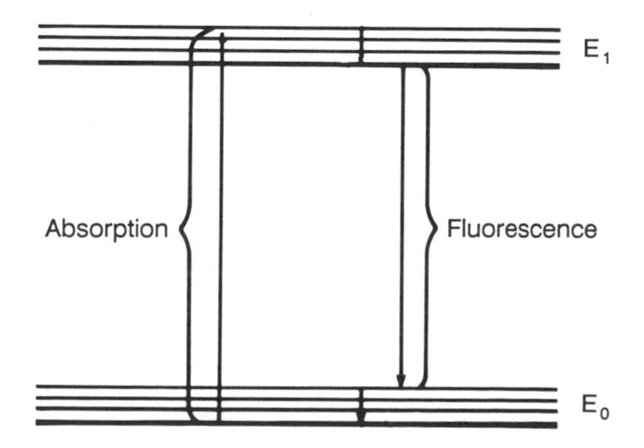

Figure 11.13. An energy level diagram showing the transitions occurring when absorption is accompanied by fluorescence.

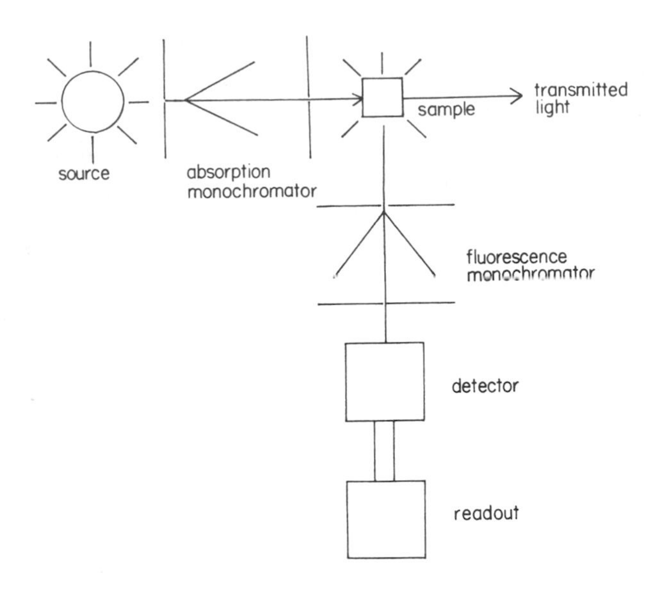

Figure 11.14. The basic fluorometer. The two monochromators can be glass filters.

The types of compounds that can be analyzed by fluorometry are limited. The drop of an electron to the ground state must be accompanied by the emission of light (it must be a direct drop). The kind of electron which is most apt to be able to do this is a π electron and, more specifically, electrons found in benzene rings. Fused benzene ring systems, such as those in Figure 11.15, are especially highly fluorescent compounds. Metals can be analyzed by fluorometry if they are able to form complex ions by reaction with a ligand having the required electrons. An example is aluminum, as shown in Figure 11.16.

Fluorometry and absorption spectrophotometry are competing techniques in the sense that both are techniques used to analyze for molecular species. Each offers its own advantages and disadvantages, however. The number of chemical species that can exhibit fluorescence is very limited. However, for those species that do fluoresce, the fluorescence is generally very intense. Thus we can say that while absorption spectrophotometry is much more universally applicable, fluorometry suffers less from interferences and is usually much more sensitive. Therefore, when an analyte does exhibit the somewhat rare quality of fluorescence, fluorometry is likely to

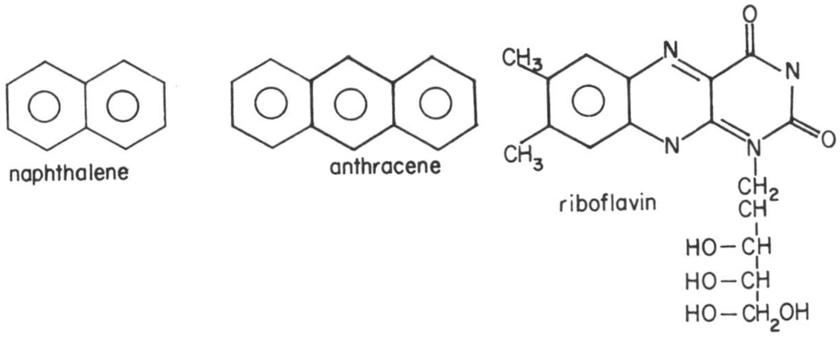

naphthalene

anthracene

riboflavin

Figure 11.15. Highly fluorescent compounds.

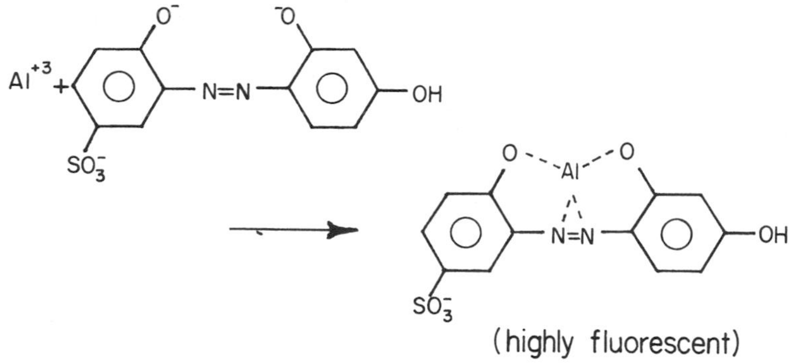

Figure 11.16. The formation of a complex ion of aluminum, which is highly fluorescent.

be chosen for the analysis. The analysis of foods for vitamin content is an example, since vitamins such as riboflavin and niacin exhibit fluorescence and fluorometry would be both relatively free of interference and very sensitive.

EXPERIMENTS

EXPERIMENT 22: COLORIMETRIC ANALYSIS OF PREPARED OR REAL WATER SAMPLES FOR IRON

NOTE: Safety glasses are required.

1. Prepare a 100 ppm Fe stock (100 mL) from the available 1000 ppm solution.
2. Prepare standards of 1.0, 2.0, 3.0, and 4.0 ppm iron from the 100 ppm stock in 50-mL volumetric flasks, but do not dilute to the mark. Have a fifth flask ready, but add no iron to it – it is for a blank. If real water samples are being analyzed and you expect the iron content to be low, prepare two additional standards that are 0.1 ppm and 0.5 ppm.
3. Prepare the real water samples by pipetting 25 mL of each into 50-mL volumetric flasks. Do not dilute to the mark. If a prepared water sample is provided, do not dilute to the mark until directed to do so in Step 4. Your instructor may also provide you with a control sample.
4. To each of the flasks (standards, samples, blank, and control) add

   (a) 0.5 mL of 10% hydroxylamine hydrochloride
   (b) 2.5 mL of 0.1% *o*-phenanthroline solution
   (c) 4.0 mL of 10% sodium acetate

   These reagents are required for proper color development. Hydroxylamine hydrochloride is a reducing agent which is required to keep the iron in the +2 state. The *o*-phenanthroline is a ligand which reacts with the Fe^{+2} to form the orange-colored complex ion which is the "absorbing species". In addition, since the reaction is pH dependent, sodium acetate is needed for buffering at the optimum pH.

Dilute each to the mark with distilled water and mix thoroughly.

5. Using one of your standards and the Spectronic 20, or another single-beam spectrophotometer, obtain an absorption (or transmittance) spectrum of the Fe/*o*-phenanthroline complex ion (instructor will demonstrate use of instrument). Determine the wavelength of maximum absorbance from the spectrum and use this wavelength to obtain absorbance readings of all the solutions. (Use the blank for 100% T setting.)
6. Plot the absorbance vs. concentration either manually or using a computer, as directed by your instructor, and obtain the concentration of the unknowns.
7. If real water samples were analyzed, remember that these samples were diluted by a factor of 2. Multiply the concentrations in Step 6 by 2 to get the final answers in that case.

EXPERIMENT 23: SPECTROPHOTOMETRIC ANALYSIS OF A PREPARED SAMPLE FOR TOLUENE

NOTE: The slightest contamination with water or other solvent will result in gross error. Safety glasses are required.

1. Prepare 100 mL of a stock standard solution of toluene in cyclohexane by pipetting 0.1 mL of toluene into the flask and diluting to the mark with cyclohexane. This solution has a concentration of 0.870 g/L.
2. Prepare four standards with concentrations of 0.0435 g/L, 0.0870 g/L, 0.174 g/L, and 0.261 g/L in cyclohexane solvent by diluting the stock solution prepared in Step 1. Use 25-mL volumetric flasks.
3. Obtain an unknown from your instructor and dilute to the mark with cyclohexane.
4. Obtain an absorbance spectrum of each of your standards and the unknown on a scanning UV/vis spectrophotometer. If directed by your instructor, use a computer for data acquisition and storage.
5. Find the maximum absorbance value for each set of data either manually or with the use of a computer.
6. Prepare a Beer's law plot and obtain the concentration of the unknown.

EXPERIMENT 24: DETERMINATION OF NITRATE IN WATER BY UV SPECTROPHOTOMETRY

NOTE: Safety glasses are required.

1. Crush and dry a small quantity of potassium nitrate for 1 h. Prepare a stock standard by weighing 36.1 mg of the dried KNO_3 into a 100-mL flask and diluting to the mark with distilled water. Shake to dissolve. This solution is 50 ppm *N*. (See Chapter 7 for the calculation for preparing a solution of a certain ppm from a pure solid chemical con-

taining the element. This is 50 ppm *N* from KNO_3. See also question #27 at the end of Chapter 7.)

2. Prepare standards that are 0.5, 1.0, 3.0, 5.0, and 10.0 ppm *N* in 50-mL volumetric flasks by diluting the stock standard with distilled water. After diluting to the mark, pipet 1.0 mL of 1 *N* HCl into each flask and shake.
3. Prepare the water samples by adding 1 mL of 1 *N* HCl to 50 mL of each sample. This can be done by filling (rinse first) 50-mL volumetric flasks with the samples to the 50-mL mark and then pipetting 1.0 mL of the HCl into these flasks. Shake well. Prepare a blank using distilled water and 1 *N* HCl in the same ratio as the others. A control may also be provided. Pipet the 1 *N* HCl into the control flask as well, in the same ratio as the others.
4. Measure all solutions, 0.5 through 10 ppm, on the UV spectrophotometer—use quartz cuvettes! Measure the absorbances at 220 nm and at 275 nm for each solution. Since dissolved organic matter may also absorb at 220 nm, it is necessary to correct for interference. Subtracting two times the absorbance at 275 from the absorbance at 220 is a sufficient correction procedure for this interference. Thus, subtract two times the absorbance at 275 from the absorbance at 220 for each. Plot the resulting absorbance vs. concentration and obtain the sample concentration. Use a computer if so directed by your instructor. If any water samples were diluted prior to the measurement, apply the dilution factor before reporting the results.

EXPERIMENT 25: THE COLORIMETRIC DETERMINATION OF OZONE IN AIR

NOTE: In this experiment, an air sample is collected by means of a midget impinger filled with an "absorbing" solution and an air sampling pump. The absorbing solution is a pH-buffered solution of potassium iodide. Upon contact with ozone the iodide converts to iodine, which is measured colorimetrically. The standards are solution of iodine. Safety glasses are required.

1. To prepare 500 mL of absorbing solution, dissolve 6.81 g of KH_2PO_4 (anhydrous), 7.10 g of Na_2HPO_4 (anhydrous), and 5.00 g of potassium iodide in distilled water in a 500-mL volumetric flask and dilute to the mark with distilled water. Shake. Keep this solution from exposure to direct sunlight.
2. Prepare a stock standard iodine solution by dissolving 3.2 g of KI and 0.635 g of iodine in distilled water in a 100-mL volumetric flask. Dilute to the mark with distilled water. This solution is 0.025 *M* in iodine, I_2.
3. Prepare a second stock standard iodine solution from the one prepared in Step 2. This solution should be 0.00125 *M* in iodine and should be prepared in a 100-mL flask. Dilute with absorbing solution.
4. Prepare a series of standards of iodine for the Beer's law plot. These should be prepared from the solution in Step 3 in 25-mL flasks and

should have the following concentrations: 0.000001 *M*, 0.00001 *M*, 0.00002 *M*, 0.00003 *M*, and 0.000045 *M*. Dilute with the absorbing solution. A control sample may also be provided.

5. For collecting the sample, pipet exactly 10 mL of the absorbing solution into the midget impinger. Connect the impinger to the pump with a short piece of flexible tubing, preferably not rubber. The pump should be precalibrated to draw about 1 L of air per minute through the impinger. This flow rate should be precisely measured. Place the sampling apparatus at the sampling site and sample for 30 min.
6. Set up a spectrophotometer or colorimeter to measure absorbance at 352 nm. Using the absorbing solution as a blank, measure the absorbance of all standards, the control, and the sample. Plot absorbance vs. concentration and obtain the concentration of the unknown from the graph. A computer may be employed for some of this. Consult your instructor.
7. The stoichiometry of the $O_3 \rightarrow I_2$ reaction is one to one. Therefore, the molarity of I_2 in the unknown is also the molarity of O_3. Calculate the concentration of O_3 in the air as follows:

$$\text{ppm } O_3 \text{ in air} = \frac{\text{M of } O_3 \times 0.010 \times 24.45 \times 10^6}{\text{sampling time (min)} \times \text{flow rate(L/min)}}$$

in which 24.45 is the molar volume of O_3 at 25°C and 760 torr. Your instructor may wish you to show a derivation of this equation (Question #38)

EXPERIMENT 26: THE COLORIMETRIC DETERMINATION OF MANGANESE IN STEEL

NOTE: In this experiment, the manganese content in a sample of steel is determined by oxidizing the manganese metal to permanganate and then measuring the purple color of permanganate by visible spectrophotometry. Standards are prepared from a stock 1000 ppm Mn solution and oxidized to permanganate alongside the sample solution. Safety glasses are required.

1. The standard solutions should be 2, 5, 10, 15, and 20 ppm Mn and will ultimately be diluted to 100 mL. Calculate the amount of 1000 ppm Mn needed for each and pipet these amounts into separate 250-mL beakers. Add 25 mL (graduated cylinder) of dilute nitric acid (25% by volume) to each. Use a sixth beaker containing only the 25 mL of HNO_3 for a blank. Cover with watch glasses.
2. Weigh a sample of the steel (0.25–0.30 g) into a seventh 250-mL beaker and add 25 mL of the dilute HNO_3. Cover this beaker with a watch glass and simmer on a hot plate in a fume hood until the steel is dissolved or until only a small amount of carbon residue remains.
3. Remove from the hot plate and allow to cool for 10 min. Then add 0.50 g of ammonium peroxydisulfate to each of the seven beakers, bring all seven solutions with watch glasses in place to boil in a fume

hood, and boil gently for 10–15 min. The purpose of this step is to oxidize the carbon compounds in the steel. The standards and blank are treated identically so that they will have a matrix identical to the sample. Allow to cool again.

4. The color developing agent, the oxidizing agent used to quantitatively oxidize the manganese to permanganate at this point, is KIO_4. Add to each beaker 25 mL of water, 5 mL of concentrated H_3PO_4 and 0.4 g of KIO_4. Return each to the hot plate and boil again for 5 min (watch glass in place). Cool again and transfer quantitatively to 100-mL volumetric flasks. (A quantitative transfer is one in which the contents of one container, including what might be adhering to a watch glass, are completely and carefully transferred, along with at least two water rinsings, to a second container.) Dilute all flasks to the mark and shake.
5. Measure the absorbance of the six solutions (standards and sample), using the blank for calibration, as usual, at λ_{max} (see Figure 11.9b). A computer may be used for data acquisition.
6. Plot A vs. C and obtain the concentration of the unknown solution, C_{unk}. A computer may be used for this. Calculate the concentration of Mn (in ppm) in the steel as follows:

$$\text{ppm Mn} = \frac{C_{unk} \times 0.100}{\text{kg of steel}}$$

Your instructor may wish you to show a derivation of this equation. See Question #33.

EXPERIMENT 27: THE DETERMINATION OF PHOSPHORUS IN WATER AND SOIL

Introduction

In this experiment, the phosphorus in soil and water samples is determined by a reaction with potassium antimonyl tartrate, ammonium molybdate, and ascorbic acid. The phosphorus in the soil samples is first extracted with a dilute HCl/NH_4F solution.

NOTE: All glassware should be either washed in phosphate-free detergent or rinsed with chromic acid prior to use to avoid phosphate contamination. *Use caution* when rinsing with chromic acid. Safety glasses are required.

Preparation of Soil Samples

1. Prepare 1000 mL of an extraction solution that is 0.025 *M* HCl and 0.03 *M* NH_4F. Adjust the pH to 2.6 ± 0.05 with either 6 *M* HCl or 6 *M* NH_4OH, whichever is required. This solution is an acidic solution of a fluoride salt, hence it will tend to etch glass (see Chapter 2, Section 2.4). You can prepare it in a volumetric flask, but store it in a plastic bottle.
2. Weigh 2 g of each soil sample (from a larger sample that has been dried and crushed ahead of time) into dry 50-mL Erlenmeyer flasks.

Pipet 20 mL of the extraction solution into each flask, stopper, and shake at a reasonably rapid rate on a shaker for 10–15 min. Filter through dry Whatman #2 (or equivalent) filter paper into dry 50-mL beakers. Do not rinse the filtered soil, since that would dilute the extract by an inexact amount. Refilter if the filtrate appears cloudy or turbid.

Preparation of Reagent Solutions

3. *Potassium antimonyl tartrate solution.* In a 100-mL volumetric flask, dissolve 0.45 g of potassium antimonyl tartrate in about 50 mL of distilled water. Swirl gently until dissolved, and then dilute to the mark and shake. Prepare just prior to the run.
4. *Ammonium molybdate solution.* Dissolve 6 g of ammonium molybdate in distilled water in a 100-mL volumetric flask, dilute to the mark with distilled water, and shake. Prepare just prior to the run.
5. *Ascorbic acid solution.* Weigh 2.7 g of ascorbic acid into a 100-mL volumetric flask, dilute to the mark with distilled water, and shake. Prepare just prior to the run.
6. *Dilute sulfuric acid.* Prepare 100 mL of 2.52 *M* H_2SO_4 from concentrated sulfuric acid.
7. *Combined reagent.* To prepare 100 mL of the color-producing reagent, mix the following portions in the following order, swirling gently after each addition: 50 mL of the dilute sulfuric acid, 5.00 mL of the potassium antimonyl tartrate solution, 15 mL of the ammonium molybdate solution, and 30 mL of the ascorbic acid solution. Use a 125-mL Erlenmeyer flask. If this solution turns blue after several minutes, phosphate contamination is indicated and all solutions used will need to be re-prepared. This combined reagent is stable for a few hours.

Preparation of Standard Solutions

8. Prepare 100 mL of a 50 ppm phosphorus solution by dissolving 0.022 g of dried KH_2PO_4 in a small amount of water in a 100-mL volumetric flask. If soil samples are to be determined, dilute to the mark with extraction solution so that the matrix of the standards will match that of the samples. Otherwise, dilute to the mark with distilled water.
9. Prepare 4–6 working standards in the range of 0 to 2 if water is analyzed, or 0–10 ppm if soil is analyzed, from the stock standard prepared in Step 8. Use 50-mL volumetric flasks. Drinking-water samples will likely have very low concentrations (2 ppm maximum), while soil extracts will likely have higher concentrations. Again, dilute to the mark with extraction solution if soil samples are to be determined; otherwise dilute with distilled water. A control sample may also be provided.

Absorbance Measurements and Results

10. For the water samples and associated standards and control (distilled water for the blank), pipet 25.00 mL into a clean, dry 125-mL Erlenmeyer flask. Add 4.0 mL of the combined reagent and mix thoroughly. After at least 10 min, but no more than 30 min, measure the absorbance of each at 880 nm.
11. For the soil extracts and associated standards and control (extraction solution for the blank), pipet 5.00 mL into a clean, dry 50-mL Erlenmeyer flask. Add 8.00 mL of the combined reagent and mix thoroughly. After at least 10 min but not more than 30 min, measure the absorbance of each at 880 nm.
12. Plot absorbance vs. concentration and obtain the concentration of the unknowns from the graph. Calculate the concentrations in the soil as follows:

$$\text{ppm P} = \text{ppm P from graph} \times 10$$

EXPERIMENT 28: FLUOROMETRIC ANALYSIS OF A PREPARED SAMPLE FOR RIBOFLAVIN

NOTE: Safety glasses are required. Riboflavin solutions are light sensitive. Store them in the dark as you prepare them.

1. Prepare 100 mL of a 10 ppm riboflavin solution from the available 50 ppm solution. Use 5% (by volume) acetic acid for dilution. Your instructor may ask you to prepare your own 50 ppm solution. In that case, use 5% acetic acid for that dilution as well.
2. Prepare a series of standard solutions which are 0.2, 0.4, 0.6, 0.8, and 1.0 ppm from this 10 ppm solution and again use 5% acetic acid for the dilution. Use 50-mL volumetric flasks for these standards.
3. Obtain your unknown from your instructor and again dilute with 5% acetic acid to the mark.
4. Obtain fluorescence measurements (F) on all of your standards and the unknown. Use 5% acetic acid for a blank. Your instructor will demonstrate the use of the instrument.
5. Make a graph of F vs. C and determine the concentration of the unknown. Use a computer for this as directed by your instructor.

EXPERIMENT 29: FLUOROMETRIC ANALYSIS OF VITAMIN TABLETS FOR RIBOFLAVIN

NOTE: Safety glasses are required. Riboflavin solutions are light sensitive; store them in the dark as you prepare them.

1. Prepare five riboflavin standards as directed in Experiment 28. A control sample may also be provided. Dilute it to the mark as with the standards.
2. Obtain the weight of one tablet of each brand to be tested. Use the analytical balance and call this weight W_T.

3. Obtain from the label on the sample bottle the weight of riboflavin in one tablet. Call this weight W_R.
4. Calculate the weight of each brand required to give a solution concentration in the range of your standards. Call this W_S.

$$W_S = \frac{0.50 \text{ mg/L} \times 0.050 \text{ L}}{W_R} \times W_T$$

(This calculation assumes that 50.0 mL of a 0.50 ppm solution is to be prepared from the tablet[s]. If W_S is so small that it cannot be measured accurately (consult your instructor), weigh 10 times the calculated amount and then dilute prior to the measurement.)

5. Crush a tablet (or several tablets if $W_S > W_T$) with a mortar and pestle.
6. Weigh W_S into a 50-mL flask (analytical balance), add 5% acetic acid to the mark, and shake well. Allow to settle before pouring into a cuvette.
7. Read the fluorescence of all standards, samples, and control and plot F vs. C, using a computer if so desired. Obtain unknown concentrations (call these C_U).
8. Calculate the milligrams of riboflavin in the tablets as follows:

$$\text{mg riboflavin} = C_U \times 0.050 \text{ L} \times \frac{W_T}{W_S}$$

9. Compare with the bottle labels.

QUESTIONS AND PROBLEMS

1. Why is a molecular absorption spectrum called a molecular "fingerprint"?
2. Briefly describe the light sources used for both visible and UV work.
3. What is a monochromator and how does it work?
4. Consider an experiment in which an analyst wants to change the wavelength used in a given colorimetry experiment from 392 nm (blue light) to 728 nm (red light). He/she turns a knob on the face of the instrument to do this. Tell exactly what is happening inside the instrument when he/she does this.
5. Describe two different types of dispersing elements used in spectrophotometers.
6. Compare the material from which cuvettes must be made for UV, visible, and IR work.
7. What is a photomultiplier tube? Describe what it does and how it works.
8. What is the mathematical definition of transmittance? Also define the parameters that are found in this mathematical definition.
9. Describe the method of calibrating a single-beam instrument.

10. Why must the 100%T adjustment (the adjustment using the blank) be made on a single-beam spectrophotometer when one changes the wavelength used?
11. What absorbance corresponds to a transmittance of
    (a) 0.821?
    (b) 0.492?
    (c) 0.244?
12. What absorbance corresponds to a percent transmittance of
    (a) 46.7% T?
    (b) 28.9% T?
    (c) 68.2% T?
13. What transmittance corresponds to an absorbance of
    (a) 0.622?
    (b) 0.333?
    (c) 0.502?
14. What is the percent transmittance given that the absorbance is
    (a) 0.391?
    (b) 0.883?
    (c) 0.490?
15. What are the three advantages of a double-beam instrument over a single-beam instrument?
16. Imagine an experiment in which the molecular absorption spectrum of a particular chemical species is needed. Which instrument is preferred – a single-beam instrument or a double-beam instrument? Why?
17. What is the difference between an absorption spectrum and a transmittance spectrum?
18. Define chromophore, bathochromic shift, hyperchromic shift, hypochromic effect, and auxochrome.
19. What is Beer's law? Describe each of the parameters with one word.
20. What is the absorbance given that the absorptivity is 2.30×10^4 L/(mol cm), the path length is 1.00 cm, and the concentration is 0.0000453 *M*?
21. A sample in a 1-cm cuvette gives an absorbance reading of 0.558. If the absorptivity for this sample is 15,000 L/(mol cm), what is the molar concentration?
22. If the transmittance for a sample having all the same characteristics as in #21 is measured as 72.6%, what is the molar concentration?
23. The transmittance of a solution, measured at 590 nm in a 1.5-cm cuvette, was 76.2%.
    (a) What is the corresponding absorbance?
    (b) If the concentration is 0.0802 *M*, what is the absorptivity of this species at this wavelength?
    (c) If the absorptivity is 10,000 L/(mol cm), what is the concentration?

24. Calculate the transmittance of a solution in a 1.00-cm cuvette given that the absorbance is 0.398. What additional information, if any, would you need to calculate the molar absorptivity of this analyte?

25. What is the molar absorptivity given that the absorbance is 0.619, the path length is 1.0 cm, and the concentration is 0.00000423 *M*?

26. Calculate the concentration of an analyte in a solution given that the measured absorbance is 0.592, the pathlength is 1.00 cm, and the absorptivity is 3.22×10^4 L/(mol cm).

27. What is the concentration of an analyte given that the %T is 70.3%, the path length is 1.0 cm, and the molar absorptivity is 8382 L/(mol cm)?

28. What is the path length in centimeters when the molar absorptivity for a given absorbing species is 1.32×10^3L/(mol cm), the concentration is 0.000923 *M*, and the absorbance is 0.493?

29. What is the path length in centimeters when the transmittance is 0.692, the molar absorptivity is 7.39×10^4 L/(mol cm), and the concentration is 0.0000923 *M*?

30. What is the transmittance when the molar absorptivity for a given absorbing species is 2.81×10^2 L/(mol cm), the path length is 1.00 cm, and the concentration is 0.000187 *M*?

31. What is the molar absorptivity when the %T is 56.2, the path length is 2.00 cm, and the concentration is 0.0000748 *M*?

32. In each of the following, enough data are given to calculate the indicated parameter(s). Show your work.
    (a) Calculate A given that T is 0.551.
    (b) Calculate the molar absorptivity given that A is 0.294, "b" is 1.00 cm, and "c" is 0.0000351 *M*.
    (c) Calculate the transmittance given that the absorbance is 0.297.
    (d) Calculate T given that %T is 42.8%.
    (e) Calculate %T given that the absorptivity is 12,562 L/(mol cm), the path length is 1.00 cm, and the concentration is 0.00000355 *M*.

33. In each of the following, enough data are given to calculate the indicated parameter(s).
    (a) Calculate A given that T is 0.651.
    (b) Calculate "a" given that A is 0.234, "b" is 1.00 cm, and "c" is 0.0000391 *M*.
    (c) Calculate T given that A is 0.197.
    (d) Calculate T given that %T is 62.8%.
    (e) Calculate %T given that "a" is 13562 L/(mol cm), "b" is 1.00 cm, and "c" is 0.00000355 *M*.

34. A standard 5 ppm iron sample gave a transmittance reading of 52.8%. What is the concentration of an unknown iron sample if its transmittance is 61.7%?

35. A series of five standard copper solutions is prepared and the absorbances measured as indicated below. Plot the data and determine the concentration of the unknown.

A	C (ppm)
0.104	1
0.198	2
0.310	3
0.402	4
0.500	5
0.334	C_u

36. Match the items in the left hand column to items in the right hand column by writing the appropriate letter in the blank. Each is used only once.

__ T	(a) the intensity of light after having passed through a solution of an absorbing species
__ A	(b) path length
__ I	(c) molar absorptivity
__ I_o	(d) I/I_0
__ a	(e) A = abc
__ b	(f) -log T
__ Beer's law	(g) the intensity of light after having passed through a blank solution
__ %T	(h) absorptivity
__ ϵ	(i) T × 100

37. Consider the experiment for determining the percent of Mn in steel (Experiment 26). Figure 11.9b shows the absorption spectrum of $KMnO_4$. What is the best choice for a wavelength for doing this analysis and why?

38. In Step 7 of Experiment 25 you were given a formula for calculating the ppm O_3 in the air sample. Show by appropriate cancellation of all units involved that the final ppm units are microliters of O_3 per liter of air.

39. In Step 6 of Experiment 26 a formula was given for calculating the ppm Mn in steel. Show by appropriate cancellation of all units involved that the final ppm units are milligrams of Mn per kilogram of steel.

40. What does it mean to say that a pair of "matched cuvettes" is needed in a given experiment? Explain why a pair of cuvettes would ever need to be "matched".

41. A certain pair of cuvettes is not a matched pair. Tell what exactly may be different about them – things over which the analyst does not have any control. Name at least two things.

42. State the importance of the wavelength of maximum absorbance, λ_{max}.

43. Is the wavelength of fluorescence longer or shorter than the wavelength of absorption? Explain your answer with the help of an energy level diagram.

44. Is the energy of absorption more or less than the energy of fluorescence? Explain your answer with the help of an energy level diagram.

45. Why are there two monochromators in a fluorometer?

46. What is meant by a "right angle configuration" in a fluorometer and why is this instrument constructed this way?

47. A fluorometer differs in basic design from an absorption spectrophotometer in two major ways.
    (a) What are they?
    (b) Explain the need for these design differences.

48. Draw a diagram of a fluorescence instrument and point out the differences between it and the basic single-beam absorption spectrophotometer.

49. When performing a quantitative analysis procedure using fluorometry, what parameter is measured by the instrument and plotted vs. concentration?

50. The fluorescence of a 0.10 ppm solution of aluminum/acid alizarin garnet R is 39.6. What is the concentration of this species if F = 46.2?

51. Why is it that fused benzene ring (polycyclic) compounds such as naphthalene, anthracene, and riboflavin can be analyzed by fluorometry while uncomplexed metal ions cannot?

52. Fluorometry is more selective and more sensitive than absorption spectrometry. What do these two terms mean?

53. Under what circumstances would you want to use a fluorometric procedure rather than an absorption spectrometry procedure? Explain briefly.

54. What two advantages does fluorometry have over absorption spectrophotometry?

55. The fact that very few chemical species fluoresce works both to the advantage and disadvantage of fluorometry as a quantitative technique. Explain this.

56. How do absorption spectrophotometry and fluorometry compare in terms of
    (a) instrument design?
    (b) sensitivity?
    (c) applicability?

57. Derive and discuss the equations given in Step 12 of Experiment 27 and Steps 4 and 8 of Experiment 29, showing especially that the units cancel appropriately.

58. The analysis of riboflavin in a vitamin tablet yielded the following data. How many milligrams of riboflavin are in one tablet?
    Weight of one tablet .... 1.1793 g
    Weight of sample taken and diluted to 50 mL .... 0.00177 g

C (ppm)	F
0.20	18.6
0.40	37.4
0.60	56.4
0.80	76.7
C_u	44.8

59. Selenium content in plants can be determined by fluorometry. The procedure involves treatment with nitric and perchloric acid followed by a quantitative reaction to form a fluorescent species and finally an extraction into an organic solvent. If 1.0277 g of plant material were weighed and the resulting solution, after a reaction to form the fluorescent species, was diluted to 25 mL, what is the ppm Se in the plant material? Assume that all of the fluorescent species gets extracted into exactly 25 ml of the organic solvent. The following is the fluorescence data for the sample and a series of standard solutions:

F	C (ppm)
98.3	0.500
77.4	0.400
56.2	0.300
37.0	0.200
17.9	0.100
40.9	unknown

Chapter 12

Infrared Spectrometry

12.1 INTRODUCTION

As discussed in Chapter 10, the absorption of infrared light causes vibrational energy transitions in molecules. The usefulness of the technique lies especially in the fact that only very specific wavelengths (energies) of infrared light are able to be absorbed when a single particular kind of molecule is in the path of the light. As with UV/vis absorption, the absorbance vs. wavelength plot is a molecular fingerprint of the molecule. The difference, however, lies in the fact that in the infrared region the absorption bands are extremely sharp and each such band is associated with a particular covalent bond present in the molecule. In Chapter 10 we referred to the ball and spring model of a molecule and how the initiation of a vibration of the spring constitutes an energy transition caused by an infrared wavelength. The point here is that the wavelength at which the sharp absorption band is observed depends on what atoms the "spring" connects and whether the bond is a single bond, a double bond, etc. Thus, a -C–H- bond absorbs a particular wavelength, a -C–O- bond a different wavelength, a -C=O yet a different wavelength, etc. Additionally, various wavelengths can be absorbed depending on what mode of vibration is involved. These can be stretching vibrations, rocking and bending motions, etc. (Refer back to Figure 10.7.) The technique is especially useful, therefore, for qualitative analysis. Not only is the infrared absorption spectrum a molecular fingerprint, but the presence of particular bonds in the structure will be manifested in the presence of the corresponding sharp absorption bands at particular wavelengths, a fact that has considerable usefulness in the narrowing down of the possibilities in a qualitative analysis scheme. This topic will be expanded upon later in this chapter.

The nature of the sample holder is important to consider. In Chapter 11 we referred to the need for the sample holder in infrared spectrometry to be composed of inorganic salt crystals. Since infrared wavelengths are of such energy so as to cause covalent bonds to vibrate, any covalent compound would not be suitable as sample holder material because such material will absorb its characteristic wavelengths of infrared light. This, in turn, obviously may cause erroneous conclusions in qualitative analysis schemes. Glass or plastic materials therefore cannot be used. Material with ionic bonds, however, is not a problem because ionic bonds do not absorb infrared wavelengths – they do not undergo vibrational energy transitions. Thus, inorganic compounds such as NaCl and KBr are commonly used as matrix material and sample "windows" in infrared spectrometry.

12.2 LIQUID SAMPLING

For pure liquids ("neat" liquids) and liquid solutions, sandwiching a thin layer of liquid between two large NaCl or KBr crystals (windows) is the classic procedure for mounting the sample in the path of the light. Typical dimensions for such windows are about 2 cm wide × 3 cm long × 0.5 cm thick. Positioning or holding the crystals in place is done using either a "sealed cell", a "demountable cell", or a combination "sealed demountable" cell. Sealed cells are permanent fixtures for the windows and cannot be disassembled. They have a fixed path length and are very useful for quantitative analysis, since the path length is reproduced from one standard to another and also for the unknown. Demountable cells can be disassembled so as to change the path length. The sample is placed on the window in a space created by a "spacer" while it is disassembled. The thickness of the spacer establishes the path length. It is then reassembled with the thin layer in place between the windows. Sealed demountable cells are demountable cells which include inlet and outlet ports for introducing and eliminating the liquid samples. All demountable cells are designed to allow easy disassembly for changing the spacer between the crystals (the path length) or for eliminating the spacer altogether for viscous samples, for example. Figure 12.1 shows a drawing of a typical sealed demountable cell. The top neoprene gasket and window have holes drilled in them to coincide with the inlet and outlet ports to facilitate filling the space, created by the spacer, with the liquid sample. The path of the liquid sample is shown as dashed lines in the figure.

Filling the cells with sample and eliminating the sample when finished can be troublesome. The sample inlet and outlet ports are tapered to receive a syringe with a Luer (tapered) tip. The usual procedure for filling is to raise the outlet end by resting it on a pencil or similar object (to eliminate air bubbles) and then to use a pressure/vacuum system with two syringes, one in the outlet port and one in the inlet port, as shown in Figure 12.2. While pushing on the plunger of the syringe containing the liquid sample in the inlet port and pulling up on the plunger of the empty syringe in the outlet port, the cell can be filled without excessive pressure on the inlet side. This reduces the possibility of damaging the cell due to the excessive pressure that may needed, especially when working with unusually viscous samples

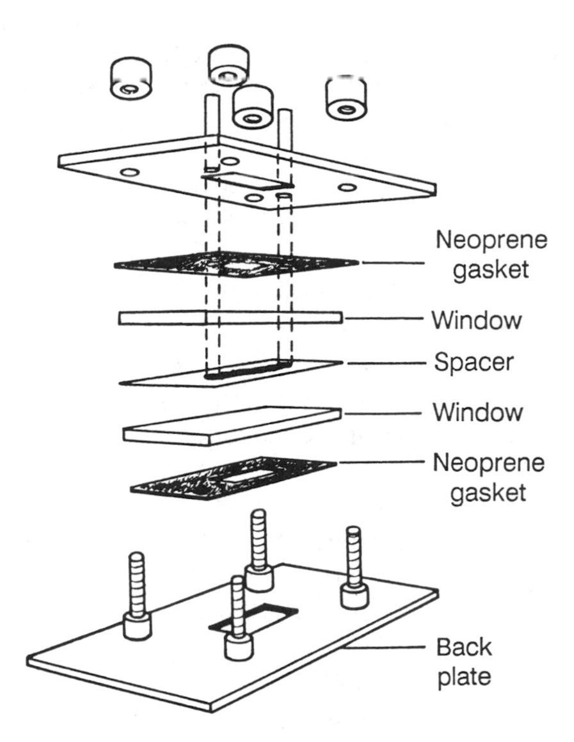

Figure 12.1. The sealed demountable cell assembly. (Adapted from Chia, L. and Ricketts, S., Basic Techniques and Experiments in Infrared and FT-IR Spectroscopy, The Perkin-Elmer Corporation, Norwalk, CT, 1988. With permission.)

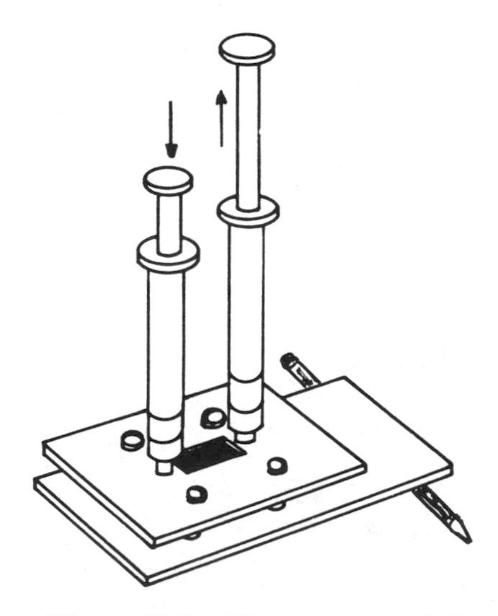

Figure 12.2. The recommended method of filling a sealed or demountable cell. (From Kenkel, J., *Analytical Chemistry Refresher Manual*, Lewis Publishers, Chelsea, MI, 1992. With permission.)

and short path lengths. Tapered Teflon® plugs are use to stopper the ports immediately after filling. The cell may be emptied and readied for the next sample by using two empty syringes and the same push-pull method. When refilling, an excess of liquid sample may be used to rinse the cell and eliminate the residue from the previous one. Alternatively, the cell may be rinsed with a dry volatile solvent and the solvent evaporated before introducing the next sample.

The analyst must be careful to protect the salt crystals from water during use and storage. Sodium chloride and potassium bromide are, of course, highly water soluble, and the crystals may be severely damaged with even the slightest contact with water. All samples introduced into the cell must be dry. This is important for another reason. Water contamination will show up on the measured spectrum and cause erroneous conclusions. If the windows are damaged with traces of water, they will become "fogged" and will appear to become non-transparent. The windows may be re-polished if this happens. Depending on the extent of the damage, various degrees of abrasive materials may be used, but the final polishing step must utilize a polishing pad and a very fine abrasive. Polishing kits are available for this. Figure 12.3 shows the correct method for polishing. Finger cots should be used to protect the windows from finger moisture.

Liquids can be sampled as either the "neat" liquid (pure) or mixed with a solvent (solution). The neat liquid is more desirable since the spectrum will show absorption due to the liquid only. However, when a solution is run both the analyte and the solvent will produce absorption bands, and they must be differentiated. Some solvents have rather simple IR spectra and are thus desirable as solvents. Examples are carbon tetrachloride (only –C–Cl bonds) and methylene chloride (CH_2Cl_2). Their infrared spectra are shown in Figure 12.4. Alternatively, a double-beam instrument can be used to cancel out the solvent absorption. These instruments will be discussed later in this section.

12.3 SOLID SAMPLING

Infrared spectrometry is one of the few analytical techniques that is routinely used to analyze solid undissolved samples. The techniques to be

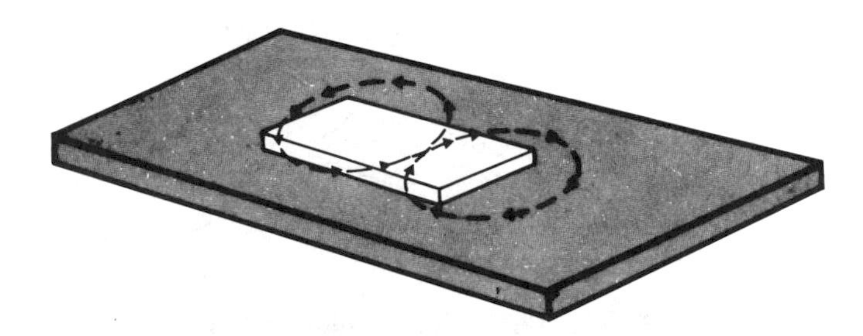

Figure 12.3. The correct method for polishing crystal materials for infrared sample cells. Note the figure 8 motion. (From Kenkel, J., *Analytical Chemistry Refresher Manual*, Lewis Publishers, Chelsea, MI, 1992. With permission.)

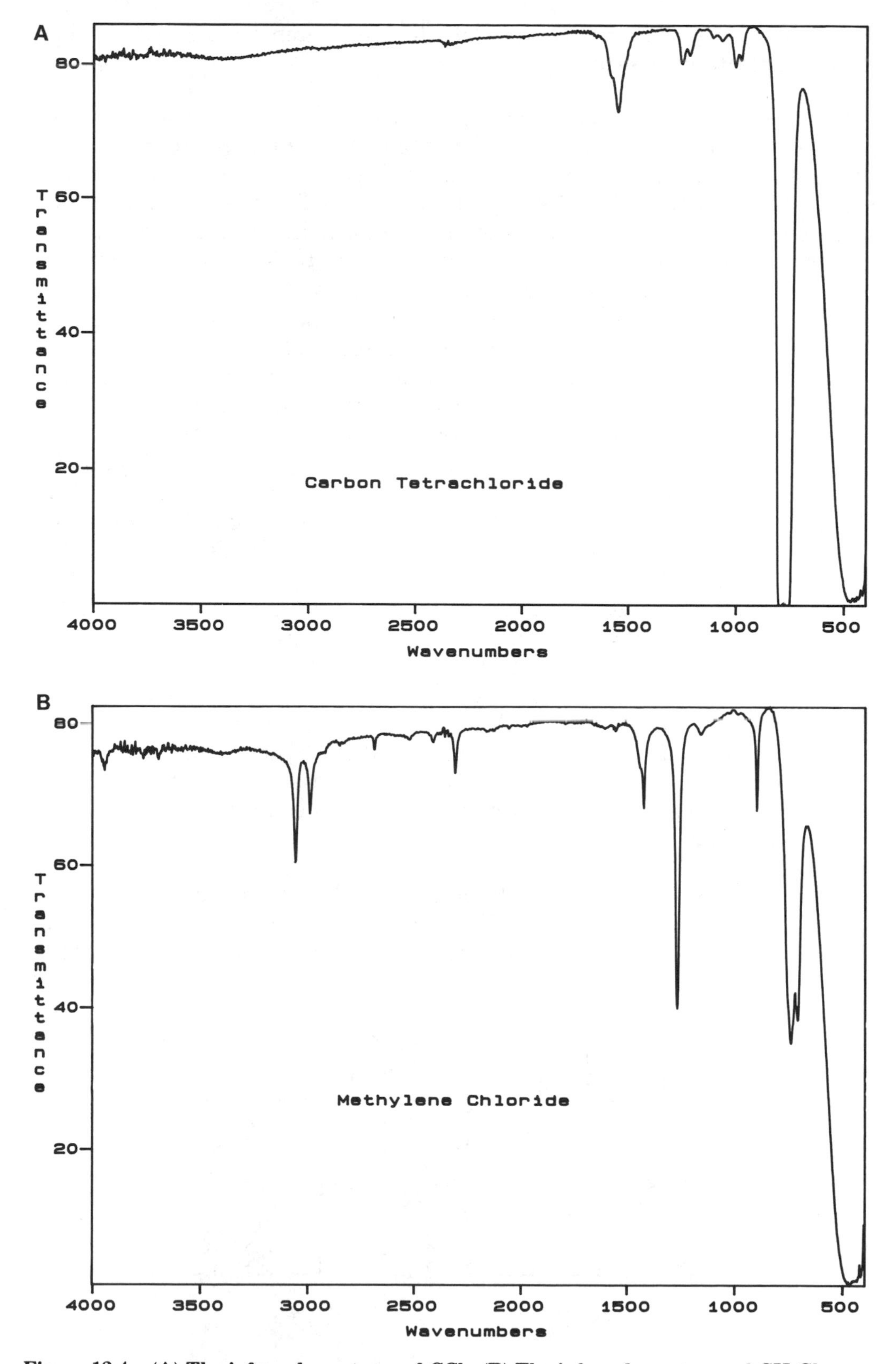

Figure 12.4. (A) The infrared spectrum of CCl_4. (B) The infrared spectrum of CH_2Cl_2.

described here include the KBr pellet, the Nujol (mineral oil) mull, and the diffuse reflectance method.

KBr Pellet. The KBr pellet technique is based on the fact that *dry*, finely powdered potassium bromide has the property of being able to be squeezed under very high pressure into transparent discs – transparent to both infrared light and visible light. It is important for the KBr to be dry both in order to obtain a good pellet and to eliminate absorption bands due to water in the spectrum. If a small amount of the dry solid analyte (0.1% to 2.0%) is added to the KBr prior to pressing, then a disc (pellet) can be formed from which a spectrum of the solid can be obtained. Such a disc is simply placed in the path of the light in the instrument and the spectrum measured.

Two methods of pressing the KBr pellet will be described here. First, a pellet die consisting of a threaded body and two bolts with polished faces may be used. One bolt is turned completely into the body of the die. A small amount of the powdered sample, enough to cover the face of the bolt inside, is added and the other bolt turned down onto the sample, squeezing it into the pellet. (See Figure 12.5.) The two bolts are then carefully removed. The body of the die is placed in the instrument so that the light beam passes directly through the center of the die and through the pellet.

The other method utilizes a hydraulic press and a cup die. The cup die consists of a base, with a protruding center, and a hollow cylinder that fits snugly over the protrusion. With the cylinder in place, the powdered sample is added to the cylinder to cover the face of the protrusion and a second metal piece with a protrusion is placed on the top of the assembly. The entire assembly is placed into a laboratory press. (See Figure 12.6.)

It is important in either case for the KBr and sample to be dry, finely powdered, and well mixed. An agate mortar and pestle is recommended for the grinding and mixing of the KBr and sample.

Nujol Mull. The Nujol (mineral oil) mull is also often used for solids. In this method, a small amount of the finely divided solid analyte (1- to 2-μ particles) is mixed together with an amount of mineral oil to form a mixture

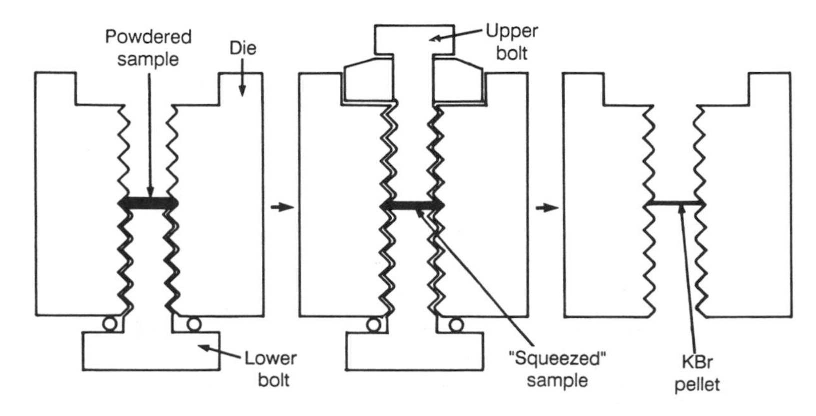

Figure 12.5. The procedure for making a KBr pellet with the use of a threaded pellet die. (From Kenkel, J., *Analytical Chemistry Refresher Manual*, Lewis Publishers, Chelsea, MI, 1992. With permission.)

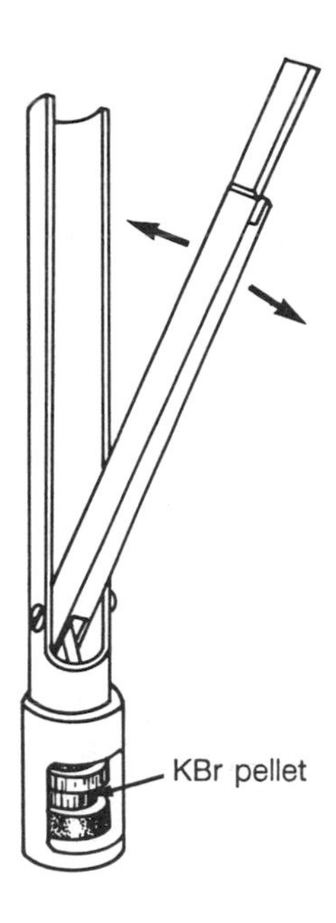

Figure 12.6. A pellet-making assembly which utilizes a laboratory hand press. (Reproduced with permission of Spectra-Tech, Inc., Stamford, CT.)

with a toothpaste-like consistency. This mixture is then placed (lightly squeezed) between two NaCl or KBr windows which are similar to those used in the demountable cell discussed previously for liquids. If the particles of solid are not already the required size when received, they must be finely ground with an agate mortar and pestle and can be ground directly with mineral oil to create the mull to be spread on the window. Otherwise, a small amount (about 10 mg) of the solid is placed on one window along with one small drop of mineral oil. A gentle rubbing of the two windows together with a circular or back and forth motion creates the mull and distributes it evenly between the windows. The windows are placed in the demountable cell fixture which is placed in the path of the light.

A problem with this method is the fact that mineral oil is a covalent substance and its characteristic absorption spectrum will be found superimposed in the spectrum of the solid analyte, as with the solvents used for liquid solutions discussed previously. However, the spectrum is a simple one (Figure 12.7) and often does not cause a significant problem.

Diffuse Reflectance. With the advent of the Fourier transform infrared spectrometry (FTIR) instrumentation (see below), a technique called diffuse reflectance has become popular for solids. In this technique, the powdered KBr/analyte mixture (about 5% analyte) is placed in a small sample cup and the light beam shines directly on this powdered sample. The diffuse reflectance, or the light returning from the surface of the sample that is scattered (and not simply reflected), is measured by the FTIR system and the spectrum displayed. The advantage of this method is that an analyte that does not form a good KBr pellet can be run with little or no problem.

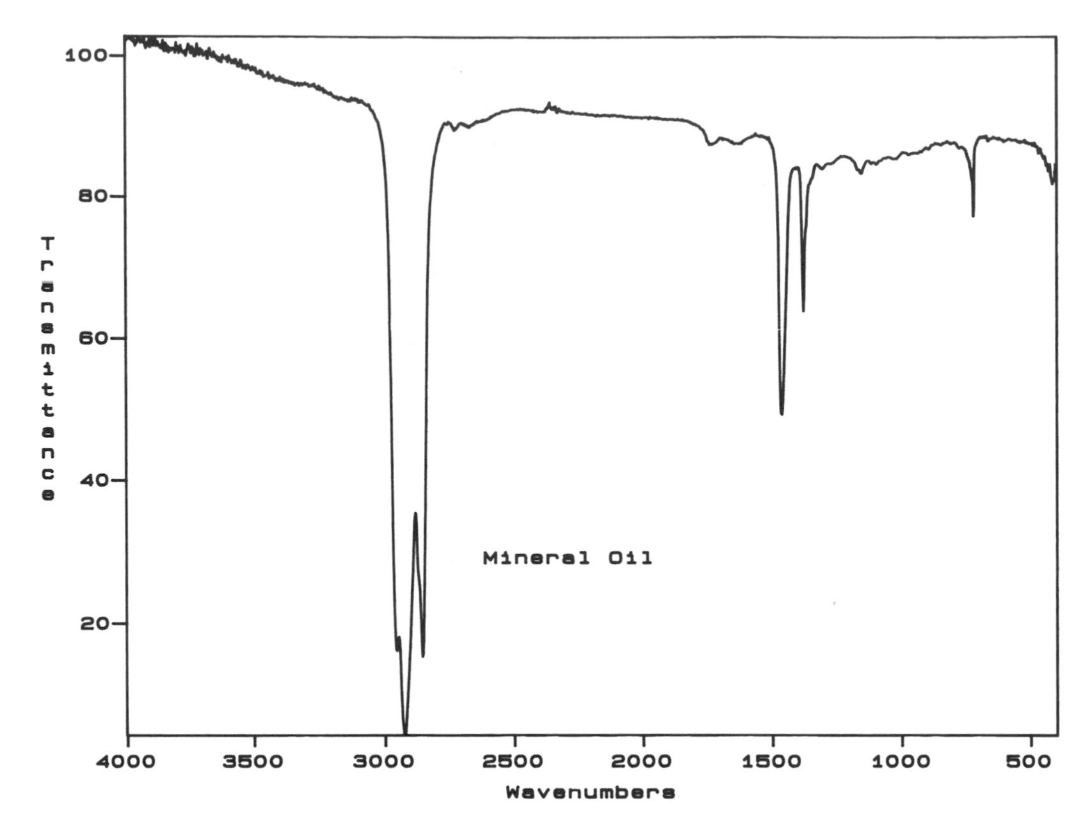

Figure 12.7. The infrared spectrum of mineral oil.

12.4 INSTRUMENT DESIGN

The traditional infrared spectrometer is very similar to the UV/vis spectrophotometer, the so-called double-beam "dispersive" instrument utilizing the traditional slit/dispersing element/slit monochromator to create a narrow band width, often referred to as the single wavelength, to shine on the sample. The modern IR instrument, however, utilizes an interferometer in which the light does not get dispersed and yet the traditional infrared spectrum is measured. It is called the FTIR. Both of these designs will now be discussed.

Double-Beam Dispersive. A diagram of the double-beam dispersive infrared spectrometer is shown in Figure 12.8. It is very similar to the UV/vis spectrophotometer described in Chapter 11. There are some differences, however. The light source is typically a nichrome wire coil that has a high electrical resistance and emits intense heat (infrared light) when an electric current is passed through it, much like the burner on an electric stove. The source may also be a "Globar", which is a silicon carbide rod, or a "Nernst Glower", which is a rare earth oxide cylinder. Both of these operate on basically the same principle as the nichrome wire. The detector is a thermocouple transducer, a device which converts heat energy into an electric signal.

The IR instrument is typically a "double-beam in space" instrument, whereas the UV/vis instrument is a "double-beam in time" instrument. "Double-beam in time" refers to the fact that the light beam from the source

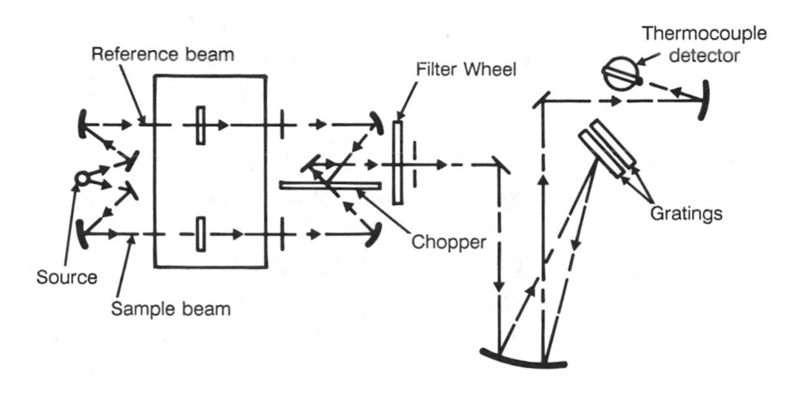

Figure 12.8. A diagram of a double-beam dispersive infrared instrument. (Reproduced from the AccuLab 5 instruction manual with permission of Beckman Instruments Inc., Fullerton, CA.)

is split into two beams existing alternately in time. At one moment the sample beam exists and passes through the sample compartment, while at the next moment the reference beam exists and passes through the sample compartment. A rotating half-mirror (chopper) ahead of the sample compartment accomplishes this splitting of the beam. (See Chapter 11.) In the "double-beam in space" instrument, both light beams exist simultaneously. Two mirrors that point in different directions, as depicted in Figure 12.8, send two light beams through the sample compartment. Thus, both beams exist simultaneously and pass through the sample compartment together. The monochromator system, located between the sample compartment and detector in such an instrument, does utilize a chopper to combine the light beams, however, and the detection system is synchronized with the chopping frequency so as to differentiate the two signals and display the percent transmittance of the sample at the recorder. As mentioned above under liquid sampling, placing the solvent of the analyte species in the reference path will cause all solvent absorption bands to disappear from the recorded spectrum just as the blank adjustment is automatically made in the UV/vis instruments discussed earlier. This does, however, require two exactly matched IR cells for accurate work. This cell matching is less important for qualitative work, which is a more common application of the technique.

The optical system for dispersive IR instruments has an additional requirement. The absorption of IR light by the mirrors, lenses, and dispersing element must be minimized. Thus, various reflection and transmittance gratings composed of non-absorbing materials are used, often in combination with transmission filters, such as the "filter wheel" in Figure 12.8. Front-reflecting mirrors are used, and glass collimating lenses are absent. In addition, more than one grating is needed to accommodate the different regions of the infrared spectrum, and this fact requires the instrument to stop scanning to allow the re-alignment of these gratings in the middle of a run and then to resume. Figure 12.8 shows two gratings. The gratings are usually made of glass or plastic and are coated with aluminum.

Fourier Transform Infrared Spectrometry (FTIR). The modern FTIR instrument is designed to perform the same functions as the dispersive instruments. Such an instrument, however, does not utilize a light-dispersing monochromator and associated optics. The light from the source, typically the same type of source as in the dispersive instruments, passes through the optical path undispersed. The result, however, is the same – the infrared absorption spectrum, the plot of %T vs. wavelength – but it can be obtained much faster than with the dispersive instrument. We will undertake a simplified discussion as to how this happens.

In the FTIR instrument, the undispersed light beam passes through the sample and all wavelengths and the corresponding absorption data are received at the detector simultaneously. A computerized mathematical manipulation known as the Fourier transform is performed on these data in order to obtain absorption data for each individual wavelength. To present the data in a form that can utilize the Fourier transform, the wave pattern created by combined constructive and destructive interference of all wavelengths of light from the source over time is utilized. This pattern, known as an interferogram, is created by moving one beam of light from the source through another. The device for doing this is known as an interferometer.

Figure 12.9 shows a diagram of an interferometer. It consists of a beam splitter and two mirrors, one fixed and one movable. Consider first a light source of a single wavelength. As light from the source strikes the beam splitter, the beam is split such that half of the intensity is transmitted to the movable mirror while half is reflected to the fixed mirror. Both beams are reflected back to the splitter, where they join again and proceed toward the sample. If the distance from the splitter to the movable mirror is exactly equal to the distance to the fixed mirror, then their rejoining will result in constructive interference and the intensity reaching the sample will be a maximum. The position of the movable mirror, however, may also be such that complete destructive interference occurs. This would occur if the movable mirror distance is equal to the fixed mirror distance plus half of the wavelength. Other movable mirror distances would result in intermediate intensities, the cycle would repeat as the distance is increased, and thus the

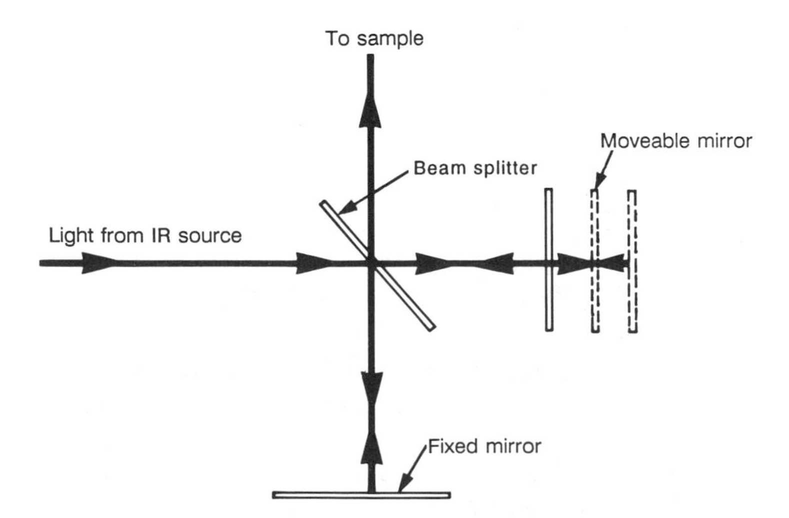

Figure 12.9. A diagram of an interferometer. See text for discussion. (From Kenkel, J., *Analytical Chemistry Refresher Manual*, Lewis Publishers, Chelsea, MI, 1992. With permission.)

"interferogram", known as a cosine wave, shown in Figure 12.10a would result. Now consider a light beam consisting of all wavelengths in the infrared region—in other words, the light from the light source in a typical infrared instrument. In this case the interferogram in Figure 12.10b would result. The strong intensity in the center occurs when the distances of both mirrors from the splitter are equal and we have complete constructive interference of all wavelengths. Motor-driving the movable mirror to small distances toward and away from the equidistance point "codes" the light and its wavelengths and creates the possibility for it to be "decoded" by the Fourier transform mathematical manipulation by computer at the detector, resulting in the infrared spectrum of the sample.

The advantages of FTIR over the dispersive technique are (1) it is faster, making it possible to be incorporated into chromatography schemes, as we will see briefly in Chapters 16 and 17; and (2) the energy reaching the detector is much greater, thus increasing the sensitivity.

12.5 QUALITATIVE AND QUANTITATIVE ANALYSIS

The ultimate goal of infrared analysis is, of course, the identification of the substance measured or the determination of the quantity of the substance measured. The usefulness of the technique lies mostly in the identification aspects, and we will emphasize this here. However, we will also briefly address the quantitative aspects as well.

Qualitative Analysis. Early on in this section, we spoke of how specific bonds present in molecules give rise to absorption bands at specific corresponding wavelengths in the infrared region. This is the basis of qualitative analysis using this technique and we will now expand on this idea, especially discussing what wavelengths are absorbed by each type of bond.

The region of the electromagnetic spectrum involved here of course is the infrared region. This spans the wavelength region from about 2.5 μm to about 17 μm or, in terms of wave number, from about 4000 cm^{-1} to about 600 cm^{-1}. Infrared spectra are transmission spectra calibrated most often in wave number, and so we will be speaking in terms of wave number for the remainder of this discussion and depicting the spectra with a 100%T base-

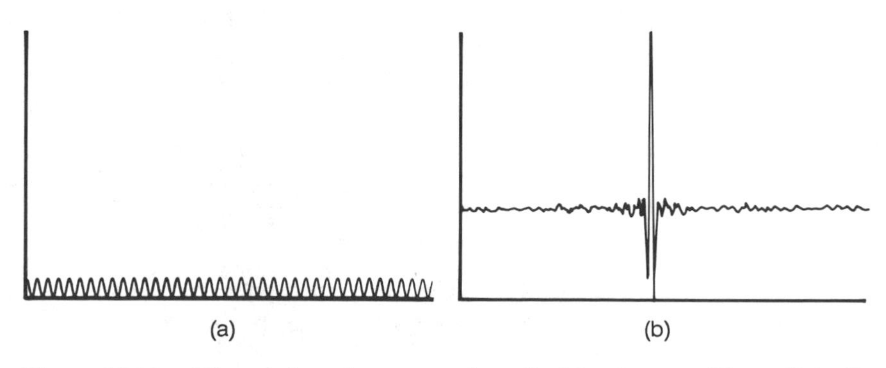

Figure 12.10. The two interferograms described in the text. (From Chia, L. and Ricketts, S., Basic Techniques and Experiments in Infrared and FT-IR Spectroscopy, The Perkin-Elmer Corporation, Norwalk, CT, 1988. With permission.)

line at the top of the chart and the peaks deflecting toward the 0%T level when an absorption occurs, as in the spectra already discussed briefly (Figures 12.4 and 12.7). (See Figure 12.11 for another example.) The absorption pattern derived from the particular molecule present in the path of the light is a molecular "fingerprint" just as it was in the UV/vis region (Chapter 11). The infrared spectrum, however, is useful for an additional reason. The absorption bands, as they are typically recorded, appear much sharper, and we have the ability to conclude that a molecule has a particular type of bond in its structure when we observe the corresponding characteristic absorption band in the spectrum. This is particularly true in the region from about 4000 cm^{-1} to about 1500 cm^{-1} and less true in the 1500 cm^{-1} to 600 cm^{-1} region. This latter region is thus often described separately as the "fingerprint" region and is used mostly to match, peak for peak, the spectrum of an unknown with a spectrum of a known, perhaps from a library of known spectra, such as the Sadtler library.* (See Figure 12.12.) Thus, we look for characteristic bands in the 4000 cm^{-1} to 1500 cm^{-1} region to perhaps assign the unknown to a particular class of compounds, i.e., to narrow down the possible structures, and then look to the fingerprint region and the overall spectrum to make the final determination, matching peak for peak.

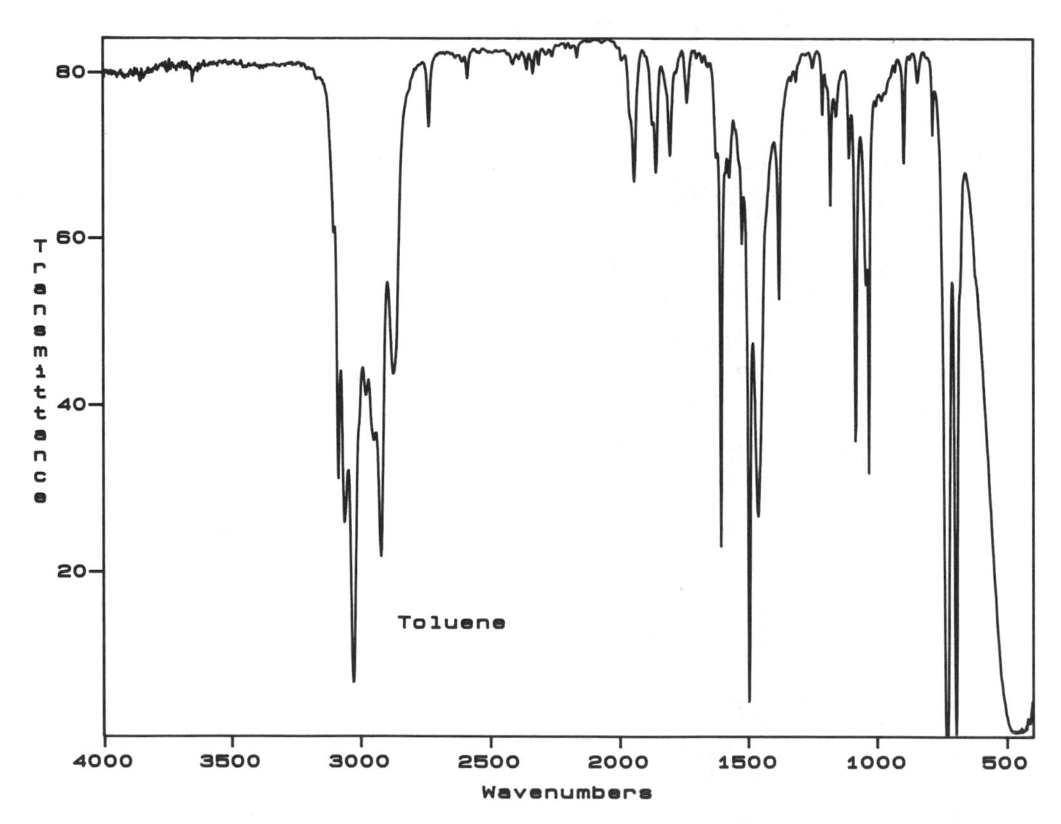

Figure 12.11. The infrared spectrum of toluene. Infrared spectra are transmission spectra with the peaks recorded from the top down as shown.

*This refers to the collection of standard spectra published by Sadtler Research Laboratories, Philadelphia, PA. A smaller volume appropriate for academic laboratories is *A Spectrum of Spectra*, Tomasai, R. A., University of Tulsa, Tulsa, OK, 1992.

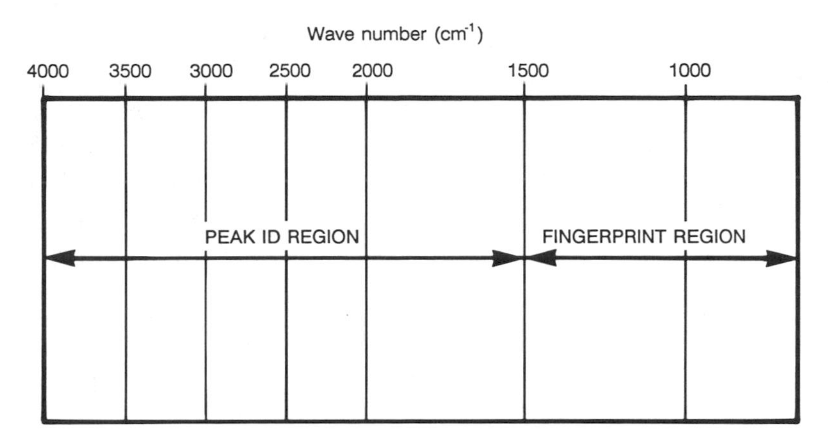

Figure 12.12. The "fingerprint" and "peak ID" regions of infrared spectra. (From Kenkel, J., *Analytical Chemistry Refresher Manual*, Lewis Publishers, Chelsea, MI, 1992. With permission.)

In addition to the location (i.e., wave number) of the absorption bands, it can also be useful to examine the width and depth of the absorption. The descriptions "broad" and "sharp" are often used to describe the width of the peaks, and "weak", "medium", and "strong" are used to describe the depth of the peaks. To understand these descriptions, refer to Figure 12.13. Spectra "A" and "C" show broad, strong peaks centered around 3300 cm^{-1}. Spectrum "B" shows a sharp, strong peak at 1685 cm^{-1}. Spectrum "C" shows a sharp, medium peak at 1220 cm^{-1} and a series of weak peaks between 1700 cm^{-1} and 2000 cm^{-1}.

Table 12.1 correlates the location, width, and depth of various IR absorption patterns for some common kinds of bonds. By correlating Figure 12.13 with Table 12.1, it should be obvious that spectrum "A" in Figure 12.13 is that of an alcohol with no benzene rings, spectrum "B" is the spectrum of a compound containing a carbonyl group (such as an aldehyde) with no benzene ring, and spectrum "C" is the spectrum of an alcohol with a benzene ring. Figure 12.14 is a correlation chart showing the location of the absorption peaks of most bonds.

Finally, earlier, the need for using a solvent such as carbon tetrachloride, dichloromethane, or Nujol (mineral oil) was indicated for some applications. The analyst needs to know what the spectra of these solvents look like in order to be able to account for the interfering solvent peaks. Figures 12.4 and 12.7 gave the IR spectra for these compounds.

Quantitative Analysis. Quantitative analysis procedures using infrared spectrometry utilize Beer's law. Once the %T or absorbance measurements are made, the data reduction procedures are identical to those outlined previously (in Chapter 11) for UV/vis spectrophotometry. This means that the analyst could prepare just one standard solution to which to compare the unknown using the ratio and proportion scheme (Equation 11.6), determine the concentration by direct calculation (Equation 11.7), or prepare a series of standards and utilize a Beer's law plot (see Figure 11.11 and accompanying discussion).

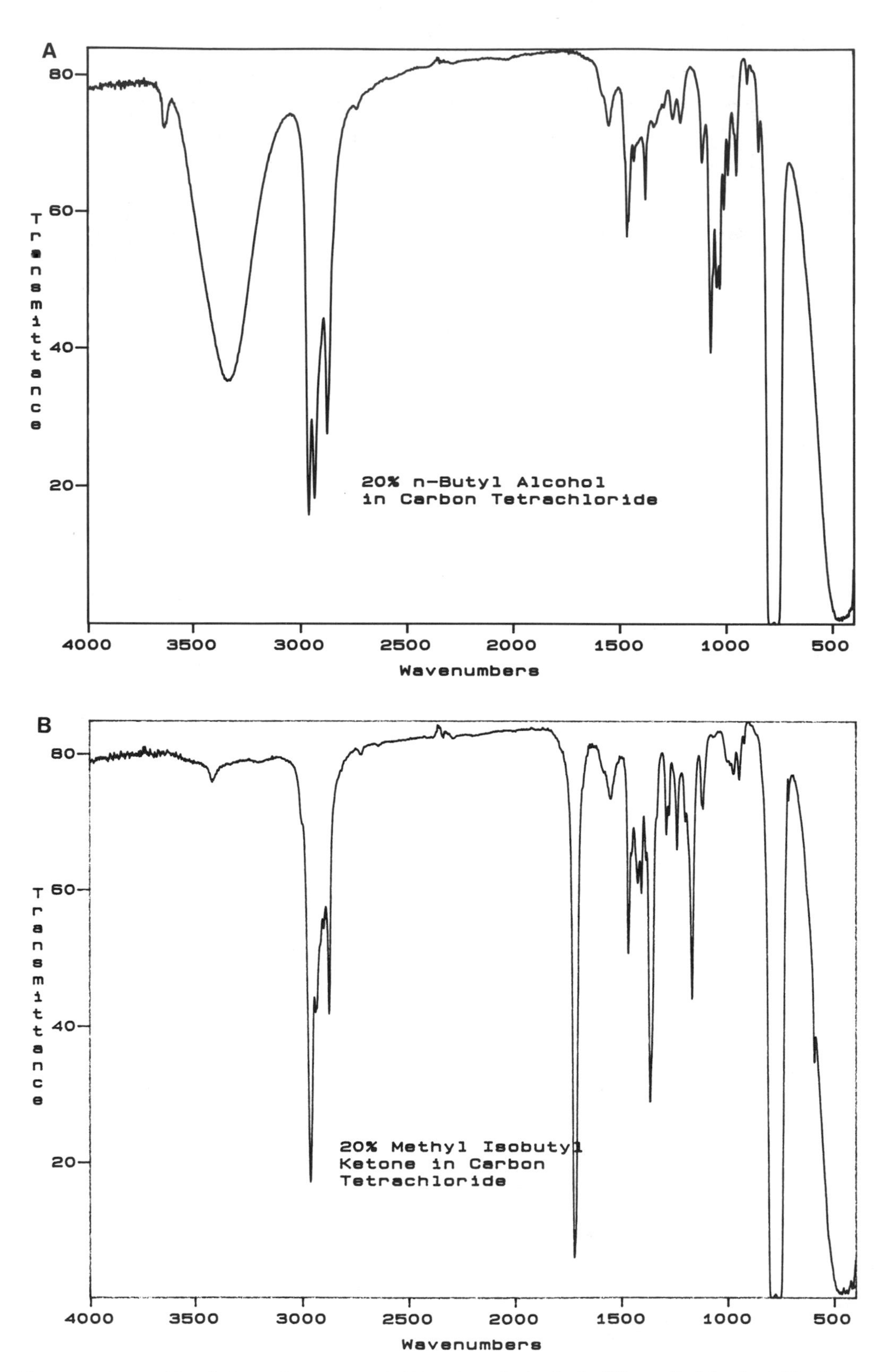

Figure 12.13 The infrared spectra of (A) *n*-butyl alcohol, (B) methyl isobutyl ketone, and (C) benzyl alcohol. See text for discussion.

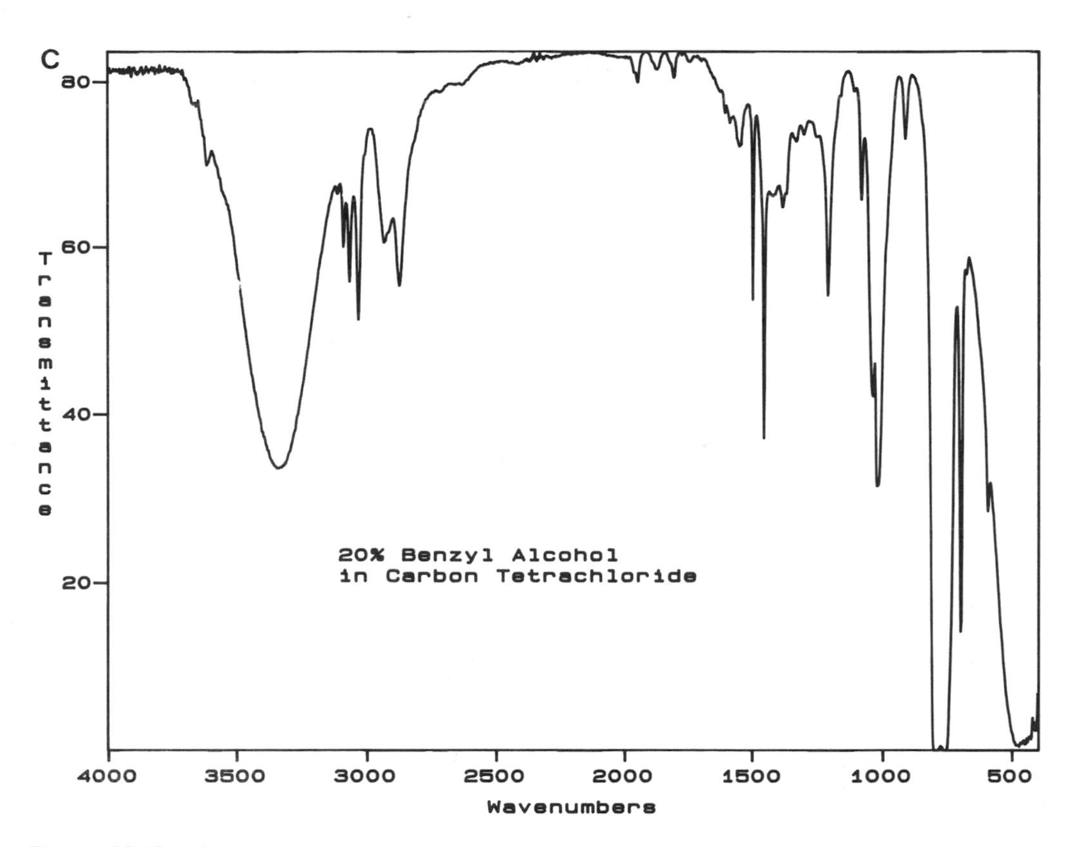

Figure 12.13. Continued.

Table 12.1. Some easily recognizable IR absorption patterns.

Bond	Description of Absorption Pattern
-C–H–, where C is not part of a benzene ring	Sharp, strong peak on the low side of 3000 cm^{-1}, between about 2850 cm^{-1} and 3000 cm^{-1}
-C–H–, where C is a part of a benzene ring or double bond	Sharp, medium peak on the high side of 3000 cm^{-1}, between about 3000 cm^{-1} and 3100 cm^{-1}
-C–H–, where C is a part of triple bond	Sharp, medium peak on the high side of 3000 cm^{-1}, between about 3250 cm^{-1} and 3350 cm^{-1}
-O–H–, in alcohols, phenols, and water, for example	Broad, strong peak centered at about 3300 cm^{-1}
-C=O, in aldehydes, ketones, etc.	Sharp, strong peak at about 1700 cm^{-1}
-C–C–, in benzene ring	Two sharp, strong peaks near 1500 cm^{-1} and 1600 cm^{-1}, and a series of weak peaks (overtones) between 1600 cm^{-1} and 2000 cm^{-1}, the latter in the case of a monosubstituted benzene ring

From Kenkel, J., *Analytical Chemistry Refresher Manual*, Lewis Publishers, Chelsea, MI, 1992. With permission.

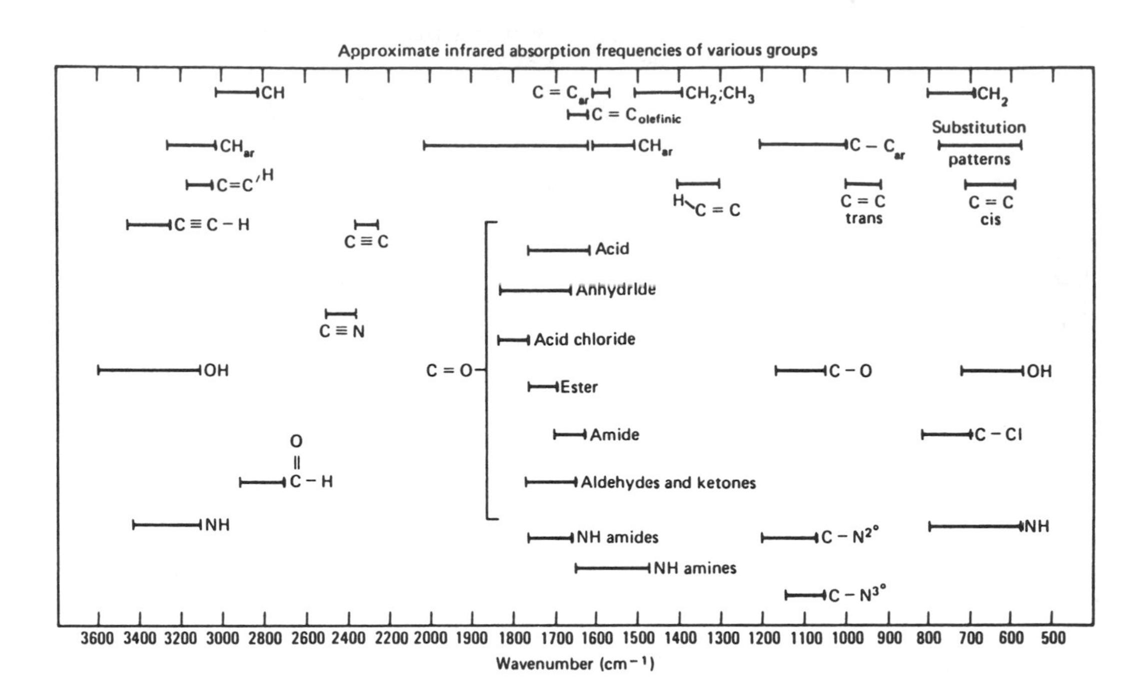

Figure 12.14. A correlation chart for infrared spectroscopy. (From Zubrick, J. W., *Organic Chemistry Survival Manual*, Copyright © John Wiley & Sons, New York, 1988. With permission.)

Reading the %T from the recorded IR spectrum for quantitative analysis can be a challenge. In the first place, there can be no interference from a nearby peak due to the solvent or other component. One must choose a peak to read that is at least nearly, if not completely, isolated.

Second, the baseline for the peak must be well defined. Since the baseline may not be completely straight across the wavelength region, the analyst does not endeavor to have the instrument trace across the 100%T line. Rather, something less than 100%T is chosen at the beginning of the run such that the baseline, even though it likely will not be straight, remains on scale (less than 100%T) for the entire scan. Thus, two %T readings must be taken, one for the baseline (corresponding to where the baseline would be if the peak were absent – a "blank" reading) and one for the minimum %T, the tip of the peak. The two readings are then converted to absorbance and the absorbance of the baseline subtracted from the absorbance of the peak. The result is the absorbance of the sample. It is recommended, for accurate work, that two or more spectra of each standard and the unknown be recorded and an average of each absorbance calculated.

It should also be mentioned that the path length of the sample cell used must be constant for all standards and the unknown, or at least known so that a correction can be applied if necessary. Using care when filling and cleaning cells is also important to avoid alterations in path length due to excessive pressure or leaking.

EXPERIMENTS

EXPERIMENT 30: QUALITATIVE ANALYSIS BY INFRARED SPECTROMETRY–LIQUID SAMPLING

DISCUSSION: In this experiment you will be given two unknown organic liquids to attempt to identify with the help of infrared spectrometry. For one of the unknowns you will be given its molecular formula. The other must be identified by matching its infrared spectrum to a spectrum in a reference catalog of spectra (sometimes called a "spectral library"). If your spectra are acquired using a computer, your instructor may ask you to use your computer software to help identify your unknowns. Also, your instructor may want to demonstrate the disassembly and reassembly of the sealed demountable cell and the changing of the spacer. One part of this experiment requires a 10% solution of the unknown while another part requires a "neat" sample; thus different spacer thicknesses will be required. Alternatively, two or more different cells may be used.

SAFETY NOTE: Avoid contact of all organic compounds with your skin and also avoid breathing their vapors as much as possible. Safety glasses are required.

1. Obtain two unknowns from your instructor. One will have an identifying number; the other will have an identifying letter. For the one identified by a number, also obtain its molecular formula from your instructor. Record this information.

2. Prepare a 10% solution of your "lettered" unknown in carbon tetrachloride. Fill a sealed demountable cell (available from your instructor) with this solution. This cell must have a path length suitable for a 10% solution. Record the IR spectrum as directed by your instructor. After obtaining the spectrum, clean the cell according to your instructor's directions.

3. While other students are using the instrument, you can either proceed with your numbered unknown (Step 4) or begin the interpretation process for your lettered unknown. This interpretation process will involve either using a computer, such as searching a computer library if your spectrum was obtained with the help of a computer, or a manual search using a catalog of spectra. In any case, a preliminary look at the spectrum to check for obvious signs of the various functional groups (refer to Table 12.1 and Figure 12.14) will help reduce the number of possibilities. Keep in mind that your spectrum has the carbon tetrachloride peaks, which your catalog's spectra may not. Report your decision as to its identity to your instructor and justify your decision.

4. Based on your molecular formula alone you can make some decisions regarding your numbered unknown. For example, there must be at least six carbons in order for it to have a benzene ring. If there is a benzene ring, chances are that it will have approximately the same number of hydrogens as carbons. If there is one oxygen, it is probably either an alcohol, ether, aldehyde, or ketone. If you are waiting to use the instrument, try to come up with possible structures based on such clues from the formula.

5. Record the infrared spectrum of your numbered unknown as a "neat" sample. Again you will need to use a demountable cell with appropriate path length. Identify functional groups and other structural features indicated in the spectrum.

6. Decide on possible structures for your numbered unknown. It is very possible that several different structures (isomers) will fit. Report all of these to your instructor and justify your decisions. Finally, your instructor may want you to find the spectra of your choices in the catalog or use the computer to help identify it as you did the lettered unknown.

EXPERIMENT 31: QUALITATIVE ANALYSIS BY INFRARED SPECTROMETRY–SOLID SAMPLING

DISCUSSION: In this experiment you will identify two organic solids which your instructor will give you. The procedure is similar to Experiment 30. For one solid you will be given the molecular formula and be asked to identify as many isomers as you can that may fit both the formula and the spectrum (without looking at a catalog of spectra). For the other you will match your spectrum with one from a catalog of spectra. Again, if a computer was used to acquire the spectra, it may also be used to help identify your unknowns at the discretion of your instructor. Your instructor will indicate which of the four solid sampling techniques (solution, KBr pellet, mineral oil mull, or diffuse reflectance) will be used and will give you specific instructions in the technique involved. Reread the "solid sampling" portion of this chapter for hints.

1. Obtain the two organic solids and record their numbers and molecular formulas if appropriate.

2. Following your instructor's directions, prepare the sample for analysis by solution, KBr pellet, mineral oil mull, or diffuse reflectance, whichever is appropriate. Record the spectra for the two compounds and attempt to identify them in the same manner as in Experiment 30.

EXPERIMENT 32: QUANTITATIVE INFRARED ANALYSIS OF ISOPROPYL ALCOHOL IN TOLUENE*

DISCUSSION: Please reread the "quantitative analysis" segment under Section 12.5 to prepare for this exercise. Isopropyl alcohol exhibits an infrared absorption peak at 817 cm^{-1}. This peak is well isolated from any other peak due to isopropyl alcohol and also is isolated from any peaks due to toluene in a mixture of isopropyl alcohol and toluene. Thus, this peak is appropriate for quantitative analysis of isopropyl alcohol dissolved in toluene. Your instructor may ask you to disassemble and re-assemble the cell so as to have the appropriate spacer in place.

1. Prepare 25 mL of four standard solutions of isopropyl alcohol in toluene that are 20%, 30%, 40%, and 50% alcohol. (Use *dry* volumetric flasks!) Obtain an unknown from your instructor and dilute to the mark with toluene.
2. Obtain a sealed demountable cell from your instructor and fill it with the 20% solution. Place this cell in the instrument and obtain at least two transmittance spectra. Record the %T values from the two spectra at the tip of the 817 cm^{-1} peak and at the baseline. If the values from the two spectra differ by more than 1%, obtain a third spectrum. If the %T values from three spectra differ by more than 2%, obtain a fourth. Repeat with the other three standards and the unknown.
3. Convert all %T readings to absorbance, A. Subtract the absorbance at the baseline from the absorbance at the tip of the 817 cm^{-1} peak in each spectrum and calculate the average for each solution. Plot the resulting absorbances vs. concentration and obtain the concentration of the unknown.

QUESTIONS AND PROBLEMS

1. Why is it that the infrared spectrum of an organic compound is more useful than the UV-vis spectrum for qualitative analysis?
2. Why are inorganic compounds useful as sample "windows" and matrix material for infrared analysis?
3. What is a "neat" liquid?
4. Name three methods for mounting a liquid sample in the path of the light in an infrared spectrometer.
5. What are the differences between a "sealed" cell, a "demountable" cell, and a "sealed demountable" cell for liquid sampling?
6. What defines the path length in a sealed demountable IR cell?

*This experiment is adapted from Chia, L. and Ricketts, S., Basic Techniques and Experiments in Infrared and FT-IR Spectroscopy, The Perkin-Elmer Corporation, Norwalk, CT, 1988. With permission.

7. Describe the pressure/vacuum method of filling an IR cell equipped with inlet and outlet ports.

8. Name two separate problems that occur when a sample for infrared analysis is contaminated with water.

9. Why is a water-contaminated sample a problem for the IR cells?

10. What problem is encountered with spectra interpretation when a solution of the analyte in a particular solvent is analyzed? Why does the use of carbon tetrachloride as the solvent minimize this problem?

11. Name four methods by which the IR spectrum of a solid can be obtained.

12. What is a KBr pellet?

13. Give two reasons why the potassium bromide used to make the KBr pellet must be dry.

14. Explain the use of a hydraulic press for IR laboratory work.

15. What is a Nujol mull?

16. What advantage do the KBr pellet method and the diffuse reflectance method have over the solution and mineral oil mull methods for solids?

17. Why is it that the presence of mineral oil in the mineral oil mull usually does not cause a problem with spectral interpretation?

18. Briefly describe the diffuse reflectance method for solids.

19. What do the letters FTIR stand for?

20. Distinguish between the "double-beam dispersive" infrared instrument and the "FTIR" instrument.

21. Fill in the blanks with either "double-beam dispersive" or "FTIR", whichever correctly completes the statement:
    (a) uses an interferometer ____________
    (b) uses a moveable mirror ____________
    (c) utilizes the slit/dispersing element/slit arrangement ____________
    (d) measures the IR spectrum in a matter of seconds ____________
    (e) disperses IR light into a "spray" of wavelengths, a narrow band of which is selected ____________
    (f) a Fourier transform is performed on the data in order to obtain the absorption data at all wavelengths ____________

22. What is an interferometer and what is its function in an FTIR instrument?

23. What are the advantages of the FTIR over the double-beam dispersive instrument?

24. Answer with either "double-beam dispersive" or "FTIR" in order to indicate which instrument design is described by each statement.
    (1) utilizes a moveable mirror to create an interference pattern
    (2) records the IR spectrum in a matter of seconds
    (3) is an older design that is nearly obsolete
    (4) utilizes an "interferometer"
    (5) utilizes a double beam that is dispersed *after* it passes through the sample
25. Describe the light source in an IR instrument.
26. Distinguish between a "double beam in space" instrument and a "double beam in time" instrument. Which is used in IR instruments?
27. How are the infrared double-beam dispersive instrument described in this chapter and the UV-vis double-beam instrument described in Chapter 11 the same and how are they different?
28. How are the double-beam dispersive and FTIR instruments the same and how are they different?
29. For each of the four IR spectra in Figure 12.15, tell which of the six bonds in Table 12.1 are present and which are absent. When finished, suggest a total structure for each that fits your observations in each case.
30. Look at the three infrared spectra in Figure 12.16 and answer the following questions.
    (a) Are any of the spectra (a, b, and c) that of an alcohol? If so, which? What absorption pattern(s) at what wavelength(s) identifies an alcohol?
    (b) Are any of the spectra (a, b, and c) that of a compound containing a benzene ring? If so, which? What three absorption patterns at what wavelengths show that a compound has a benzene ring?
    (c) Are any of the spectra (a, b, and c) that of a compound containing only carbons and hydrogens? If so, which? Benzene rings contain only carbons and hydrogens. Might the spectrum or spectra you chose for your answer above indicate a benzene ring? (Tell what absorption patterns are present or not present that would support your answer.)
    (d) Do any of the three compounds have a carbonyl group (-C=O)? If so, which – a, b, and/or c? What absorption pattern, or lack thereof, at what wavelength supports your answer?
    (e) One of the following is compound a, one is compound b, and one is compound c: *n*-pentane, 3-methylphenol, benzophenone. Identify which is a, which is b, and which is c based on the spectra provided.
31. Describe an experiment in which the *quantity* of an analyte is determined by infrared spectrometry.

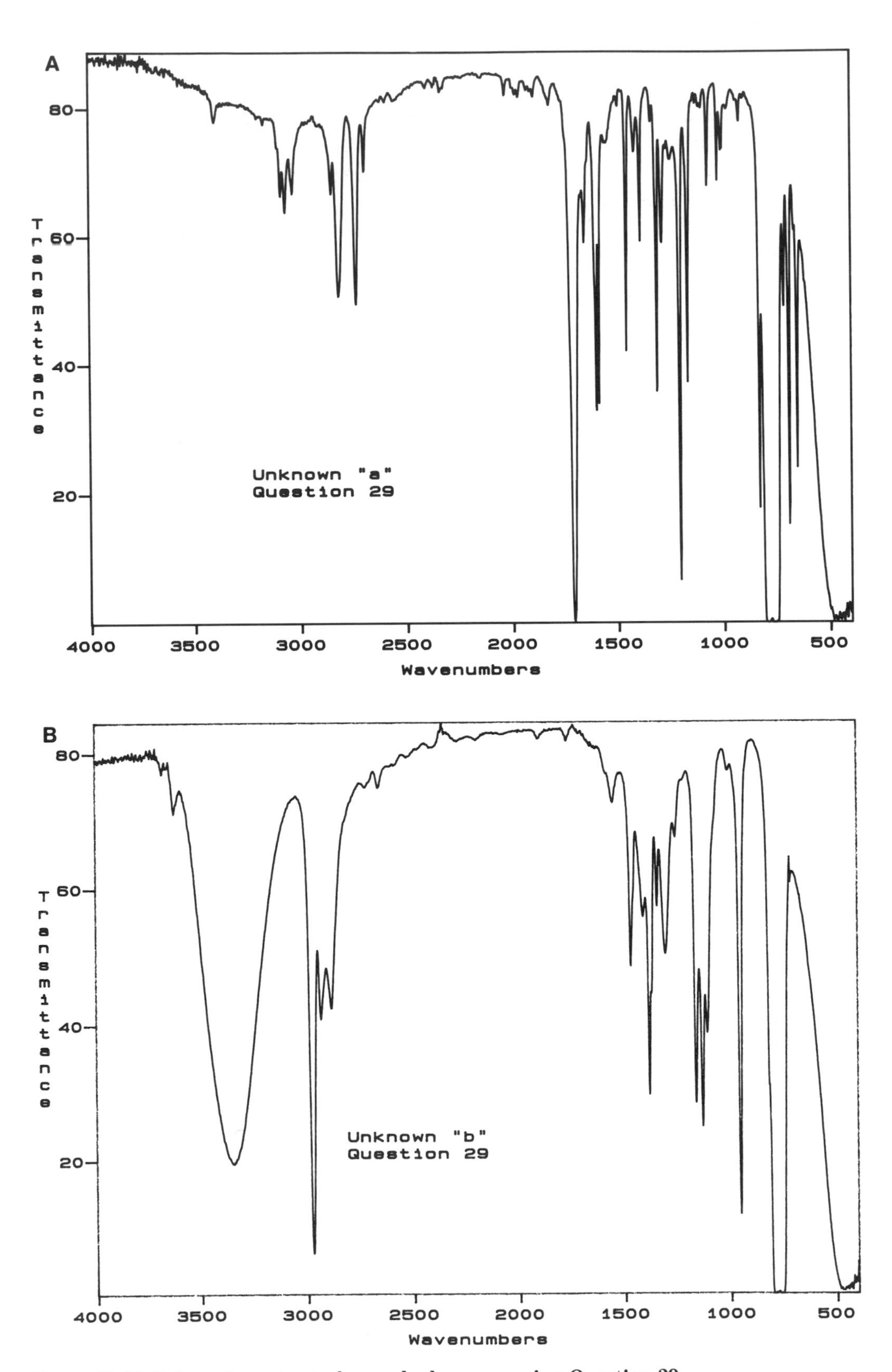

Figure 12.15. Infrared spectra to be used when answering Question 29.

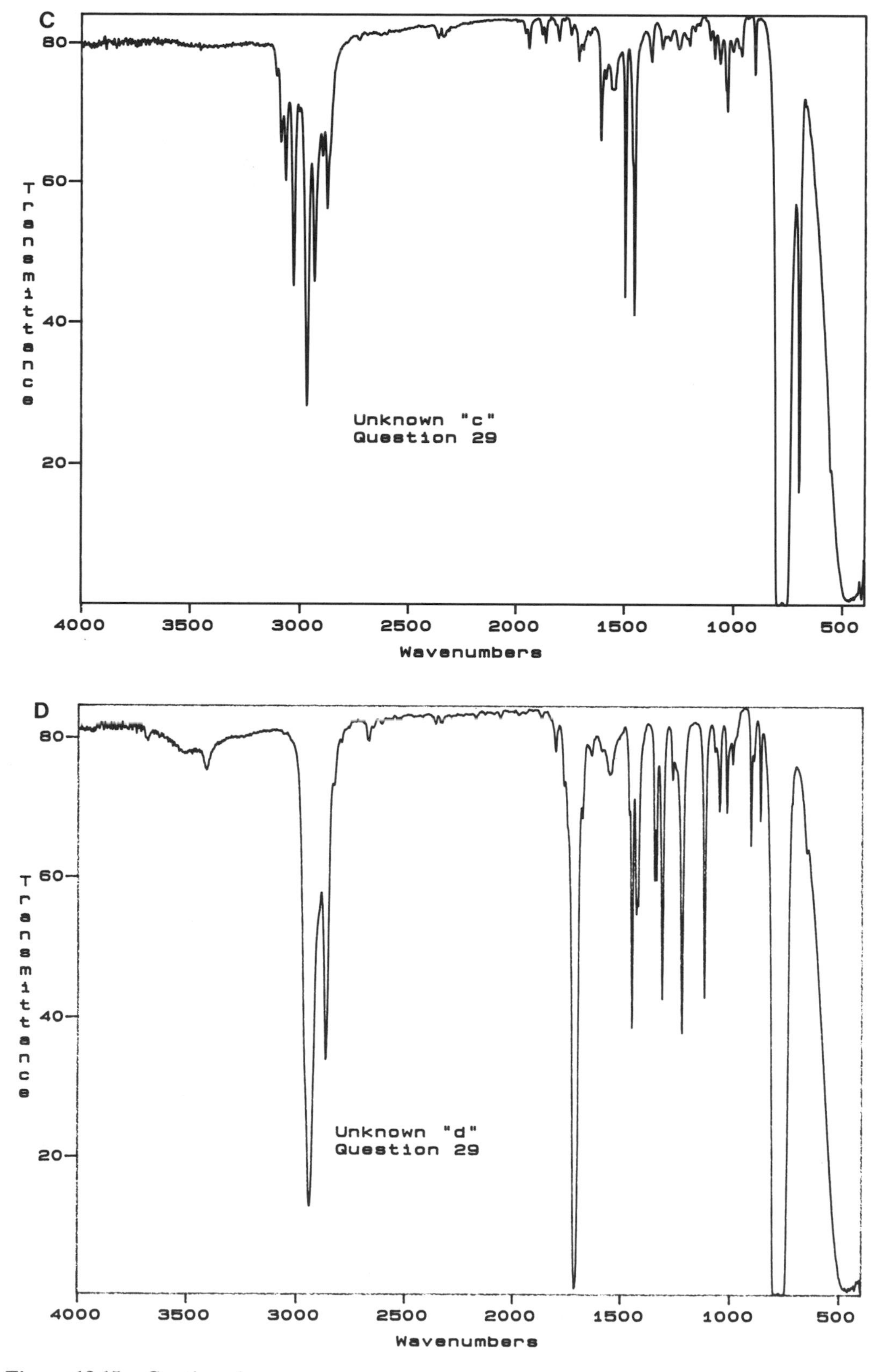

Figure 12.15. Continued.

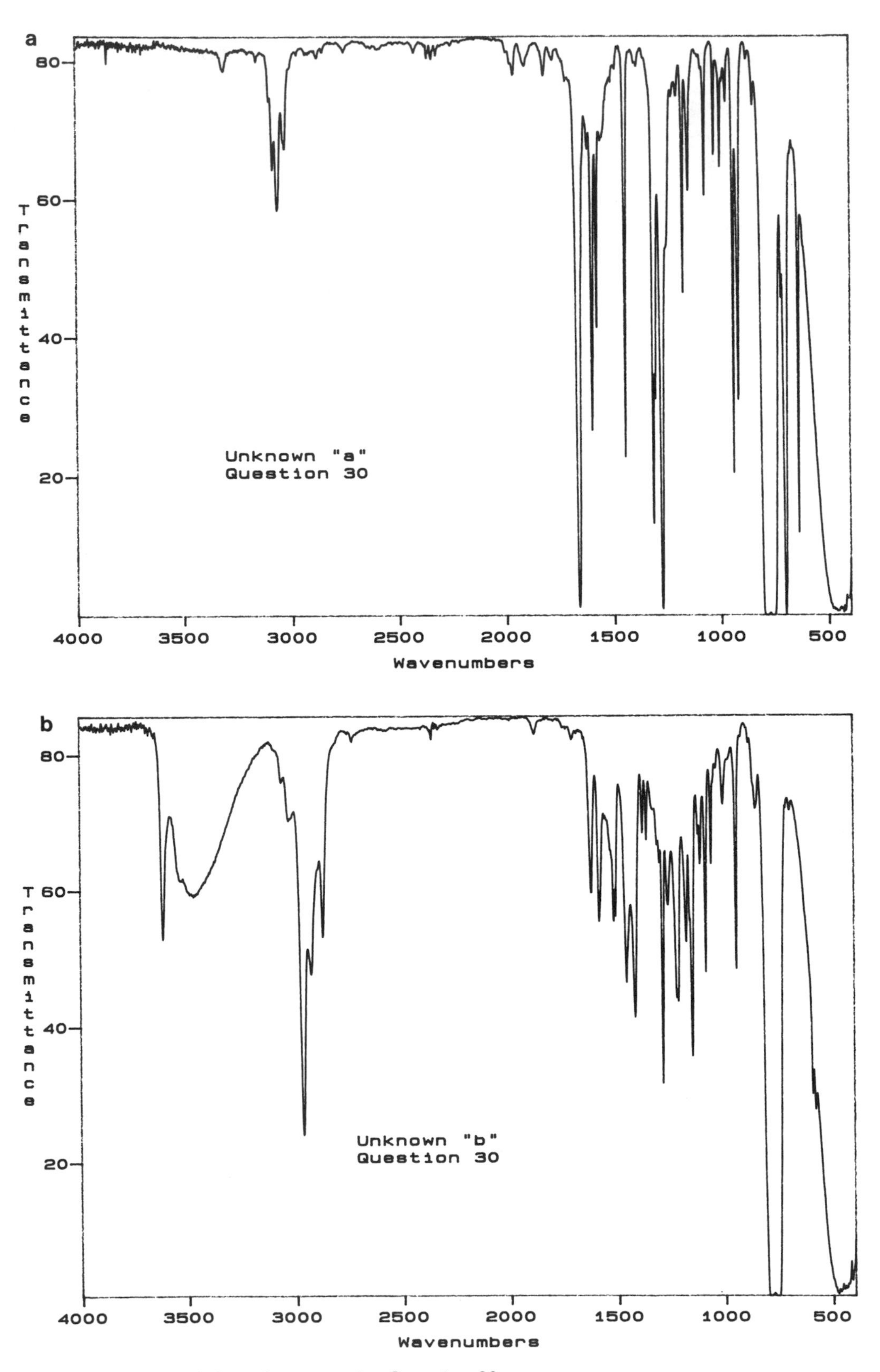

Figure 12.16. The infrared spectra for Question 30.

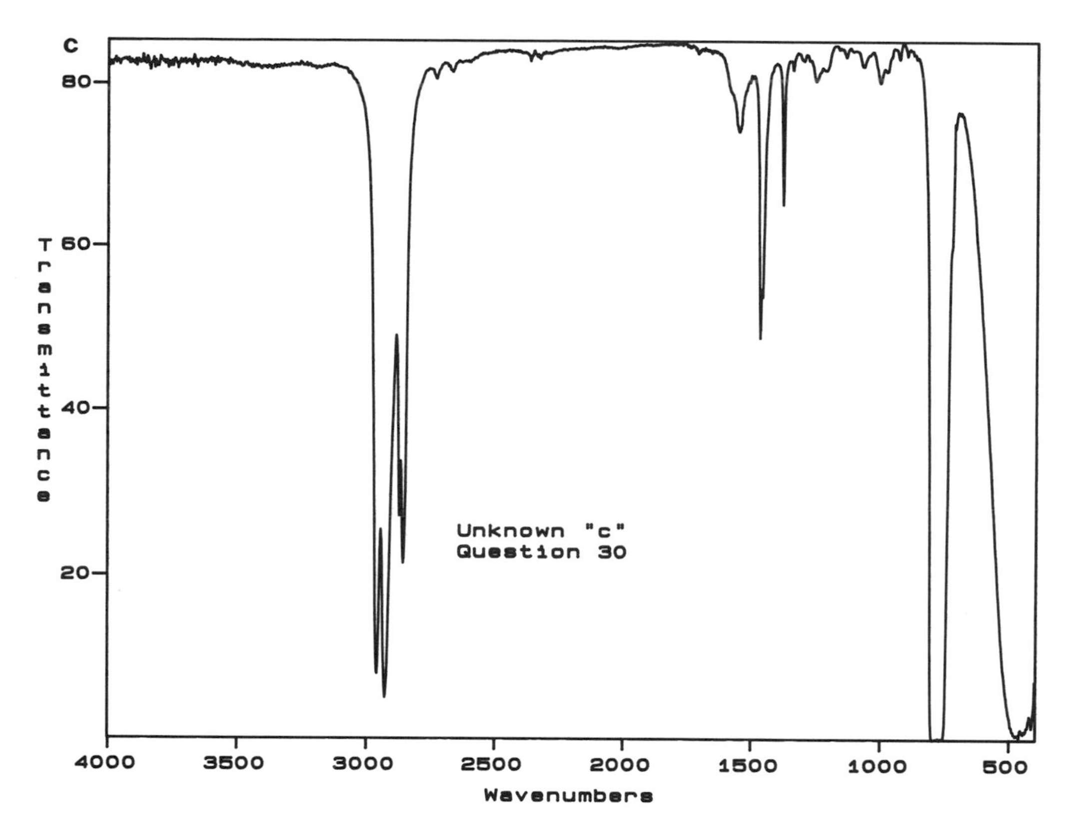

Figure 12.16. Continued.

Chapter 13

Nuclear Magnetic Resonance Spectroscopy and Mass Spectrometry

13.1 NUCLEAR MAGNETIC RESONANCE SPECTROSCOPY

Introduction

In the last two chapters we have studied instrumental analysis techniques which are based on the phenomenon of light absorption. We have discussed techniques utilizing light in the UV/vis region involving electronic energy transitions – the elevation of electrons to higher energy states. We have also discussed techniques utilizing light in the infrared region involving molecular vibrational and rotational energy transitions. In this chapter, we introduce the concepts of nuclear magnetic resonance spectrometry (NMR) and mass spectrometry (Mass Spec). First, let us discuss NMR. This technique utilizes light in the radio wave region of the spectrum and involves nuclear spin energy transitions which occur in a magnetic field.

The absorption to be described is based on the theory and experimental evidence that the nuclei of the atoms bonded to each other in molecules spin on an axis like a top. Since any given nucleus is positively charged, a small magnetic field exists around it. If we were to bring a spinning nucleus into an external magnetic field, such as between the poles of a magnet, the nucleus, representing the smaller magnetic field, would align itself to the external field. It is possible to become aligned either in the same direction as the external field or in the opposite direction. Alignment in the opposite direction represents a slightly higher energy state. (See Figure 13.1.) The energy difference between the two different alignments is on the order of a

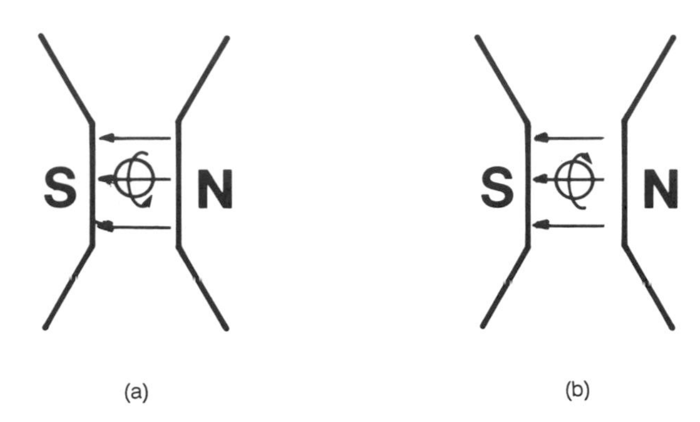

Figure 13.1. Alignment of a spinning nucleus (a) with a magnetic field and (b) opposed to a magnetic field. (From Kenkel, J., *Analytical Chemistry Refresher Manual*, Lewis Publishers, Chelsea, MI, 1992. With permission.)

radio-frequency (RF) wavelength. Thus, light in the radio-frequency region of the electromagnetic spectrum can be absorbed by molecules in a magnetic field so as to cause the nuclear spin energy transition. (See Figure 13.2.) While the phenomenon just described applies to nuclei of all elements, NMR has found its most useful application in the measurement of the hydrogen nucleus, probably because the vast majority of molecular structures contain hydrogen atoms. For this reason, it is sometimes also referred to as proton magnetic resonance (PMR). The application lies mostly in the determination of the structure of organic compounds; thus, it is mostly a qualitative analysis tool for such compounds.

The NMR spectroscopy technique has been used in organic and analytical laboratories for many years. Fairly recently, however, the Fourier transform technology, similar to that described for infrared spectrometry in the pre-

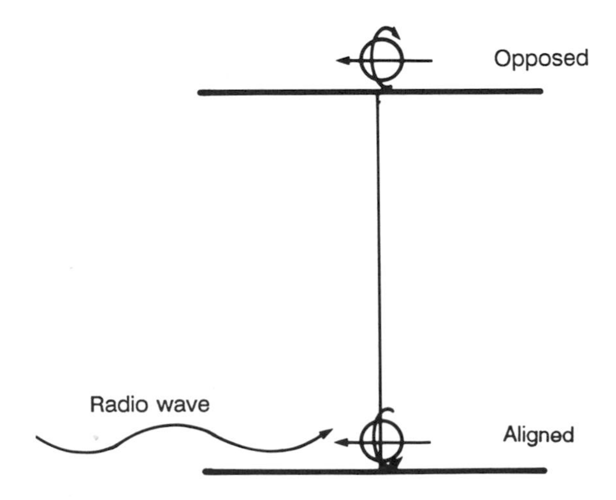

Figure 13.2. An energy level diagram showing the transition from one nuclear spin energy state to another. (From Kenkel, J., *Analytical Chemistry Refresher Manual*, Lewis Publishers, Chelsea, MI, 1992. With permission.)

vious chapter, has been applied to NMR (FTNMR) and has altered the manner in which the NMR spectrum is obtained. As with FTIR, FTNMR requires much less time (just a few seconds, as opposed to 2–5 min) to obtain a spectrum and is now used routinely in most NMR laboratories. We will now give brief descriptions of these instruments beginning with the traditional instrument.

The Traditional Instrument

Central to the instrumental design is a large magnet with the north and south poles facing each other as shown in Figure 13.1. There are several different types of magnets in common use, but stability and the ability to produce precise magnetic field strengths are common requirements. The ability to carefully vary the strength of the field between the two poles and the ability to output the magnitude of the field over time to a recorder are also requirements. A pair of coils positioned parallel to the magnet poles and connected to a sweep generator permits precise scanning of the magnetic field. Additionally, incorporated into the unit are an RF transmitter capable of emitting a precise frequency, an RF receiver/detector for detecting absorption, and an x-y recorder or computer to plot the output of the detector vs. the applied magnetic field. The sample is held in a 5 mm O.D. glass tube containing less than 0.5 mL of liquid, which in turn is held in a fixture called the "sample probe". A schematic diagram of the traditional NMR spectrometer is shown in Figure 13.3.

A precise radio frequency is emitted by the transmitter, so there is no component needed to act as a monochromator. However, the receiver/detector warrants additional comment. There are two designs for detectors. One utilizes a coil wrapped around the sample tube as the transmitter and a second coil arranged at right angles to the transmitter coil as the detector. This unique design will detect a signal only if absorption has taken place. The other design utilizes a single coil wrapped around the sample which, with the use of an appropriately designed electronic circuit, acts both as the transmitter and the receiver. The x-axis of the recorder is connected to the scanning mechanism as described previously and thus, with absorption plotted on the y-axis, the magnetic field strength is plotted on the x-axis.

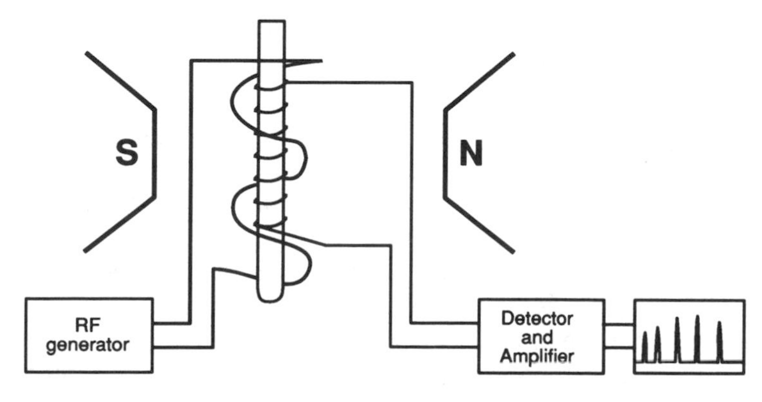

Figure 13.3. Schematic diagram of an NMR spectrometer as described in the text.

The result is a plot of absorption vs. field strength, the so-called NMR spectrum. Traditional NMR thus represents a departure from other absorption instruments in that the wavelength never changes. Rather, the applied field strength changes and the absorption of the fixed wavelength that occurs as the field strength is varied is recorded. (See the discussion of chemical shift below for a more precise discussion of what is plotted on the x-axis.)

The magnitude of the RF frequency used is variable, however. Some instruments are 60 MHz (megahertz, or 1 million cycles per seconds). Others are 100 MHz or higher. The magnitude of the frequency dictates the magnitude of the magnetic field strength required. The spread between the two energy levels in question (which corresponds to the energy of the RF frequency) depends on the strength of the field. A 60-MHz instrument requires a field strength of 14,092 gauss (the gauss is a unit of field strength), while a 100-MHz instrument requires 23,486 gauss.

Fourier Transform NMR

In FTNMR instruments, the instrument design is identical except that very brief repeating pulses of RF energy are applied to the sample while holding the magnetic field constant. As with FTIR, the resulting detector signal contains the absorption information for all transitions occurring in the sample, which in this case are all the nuclear energy transitions of all nuclei. The Fourier transformation, performed by computer, then sorts out and plots the absorption as a function of frequency, giving the same result, the NMR spectrum, as the traditional instrument but in much less time. In addition, it is more sensitive and gives better resolution.

Chemical Shifts

Electrons are, of course, present around and near the hydrogen nuclei in a molecule, and they are generating individual magnetic fields too. These very small fields oppose the applied field, giving an effective applied field somewhat smaller than expected and therefore a slightly shifted absorption pattern, the so-called chemical shift. If all hydrogen nuclei in a molecule were shielded by electrons equally, they would all give an absorption peak at the same frequency/field combination. Due to the different environments surrounding the different hydrogens in an organic molecule, however, we find absorption peaks at locations representing each of these environments. In other words, each type of hydrogen in a structure gives a characteristic absorption peak at a specific field value. Methyl alcohol, CH_3OH, for example, would give two peaks, one representing the methyl hydrogens and one representing the hydroxyl hydrogen, while cyclohexane, which has 12 equivalent hydrogens, would give just one peak, etc. The importance of this information is that (1) from the number of different peaks we can tell how many different kinds of hydrogen there are and (2) from the amount of shielding shown we can determine the structure nearby.

Since the shifts can be extremely slight, making the actual field strength of the peak difficult to measure precisely, a reference compound, typically tetramethylsilane, TMS, is often added to the compound measured. All the hydrogens in the TMS structure are equivalent (Figure 13.4). It is then

$$
\begin{array}{c}
CH_3 \\
| \\
CH_3 - Si - CH_3 \\
| \\
CH_3
\end{array}
$$

Figure 13.4. The structure of tetramethylsilane (TMS).

convenient to modify the x-axis to show the difference between the single TMS peak and the compound's peaks. The x-axis thus typically represents a "difference factor" symbolized by the Greek letter lower-case delta (δ), in which the single TMS peak is 0.0 parts per million (ppm) field strength difference and other peaks then are so many ppm away from the TMS peak. Figure 13.5 shows such an NMR spectrum for methyl alcohol.

Peak Splitting and Integration

Additional qualitative information is possible with NMR spectra. First, a given absorption peak, while apparently arising from one particular kind of hydrogen, may appear to be split into two or more peaks, giving what are termed "doublets", "triplets", etc. Also, two obviously different kinds of hydrogens may give rise to just one peak (a singlet). Each of these phenomena is caused by effects of the immediate environment of the hydrogen and will present specific qualitative information. Second, the size of a peak is indicative of the number of hydrogens it represents. This, too, is important qualitative information, and so many NMR spectrometers are equipped with integrators for determining peak size. A second pen on the recorder or a computer is used to give an integration trace which is then used to determine this number of hydrogens. Figure 13.6 is the spectrum of ethylbenzene, showing a quartet and a triplet as well as the integrator trace.

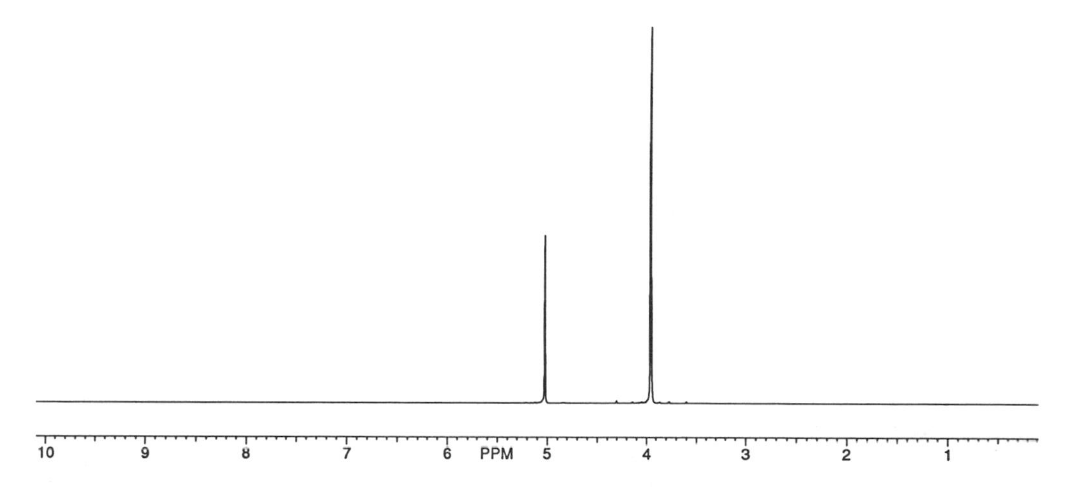

Figure 13.5. The NMR spectrum of methyl alcohol. (Courtesy of Dr. Richard Shoemacher, Chemistry Department, University of Nebraska, Lincoln.)

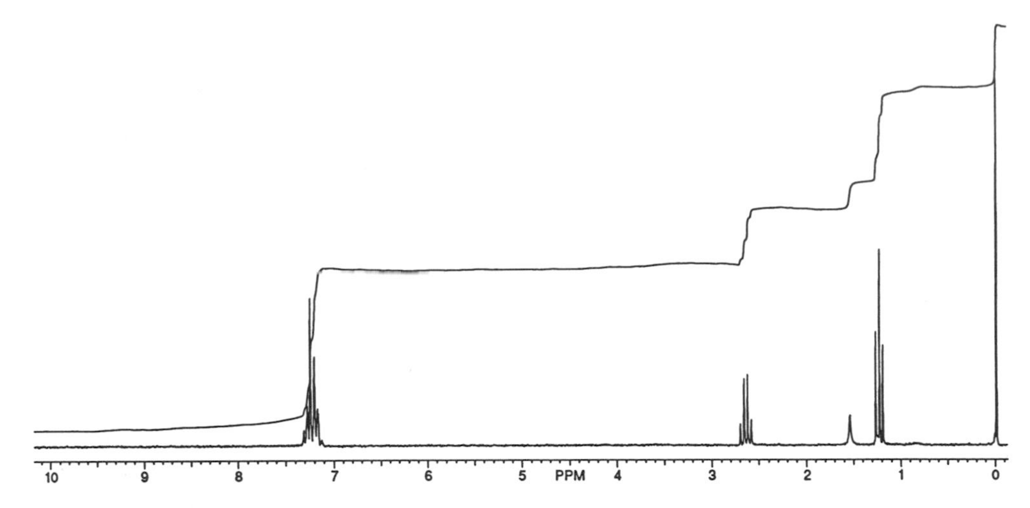

Figure 13.6. The NMR spectrum of ethylbenzene. (Courtesy of Dr. Richard Shoemacher, Chemistry Department, University of Nebraska, Lincoln.)

For more details on these and other concepts of NMR, please refer to comprehensive organic chemistry texts such as those listed under "References Sources" in Appendix 10.

13.2 MASS SPECTROMETRY

Introduction

The final molecular spectroscopy technique we will discuss is mass spectrometry. Unlike all the others, this technique does not use light at all. Very briefly, the instrument known as a mass spectrometer utilizes a high-energy electron beam to cause total destruction and fragmentation of the molecules of the sample. This fragmentation results in small charged pieces or fragments of the molecules which are then made to move through a magnetic field. The magnetic field affects each of the fragments differently according to their mass and charge, and thus they become separated. Finally, a detector sensitive to these fragments is placed in their path which, in combination with a readout device, such as a recorder or monitoring screen, allows the operator to determine the mass to charge ratio of each fragment and to identify not only the fragment, but the entire molecule. We will briefly discuss the details.

Instrument Design

There are two different instrument designs we will discuss. These are the magnetic sector mass spectrometer and the quadrupole mass spectrometer. Both designs consist of a system for sample introduction, the electron beam to create the fragmentation, a magnet to create the magnetic field, and a detection system. The entire path of the fragments, including the inlet system, must be evacuated to 10^{-4} to 10^{-8} torr. This requirement means that a sophisticated vacuum system must also be part of the setup. The reason for the vacuum is to avoid collisions of both the electron beam and the sample ions with contaminating particles which would alter the results.

The difference between the magnetic sector mass spectrometer and the quadrupole mass spectrometer lies in the design of the magnet. A diagram of a magnetic sector mass spectrometer is shown in Figure 13.7. In this instrument the magnet is a powerful, variable-field electromagnet, the poles of which are shaped to cause a bending of the path of the fragments through a specific angle, such as 90°. It is possible to vary the field strength in such a way as to scan the magnetic field and to "focus" the ion fragments of variable mass-to-charge ratio onto the detector slit one at a time. In this way, specific fragments created at the electron beam can be separated from other fragments and detected individually.

A diagram of the quadrupole mass spectrometer is shown in Figure 13.8. Here, four short parallel metal rods ("poles") with a diameter of about 0.5 cm each are utilized. These rods are aligned parallel to and surrounding the fragment path as shown. Two nonadjacent rods, such as those in the vertical plane, are connected to the positive pole of a variable power source while the other two are connected to the negative pole. Thus, a variable electric field is created, and as the fragments enter the field and begin to pass down the center area they deflect from their path. Varying the field creates the ability to focus the fragments one at a time onto the detector slit, as in the

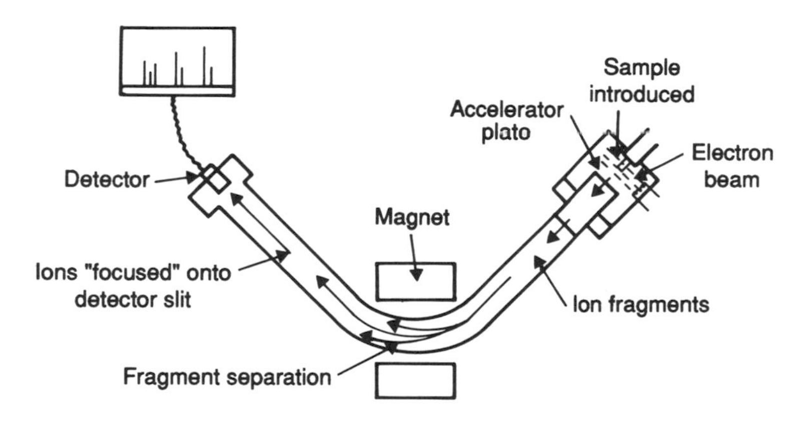

Figure 13.7. A magnetic sector mass spectrometer.

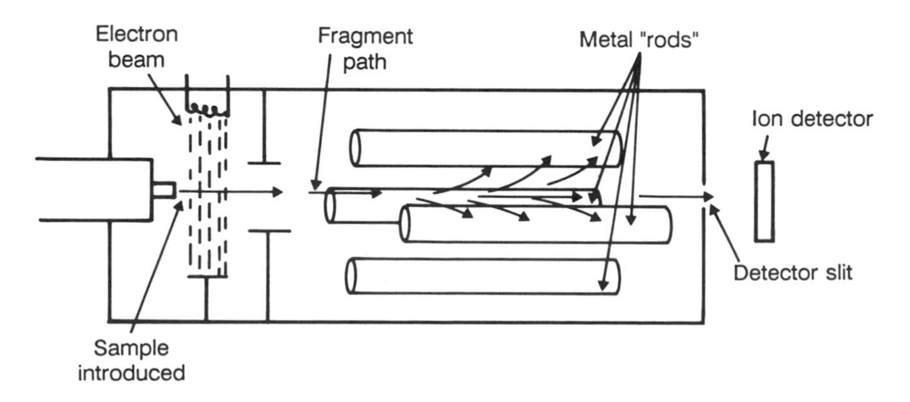

Figure 13.8. The quadrupole mass spectrometer. (From Kenkel, J., *Analytical Chemistry Refresher Manual*, Lewis Publishers, Chelsea, MI, 1992. With permission.)

magnetic sector instruments. The quadrupole instrument is newer and more popular since it is much more compact and provides a faster scanning capability.

Mass Spectra

The scanning of the magnetic field and the detection of fragments with a specific mass to charge ratio creates the possibility of manufacturing a plot of the fragment count vs. mass to charge ratio. In other words, the fractions of all specific types of fragments (of particular mass and charge) resulting from a given sample can be determined. Such a plot is called the mass spectrum. An example, the mass spectrum of aspirin, is shown in Figure 13.9.* Each vertical line evident in this figure represents a fragment of a particular mass to charge ratio given on the x-axis. The intensity of the lines represents the count, or the number of fragments detected with that ratio. The ion with a mass to charge ratio of 180 in Figure 13.9 is the "molecular ion". It loses an acetyl group (CH_2CO) to give an ion with a mass to charge ratio of 138. The mass to charge ratio of 120 is called the "base peak" because it is the most intense. The loss of CO gives the mass to charge ratio of 92.

When a certain molecule is introduced into a mass spectrometer, it will be fragmented in a certain way and its characteristic mass spectrum will al-

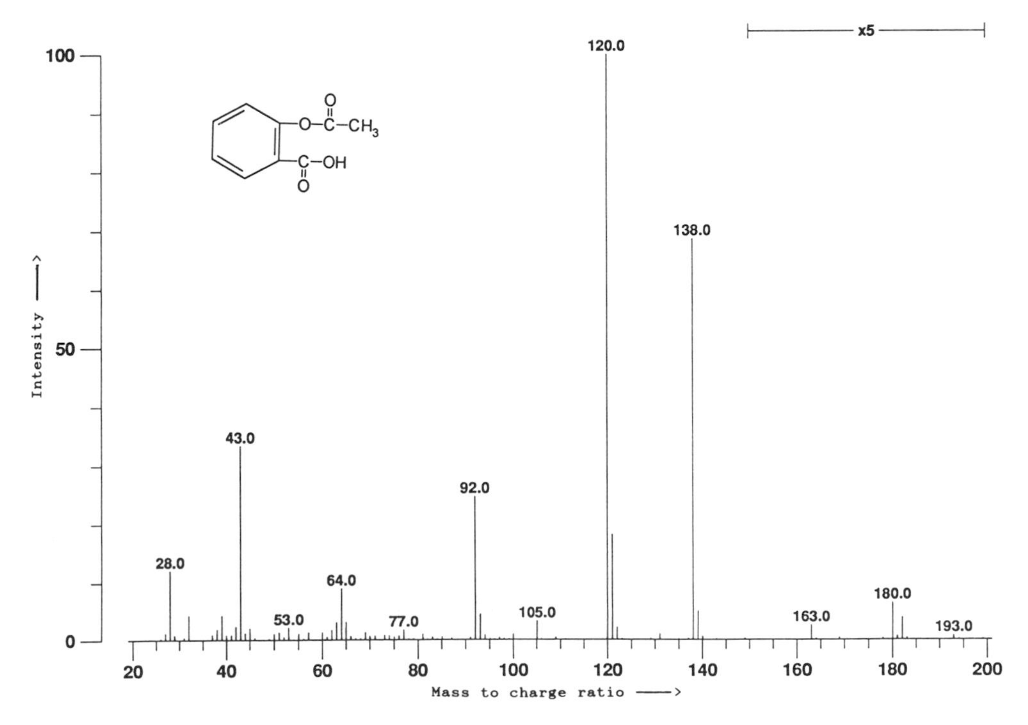

Figure 13.9. The mass spectrum of aspirin, acetylsalicylic acid.

*All mass spectra and related text and homework questions used in this text were provided by the Midwest Center for Mass Spectrometry, University of Nebraska, Lincoln, Dr. Michael Gross, Director, with partial support by the National Science Foundation, Biology Division (Grant No. DIR9017262).

ways be produced. Because the fragmentation pattern produced by a mass spectrometer can be used as a "fingerprint" of a molecule, the mass spectrum reveals, for example, if the correct compound has been synthesized and if contaminants are present. One can see that it is a "molecular fingerprint", just as absorption spectra are molecular fingerprints, and that it is a powerful tool for identification purposes.

Mass spectrometry has become an essential tool for the detection of compounds in the environment, such as the commonly used herbicide atrazine (Figure 13.10). Even relatively inexpensive commercial mass spectrometers can reach detection limits in the parts-per-billion range, while research-grade instruments can determine atrazine levels in the sub-parts-per-trillion range. Figure 13.10 shows a cluster of mass to charge ratios 215 through 218 due to the molecular ion and those molecules containing various carbon and chlorine isotopes. The molecule loses one of the CH_3 groups and one C_3H_6 group from the isopropyl amine group to give ions at mass to charge ratios of 200 and 173, respectively.

Modern mass spectrometer laboratories are linked to computer banks containing massive numbers of mass spectra obtained over the years. Specific identification, or at least a narrowing to specific possiblities, is often done by a computer that accesses these spectral files.

The mass spectrometer has been used as a detector in gas chromatography instruments and, more recently, in HPLC instruments. These combinations, referred to as GC-MS and HPLC-MS, are discussed briefly in Chapters 16 and 17.

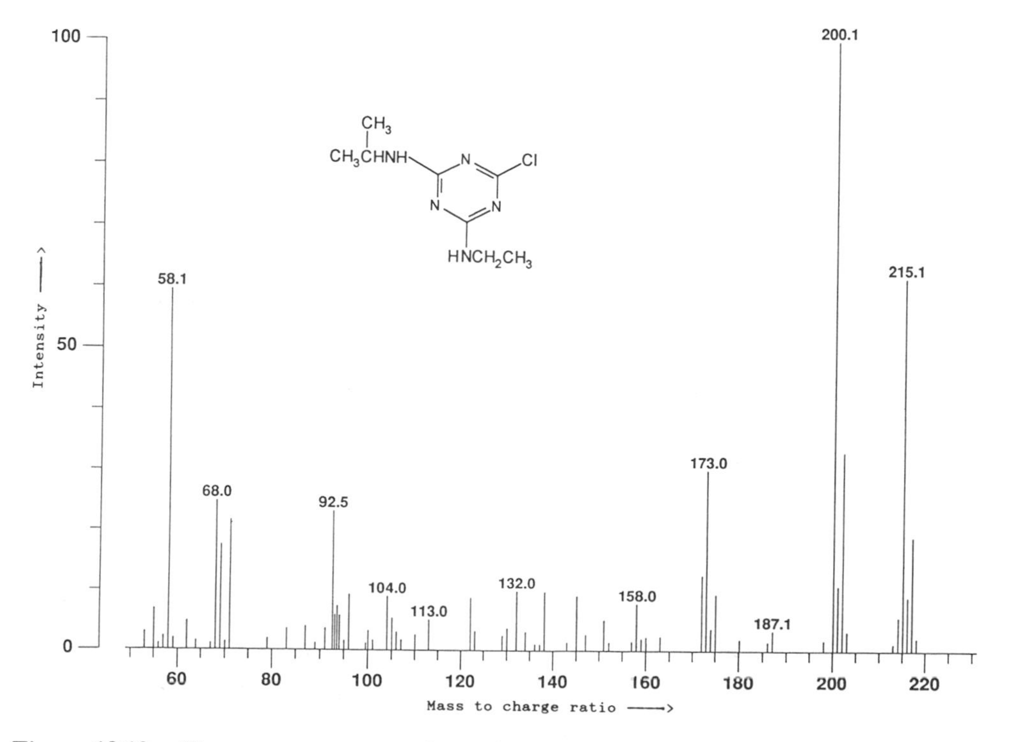

Figure 13.10. The mass spectrum of atrazine.

QUESTIONS AND PROBLEMS

1. The letters "NMR" stand for ________________ ________________ ______________________________.
2. Why is a large magnet needed as part of an NMR apparatus?
3. What wavelength region of the electromagnetic spectrum is needed for an NMR experiment? Why?
4. What two energy states of spinning nuclei present in a magnetic field are involved in an NMR experiment? Which state represents the higher energy?
5. Why is radio-frequency radiation appropriate for an NMR experiment?
6. Why is NMR sometimes referred to as PMR?
7. The letters "FTNMR" stand for ________________ ________________ ________________ ________________ ________________.
8. Briefly explain the use of each of the following NMR instrument components: sweep generator, RF transmitter, RF receiver/detector, recorder, sample probe.
9. What parameters are represented on the y-axis and on the x-axis of an NMR spectrum?
10. What is a "gauss"?
11. What is meant by the terms "hertz" and "megahertz"?
12. What is a typical value of the RF frequency (in megahertz) used in an NMR experiment? What is the magnitude of the magnetic field (in gauss) needed for an instrument utilizing this RF frequency?
13. How does a modern FTNMR experiment differ from an experiment using a traditional instrument?
14. What is the "chemical shift", and what causes it?
15. How is it that peaks representing different hydrogens in an organic molecule appear at different locations in an NMR spectrum?
16. Explain the use of tetramethylsilane (TMS) in an NMR experiment.
17. What is it about the structure of methyl alcohol that would result in two peaks in its NMR spectrum (Figure 13.5)?
18. Look at the NMR spectrum in Figure 13.11. How many different kinds of hydrogen are represented? Explain your answer.
19. What does the integrator trace in Figure 13.11 tell you about the number of different kinds of hydrogen present in the structure?
20. If you were told that the spectrum in Figure 13.11 was either ethyl alcohol or diethyl ether, what evidence would you cite that would lead you to conclude that it is ethyl alcohol and not diethyl ether?
21. If you were told that the spectrum in Figure 13.11 resulted from either ethyl alcohol or methyl ethyl ether, what evidence would you cite that would lead you to conclude that it is ethyl alcohol and not methyl ethyl ether?

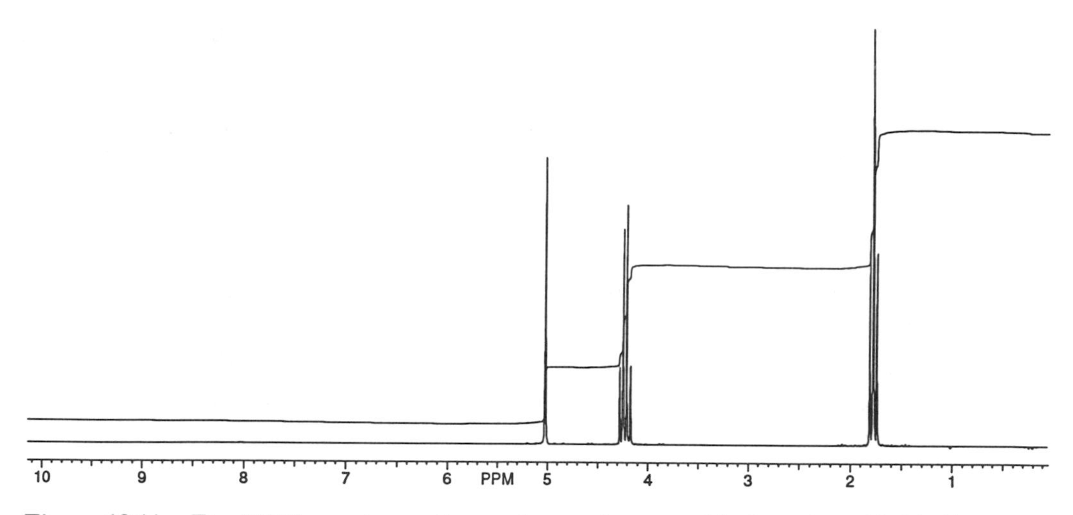

Figure 13.11. The NMR spectrum of an unknown for use with Questions 18–21. (Courtesy of Dr. Richard Shoemacher, Chemistry Department, University of Nebraska, Lincoln.)

22. The spectrum in Figure 13.12 is either that of *n*-propyl alcohol, isopropyl alcohol, or acetone. Decide which and justify your decision.
23. What causes NMR peaks to be "split" into "doublets", "triplets", etc.? How might the presence of split peaks assist with qualitative analysis?
24. What is the purpose of the high-energy electron beam utilized in a mass spectrometer?
25. How is it that a mass spectrometer can determine that the different molecular fragments created have different charges and masses (or mass to charge ratios)? Why is this fact important in its use in qualitative analysis?
26. Describe the differences between a magnetic sector mass spectrometer and a quadrupole mass spectrometer.

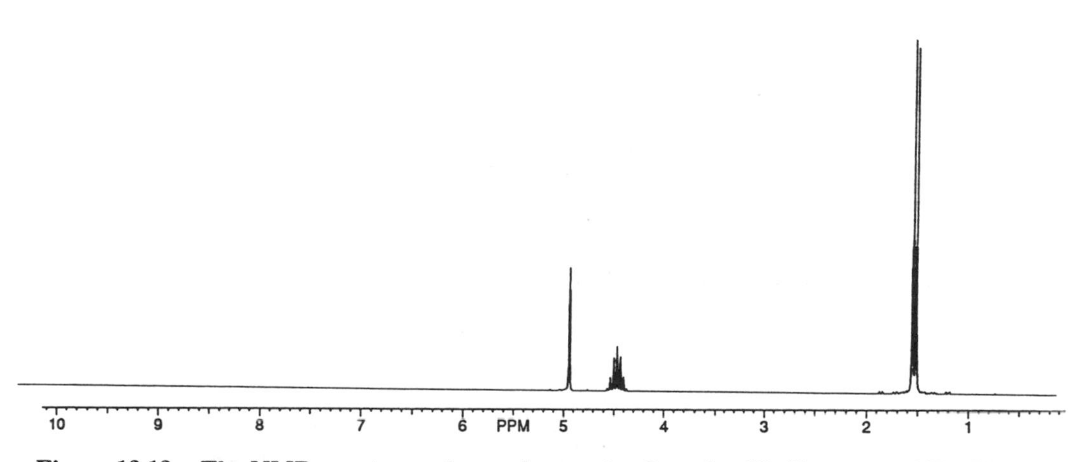

Figure 13.12. The NMR spectrum of an unknown for Question 22. (Courtesy of Dr. Richard Shoemacher, Chemistry Department, University of Nebraska, Lincoln.)

27. Why must a mass spectrometer utilize a sophisticated high-vacuum system?

28. What is plotted on the y-axis and the x-axis of a mass spectrum?

29. Briefly discuss the value of the mass spectrum as a tool for qualitative analysis.

30. The molecule which produced the mass spectrum in Figure 13.13 has the formula $C_6H_4Br_2$. Identify the compound and propose losses which would give the ion at a mass to charge ratio of 154.9. Which fragments contain bromine and how can you tell? Give the atomic composition for each ion in the molecular cluster (mass to charge ratios of 233.9 to 238.9).

31. The compound that produced the mass spectrum in Figure 13.14 is a symmetrical molecule which contains carbon, hydrogen, and oxygen. Identify the compound and explain the ions with mass to charge ratios of 182, 105, 77, and 28.

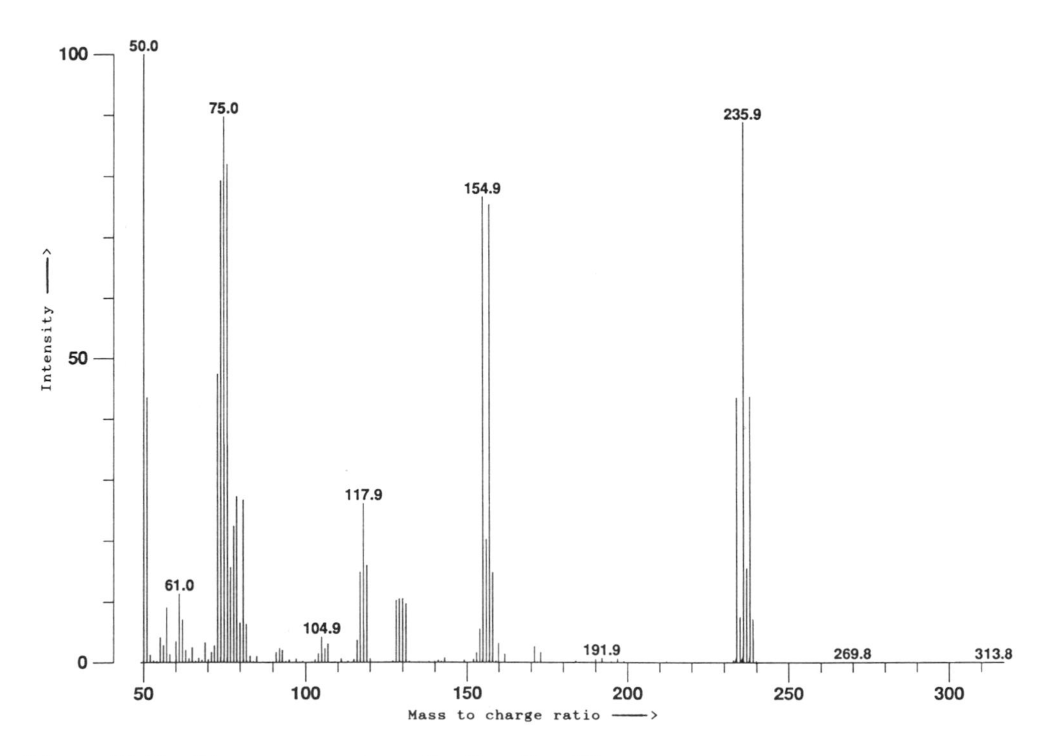

Figure 13.13. The mass spectrum for Question 30.

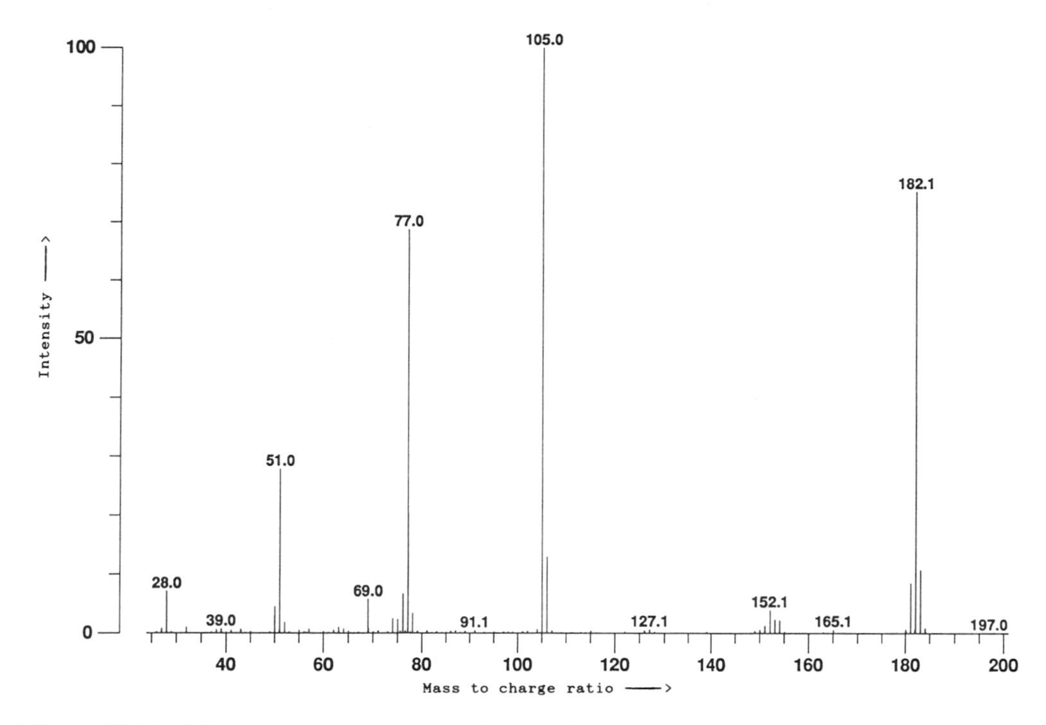

Figure 13.14. The mass spectrum for Question 31.

Chapter 14

Atomic Spectroscopy

14.1 INTRODUCTION

This chapter discusses the techniques involving the absorption and emission of light by atoms (as opposed to molecules) and builds to some extent on the discussions of light, parameters of light, light absorption and light emission, and other concepts presented in Chapters 10 and 11. If these concepts are not well understood, we suggest that you first study these chapters before beginning this chapter.

The analytical techniques known as atomic absorption (AA), flame photometry (FP), inductively coupled plasma (ICP), atomic fluorescence, and atomic emission spectrography are all included under the heading of atomic techniques or "atomic spectroscopy". Since the chemical species that absorb or emit light in these cases are atoms, the techniques are limited to sample systems from which atoms can be generated, including especially solutions of metal ions and excluding solutions of molecular species. Techniques for molecular species come under the heading of "molecular spectroscopy" and are discussed mostly in Chapters 11 and 12. Atomic spectroscopy has been applied to a wide range of metals and also to some nonmetals. In this chapter we will describe the theory, instrumentation, and application of each of the above-listed techniques.

14.2 ATOMIZATION AND EXCITATION

A device which converts metal ions into atoms is called an "atomizer". The earliest discovered atomizer was a flame. When solutions of metal ions are placed in a flame, the solvent evaporates, leaving behind crystals of the formerly dissolved salt. Dissociation into atoms then occurs – the metal ions "atomize", or are transformed into atoms. Flames are used for this purpose in most atomic absorption instruments and in all flame photometry and atomic fluorescence instruments. Such instruments, especially the AA, are easily recognized because of the centralized hooded area in which a large flame, often 6 in. wide by 6 in. or more high, is located. All atomizers, including the flame, are similar energy sources. Some are of the "atomic vapor" generator variety. Examples of these include the graphite furnace (Section 14.5), the Delves cup, and the borohydride vapor generator (Section 14.6). Another type uses an inductively coupled plasma (Section 14.7) and another uses a spark or arc across a pair of electrodes (Section 14.8). In each case, in addition to atomization, excitation of the atoms also occurs.

Fuels and Oxidants. All flames require both a fuel and an oxidant in order to exist. Bunsen burners and Meker (Fisher) burners utilize natural gas for the fuel and air for the oxidant. The temperature of such a flame is 1800 K maximum. In order to atomize and excite most metal ions and achieve significant sensitivity for quantitative analysis by atomic spectroscopy, however, a hotter flame is desirable. Most AA and FP flames today are air-acetylene flames – acetylene is the fuel, air the oxidant. A maximum temperature of 2300 K is achieved in such a flame. Ideally, pure oxygen with acetylene would produce the highest temperature (3100 K), but such a flame suffers from the disadvantage of a high burning velocity, which decreases the completeness of the atomization and therefore lowers the sensitivity. Nitrous oxide (N_2O) used as the oxidant, however, produces a higher flame temperature (2900 K) while burning at a low rate. Thus, N_2O-acetylene flames are fairly popular. The choice is made based on which flame temperature/burning velocity combination works best with a given element. Since all elements have been studied extensively, the recommendations for any given element are available. Table 14.1 lists most metals and the recommended flame for each. Air-acetylene flames are the most commonly used.

Burner Designs. There are two designs of burners for the flame atomizer that are in common use. These are the so-called "total consumption burner" and the "premix burner". In the total consumption burner (Figure 14.1),

Table 14.1. A listing showing which oxidant, air or nitrous oxide, is recommended for the various metals and nonmetals analyzed by AA.

Oxidant	Elements
Air:	Lithium, Sodium, Magnesium, Potassium, Calcium, Chromium, Manganese, Iron, Cobalt, Nickel, Copper, Zinc, Arsenic, Selenium, Rubidium, Ruthenium, Rhodium, Palladium, Silver, Cadmium, Indium, Antimony, Tellurium, Cesium, Iridium, Platinum, Gold, Mercury, Thallium, Lead, Bismuth
Nitrous Oxide:	Beryllium, Boron, Aluminum, Silicon, Phosphorus, Scandium, Titanium, Vanadium, Gallium, Germanium, Strontium, Yttrium, Zirconium, Niobium, Molybdenum, Tin, Barium, Lanthanum, Hafnium, Tantalum, Tungsten, Rhenium, Praseodymium, Neodymium, Samarium, Europium, Gadolinium, Terbium, Dysprosium, Holmium, Erbium, Thulium, Ytterbium, Uranium

From The Perkin-Elmer Corporation product literature, Norwalk, CT. With permission.

the fuel, oxidant, and sample all meet for the first time at the base of the flame. The fuel (usually acetylene) and oxidant (usually air) are forced, under pressure, into the flame, whereas the sample is drawn by aspiration into the flame through a small-diameter plastic tube. The rush of the fuel and oxidant through the burner head creates a vacuum in the sample line and draws the sample from the sample container into the flame. This type of burner head is used in flame photometry and is not useful for atomic absorption. The reason for this is that the resulting flame is turbulent and nonhomogeneous – a property that negates its usefulness in AA, since the flame must be homogeneous for the same reason that different sample cuvettes in molecular spectroscopy must be closely matched. One would not want the absorption properties to change from one moment to the next because of the lack of homogeneity in the flame. The lack of homogeneity, however, does not affect the quality of the data obtained with a flame photometer. The reason will become clear in Section 14.4.

The premix burner does away with the homogeneity difficulty and is the burner typically used in flame AA. The sample is again drawn from the sample container by aspiration through a small-diameter flexible plastic tube, nebulized (split into a fine mist), and mixed with the fuel and oxidant with the use of a flow spoiler (such as a set of baffles) prior to introduction into the flame. The burner head typically used is rectangular and has a 4- to 6-in.-long slot through which the premixed fuel, oxidant, and sample emerge and are ignited, creating the flame. Figure 14.2 is a diagram of this design. In the nebulizer, the sample enters an even smaller-diameter tube and then impacts a glass bead, creating the fine mist. There is an adjustment on the nebulizer which controls the aspiration rate and thus the amount of sample

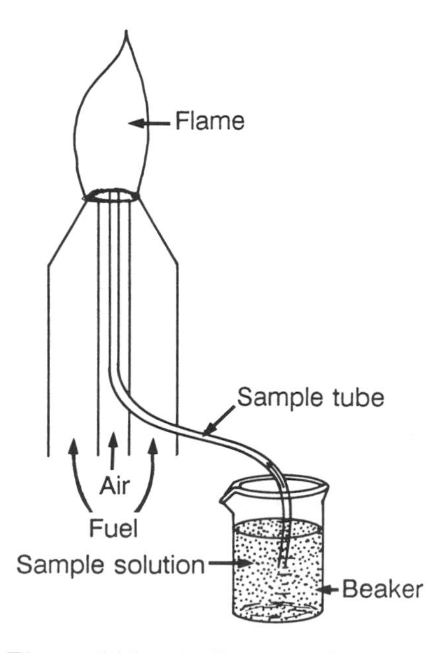

Figure 14.1. A diagram of a total consumption burner. (From Kenkel, J., *Analytical Chemistry Refresher Manual*, Lewis Publishers, Chelsea, MI, 1992. With permission.)

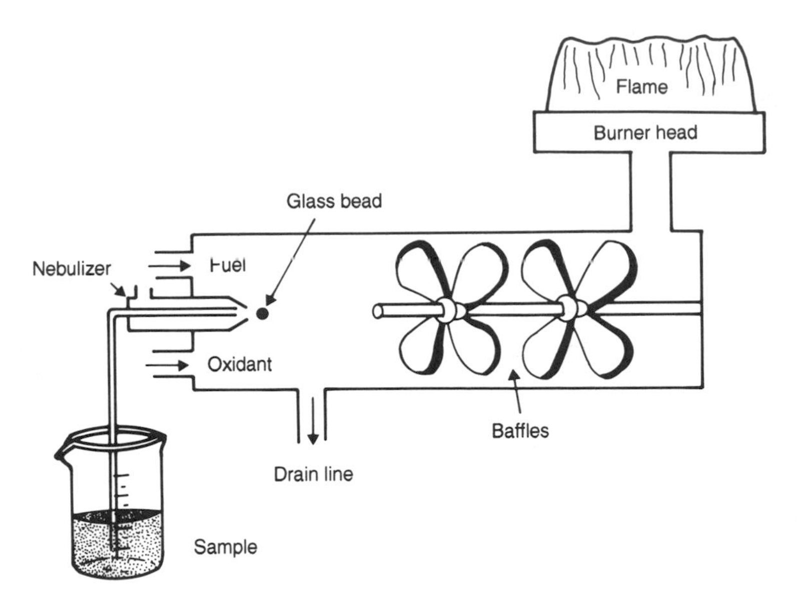

Figure 14.2. A diagram of a premix burner.

reaching the flame. This adjustment is usually set so as to obtain the maximum absorbance on the readout. Most instruments are equipped to accept a variety of fuels and oxidants. As the gas combinations are varied, it is usually necessary to change the burner head to one suitable for the particular combination chosen. A faster burning mixture would require a burner head with a smaller slot so as to discourage drawing the flame inside the burner head causing a flashback, a minor explosion.

Flashbacks can also occur when air is drawn back through the drain line, illustrated at the bottom of the premix chamber in Figure 14.2. The drain line is necessary to allow droplets of solution that don't make it to the flame to drain out. This problem is solved by forming a trap in the drain line (a flexible plastic tube) and keeping the end of the drain immersed in the waste solution contained in a bottle below the instrument.

Review of Theory. The theory of light absorption by atoms and molecules and the concepts of "line" and "continuous" spectra that result were presented in Chapter 10. What follows is a review of these subjects for atoms.

Following atomization, as indicated above, excitation of the atoms occurs to a small extent in the flame atomizer. A very small percentage of the atoms absorb energy from the flame and are elevated to an excited state. The technique of flame photometry is derived from this process, as we will see. The remaining high percentage of ground-state atoms that are therefore also present in the flame are subject to excitation with a light source. Flame atomic absorption is derived from *this* process.

As discussed in Chapter 10, basic atomic theory holds that electrons in atoms exist in energy levels around the nucleus. This theory also holds that electrons can be moved from one energy level to a higher one if conditions are right. These conditions consist of (1) the absorption of sufficient energy by the electrons and (2) a vacancy for the electron with this greater energy

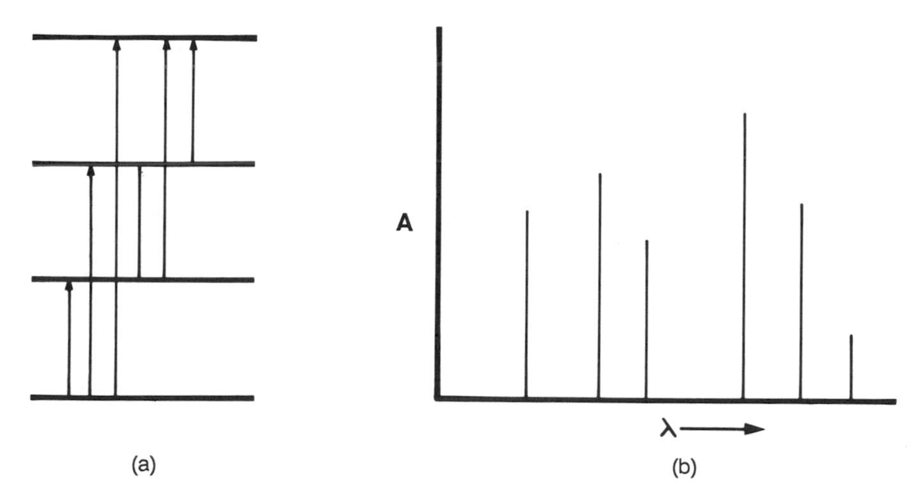

Figure 14.3. (a) A hypothetical energy level diagram showing four electronic levels and the transitions that are possible within these levels. (b) A line absorption spectrum. Each transition indicated in (a) corresponds to a line in the spectrum in (b).

in a higher level. In other words, if an electron absorbs the energy required to be promoted to a higher vacant energy level, then it will be promoted to that level and the atom will have undergone a transition from the ground electronic state to an excited electronic state. These "electronic energy levels" are the only ones that exist in atoms (there are no vibrational levels as with molecules), and thus electronic "transitions" are the only kind of energy transitions that can occur. This fact accounts for many of the differences between molecular spectrophotometers and atomic spectrophotometers and between the theories associated with each.

The individual, discrete, electronic energy transitions possible in atoms are depicted in Figure 14.3a. Since no vibrational transitions take place, only a limited number of energies, those corresponding to the very specific electronic transitions, have a chance of being absorbed. For flame atomic absorption, in which wavelengths of light get absorbed by the atoms, the result is a "line" absorption spectrum (Figure 14.3b) rather than a "continuous" absorption spectrum as found for molecules. The term "line" is used here in reference to the individual lines, or wavelengths, of absorption evident in Figure 14.3b.

Those atoms excited by the light source in flame atomic absorption are those that are measured by this technique. Those atoms excited by the flame are those that are measured by flame photometry. Flame AA utilizes a light source and absorption is measured, while flame photometry does not use a light source and light emission is measured. Details of these two techniques are given in the following two sections. As mentioned earlier, atomization can be caused by methods other than a flame, and thus excitation can also be caused by energy sources other than a flame or light source. Details of techniques associated with these are given in Sections 14.6, 14.7, and 14.8.

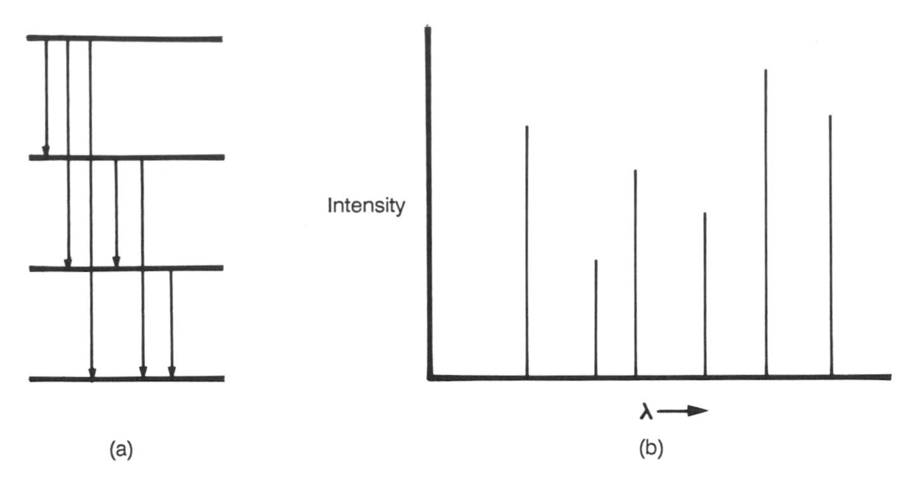

Figure 14.4. (a) A hypothetical set of transitions from higher electronic states to lower electronic states. (b) The line emission spectrum that results from the transitions in (a).

14.3 FLAME EMISSION

The Flame Test

Flame photometry is the technique which measures the atoms excited by a flame (and not by a light source). This measurement is possible because atoms that find themselves in an excited state (such as those found naturally from a solution aspirated into a flame) will readily lose the gained energy in order to revert back to the ground state. Since the magnitude of the energy lost is on the order of light energy, light is emitted. The wavelengths of the emitted light correspond to those same wavelengths as those that were absorbed in the flame atomic absorption technique discussed briefly in the last section, since exactly the same energy transitions occur, except in reverse. Figure 14.4 illustrates this phenomenon. The spectrum shown in Figure 14.4b is a "line emission" spectrum. Thus, a line spectrum can be either an absorption spectrum or an emission spectrum depending on the process measured.

Each individual metal has its own characteristic emission and absorption pattern–its own unique set of wavelengths emitted or absorbed–its own unique line emission or absorption spectrum. This is because each individual metal atom has its own unique set of electronic levels. This fact is demonstrated in a simple laboratory test known as the "flame test". Sodium atoms present in a simple low-temperature Bunsen burner flame will emit a characteristic yellow light. Potassium atoms present in such a flame will emit a violet light. Lithium and strontium atoms emit a red light. The transitions occurring in the sodium atoms are such that the line spectrum that is emitted corresponds to yellow light, while those occurring in the potassium atom correspond to violet light, etc. (See Figure 14.5.) The usefulness of emission spectroscopy for qualitative analysis is thus apparent, especially given the fact that we can utilize monochromators and detectors to precisely measure the wavelengths involved. Indeed, all emission spectroscopy instruments, including flame photometers, ICPs, and emission spectrographs, are useful

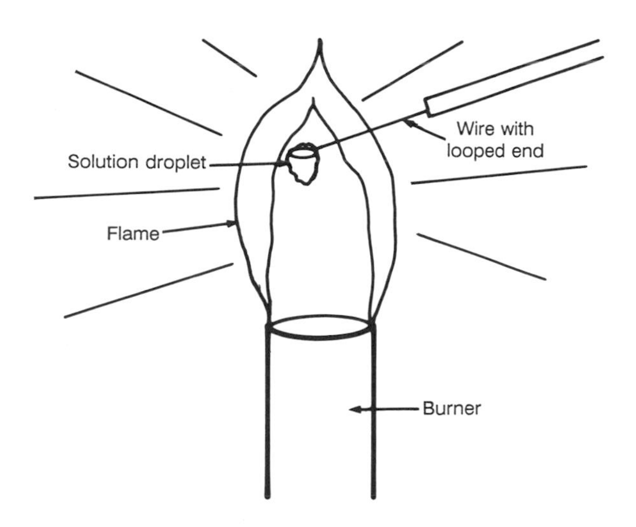

Figure 14.5. The flame test. (From Kenkel, J., *Analytical Chemistry Refresher Manual*, Lewis Publishers, Chelsea, MI, 1992. With permission).

for qualitative analysis. Flame photometry and ICP, however, find their greatest application in quantitative analysis, as we will see.

Flame Photometry

The simple flame photometer consists of three parts: the flame, a monochromator (Section 11.3), and a detector/readout (Section 11.3). As mentioned earlier, the flame originates from the total consumption burner. Of course, a premix burner may also be used; indeed, instruments (i.e., flame atomic absorption instruments) that utilize the premix burner as a matter of necessity can also function as flame photometers, as we will discuss shortly. Instruments, however, that are manufactured and sold as flame photometers cannot be used as flame AA instruments, since they are designed with a total consumption burner and no light source to excite the atoms in the flame. Why is the total consumption burner a satisfactory burner for FP? While the flame from this burner is not homogeneous in its makeup, the intensity of the light emitted by the atoms in the flame is not affected. Such emission is normally quite stable. It is, of course, the intensity of this emission that is measured for quantitative analysis, as we will see in the next section.

The second major component is the monochromator. In molecular absorption instruments (Section 11.3) a monochromator is needed to isolate the wavelength of maximum absorbance from all other wavelengths coming from the light source. In flame photometry there is no light source (independent of the flame itself), and thus the monochromator is needed for a different reason. For quantitative analysis, the intensity of the emitted light must be measured. For maximum sensitivity, we want to measure the intensity of the most intense wavelength emitted by the flame. The monochromator is thus tuned to this wavelength. Which wavelength is it? The flame emits the line spectrum of the element to be measured. The most intense line in the spectrum is generally the wavelength to zero in on. It is possible, however, that this line is not the optimum line due to interferences. If this is the case, then a different wavelength may be optimum. "Primary" and "secondary" lines, etc. are thus often defined. Generally, a flame photometry procedure has been well researched and the wavelength suggested in such a procedure will be the best wavelength.

The monochromator is useful for another reason. In the molecular spectroscopy techniques discussed in Chapter 11, the sample cuvette is located in a light-tight box (sample "compartment") such that room light is not a problem. With flame techniques, however, the flame must be in an open area of the instrument. The monochromator thus also serves to isolate the desired wavelength from room light.

The final component of a flame photometer is the detector/readout and associated electronics. The detector is essentially the same component described in Section 11.3. It is a device that generates an amplified electronic signal when light strikes it. In this case, however, the associated electronics are designed to measure and output the intensity of the light rather than a transmittance or absorbance. Thus, the readout displays relative intensity. Figure 14.6 shows a schematic diagram of the flame photometer.

Application

Application of flame photometry can be of both a qualitative and quantitative nature. For qualitative analysis the determination of which wavelengths are emitted is the key, since each element emits its own characteristic line spectrum. Thus, one would check for the emission of an element's wavelengths by tuning the monochromator to the expected wavelengths for the element in question and checking for a readout of intensity. The expected wavelengths can be obtained from tables of emission lines such as Table 14.2.

Quantitative analysis by flame photometry is encountered more frequently than qualitative analysis in an analytical laboratory, however. As implied previously, the intensity of the emitted light is directly and linearly proportional to the concentration. Thus, the standard curve for quantitative analysis is a plot of intensity vs. concentration (Figure 14.7). As discussed

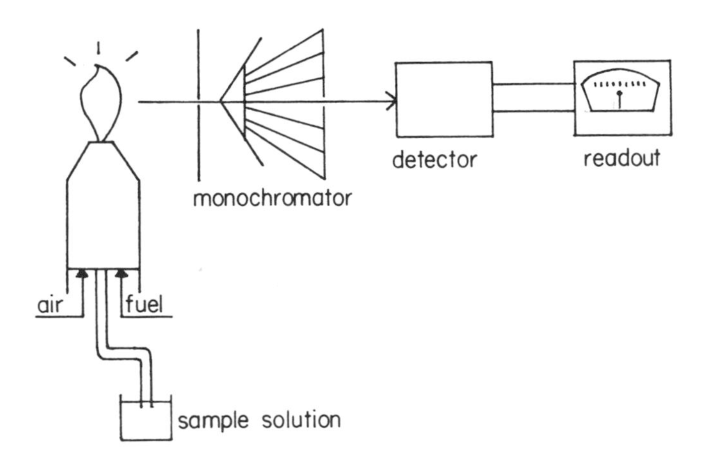

Figure 14.6. A diagram of a flame photometer.

Table 14.2. The elements typically determined by flame photometry, their corresponding primary lines, their detection limit, and applications.

Element	Primary Line	Detection Limit (ppm)	Applications
Lithium	670.8 nm	0.00003	Water, wastewater, biological fluids, soil extracts, etc.
Potassium	766.5 nm	0.0005	Water, wastewater, biological fluids, soil extracts, etc.
Sodium	589.0 nm	0.0005	Water, wastewater, biological fluids, soil extracts, etc.
Strontium	460.7 nm	0.09	Water, wastewater, soil extracts, etc.

in this book for other techniques, the standard curve is prepared by measuring the readout for a series of standard solutions of the element of interest. Table 14.2 also gives the detection limit and application of the elements typically determined by flame photometry.

It should be pointed out that other atomic techniques, especially flame and graphite furnace AA and ICP, have distinct advantages over flame photometry in many respects, especially sensitivity and linear concentration ranges. However, the flame photometry technique has been around longer and some routine analyses are well established, especially in clinical laboratories. Also, the instruments are much less expensive.

The flame photometer also has a specific application as a detector for gas chromatography. (See Section 16.7.)

14.4 FLAME ATOMIC ABSORPTION

Introduction

Of the various atomization techniques (flame, graphite furnace, Delves cup, and borohydride vapor generator) with which we combine the use of a light source for excitation, producing an absorption technique, the flame atomizer is the oldest and the most common. We refer to these techniques collectively as atomic absorption (AA) techniques, but the one which uses the flame atomizer is also often referred to as simply atomic absorption. In

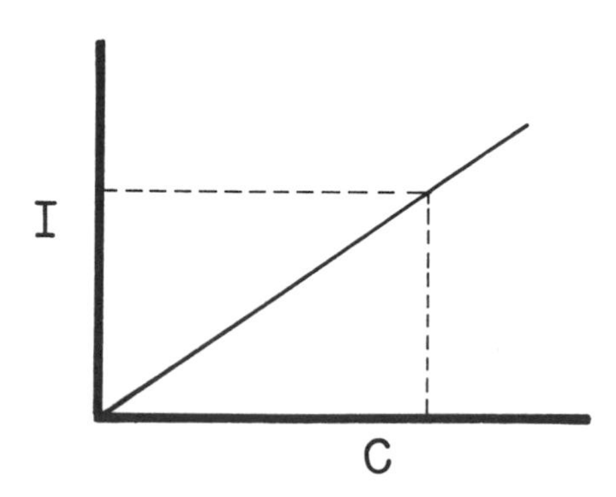

Figure 14.7. The standard curve for quantitative analysis in flame photometry. I = intensity, C = concentration.

order to make a distinction, however, between the use of the flame and the others, we refer to the absorption technique which utilizes a flame as the atomizer as flame atomic absorption. This is the technique discussed in this section. The others will be discussed in Section 14.6.

As stated previously, only a very small percentage of the atoms in the flame are actually present in an excited state at any given instant. The exact percentage depends on the flame temperature, but at the hottest temperature of any flame atomizer used for AA (2900 K), the proportion of atoms actually in the excited state at a given instant is much less than 0.01% of the total. Thus, there is a large percentage of atoms that are in the ground state and available to be excited by some other means, such as a beam of light from a light source. Flame AA takes advantage of this fact and uses a light beam to excite these ground-state atoms in the flame. Thus, AA is very much like molecular absorption spectroscopy in that light absorption (by these ground-state atoms) is measured and related to concentration. The major differences lie in instrument design, especially with respect to the light source, the sample "container", and the placement of the monochromator.

Instrumentation

The basic flame AA instrument is shown in Figure 14.8. The light source, in most cases a hollow cathode tube, is a lamp that emits exactly the wavelength required for the analysis (without the use of a monochromator). This light beam is directed at the flame containing the sample. The flame standing atop the slot of the premix burner (Section 14.2) is wide (the width of the burner head slot). This width represents the path length for Beer's law (Section 11.3) considerations. The 4- to 6-in. width thus aids in determining small concentrations of the metal being analyzed. The light beam then enters the monochromator, which is tuned to the recommended line, the so-called "primary" line, from the metal's line spectrum. This line emerges from an adjustable slit opening and is thus the wavelength that strikes the detector. Since this is an absorption technique, the electronics associated with the detector/readout are designed to display either absorbance or transmittance on the readout.

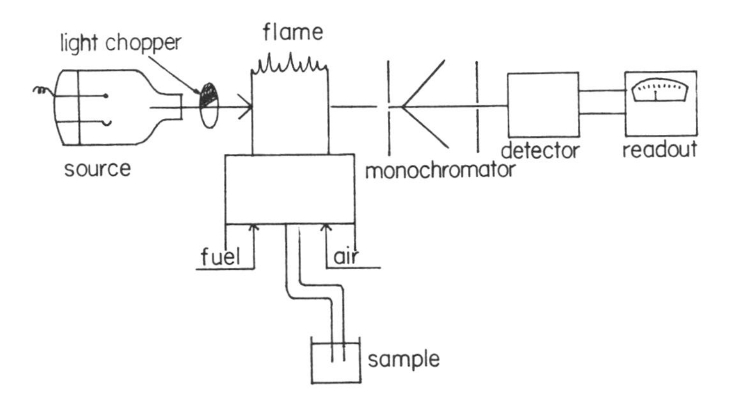

Figure 14.8. The basic flame atomic absorption instrument.

The Hollow Cathode Lamp. How is it that the hollow cathode lamp emits exactly the wavelength required without the use of a monochromator? The reason is that atoms of the metal to be tested are present within the lamp, and when the lamp is on, these atoms are supplied with energy which causes them to elevate to the excited states. Upon returning to the ground state, exactly the same wavelengths that are useful in the analysis are emitted since it is the analyzed metal with exactly the same energy levels that undergoes excitation. Figure 14.9 is an illustration of this. The hollow cathode lamp therefore must contain the element being determined. A typical atomic absorption laboratory has a number of different lamps in stock which can be interchanged in the instrument depending on what metal is being determined. Some lamps are "multielement", which means that several different specified kinds of atoms are present in the lamp and are excited when the lamp is on. The light emitted by such a lamp consists of the line spectra of all the kinds of atoms present. No interference will usually occur, however, since the monochromator isolates a wavelength of our own choosing.

The exact mechanism of the excitation process in the hollow cathode lamp is of interest. Figure 14.10 is a close-up view of this lamp and of the mechanism. The lamp itself is a sealed glass envelope containing either argon or neon gas (neon is shown in the figure). When the lamp is on, neon atoms are ionized, as shown, with the electrons drawn to the anode (positively charged electrode), while the neon ions, Ne^+, "bombard" the surface of the cathode

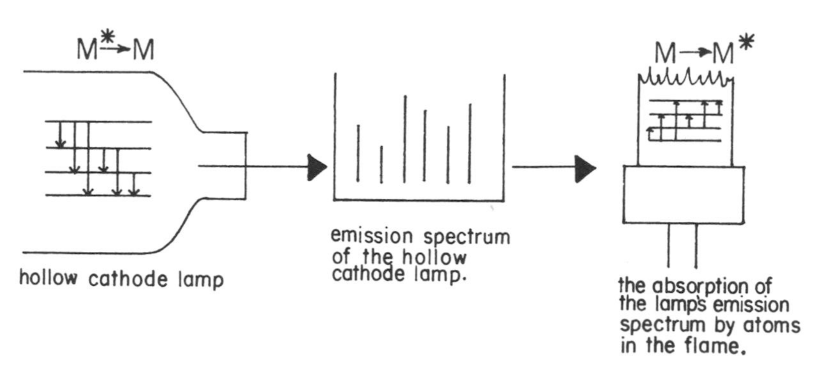

Figure 14.9. The light emitted by the hollow cathode lamp is exactly the wavelength needed by the atoms in the flame.

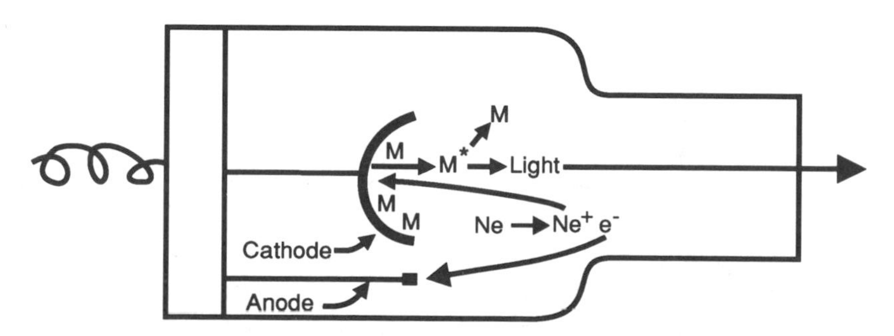

Figure 14.10. The hollow cathode lamp and the process of metal atom excitation and light emission.

(negatively charged electrode). The metal atoms, M, in the cathode are sputtered from the surface and raised to the excited state as a result of the bombardment. When the atoms return to the ground state, the characteristic line spectrum of that atom is emitted. It is this light which is directed at the flame, where unexcited atoms of the same element absorb the radiation and are themselves raised to the excited state. As indicated previously, the absorbance is measured and related to concentration.

Electrodeless Discharge Lamp. A light source known as the electrodeless discharge lamp (EDL) is sometimes used. In this lamp there is no anode or cathode. Rather, a small, sealed quartz tube containing the metal or metal salt and some argon at low pressure is wrapped with a coil for the purpose of creating an RF field. The tube is thus inductively coupled to an RF field and the coupled energy ionizes the argon. The generated electrons collide with the metal atoms, raising them to the excited state. The characteristic line spectrum of the metal is thus generated and is directed at the flame just as with the hollow cathode tube. EDLs are available for 17 different elements. Their advantage lies in the fact that they are capable of producing a much more intense spectrum and thus are useful for those elements whose hollow cathode lamps can produce only a weak spectrum.

Light Chopper. Another item relating to instrument design that is different from the molecular absorption spectrophotometers is the fact that there are two sources of light that simultaneously enter the monochromator and strike the detector. How can the detector measure only the intensity of the light that does not get absorbed and not measure the light emitted by the flame, since both sources of light are of the same wavelength? Notice in Figure 14.8 that there is a light chopper placed between the lamp and the flame. The light is "chopped" with a component similar to the rotating half-mirror beam-splitting device discussed for double-beam instruments in Chapter 11. The detector thus sees alternating light intensities. At one moment only the light emitted by the flame is read since the light from the lamp is cut off, while at the next moment the light from both the flame emission and the transmission of the lamp's light is measured since the lamp's light is allowed to pass. The electronics of the detector are such that the emission signal is subtracted from the total signal and this difference then is what is measured. Absorbance is usually displayed on the readout.

Double-Beam AA. There is also a difference in the design of double-beam instruments. In flame AA the second beam does not pass through a second sample container as it does for the molecular instruments, since a second flame would be required and it would need to be identical to the sample flame. Rather, the second beam simply bypasses the flame and is relayed to the detector directly. This design eliminates variations due to fluctuations in source intensity (the major objective of a double beam), but does not eliminate effects due to the flame or other components in the sample (blank components). These must still be adjusted for by reading the blank at a separate time. Also, the flame's emission continues to be accounted for via the chopper as with the single-beam instrument. (See Figure 14.11.)

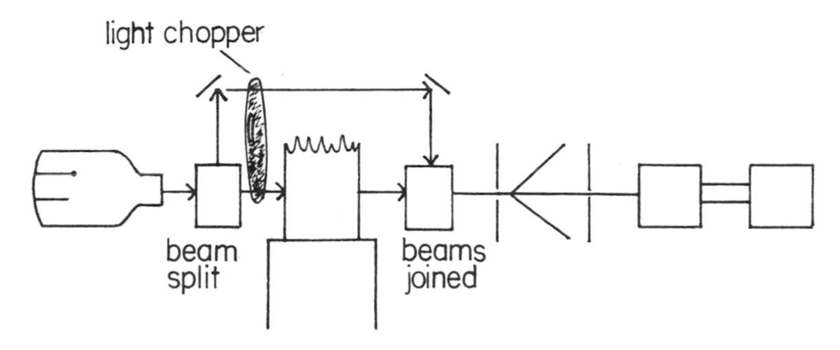

Figure 14.11. A representation of a double-beam atomic absorption instrument.

Monochromator, Detector, Readout. Descriptions of the monochromator and detector/readout are similar to those presented for molecular instruments in Chapter 11. One important difference is the placement of the monochromator. It is situated between the light source and the sample in UV/vis molecular instruments, but between the sample and the detector in atomic instruments, an arrangement similar to infrared molecular instruments. In AA instruments, the light source (the hollow cathode lamp) does not require a monochromator in order to obtain the desired wavelength, as indicated previously. However, once the light has passed through the flame, which, as with flame photometry, is located in an open, exposed portion of the instrument, it is desirable to screen out light from the flame and light from the room as well as extraneous lines from the source before the light strikes the detector. Thus, the monochromator is placed just before the detector. Another difference between the AA monochromator and the monochromator in molecular instruments is that there is a manually adjustable exit slit opening on the AA instrument. One can thus widen or narrow the slit manually to optimize the optics for the metal being analyzed.

The readout is in the form of a meter display on the instrument, a recorder, or through data acquisition with a computer. In a typical experiment, the readout consists of a series of absorbance readings with blank readings in between, since the blank is ordinarily read before each sample or standard. A strip-chart recording, for example, of the absorbance values for a series of five standard solutions would appear as in Figure 14.12.

Application

Quantitative analysis in atomic absorption spectroscopy utilizes Beer's law. The standard curve is a Beer's law plot, a plot of absorbance vs. concentration. The usual procedure, as with other quantitative instrumental methods, is to prepare a series of standard solutions over a concentration range suitable for the samples being analyzed, i.e., such that the expected sample concentrations are within the range established by the standards. The standards and the samples are then aspirated into the flame and the absorbances read from the instrument. The Beer's law plot will reveal the useful linear range and the concentrations of the sample solutions. In addition, information on useful linear ranges is often available for individual elements and instrument conditions from manuals of analytical methods available from instrument manufacturers such as Perkin-Elmer.

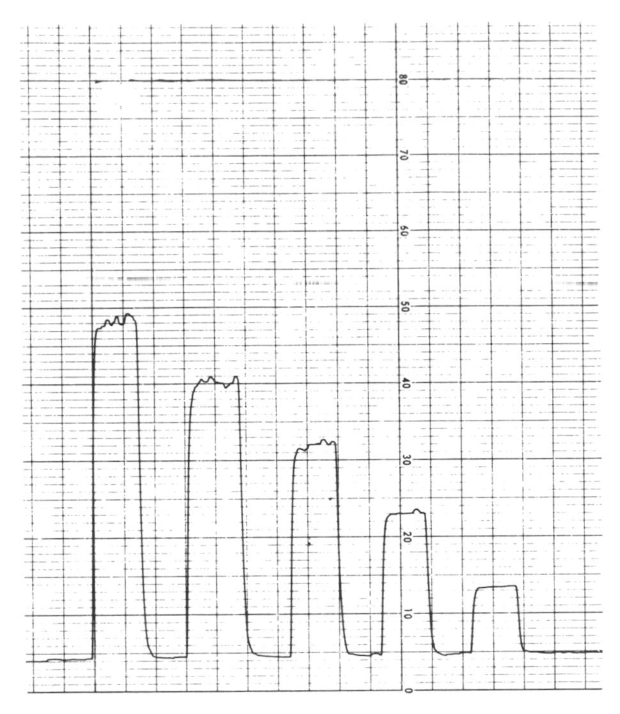

Figure 14.12. A strip-chart recording of the absorbance values of a series of standard solutions measured by an atomic absorption instrument.

Interferences. Interferences can be a problem in the application of AA. Interferences can be caused by chemical sources (chemical components present in the sample matrix) or instrumental sources (so-called "spectral" interferences arising from an instrument condition that turns out to be less than optimal for a particular sample). Chemical interferences usually result from incomplete atomization caused by an unusually strong ionic bond. An example is in the analysis of a sample for calcium. The presence of sulfate or phosphate in the sample matrix along with the calcium suppresses the reading for calcium because of incomplete atomization due to the strong ionic bond between calcium and the sulfate and phosphate. This results in a low reading for the calcium in the sample in which this interference exists. The usual solution to this problem is to add a substance to the sample which would chemically free the element being analyzed, calcium in our example, from the interference. With our calcium example, the substance that accomplishes this is lanthanum. Lanthanum sulfates and phosphates are more stable than the corresponding calcium salts, and thus the calcium is free to atomize when lanthanum is present.

In addition to the above method for removing a chemical interference, another possible solution may be to exactly match the matrix of the standard and blank to that of the sample; thus the interference would be present in all solutions tested (at the same concentration) and this would negate the problem, although sensitivity may be decreased due to a smaller concentration of atomized metal ions. The method of standard additions described in Chapter 9 is a way in which this may be accomplished.

An example of a spectral interference is when the spectral line of the element being determined nearly overlaps the line of another element in the sample, such that some of the light from the hollow cathode lamp will be absorbed by this interfering element, creating an absorbance reading that is high. We call this an instrumental interference because the slit opening setting suggested by the instrument manufacturer is too wide to totally isolate the desired wavelength emerging from the monochromator. The solution is to use a narrower slit width or to zero in on a different line, a so-called "secondary" line for the analyte element rather than the "primary" line. Recommended secondary lines are also found in instrument manufacturers' methods manuals and other literature sources.

Safety and Maintenance. There are a number of important safety considerations regarding the use of AA equipment. These center around the use of highly flammable acetylene, as well as the use of a large flame, and the possible contamination of laboratory air by combustion products. The acetylene is stored in a compressed gas cylinder as in a welding lab (although the gas is of a special purity for use in AA instruments). All precautions relating to compressed gas cylinders must be enforced–the cylinders must be secured to an immovable object, such as a wall, they must have approved pressure regulators in place, etc. Tubing and connectors must be free of gas leaks. There must be an independently vented fume hood in place over the flame to take care of toxic combustion products. Volatile flammable organic solvents, such as ether and methylene chloride, and their vapors must not be present in the lab when the flame is lit.

Precautions should be taken to avoid flashbacks. Flashbacks result from improperly mixed fuel and air, such as when the flow regulators on the instrument are improperly set or when air is drawn back through the drain line of the premix burner (see Section 14.2). Manuals supplied with the instruments when they are purchased give more detailed information on the subject of safety.

Finally, periodic cleaning of the burner head and nebulizer is needed to ensure a minimal noise level due to impurities in the flame. Scraping the slot in the burner head with a sharp knife or razor blade to remove carbon deposits and/or removing the burner head for the purpose of cleaning in an ultrasonic bath are two commonplace maintenance chores. The nebulizer should be dismantled, inspected, and cleaned periodically to remove impurities that may be collected there.

Sensitivities, Detection Limits, and Analytical Uses. Sensitivity and detection limit have specific definitions in AA. Sensitivity is defined as the concentration of an element which will produce an absorption of 1%. It is the smallest concentration that can be determined with a reasonable degree of accuracy. Detection limit is the concentration which gives a readout level that is double the noise level inherent in the experiment. It is a qualitative parameter in the sense that it is the minimum concentration that can be detected, but not accurately determined, like a "blip" on a noisy baseline. It would tell the analyst that the element is present, but not necessarily at an accurately determinable concentration level. Some sensitivities and detection limits for selected elements are given in Table 14.3.

Table 14.3. Some sensitivities and detection limits of selected elements in flame atomic absorption at the primary wavelength using an air/acetylene flame and both a flow spoiler and a glass impact bead in the premix chamber.

Element	Sensitivity (ppm)	Detection Limit (ppm)
Silver	0.03	0.0009
Arsenic	0.51	0.14
Calcium	0.08	0.001
Cadmium	0.02	0.0005
Chromium	0.04	0.002
Copper	0.03	0.001
Iron	0.04	0.003
Magnesium	0.003	0.0001
Manganese	0.03	0.001
Mercury	2.2	0.17
Nickel	0.04	0.004
Lead	0.19	0.01
Zinc	0.011	0.0008

Courtesy of The Perkin-Elmer Corporation, Norwalk, CT.

The analytical uses of flame AA are very numerous. As indicated in Section 14.1, the vast majority of applications of atomic techniques are to solutions of metal ions. Thus, whenever metals are to be determined in a sample, atomic techniques are likely to be chosen. Flame AA is the most widely used of these at the present time, probably because it has been around a long time and the instruments are less expensive than those which perhaps offer important advantages (see later sections). Table 14.4 lists some of the more common analytical uses of flame AA as well as other AA techniques.

14.5 GRAPHITE FURNACE ATOMIC ABSORPTION

There are some limitations to the use of flames as atomizers compared to other atomizers that have been invented. First, they require relatively large sample volumes (at least several milliliters). Second, they produce fewer atoms in the path of the light, making them less sensitive. Third, they are not applicable to samples in certain matrices, due to undesirable matrix effects. This section discusses the alternative methods and especially emphasizes the graphite furnace.

The graphite furnace method of atomization is useful and well established. The conversion of ions to atoms occurs in a small hollow graphite cylinder (tube) which replaces the burner head in flame AA. (The flame and graphite furnace are interchangeable in most instruments.) This cylinder is positioned horizontally such that the light beam passes directly through the center of it. There is a small hole in the center of the top side of the cylinder for introducing the sample. The sample is not introduced through any form of aspiration, but rather by means of an injection through this small hole, often onto a platform inside the tube. Thus, atoms are in the path of the light for a limited time defined by the sample volume. Very small and carefully measured sample volumes, 5–50 μL, are used. This small size can be a distinct advantage when only small volumes are available. The tube is encased in a larger tube in order to facilitate protection against air oxidation of

Table 14.4. Examples of analytical uses for atomic absorption.

Field of Endeavor	Examples of Samples and Analytes
Agriculture	Calcium, copper, iron, magnesium, manganese, potassium, sodium and zinc in soils, plant tissue, feeds and fertilizers
Biochemistry	Sodium, potassium, iron, lithium, copper, zinc, gold, lead, calcium, and magnesium in biological fluids; tissue, fingernails, and hair
Environment	Calcium, copper, lithium, magnesium, manganese, potassium, sodium, strontium, and zinc in seawater and natural water, and other metals in airborne particulates
Food	Various metals in food and food additives, meat, seafood, cooking oils, and beverages
Heavy industry	Various metals in cement, coal ash, glass, ceramics, paint and paint additives, etc.
Metallurgy	Various minor elements in alloys, such as steels, brasses, alloys of aluminum, magnesium, lead, tin, copper, titanium, nickel, and iron, as well as plating solutions
Petroleum	Various metals in lubricating oils and additives, various wear metals in used lubricating oils, lead in gasoline, and metals in other fuels, oils, and additives
Pharmaceuticals	Various metals in pharmaceutical preparations

Compiled from The Perkin-Elmer Corporation, Analytical Methods for Atomic Absorption Spectrophotometry, Norwalk, CT, 1982. With permission.

the graphite at the high temperature needed for atomization. An inert gas (argon or nitrogen) is fed both into the larger tube (but outside the inner tube) and into the inner tube. To further protect the tube and to prevent the sample from permeating through, a coating of pyrolytic graphite is often applied. Contact rings and quartz windows are attached at both ends of the tubes so as to seal the entire system. (See Figure 14.13.)

The tube itself acts as a heating element. The contact rings referred to above serve to make electrical contact to a power source. The power is controlled according to a three-step temperature program such that the

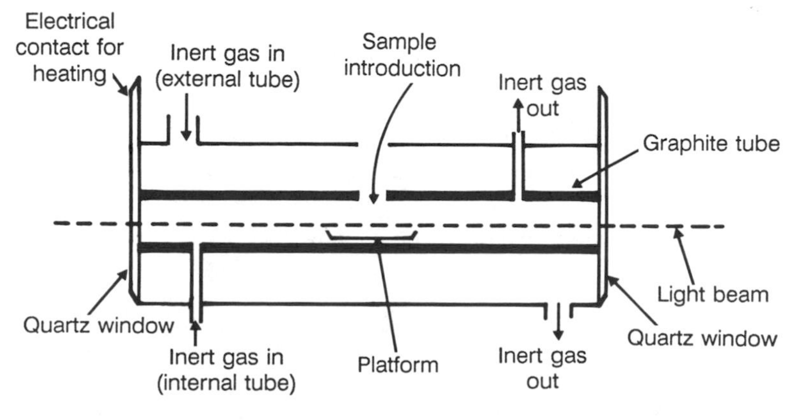

Figure 14.13. The graphite furnace atomizer. (From Kenkel, J., *Analytical Chemistry Refresher Manual*, Lewis Publishers, Chelsea, MI, 1992. With permission).

sample is first dried, then ashed, and finally atomized at a temperature of about 2500 K. The atomic vapor fills the tube at this temperature and light from the hollow cathode lamp is absorbed, with the absorbance measured as in flame AA. The drying and ashing steps are required to destroy organic matter which produces smoke inside that would otherwise scatter the light from the source, unless it is first flushed from the system by the flowing inert gas. The programming of the temperature then provides the time required for this to take place prior to the atomization.

Since there is a limited amount of sample present and the atomic vapor is also eventually swept from the tube by the flowing inert gas, the absorbance signal is transient. This means that the recording of the absorbance is usually done as a function of time, such as on a recorder or by computer data acquisition, and the transient signal then appears as a sharp peak at a particular time. The entire process of drying, ashing, and absorbance measurement is completed in a matter seconds.

The advantages of the graphite furnace are (1) the fact that only small sample volumes are needed, as previously stated, and (2) the sensitivity is substantially increased. Atomization in this furnace is nearly 100% efficient, which is a tremendous improvement over the flame (0.1%). Detection limits are improved by as much as 1000 times. The major disadvantages are (1) there often can be matrix and background absorption effects, and (2) reproducibility is not as good as with flames. Standard additions (Chapter 9) can be a solution to the matrix effect. Continuum source correction and Zeeman effect correction (utilizing a magnetic field) have been used effectively to reduce background effects. Detailed discussions of these are beyond the scope of this book.

14.6 VAPOR GENERATION METHODS

An alternative room-temperature atomization method has been devised for mercury known as the cold vapor mercury technique. Also, an alternative technique for the difficult elements arsenic, bismuth, germanium, lead, antimony, selenium, tin, and tellurium called the hydride generation technique has been devised. These are collectively referred to as vapor generation techniques. They utilize a chemical reduction (of the metal ion to the metal atom or to the metal hydride) process rather than a direct atomization in a flame or furnace.

In the cold vapor technique for mercury, mercury atoms are generated via chemical reduction with a strong reducing agent, such stannous chloride. A stream of nitrogen or air sweeps the mercury atoms into a sample absorption cell that is positioned in the path of the light. A transient absorption signal, similar to the graphite furnace signal, is observed and measured.

In the hydride generation technique, sodium borohydride is the usual reducing agent. Again the reaction product is swept into the path of the light, but, unlike the mercury technique, the analyte is not yet in the free atomic state at this stage, but rather is in the molecular hydride state. Once in the sample absorption cell, this vapor is heated with either a cool air/acetylene flame or with an electrical heating unit to create the atoms.

Advantages of vapor generation for the above-listed elements can include (1) a decrease in the effects of sample matrices, (2) improved precision, and (3) higher sensitivity.

The Delves Cup

Improvements for some elements can be achieved by simply holding the sample in a cup in the flame (no aspiration). This cup, called the "Delves cup", is composed of nickel. The atomic vapor generated is channeled into a sample absorption cell and the absorbance measured. Such a technique can utilize a sample as small as 100 μL (an advantage), but often suffers from poorer precision.

14.7 INDUCTIVELY COUPLED PLASMA

A technique that is gaining in popularity due to some important advantages over the techniques mentioned so far in this chapter is the inductively coupled plasma technique, better known as ICP. ICP is strictly an atomic emission technique most closely related to the flame photometry technique described earlier. The ICP emission source is not a flame, however, but a plasma. A plasma source is a flame-like system of ionized, very hot, flowing argon gas that is directed through a quartz tube wrapped with a copper wire (coil). Radio-frequency energy is applied to the coil, creating an intense magnetic field inside the tube. The ionization of the gas is initiated by a Tesla coil, or spark, upstream from the quartz tube, causing the argon to become conductive, creating argon ions and electrons in the flow path. As the partially ionized argon flows through the quartz tube and the RF-generated magnetic field, more ionization occurs and the gas becomes extremely hot – 9000 K to 10,000 K. What emerges from the quartz tube then is a very hot spray of ionized argon that is the ICP source, giving the appearance of a flame.

The sample and standards, in the form of solutions, are introduced into the flowing argon by aspiration just prior to the initial ionization. The entire system is pictured in Figure 14.14.

As stated previously in this chapter, a hotter source increases both atomization efficiency and excitation efficiency. Thus, the hotter an excitation source the more intense one would expect the emission to be and the smaller the concentrations of metal ions that can be detected and accurately measured. ICP is therefore much more sensitive than all other atomic techniques. In addition, the concentration range over which the emission intensity is linear is broader and simultaneous "multielement" analysis of a sample is possible. The instruments are more costly than AA and other systems.

14.8 OTHER ATOMIC EMISSION TECHNIQUES

Arc or Spark Emission Spectrography. A technique which utilizes a solid sample for light emission is arc or spark emission spectrography. In this technique, a high voltage is used to excite a solid sample held in an electrode cup in such a way that when a spark or arc jumps from this electrode to another electrode in this arrangement, atomization, excitation, and emission occur and the emitted light again is measured. The usual configuration

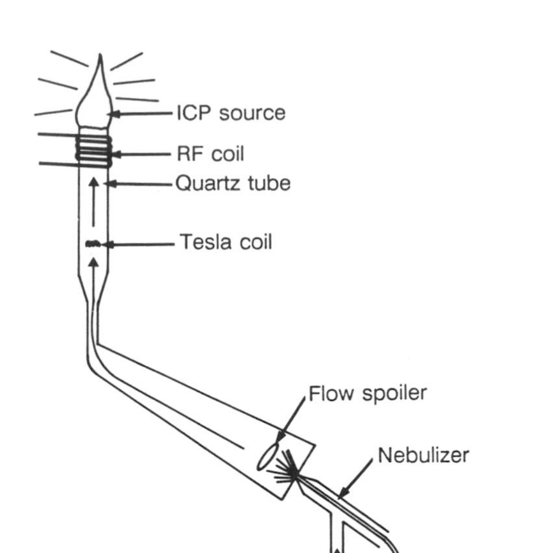

Figure 14.14. The inductively coupled plasma system described in the text. (From Kenkel, J., *Analytical Chemistry Refresher Manual*, Lewis Publishers, Chelsea, MI, 1992. With permission).

is such that the emitted light is dispersed and detected with the use of photographic film, hence the name "spectrography". The "picture" that results is that of a combined line spectrum of all the elements in the sample. Identification (qualitative analysis) is then possible by comparing the locations of the lines on the film to locations of lines on a standard film. Figure 14.15 shows the instrumental arrangement.

Atomic Fluorescence. Finally, we briefly describe a technique based on both absorption and emission – atomic fluorescence. When atoms that have

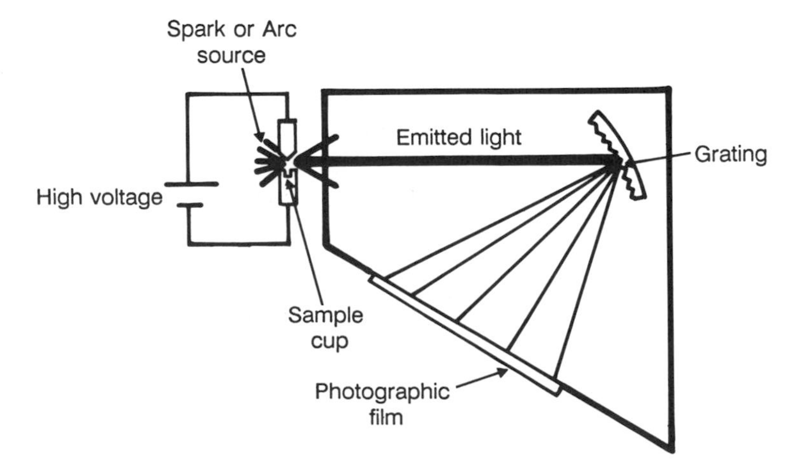

Figure 14.15. A diagram of a spark emission instrument. (From Kenkel, J., *Analytical Chemistry Refresher Manual*, Lewis Publishers, Chelsea, MI, 1992. With permission).

been elevated to higher energy levels return to the ground state, the pathway could take them to some intermediate electronic states prior to the final drop. Such a series of drops back to the ground state, if accompanied by light emission, is a form of fluorescence – in this case, atomic fluorescence. (See Chapter 11 for a discussion of molecular fluorescence.) As with molecular fluorescence, the intensity of this emitted light is measured at right angles to the incident light and related to concentration.

14.9 SUMMARY OF ATOMIC TECHNIQUES

Table 14.5 summarizes the description and application of the techniques discussed in this chapter.

EXPERIMENTS

EXPERIMENT 33: VERIFYING OPTIMUM INSTRUMENT PARAMETERS FOR FLAME AA

Introduction

All atomic absorption analyses require that a number of parameters and instrument settings be optimized prior to taking readings. These include slit setting, wavelength setting, lamp current, lamp alignment, gain setting, aspiration rate, burner head position, fuel flow rate, and oxidant flow rate. In this experiment, all of these will be tentatively set for optimum use according to manufacturer's and/or instructor's suggestions; then each will be varied slightly to observe the effect on the measured absorbance. Safety glasses are required at all times while operating the instrument, and the flame must *never* be left unattended.

Procedure

1. Prepare a standard solution of a metal by diluting a stock standard with water to a concentration that will give a reasonable absorbance when conditions are optimal, such as 5 ppm copper or zinc. Your instructor may have a suggestion for which metal and what concentration.
2. Set up the AA instrument according to your instructor's and/or the manufacturer's suggestions so that all settings are presumably optimum for the metal chosen (see Steps 3 and 4 of Experiment 34). This includes all parameters mentioned in the introduction. After aspirating a distilled water blank to zero the readout, obtain an absorbance reading for the standard solution prepared in Step 1. The reading should be between 0.5 and 0.8. If it isn't in this range, either adjust the concentration or expand the absorbance readout scale.
3. Now vary the parameters below one at a time as directed. After each variation, zero the readout and measure the absorbance. Return to the optimum setting before going on to the next parameter.

   Slit setting. Vary the slit setting as many times as possible. Record the absorbance values in a table along with the corresponding slit settings.

Table 14.5. A summary of the techniques described in this chapter.

Technique	Principle	Comments
Flame photometry	Intensity of light emitted by atoms in a flame is measured and related to concentration.	A well-established technique that is used for only a limited number of analytes and samples due to the applicability and sensitivity of other techniques.
Flame atomic absorption	Light absorbed by atoms in a flame is measured and related to concentration.	A well-established technique that remains very popular for a large number of samples and analytes. Instruments are relatively inexpensive.
Graphite furnace	Transient light absorbance signal from sample atomized in a graphite furnace (no flame) is measured and related to concentration.	A newer technique that has become popular due to its greater sensitivity and applicability to smaller sample size. Suffers from lack of precision.
Vapor generation	Light absorbed by chemically generated atomic vapor is measured and related to concentration.	Excellent technique for a limited number of analytes that are difficult to measure otherwise.
Inductively coupled plasma	Light emitted by atoms in an inductively coupled plasma is measured and related to concentration.	A newer technique that has become popular due to advantages in sensitivity, linear range, and multielement analysis. Instruments are costly.
Spark or arc emission	Light emitted by powdered solid sample caught in an arc or spark between a pair of electrodes is measured for qualitative analysis.	An excellent technique for qualitative analysis of metals in solids.
Atomic fluorescence	Light emitted by atoms in a flame excited by the absorption of light is measured and related to concentration	Not a popular technique, but it can offer sensitivity advantages.

Wavelength setting. Carefully change the wavelength setting, by say 0.1 nm each time, to several values both less than and greater than the optimum and measure the corresponding absorbances. Record the absorbance values in a table along with the wavelength settings if possible.

Lamp current. Vary the hollow cathode lamp current 1 mA at a time in both directions from the optimum and record the absorbance reading each time. Be careful not to exceed the maximum current rating indicated on the label of the lamp. Record the absorbance values in a table along with the corresponding values of the lamp current.

Lamp alignment. Vary the position of the lamp alignment screws, both the vertical and the horizontal (individually), half a turn in each direction from the optimum and record the absorbance with each half-turn. After recording the data for one, either the vertical or the horizontal, reset to the optimum and then vary the other. This will result in two sets of data, one for the vertical and one for the horizontal, with different absorbance values recorded for each half-turn away from the optimum in both directions.

Gain setting. If your instrument has separate readout meters for the light energy intensity and absorbance, it will be easy to perform this portion of the experiment. Simply turn the gain setting to vary the energy readout by five units at a time, then aspirate the sample and record the absorbance readout each time. If your instrument does not have separate meters, you will need to switch back and forth from energy display to absorbance display in order to get both readings. In either case, you should record a set of absorbance readings that correspond to a set of energy readings resulting from varying the gain setting.

Aspiration rate. The aspiration rate is varied with an adjustment on the nebulizer. Your instructor will demonstrate how to do this. It is possible to change this setting to the extent that air will actually be blown out of the sample tube rather than the solution being sucked in. Check this out ahead of time without the flame being lit so that the adjustment will not be made to that extreme when the flame is lit. It is possible to measure the aspiration rate by using distilled water in a 10-mL graduated cylinder and a stop watch. Fill a 10-mL graduated cylinder to the 10-mL line for each measurement. Then aspirate for 20 sec and determine how many milliliters are consumed. In this way you can calculate a "milliliters per minute" aspiration rate for each absorbance reading. NOTE: If during other experiments a slower than expected aspiration rate is observed, this may signal an obstruction in the sample tube or nebulizer orifice and indicate that a cleaning step is in order.

Burner head position. The burner head can be adjusted up-and-down and in-and-out. The optimum position is such that the light beam passes squarely through the flame about 1 cm above the burner head. This can be checked out before lighting the flame using an opaque white card, or similar material, placed perpendicular to the light beam. For this experiment, vary the position of the burner head in both directions from the optimum, both up-and-down and in-and-out, by a quarter turn of the control knob each time. You should then have two sets of data, one for the up-and-down variable and one for the in-and-out variable, showing absorbance readings varying with the number of quarter turns both clockwise and counterclockwise from the optimum.

Fuel flow rate. Your instructor will show you how to vary the rate of fuel flow to the burner head and also how to read the flow rate on the flow meter. Measure absorbance values at ten different flow rates, selecting a flow rate range that will maintain a flame while bracketing what the manufacturer's literature or your instructor may suggest as the optimum.

Oxidant flow rate. Repeat the above, but for the oxidant flow rate rather than the fuel flow rate. Use extreme caution here, since a flashback may result from too much or too little oxidant flow.

4. Make nine graphs, one for each of the varied parameters, plotting absorbance vs. the parameter setting or reading, and comment on what was discovered in each case. Also comment on what would happen in each case if the analyte metal were changed to some other metal. Would the optimum settings found be different or the same? Explain.

EXPERIMENT 34: QUANTITATIVE ATOMIC ABSORPTION ANALYSIS OF A PREPARED SAMPLE

NOTE: Your instructor will select the metal to be the analyte in this experiment. Safety glasses are required.

1. Prepare in a series of 100-mL volumetric flasks four standard solutions of the selected metal, with concentrations suggested by your instructor, from a 1000 ppm stock solution. Dilute each to the mark and shake.

2. Obtain an unknown from your instructor and dilute to the mark with distilled water.

3. General preparation of a flame AA instrument takes place as indicated below (a–d). Your instructor will demonstrate specifics.
   (a) If necessary, remove the hollow cathode lamp currently in the instrument and replace it with the lamp for your analyte.
   (b) Turn on the instrument so that it may warm up prior to the first measurement. Adjust the lamp current to the suggested operating current.
   (c) Set the monochromator and the slit opening to the proper value.
   (d) Fine tune the wavelength setting, and optimize the lamp alignment and gain setting.
   (e) Your instructor will indicate whether other parameters will need to be set by you (see the introduction to Experiment 33).

4. Turn on the fume hood and open the main valve on the acetylene gas bottle. See your instructor concerning the readings on the pressure gauges on the bottle's regulator. Turn on the air valve and acetylene valve on the instrument, and light the flame. Your instructor may want to demonstrate this as well as other adjustments.

5. Aspirate a blank (distilled water in this case) into the flame and set the absorbance meter reading to zero.

6. Aspirate your most concentrated sample. Set the "expansion" so that the absorbance reading is between 0.8 and 1.0. Obtain absorbance readings for all your standards and the unknown. Your instructor may suggest using a computer for data acquisition.

7. When finished, shut down the instrument as follows:
   (a) Turn off the acetylene valve on the instrument.
   (b) Turn off the air valve.
   (c) Turn off the power switch.
   (d) Close the acetylene bottle valve. Bleed the line free of acetylene by opening the acetylene valve on the instrument until the pressure gauges read zero. Close the valve.
   (e) Turn off the fume hood.
8. Plot absorbance vs. concentration, and determine the concentration of the unknown. Your instructor may want you to use a computer for this. Report the unknown's concentration.

EXPERIMENT 35: THE ANALYSIS OF SOIL SAMPLES FOR POTASSIUM BY ATOMIC ABSORPTION

NOTE: Soil samples should be dried and crushed ahead of time. Safety glasses are required.

1. Prepare an extracting solution as follows. Dilute 28.50 mL of glacial acetic acid to about 300 mL with distilled water. Add 34.5 mL of concentrated ammonium hydroxide and 2.54 g of NaCl, and adjust to pH = 7 with 3 *M* ammonium hydroxide or 3 *M* acetic acid, whichever is needed. Dilute to 500 mL with distilled water and mix thoroughly.
2. Prepare standards of 3, 5, 8, and 10 ppm in 25-mL volumetric flasks, using 1000 ppm K stock. A control sample may also be provided. Dilute all to the mark with extracting solution.
3. Weigh 1 g of each soil sample and place each in separate, dry 50-mL Erlenmeyer flasks.
4. Pipet 10.00 mL of extracting solution into each, then stopper and shake each vigorously by hand for 1 min. Filter each using *dry* Whatman #2 or Schleicher and Schuell 595 filter paper and a *dry* funnel. Do not rinse the soil samples in the filters, since that would dilute the sample extracts by an inexact amount.
5. Set up the atomic absorption instrument as in Experiment 34, but for potassium measurement, and obtain absorbance values for all standards, control, and samples, using the extracting solution for a blank. Your instructor may suggest using a computer for data acquisition. When finished, shut down the instrument as in Experiment 34.
6. Plot standard curve by plotting absorbance vs. concentration of standards. Determine sample concentrations from the graph. Your instructor may direct you to use a computer for this.
7. Multiply the figured ppm of sample by 10 to get ppm K in the soil. (Refer to QUESTIONS AND PROBLEMS #43.)

EXPERIMENT 36: THE ANALYSIS OF SOIL SAMPLES FOR IRON USING ATOMIC ABSORPTION

NOTE: Soil samples should be dried and crushed ahead of time. Safety glasses are required.

1. Prepare 500 mL of extracting solution by diluting 25 mL of 1 *N* HCl and 12.5 mL of 1 *N* H_2SO_4 to 500 mL with distilled water.
2. Weigh 5.00 g of each soil sample to be tested into *dry* 50-mL Erlenmeyer flasks. Pipet 20 mL of extracting solution into each flask. Stopper and shake on a shaker for at least 20 min.
3. Filter all samples simultaneously into small *dry* beakers using *dry* Whatman #42 filter paper, transferring only the supernatant. Do not rinse the soil samples in the filters, since that would dilute the sample extracts by an inexact amount.
4. While waiting for the samples to be shaken and/or filtered, prepare standards that are 0.5, 1.0, 3.0, and 5.0 ppm iron. Use a 1000 ppm iron solution and prepare one 100 ppm intermediate stock solution. Use 50-mL volumetric flasks, and use the extracting solution as the diluent. A control may also be provided. Dilute it also to the mark with the extracting solution.
5. Set up the atomic absorption instrument and obtain absorbance readings for all solutions using the extracting solution for the blank. Your instructor may suggest using a computer for data acquisition.
6. Plot absorbance vs. concentration and obtain the concentration of all sample solutions. Your instructor may direct you to use a computer for this. Calculate the ppm Fe in the soil by the following calculations:

$$\text{ppm Fe (extractable)} = \text{ppm (in solution)} \times 4$$

(Refer to QUESTIONS AND PROBLEMS #43.)

EXPERIMENT 37: THE ANALYSIS OF SNACK CHIPS FOR SODIUM BY ATOMIC ABSORPTION OR FLAME PHOTOMETRY

NOTE: Safety glasses are required. The procedure is written for AA. Your instructor may direct you to use flame photometry.

1. Prepare an HCl solution by diluting 100 mL of concentrated HCl with 44 mL of water.
2. Prepare sodium standards (by diluting 1000 ppm Na) that are 1, 3, 5, and 7 ppm. A control may also be provided. These should be approximately 0.5% HCl by volume.
3. Prepare samples (maximum of two) by grinding and weighing 5 g of each into separate 500-mL Erlenmeyer flasks. Add 50 mL of the HCl solution prepared in Step 1 to each. Bring each to a boil on a hot plate in a good fume hood, and then simmer for 5 min. Cool and transfer the supernatants for each to separate 100-mL volumetric flasks. Dilute to the mark with distilled water and shake. Filter each through Whatman #1 filter paper into *dry* 250-mL Erlenmeyer flasks. Pipet 1 mL of each of these extracts into other clean 100-mL volumetric flasks, and dilute to the mark with water. Save the original extracts in case more dilutions are needed.

4. Obtain absorbance readings on the AA and plot absorbance vs. concentration. Use a computer if so directed.
5. Calculate the milligrams of Na per 5 g of sample as follows:

$$\text{ppm found} \times 0.1 \times 100$$

(Refer to QUESTIONS AND PROBLEMS #43.)

EXPERIMENT 38: THE ATOMIC ABSORPTION ANALYSIS OF WATER SAMPLES FOR IRON USING THE STANDARD ADDITIONS METHOD

NOTE: Refer to the text (Chapter 9) for what "standard additions" entails. Safety glasses are required.

1. Obtain a water sample thought to contain iron. You will use this to dilute your standards to the mark.
2. Prepare 100 mL of a 100 ppm iron solution from the available 1000 ppm solution.
3. Prepare a set of four standards in 25-mL flasks such that the iron concentrations are 1, 3, 5, and 7 ppm. A control may also be provided.
4. Dilute each standard and the control to the mark with the sample and shake.
5. Prepare the AA instrument for iron analysis, and measure all standards, the control, and also the water sample. Use iron-free hard water for a blank. Your instructor may suggest using a computer for data acquisition.
6. Plot absorbance vs. concentration added for each sample, and obtain the concentration of the samples by extrapolating to zero. A computer may be used for this.

EXPERIMENT 39: THE ATOMIC ABSORPTION OR FLAME PHOTOMETRIC DETERMINATION OF SODIUM IN SODA POP

NOTE: Safety glasses are required. The procedure is written for AA. Your instructor may direct you to use flame photometry.

1. Prepare 100 mL of a 100-ppm sodium solution from the available 1000 ppm solution.
2. From the 100 ppm Na stock, prepare standards of 1, 3, 5, and 7 ppm in 25-mL volumetric flasks. A control sample may also be provided. Dilute to the mark with distilled H_2O.
3. Pipet 1 mL of each soda pop sample (degassed) into separate 25-mL volumetric flasks and dilute to the mark with distilled water. Shake.
4. Obtain absorbance values for all standards, samples, and control using an AA instrument. Your instructor may suggest using a computer for data acquisition.
5. Plot absorbance vs. concentration and obtain the concentration of Na in each diluted sample. A computer may be used for this. Multiply the

concentrations by 25 to get the ppm Na in the soda pop. (Refer to Problem #43.)

QUESTIONS AND PROBLEMS

1. What does it mean to say that a flame is an "atomizer"?
2. (a) What is an "atomizer"?
   (b) Some different types of atomizers are used in the different atomic techniques we have studied. One is a flame. Name two others.
   (c) Atomization occurs in a flame as one of several processes after a solution of a metal ion is aspirated. What other processes occur and in what order?
3. What temperature can be achieved by each of the following flames?
   (a) air/natural gas
   (b) air/acetylene
   (c) N_2O/acetylene
   (d) oxygen/acetylene
4. Even though an oxygen/acetylene flame can produce the hottest temperature, what disadvantage does it possess that limits its usefulness in practice?
5. What is the difference between a total consumption burner and a premix burner? Which is used for which technique?
6. Why must a premix burner have a drain line attached? What safety hazard exists because of this drain line and how do we deal with it?
7. (a) With help of an energy level diagram, tell what is the source of the "lines" in the line spectra of metals.
   (b) We speak of line spectra often as both emission spectra and as absorption spectra. Name two common sources of atomic emission spectra.
8. Describe the traditional "flame test" and tell how it is useful as a qualitative tool.
9. Why is the color of a flame containing sodium atoms different from that of a flame containing potassium atoms?
10. Describe the instrument known as a flame photometer and explain its usefulness for qualitative and quantitative analysis.
11. The percentage of atoms in a typical air/acetylene flame that are in the excited state at any point in time is less than 0.01%. What does this have to do with the fact that flame AA is a useful technique?
12. Define what is meant by the "primary line" and the "secondary lines" in a line spectrum.
13. There is no monochromator placed between the light source and the flame in an AA experiment. Why is this?
14. The hollow cathode lamp doesn't need a monochromator and must contain the metal to be analyzed in the cathode. Explain.

15. A typical AA laboratory has many hollow cathode lamps available and interchanges them in the instrument frequently. Explain why this is done in terms of the processes occurring in the lamp and in the flame.
16. What is an EDL? Describe its internal workings.
17. What purpose does the light chopper serve in an AA instrument?
18. How does a double-beam atomic instrument differ from a double-beam molecular instrument?
19. Since the sample "cuvette" is the flame located in an open area of the AA instrument, rather than a glass container held in the light-tight box as in the case of molecular instruments, how is room light prevented from striking the detector and causing an interference?
20. Why does Beer's law apply in the case of AA, but not in the case of flame photometry?
21. What are three essential differences between atomic absorption and flame photometry?
22. Answer the following questions "yes" or "no" for both flame photometry (FP) and flame atomic absorption:
    (a) Is a flame used for atomization?
    (b) Is Beer's law applicable?
    (c) Is the flame needed for the purpose of exciting atoms?
    (d) Is an external light source used to excite atoms?
    (e) Is the intensity of light emitted by a flame measured?
    (f) Is the population of unexcited atoms in the flame important?
23. Why is lanthanum used in an analysis for calcium?
24. Why is the standard additions technique a useful technique for AA?
25. What is the method of standards additions and under what circumstances would one want to use this method?
26. Plot the data given in Figure 9.5 for a standard additions experiment and determine the concentration of the sample solution.
27. List at least three safety precautions that must be taken in an AA laboratory.
28. What is the difference between "sensitivity" and "detection limit" in AA?
29. Describe the graphite furnace method of atomization.
30. Why is the absorbance signal developed in the case of the graphite furnace AA technique said to be "transient"?
31. What are the advantages of the graphite furnace method over traditional flame AA?
32. What are the advantages and disadvantages of the graphite furnace atomizer?
33. What is meant by the "vapor generation" method of atomization? Describe two different vapor generation methods in use.

34. What elements are measured by the hydride vapor generation method?

35. What are the advantages of vapor generation methods of atomization?

36. What do the letters ICP stand for? Is the ICP technique more closely related to AA or flame photometry? Explain.

37. Describe the ICP analysis method in detail.

38. What are the advantages of the ICP technique?

39. What are the advantages of the following atomic techniques over the standard flame atomic absorption?
    (a) graphite furnace AA
    (b) ICP

40. Compare flame photometry, atomic absorption (both flame and graphite furnace), ICP, atomic fluorescence, and spark emission in terms of (a) the process measured, (b) instrumental components and design, and (c) data obtained.

41. Match each statement with a choice or choices from the following list:

flame AA	flame photometry	atomic fluorescence
ICP	graphite furnace AA	spark emission

    (a) The technique which uses a partially ionized stream of argon gas called a plasma.
    (b) The technique in which a light source is used to excite atoms, the emission from which is then measured.
    (c) The three techniques which require light to be directed at the atomized ions.
    (d) The technique that utilizes a solid sample.
    (e) The four techniques that measure some form of light emission by excited atoms present in an atomizer.
    (f) The absorption technique in which the atoms are in the path of the light for a relatively short time.
    (g) The technique in which the emitting source may reach a temperature of 10,000 K.

42. Answer the following questions True or False:
    (a) In atomic absorption, the flame serves solely as an atomizer (in addition to being the sample "container").
    (b) Line spectra are emitted by atoms in a flame.
    (c) In flame photometry, one needs a light source to excite the atoms in a flame.
    (d) In flame photometry, quantitative analysis is accomplished by plotting emission intensity vs. concentration.
    (e) In atomic absorption, the monochromator is placed between the light source and the flame.
    (f) The population of atoms excited by the flame is what is measured in atomic absorption.

(g) In AA, the path length, b, is the width of the flame.
(h) The hollow cathode lamp has a tungsten filament.
(i) Atoms are raised to the excited state within a hollow cathode lamp.
(j) A total consumption burner is used in flame photometry.
(k) A premix burner has a drain line attached.
(l) In a premix burner, the fuel, oxidant, and sample solution meet at the base of the flame.
(m) The graphite furnace is an example of a nonflame atomizer.
(n) A flashback results from an improperly mixed sample solution.
(o) The fuel used in an AA unit is typically natural gas.
(p) Lanthanum is frequently used to prevent chemical interferences when analyzing for calcium.
(q) In the method of standard additions, standard quantities of the analyte are added to the blank in increasing amounts.
(r) The method of standard additions utilizes a plot of A vs. C in AA.
(s) The atomic fluorescence technique requires a light source.
(t) The spark emission technique requires a dispersing element, but not a monochromator.

43. Discuss and derive the calculations used to determine the final results in each of the experiments in this chapter. Show especially that the units cancel appropriately. Specifically this refers to calculations found in:
(a) Step 7 of Experiment 35
(b) Step 6 of Experiment 36
(c) Step 5 of Experiment 37
(d) Step 5 of Experiment 39

44. Consider the analysis of fish tissue for potassium as follows. A 3.9877-g sample is charred, ashed, and dissolved in nitric acid. After filtering and diluting the filtrate to the mark of a 100-mL volumetric flask, a further dilution was performed, 10 mL diluted to 100 mL. The follow AA data were then obtained for sample and standard:

C (ppm)	A
5.0	0.097
10.0	0.201
15.0	0.314
20.0	0.411
25.0	0.520
Sample	0.395

What is the ppm K in the original fish sample?

Chapter 15

Analytical Separations

15.1 INTRODUCTION

Modern-day chemical analysis can involve very complicated material samples—complicated in the sense that there can be many, many substances present in the sample creating a myriad of problems with interferences when the lab worker attempts the analysis. These interferences can manifest themselves in a number of ways. The kind of interference that is most familiar is probably one in which substances other that the analyte generate an instrumental readout similar to the analyte, such that the interference adds to the readout of the analyte, creating an error. However, an interference can also suppress the readout for the analyte (e.g., by reacting with the analyte). An interference present in a chemical to be used as a standard (such as a primary standard) would cause an error unless its presence and concentration were known. Analytical chemists must deal with these problems, and chemical procedures designed to effect separations or purification are now commonplace.

This chapter and also Chapters 16 and 17 describe modern analytical separation science. First, purification procedures known as recrystallization and distillation will be described. Then, the separation techniques of extraction and chromatography are discussed. This is followed by, in Chapters 16 and 17, instrumental chromatography techniques which can resolve very complicated samples and quantitate, usually in one easy step.

15.2 RECRYSTALLIZATION

Recrystallization is a purification technique for a solid, usually organic. The separation is based on the solid's solubility in a liquid solvent. A solvent is chosen such that the solid is sparingly soluble at room temperature, but considerably more soluble at higher temperatures. Both soluble and insoluble impurities are considered to be present, and the procedure removes both if their concentrations are not too large.

The key to the procedure is to use a *minimum* amount of solvent, such that the solid will just dissolve at the elevated temperature (usually the boiling point, if the solid is stable at that temperature). While maintaining this elevated temperature, any impurity that has not dissolved can be filtered out. The "insoluble" impurities are thus removed.

Soluble impurities, however, are still present in the filtrate. This is where the minimum amount of solvent comes into play. The procedure calls for the temperature to be lowered back to the original value. The soluble impurities will stay dissolved if their solubility has not been exceeded (if they are present in a small amount). However, the solid being purified will have its solubility exceeded, since the minimum amount of solvent was used to just dissolve the solid at the elevated temperature, and thus will precipitate from the solution. The solid, presumably purified, can then be filtered and dried. It may be necessary to perform the recrystallization several times in order to get the desired purity.

15.3 DISTILLATION

Distillation is a method of purification of liquids contaminated with either dissolved solids or miscible liquids. The method consists of boiling and evaporating the mixture followed by re-condensation of the vapors in a "condenser", which is a tube cooled by isolated, cold tap water. The theory is that the vapors (and thus recondensed liquids) will be purer than the original liquid. The separation is based on the fact that the contaminants have different boiling points and vapor pressures than the liquid to be purified. Thus, when the liquid is boiled and evaporated, the vapors (and recondensed liquids) created have a composition different from the original liquid. The substances with lower boiling points and higher vapor pressures are therefore separated from substances that have high boiling points and low vapor pressures.

Distillation of water to remove hardness minerals is an example and is probably the most common application in an analytical laboratory. Of all the applications of distillation, it is one of the easiest to perform. While water is known to have a relatively high boiling point and low vapor pressure, the dissolved minerals are ionic solids that generally have extremely high boiling points (indeed, extremely high *melting* points) and extremely low vapor pressures. Thus, a simple distilling apparatus and a single distillation or, at most, two (doubly distilled) or three (triply distilled) distillations will produce very pure water.*

*Water is often "deionized" using an ion-exchange (Section 15.6) cartridge to remove hardness minerals. Such deionization is often done in conjunction with distillation such that the water is both deionized and distilled prior to use. Also, if the water is contaminated with organics or other low-boiling substances, a charcoal filter is often used as well.

Organic liquids that are contaminated with other organic liquids usually constitute a much more difficult situation. Such liquids probably have such similar boiling points and vapor pressures that a distillation of a mixture of two or more would result in all being present in the distillate (the condensed vapors) – an unsuccessful purification. However, the liquid that has the highest vapor pressure and/or lowest boiling point, while not being completely purified, would be present in the distillate at a higher concentration level than the other components. It follows that if the distillate were then to be re-distilled, perhaps over and over again, further enrichment of this component would take place such that an acceptable purity would eventually be obtained. However, the time involved in such a procedure would be prohibitive. A procedure known as "fractional distillation" solves the problem.

Fractional distillation involves repeated evaporation/condensation steps before the distillate is actually collected. These repeated steps occur in a "fractionating column" (tube) – a column that contains a high surface area inert material for condensing the vapors above the original heated container. As the vapors condense on this material, the material itself heats up and the condensate re-evaporates. The re-evaporated liquid then moves further up the column, contacts more cold inert material and the process occurs again – and again and again as the liquid makes its way up the column. If a fractionating column were used that is long enough and contains a sufficient quantity of the high surface area material, any purification based on differences in boiling point and vapor pressure can be effected. A diagram of a distillation apparatus fitted with a fractionating column is shown in Figure 15.1. The high surface area packing material in a fractionating column typically consists of glass beads, glass helices, or glass wool.

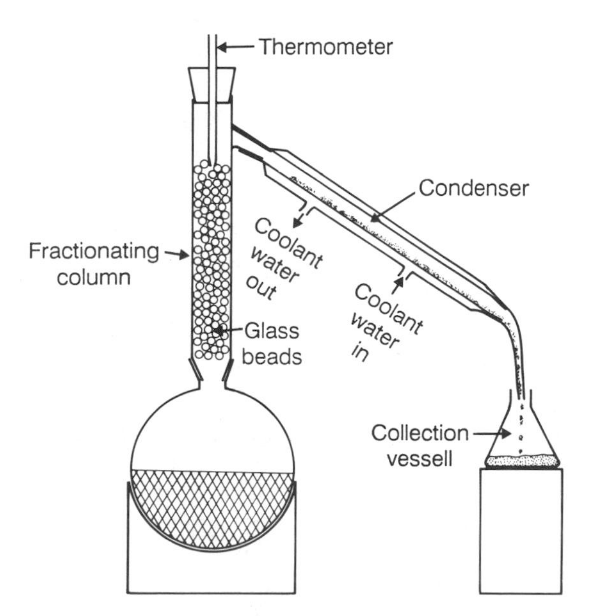

Figure 15.1. A diagram of a distillation assembly complete with a fractionating column, condenser, and collection vessel. (From Kenkel, J., *Analytical Chemistry Refresher Manual*, Lewis Publishers, Chelsea, MI, 1992. With permission.)

Each time a single evaporation/condensation step occurs in a fractionating column, the condensate has passed through what has been called a "theoretical plate". A theoretical plate is thus that segment of a fractionating column in which one evaporation/condensation step occurs. The name is derived from the concept in which the condensate is thought to be captured on small "plates" inside the fractionating column from which it was again boiled and evaporated. A fractionating column used for a given liquid mixture is then identified as having a certain number of theoretical plates, and given liquid mixtures are known to require a certain number of theoretical plates in order to achieve a given purity. The "height equivalent to a theoretical plate", or HETP, is the length of fractionating column corresponding to one theoretical plate. If the number of theoretical plates required is known, then the analyst can select a height of column that would contain the proper number of plates according to manufacturer's specifications or according to his/her own measurements of a homemade column. Height selection isn't entirely experimental, however. The use of liquid-vapor composition diagrams to predict the theoretical plates required can help. These diagrams are based on boiling point and vapor pressure differences in a pair of liquids. Further discussion of the use of these diagrams is beyond the scope of this book.

15.4 LIQUID-LIQUID EXTRACTION

Introduction

One popular method of separating an analyte species from a complicated liquid sample is the technique known as "liquid-liquid extraction" or "solvent extraction". In this method, the sample containing the analyte is a liquid solution, typically a water solution, that also contains other solutes. The need for the separation usually arises from the fact that the other solutes, or perhaps the original solvent, interfere in some way with the analysis technique chosen. An example is a water sample that is being analyzed for a pesticide residue. The water may not be a desirable solvent and there may be other solutes that may interfere. It is a "selective dissolution" method – a method in which the analyte is removed from the original solvent and subsequently dissolved in a different solvent (extracted) while most of the remainder of the sample remains unextracted, i.e., remains behind in the original solution.

The technique obviously involves two liquid phases – one the original solution and the other the extracting solvent. The important criteria for a successful separation of the analyte are (1) that these two liquids be immiscible and (2) that the analyte be more soluble in the extracting solvent than in the original solvent.

The Separatory Funnel

The extraction takes place in a specialized piece of glassware known as a separatory funnel. The separatory funnel is manufactured especially for solvent extraction. It has a teardrop shape with a stopper at the top and a stopcock at the bottom (Figure 15.2). The sample and solvent are placed together in the funnel; the funnel is then tightly stoppered and, while the stopper is held in with the index finger, shaken vigorously for a moment.

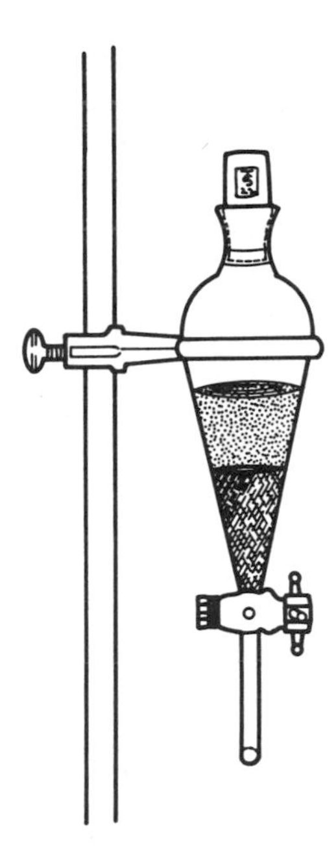

Figure 15.2. A separatory funnel containing two immiscible liquids. (From Kenkel, J., *Analytical Chemistry Refresher Manual*, Lewis Publishers, Chelsea, MI, 1992. With permission.)

Following this, the funnel may need to vented, since one of the liquids is likely to be a volatile organic solvent, such as methylene chloride. Venting is accomplished by opening the stopcock when inverted. This shaking/venting step is then repeated several times such that the two liquids have plenty of opportunity for the intimate contact required for the analyte to pass into the extracting solvent to the maximum possible extent. (See Figure 15.3 for shaking/venting illustrations.) Following this procedure, the funnel is positioned in a padded ring in a ring stand (Figure 15.2) and left undisturbed for a moment to allow the two immiscible layers to once again separate. The purpose of the specific design of the separatory funnel is mostly to provide for easy separation of the two immiscible liquid layers after the extraction

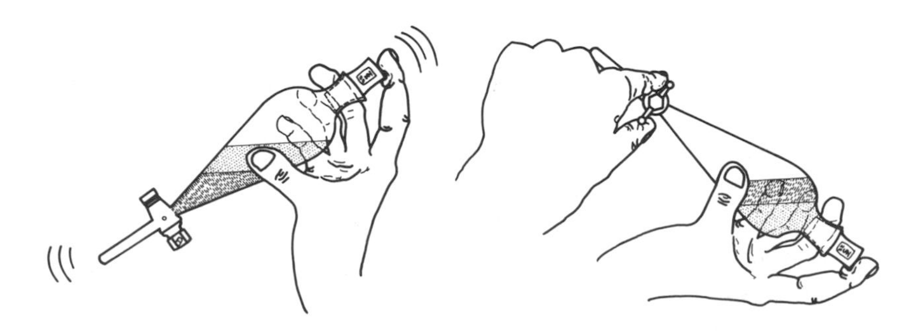

Figure 15.3. Illustrations of the positioning of the hands and the procedure for the shaking and venting mentioned in the text. (From Kenkel, J., *Analytical Chemistry Refresher Manual*, Lewis Publishers, Chelsea, MI, 1992. With permission.)

takes place. All one needs to do is remove the stopper, open the stopcock, allow the bottom layer to drain, and then close the stopcock when the interface between the two layers disappears from sight in the stopcock. The denser of the two liquids is the bottom layer and will be drained through the stopcock first. The entire process may have to be repeated several times, since the extraction is not likely to be quantitative. This means that another quantity of fresh extracting solvent may have to be introduced into the separatory funnel with the sample and the shaking procedure repeated. Even so, the experiment may never be completely quantitative. See the next section for the theory of extraction and a more in-depth discussion of this problem.

Theory

The process of a solute dissolved in one solvent being "pulled out", or "extracted", into a new solvent actually involves an equilibrium process. At the time of initial contact the solute will move from the original solvent to the extracting solvent at a particular rate, but after a time it will begin to move back to the original solvent at a particular rate. When the two rates are equal, we have equilibrium. We can thus write the following:

$$A_{orig} \rightleftarrows A_{ext} \tag{15.1}$$

in which "A" refers to "analyte" and "orig" and "ext" refers to "original solvent" and "extracting solvent", respectively. If the analyte is more soluble in the extracting solvent than the original solvent, then at equilibrium a greater percentage will be found in the extracting solvent and less in the original solvent. If the analyte is more soluble in the original solvent, then the greater percentage of analyte will be found in the original solvent. Thus, the amount that gets extracted depends on the relative distribution between the two layers, which, in turn, depends on the solubilities in the two layers. A distribution coefficient analogous to an equilibrium constant (also called the "partition coefficient") can be defined as follows:

$$K = \frac{[A]_{ext}}{[A]_{orig}} \tag{15.2}$$

Often, the value of K is approximately equal to the ratio of the solubilities of "A" in the two solvents. If the value of K is very large, the transfer of solute to the extracting solvent can be considered to be quantitative. A value around 1.0 would indicate equal distribution and a small value would indicate very little transfer.

Uses of the distribution coefficient include (1) the calculation of the amount of a solute that is extracted in a single extraction step, (2) the determination of the weight of the solute in the original solvent (important if you are quantitating the solute in this solvent), (3) calculation of the optimum volume of both the extracting solvent and the original solution to be used, (4) the number of extractions needed to obtain a particular quantity or concentration in the extracting solvent, and (5) the percent extracted. The following expansion of Equation 15.2 is useful for these:

$$K = \frac{\frac{W_{ext}}{V_{ext}}}{\frac{W_{orig}}{V_{orig}}} \qquad (15.3)$$

in which W_{ext} is the weight of the solute extracted into the extracting solvent, V_{ext} is the volume of extracting solvent used, W_{orig} is the weight of the solute in the original solvent, and V_{orig} is the volume of the original solvent used.

Example 1

The distribution coefficient for a given extraction experiment is 98.0. If the concentration in the extracting solvent is 0.0127 *M*, what is the concentration in the original solvent?

Solution

$$K = \frac{[A]_{ext}}{[A]_{orig}} = 98.0$$

$$98.0 \times [A]_{orig} = [A]_{ext}$$

$$[A]_{orig} = \frac{[A]_{ext}}{98.0}$$

$$[A]_{orig} = \frac{0.0127}{98.0} = 0.000130\ M$$

Example 2

What weight of analyte is found in 50.0 mL of an extracting solvent when the distribution coefficient is 231 and the weight of analyte found in 75.0 mL of the original solvent after extraction was 0.00723 g?

Solution

$$K = \frac{\frac{W_{ext}}{V_{ext}}}{\frac{W_{orig}}{V_{orig}}} = \frac{\frac{W_{ext}}{50.0}}{\frac{0.00723}{75.0}} = \frac{\frac{W_{ext}}{50.0}}{0.0000964} = 231$$

$$\frac{W_{ext}}{50.0} = 231 \times 0.0000964$$

$$W_{ext} = 231 \times 0.0000964 \times 50.0 = 1.11\ g$$

More complicated chemical systems may require a more universally applicable quantity called the "distribution ratio" to describe the system. These involve situations in which the analyte species may be found in different chemical states and different equilibrium species, some of which may be extracted while others are not extracted. An example is an equilibrium system involving a weak acid. In such a system there may be one (or several) protonated species and one unprotonated species. The distribution ratio, D, then takes into account all analyte species present:

$$D = \frac{C_A\ (ext)}{C_A\ (orig)} \tag{15.4}$$

in which C_A (ext) and C_A (orig) represent the total concentration of all analyte species present in the two phases regardless of chemical state. Further treatment of this situation is beyond the scope of this text.

Countercurrent Distribution

Just one extraction performed on a solution of a complicated sample will likely not result in total or at least sufficient separation of the analyte from other interfering solutes. Not only will these other species also be extracted to a certain degree along with the analyte, but some of the analyte species will likely be left behind in the original solvent as well. Thus, the analyst will need to perform additional extractions on both the extracting solvent, to remove the other solutes that were extracted with the analyte, and the original solution, to remove additional analyte that was not extracted the first time. One can see that dozens of such extractions may be required to achieve the desired separation. Eventually there would be a separation, however. The process is called countercurrent distribution.

In countercurrent distribution the extracting solvent, after first being in contact with the original solution, is moved to another separatory funnel in which there is fresh original solvent, while fresh extracting solvent is brought into the original funnel and the extractions are performed. Then the extracting solvent from the second funnel is moved to a third funnel containing fresh original solvent, the extracting solvent from the first is moved to the second funnel, and fresh extracting solvent is introduced into the first funnel. The process continues in this manner until the desired separation occurs. The concept is illustrated in Figure 15.4. The top half in each segment (a, b, etc.) represents extracting solvent while the lower half represents original solvent. A mixture of "x" and "·" is being separated. In each segment, fresh extracting solvent is introduced on the upper left while fresh original solvent is introduced on the lowered right. This illustration shows a complete separation in segments "g" and "h". Some chromatography methods are based on this concept and will be discussed in Section 15.5.

Concentrators

Following an extraction procedure, it is often necessary to evaporate a quantity of the extracting solvent in order to increase the concentration of the analyte prior to gas chromatographic analysis (Chapter 16), for example. Two so-called "concentrators" in common use are the rotary evaporator and the Kuderna-Danish evaporative concentrator. With the rotary evaporator, the analyte/solvent mixture is heated in a round-bottom flask which is rotated while lying nearly on its side in a steam bath. Solvent vapors are condensed and collected in a separate flask under a condenser off to the side while the analyte remains in the rotating flask. The result is a solution of the analyte that has a higher concentration. The procedure takes place while the system is under reduced pressure. Bumping and superheating are avoided because of the rotation.

The Kuderna-Danish apparatus is a distillation/reflux assembly that utilizes a special fractionating column called the Snyder column. The lighter,

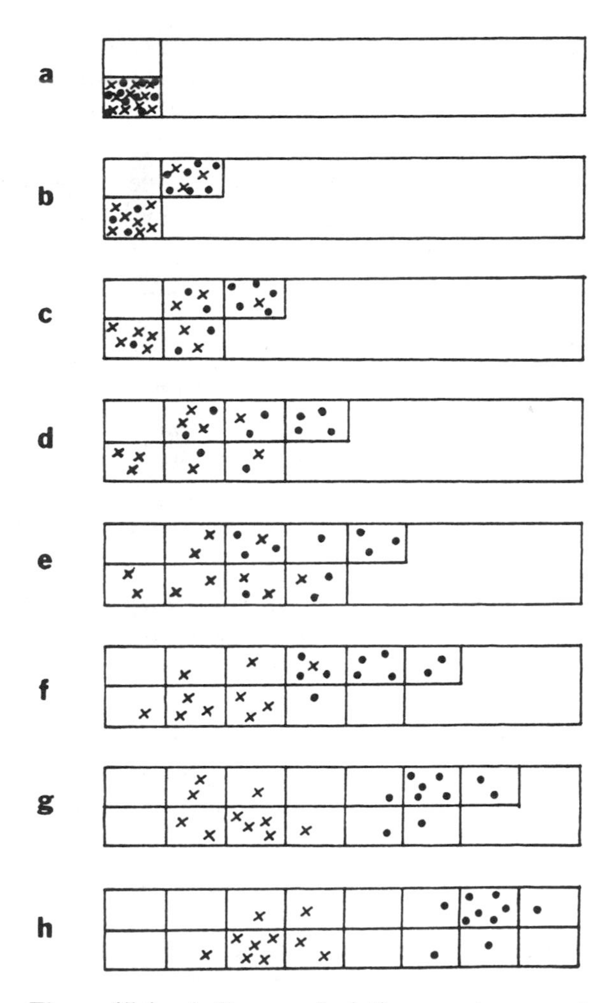

Figure 15.4. A diagram depicting countercurrent distribution as discussed in the text.

lower-boiling solvent(s) escapes the Snyder column to a solvent recovery condenser while the analyte refluxes back to the original container, the Kuderna-Danish flask and graduated concentrator tube. The end result is a 5- to 10-mL fraction of analyte concentrated in this tube.

Both of these units are used when the analyte concentration is too low to be detected otherwise.

15.5 LIQUID-SOLID EXTRACTION

There are instances in which the analyte has to be extracted from a solid material sample rather than a liquid. As in the above discussion for liquid samples, such an experiment is performed either because it is not possible or necessary to dissolve the entire sample or because it is undesirable to do so due to interferences that may also be present. In these cases, the weighed solid sample, preferably finely divided, is brought into contact with the extracting liquid in an appropriate container (not a separatory funnel) and usually shaken or stirred for a period of time such that the analyte species is removed from the sample and dissolved in the liquid. The time required for this shaking is determined by the rate of the dissolving. A separatory funnel is not used since two liquid phases are not present, but rather a liquid and a solid phase. A simple beaker, flask, or test tube usually suffices.

Following the extraction, the undissolved solid material is then filtered out and the filtrate analyzed. Examples of this would be soil samples to be analyzed for metals, such as potassium or iron (see Experiments 35 and 36 in Chapter 14), and cellophane or insulation samples to be analyzed for formaldehyde residue. The extracting liquid may or may not be aqueous. Soil samples being analyzed for metals, for example, utilize aqueous solutions of appropriate inorganic compounds, sometimes acids, while soil samples, or the cellophane or insulation samples referred to above, that are being analyzed for organic compounds utilize organic solvents for the extraction. As with the liquid-liquid examples, the extract is then analyzed by whatever analytical technique is appropriate – atomic absorption for metals, and spectrophotometry or gas or liquid chromatography for organics. Concentration methods, as with the Kuderna-Danish apparatus described in Section 15.4, may also be required, especially for organics.

It may be desirable to try to keep the sample exposed to fresh extracting solvent as much as possible during the extraction in order to maximize the transfer to the liquid phase. This may be accomplished by pouring off the filtrate and reintroducing fresh solvent periodically during the extraction and then combining the solvent extracts at the end. There is a special technique and apparatus, however, that has been developed, called the Soxhlet extraction, which accomplishes this automatically. The Soxhlet apparatus is shown in Figure 15.5. The extracting solvent is placed in the flask at the bottom, while the weighed solid sample is placed in the solvent-permeable thimble in the compartment directly above the flask. A condenser is situated directly above the thimble. The thimble compartment is a sort of cup that fills with solvent when the solvent in the flask is boiled, evaporated, and condensed on the condenser. The sample is thus exposed to freshly distilled solvent as the cup fills. When the cup is full the glass tube next to the cup is also full, and when it (the tube) begins to overflow the entire contents of the cup are siphoned back to the lower chamber and the process repeated. The advantages of such an apparatus are (1) fresh solvent is continuously in contact with the sample (without having to introduce more solvent, which would dilute the extract) and (2) the experiment takes place unattended and can conveniently occur overnight if desired.

15.6 CHROMATOGRAPHY

Introduction

A myriad of techniques used to separate complex samples come under the general heading of "chromatography". The nature of chromatography allows much more versatility, speed, and applicability than any of the other techniques discussed in this chapter, particularly when the modern instrumental techniques of gas chromatography (GC) and high-performance liquid chromatography (HPLC) are considered. These latter techniques are covered in detail in Chapters 16 and 17. In this chapter we introduce the general concepts of chromatography and give a perspective on its scope. Since there are many different classifications, this will include an organizational scheme covering the different types and configurations that exist.

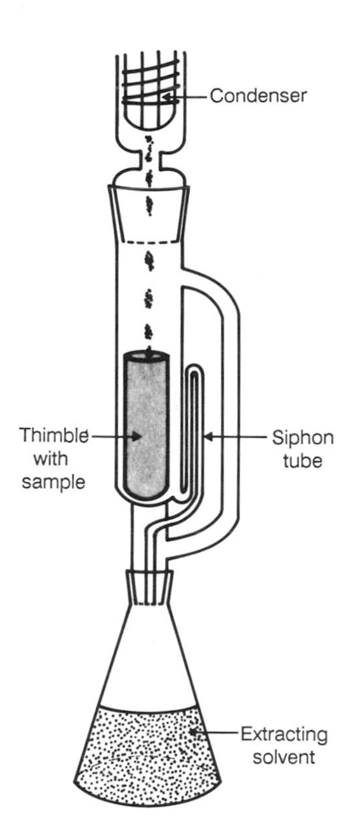

Figure 15.5. A drawing of a Soxhlet extraction apparatus. (From Kenkel, J., *Analytical Chemistry Refresher Manual*, Lewis Publishers, Chelsea, MI, 1992. With permission.)

Chromatography is the separation of the components of a mixture based on the different degrees to which they interact with two separate material phases. The nature of the two phases and the kind of interaction can be varied, and this gives rise to the different types of chromatography which will be described in the next section. One of the two phases is a moving phase (the "mobile" phase) while the other does not move (the "stationary" phase). The mixture to be separated is usually introduced into the mobile phase, which then is made to move or percolate through the stationary phase either by gravity or some other force. The components of the mixture are attracted to and slowed by the stationary phase to varying degrees, and as a result they move along with the mobile phase at varying rates and are thus separated. Figure 15.6 illustrates this concept.

The mobile phase can be either a gas or a liquid, while the stationary phase can be either a liquid or solid. One classification scheme is based on the nature of the two phases. All techniques which utilize a gas for the mobile phase come under the heading of "gas chromatography" (GC). All techniques that utilize a liquid mobile phase come under the heading of "liquid chromatography" (LC). Additionally, we have gas-liquid chromatography (GLC), gas-solid chromatography (GSC), liquid-liquid chromatography (LLC), and liquid-solid chromatography (LSC) if we wish to stipulate the nature of the stationary phase as well as the mobile phase. It is more

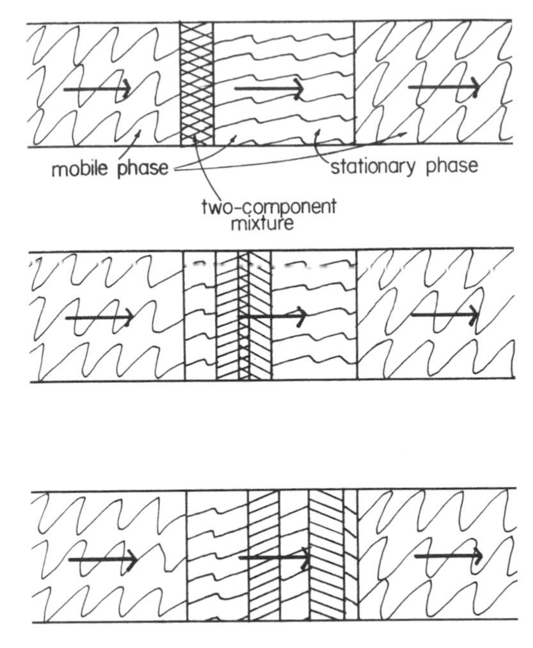

Figure 15.6. Mixture components separate as they move through the stationary phase with the mobile phase.

useful, however, to classify the techniques according to the nature of the interaction of the mixture components with the two phases. These classifications are referred to in this text as "types" of chromatography.

Types of Chromatography

Partition. In Section 15.4 we stated that some chromatography methods are based on the concept of countercurrent solvent extraction. You will recall that this is the technique in which a large number of extractions are performed, with fresh extracting solvent being brought into contact with previously extracted samples while fresh sample solvent is brought into contact with solvent extract from previous extractions (see Figure 15.4 and accompanying discussion). The extracting solvent can be thought of as continuously moving across the sample solvent while the latter remains stationary. The mixture components, which were initially found in the first segment of sample solvent, then distribute back and forth between the two phases as the extracting liquid moves and are found individually separated at different points along the way according to their individual solubilities in the two solvents. (See Figure 15.7.)

The extracting solvent in this scenario is the chromatographic mobile phase, while the sample solvent is the stationary phase. Liquid-liquid partition chromatography (LLC) is based on this idea. The mobile phase is a liquid which moves through a liquid stationary phase as the mixture components "partition" or distribute themselves between the two phases and become separated. The separation mechanism is thus the dissolving of the mixture components to different degrees in the two phases according to their individual solubilities in each.

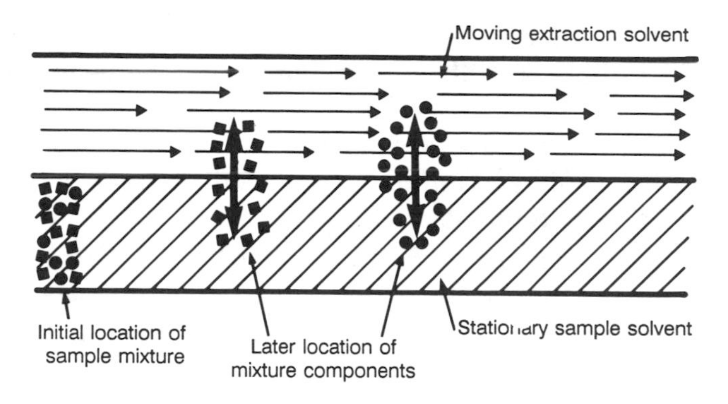

Figure 15.7. Illustration of the separation of two compounds by the movement of an extracting solvent across a sample solvent. (From Kenkel, J., *Analytical Chemistry Refresher Manual*, Lewis Publishers, Chelsea, MI, 1992. With permission.)

It may be difficult to imagine a liquid mobile phase used with a liquid stationary phase. What experimental setup allows one liquid to move through another liquid (immiscible in the first), and how can one expect partitioning of the mixture components to occur? The stationary phase actually consists of a thin liquid film either adsorbed or chemically bonded to the surface of finely divided solid particles, as shown in Figure 15.8. Chemically bonded liquid stationary phases represent a fairly recent development in this area, as opposed to adsorbed liquid phases. The latter is considered to be a more classical chromatography technique. Bonded-phase chromatography (BPC) has a distinct stability advantage. A bonded phase is not removed from the solid substrate either by reaction or by heat. BPC has become the most popular of the two by far.

Since the separation depends on the relative solubilities of the components in the two phases, the polarities of the components and those of the

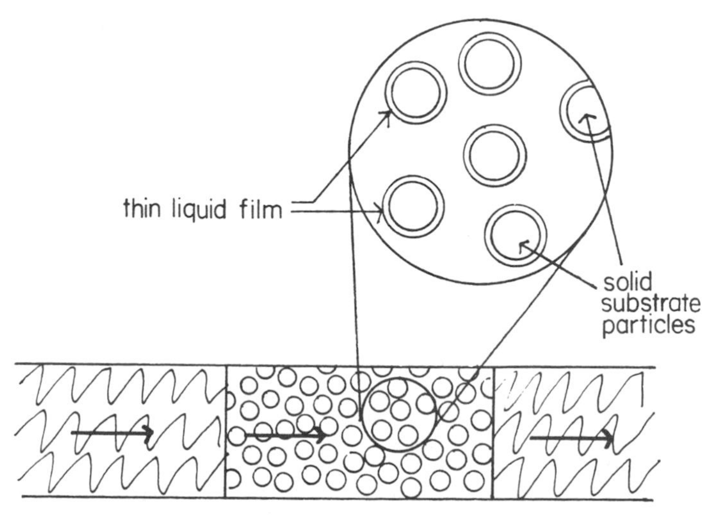

Figure 15.8. Partition chromatography, with a thin liquid film adsorbed or chemically bonded to the surface of finely divided solid particles.

stationary and mobile phases are important to consider. If the stationary phase is somewhat polar, it will retain polar components more than it will non-polar components, and thus the non-polar components will move more quickly through the stationary phase than the polar components. The reverse would be true if the stationary phase were non-polar. Of course, the polarity of a liquid mobile phase plays a role, too.

The mobile phase for partition chromatography can also be a gas (GLC). In this case, however, the mixture components' solubilities in the mobile phase are not an issue—rather their relative vapor pressures are important. This idea will be expanded in Chapter 16.

In summary, partition chromatography is a type of chromatography in which the stationary phase is a liquid adsorbed or chemically bonded to the surface of a solid substrate while the mobile phase is either a liquid or gas. The mixture components dissolve in and out of the mobile and stationary phases as the mobile phase moves through the stationary phase, and the separation occurs as a result. Examples of mobile and stationary phases will be discussed in Chapters 16 and 17.

Adsorption. Another chromatography type is adsorption chromatography. As the name implies, the separation mechanism is one of adsorption. The stationary phase consists of finely divided solid particles packed inside a tube but with no stationary liquid substance present to function as the stationary phase, as is the case with partition chromatography. Instead, the solid itself is the stationary phase and the mixture components, rather than dissolving in a liquid stationary phase, adsorb or "stick" to the surface of the solid. Different mixture components adsorb with different degrees of strength, which also depends on the mobile phase, and thus they again become separated as the mobile phase moves. The nature of the adsorption involves the interaction of polar molecules, or molecules with polar groups, with a very polar solid stationary phase. Thus, hydrogen bonding or similar molecule-molecule interactions are involved.

This very polar solid stationary phase is typically silica gel or alumina. The polar mixture components can be organic acids, alcohols, etc. The mobile phase can be either a liquid or a gas. This type of chromatography is depicted in Figure 15.9.

Ion Exchange. A third chromatography type is ion-exchange chromatography (IEC). As the name implies, it is a method for separating mixtures of ions, both inorganic and organic. The stationary phase consists of very small polymer resin beads which have many ionic bonding sites on their surfaces. These sites selectively exchange ions with certain mobile-phase compositions as the mobile phase moves. Ions that bond to the charged sites on the resin beads are thus separated from ions that do not bond. Repeated changing of the mobile phase can create conditions which will further selectively dislodge and exchange bound ions which are then also separated. This stationary-phase material can be either an anion exchange resin, which possesses positively charged sites to exchange negative ions, or a cation exchange resin, which possesses negatively charge sites to exchange positive ions. The mobile phase can only be a liquid. Further discus-

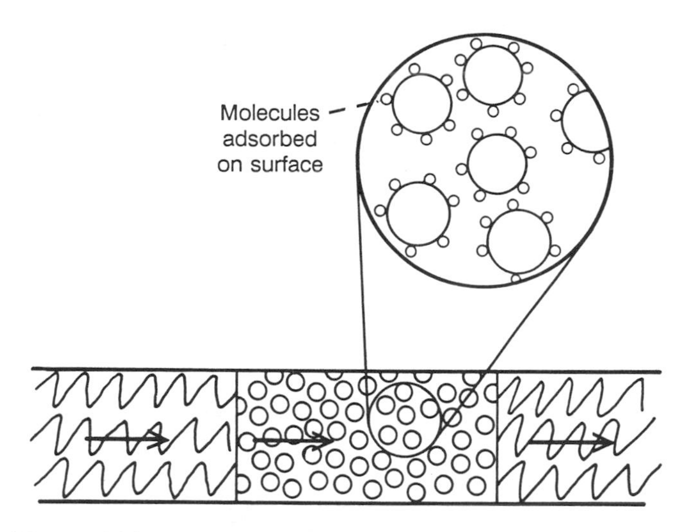

Figure 15.9. A depiction of adsorption chromatography. (From Kenkel, J., *Analytical Chemistry Refresher Manual*, Lewis Publishers, Chelsea, MI, 1992. With permission.)

sion of this type can be found in Chapter 17. Figure 15.10 depicts ion-exchange chromatography.

Size Exclusion. Size exclusion chromatography (SEC), also called gel permeation chromatography (GPC) or gel filtration chromatography (GFC), is a technique for separating dissolved species on the basis of their size. The stationary phase consists of porous polymer resin particles. The components to be separated can enter the pores of these particles and be slowed from progressing through this stationary phase as a result. Thus, the separation depends on the sizes of the pores relative to the sizes of the molecules to be separated. Small particles are slowed to a greater extent than larger

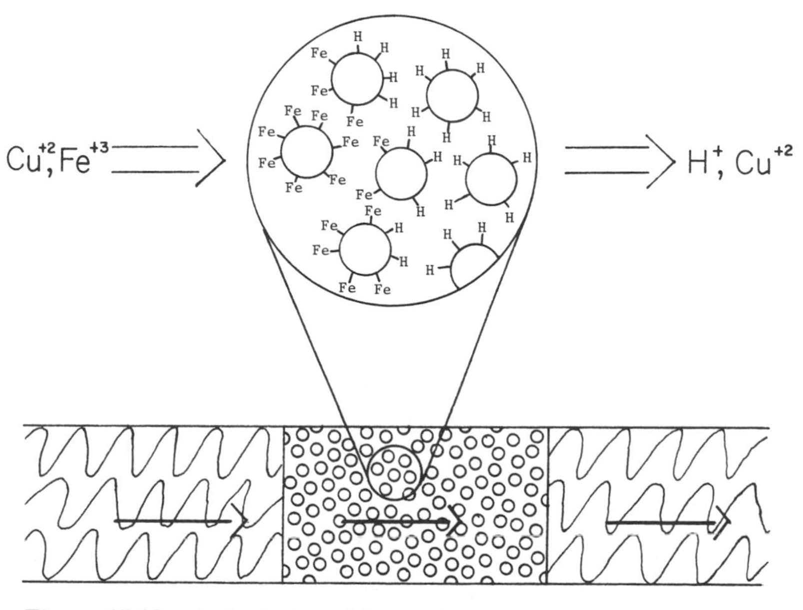

Figure 15.10. A depiction of ion-exchange chromatography.

particles, some of which may not enter the pores at all, and thus the separation occurs. The mobile phase for this type can also only be a liquid. SEC, too, is discussed further in Chapter 17. The separation mechanism is depicted in Figure 15.11.

Chromatography Configurations

Chromatography techniques are further classified in this text according to "configuration" – how the stationary phase is contained, how the mobile phase is configured with respect to the stationary phase in terms of physical state (gas or liquid), positioning, and how and in what direction the mobile phase travels in terms of gravity, capillary action, or other forces.

Configurations can be broadly classified into two categories: the "planar" methods and the "column" methods. The planar methods utilize a thin sheet of stationary-phase material and the mobile phase moves across this sheet, either upward ("ascending" chromatography), downward ("descending" chromatography), or horizontally ("radial" chromatography). Column methods utilize a cylindrical tube to contain the stationary phase, and the mobile phase moves through this tube either by gravity, with the use of a high-pressure pump, or by gas pressure. Additionally, with the exception of paper chromatography, those that utilize a liquid for the mobile phase are capable of utilizing all of the types (partition, adsorption, etc.) reviewed above. Paper chromatography utilizing unmodified cellulose sheets is strictly partition chromatography (see the next section). If the mobile phase is a gas (gas chromatography), the type is limited to adsorption and partition methods. Table 15.1 summarizes the different configurations. Let us consider each individually.

Paper and Thin-Layer Chromatography. Paper chromatography and thin-layer chromatography (TLC) constitute the planar methods mentioned

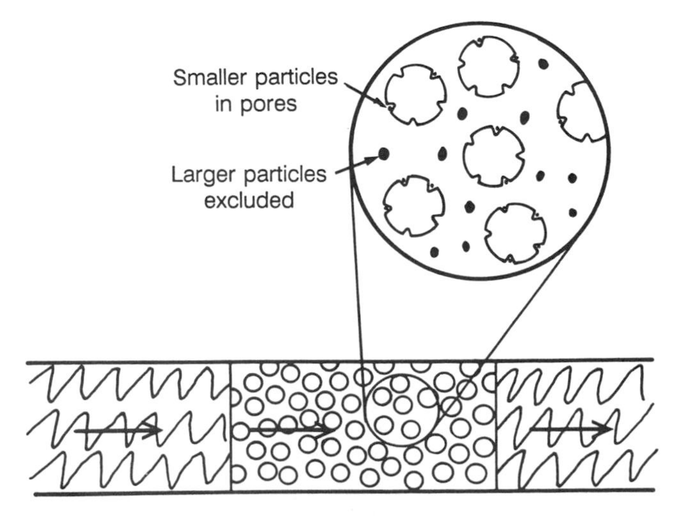

Figure 15.11. A depiction of size exclusion chromatography. (From Kenkel, J., *Analytical Chemistry Refresher Manual*, Lewis Publishers, Chelsea, MI, 1992. With permission.)

Table 15.1. Chromatography configurations and their applicable options.

Geometry	Configuration	Migration Direction	Applicable Types
Planar	Paper	Ascending, descending, radial	Partition
Planar	Thin layer	Ascending, descending, radial	Adsorption, partition, ion exchange, size exclusion
Column	Open column	Descending	Adsorption, partition, ion-exchange, size exclusion
Column	GC	N/A	Adsorption, partition
Column	HPLC	N/A	Adsorption, partition, ion-exchange, size exclusion

From Kenkel, J., *Analytical Chemistry Refresher Manual*, Lewis Publishers, Chelsea, MI, 1992. With permission.

above. Paper chromatography makes use of a sheet of paper having the consistency of filter paper (cellulose) for the stationary phase. Since such paper is hydrophilic, the stationary phase is actually a thin film of water unintentionally adsorbed on the surface of the paper. Thus, paper chromatography represents a form of partition chromatography only. The mobile phase is always a liquid.

With thin-layer chromatography, the stationary phase is a thin layer of material spread across a plastic sheet or a glass or metal plate. Such plates or sheets either may be purchased commercially already prepared or they may be prepared in the laboratory. The thin-layer material can be any of the stationary phases described earlier, and thus TLC can be any of the four types, including adsorption, partition, ion exchange, and size exclusion. Perhaps the most common stationary phase for TLC, however, is silica gel, a highly polar stationary phase for adsorption chromatography, as mentioned earlier. Also common is pure cellulose, the same material used for paper chromatography, and here also we would have partition chromatography. The mobile phase for TLC is always a liquid.

The most common method of configuring a paper or thin-layer experiment is the ascending configuration shown in Figure 15.12. The mixture to be separated is first "spotted" (applied as a small "spot") within 1 in. of one edge of a 10-in. square paper sheet or TLC plate. A typical experiment may be an attempt to separate several spots representing different samples and standards on the same sheet or plate. Thus, as many as eight or more spots may be applied on one sheet or plate. So that all spots are aligned parallel to the bottom edge, a light pencil mark can be drawn prior to spotting. The size of the spots must be such that the mobile phase will carry the mixture components without streaking. This means that they must be rather small – they

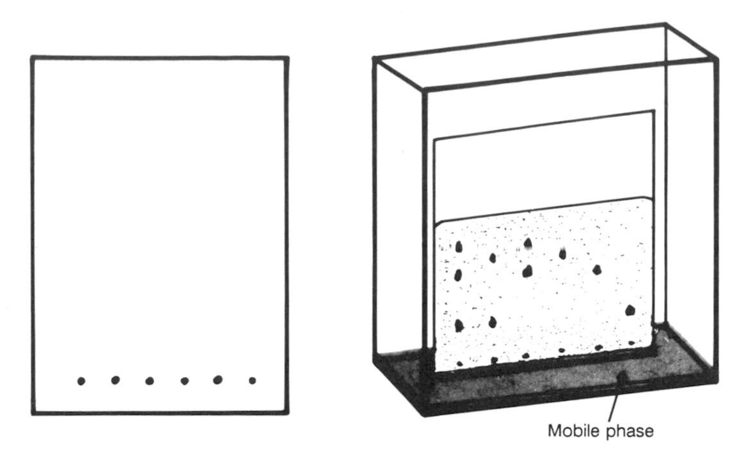

Figure 15.12. The paper or thin-layer chromatography configuration. The drawing on the left shows the paper or TLC plate with the spots applied. The drawing on the right shows the chromatogram in the developing chamber nearing complete development.

must be applied with a very small diameter capillary tube or micropipet. An injection syringe with a 25-μL maximum capacity is usually satisfactory.

Following spotting, the sheet or plate is placed spotted edge down in a developing chamber that has the liquid mobile phase in the bottom to a depth lower than the bottom edge of the spots. The spots must not contact the mobile phase. The mobile phase proceeds upward by capillary action and sweeps the spots along with it. At this point, chromatography is in progress, and the mixture components will move with the mobile phase at different rates through the stationary phase; if the mixture components are colored, evidence of the beginning of a separation is visible on the sheet or plate. The end result, if the separation is successful, is a series of spots along a path immediately above the original spot locations, each representing one of the components of the mixture spotted there. (See Figure 15.13.)

If the mixture components are not colored, any of a number of techniques designed to make the spots visible may be employed. These include iodine staining, in which iodine vapor is allowed to contact the plate. Iodine will absorb on most spots, rendering them visible. Alternatively, a fluorescent substance may be added to the stationary phase prior to the separation (available with commercially prepared plates) such that the spots, viewed under an ultraviolet light, will be visible because they do not fluoresce while the stationary phase surrounding the spots does fluoresce.

The visual examination of the chromatogram can reveal the identities of the components, especially if standards were spotted on the same paper or plate. Retardation factors (so-called R_f factors) can also be calculated and used for qualitative analysis. These factors are based on the distance the mobile phase has traveled on the paper (measured from the original spot of the mixture) relative to the distances the components have traveled, each measured from either the center or leading edge of the original spot to the center or leading edge of the migrated spot:

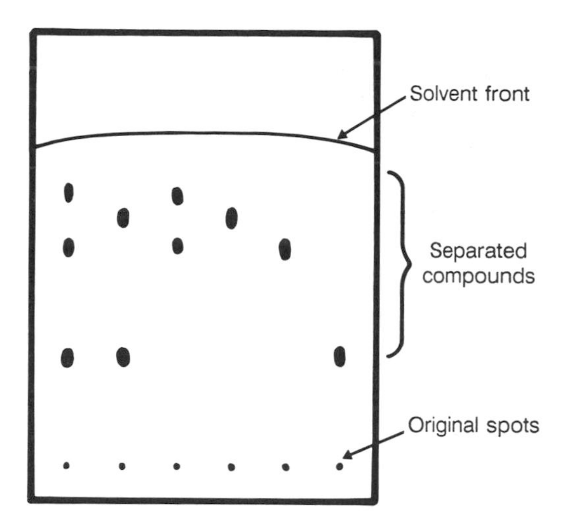

Figure 15.13. A developed paper or thin-layer chromatogram.

$$R_f = \frac{\text{distance mixture component has traveled}}{\text{distance mobile phase has traveled}} \qquad (15.5)$$

These factors are compared to those of standards to reveal the identities of the components.

Quantitative analysis is also possible. The spot representing the component of interest can be cut (in the case of paper chromatography) or scraped from the surface (TLC), dissolved, and quantitated by some other technique, such as spectrophotometry. Alternatively, modern scanning densitometers, which utilize the measurement of the absorbance or reflectance of ultraviolet or visible light at the spot location, may be used to measure quantity.

Using the TLC concept to prepare pure substances for use in other experiments, such as standards preparation or synthesis experiments, is possible. This is called preparatory TLC and involves a thicker layer of stationary phase so that larger quantities of the mixture can be spotted and a larger quantity of pure component obtained.

Additional details of planar chromatography, such as methods of descending and radial development, how to prepare TLC plates, tips on how to apply the sample, what to do if the spots are not visible, and the details of preparatory TLC, etc., are beyond the scope of this text.

Classical Open-Column Chromatography. Another configuration for chromatography consists of a vertically positioned glass tube in which is placed the stationary phase. It is typical for this tube to be open at the top, to have an inner diameter on the order of a centimeter or more, and to have a stopcock at the bottom, making it similar to a buret in appearance. With this configuration, the mixture to be separated is placed at the top of the column and allowed to pass onto the stationary phase by opening the stopcock. The mobile phase is then added, continuously fed into the top of the column, and flushed through by gravity flow. The mixture components separate on the stationary phase as they travel downward and, unlike the planar methods, are then collected as they elute from the column. In the classical experiment, a "fraction collector" is used to collect the eluting solution. A typical fraction collector consists of a rotating carousel of test tubes positioned under the column such that fractions of eluate are collected over a period of time, such as overnight or a period of days, in individual test tubes. (See Figure 15.14.) This then makes qualitative or quantitative analysis possible through the analysis of these fractions by some other technique, such as spectrophotometry.

The length of the column is determined according to the degree to which the mixture components separate on the stationary phase chosen. Difficult separations would require more contact with the stationary phase and thus may require longer columns. Again, all four types (adsorption, partition, ion exchange, and size exclusion) can be used with this technique.

It is well known that classical open-column chromatography has been largely displaced by the modern instrumental techniques of liquid chromatography. However, open columns are still used where extensive sample "cleanup" in preparation for the instrumental method is necessary. One can imagine that "dirty" samples originating, for example, from animal feed extractions or soil extractions, etc. may have large concentrations of undesirable components present. Since only very small samples (on the order of 1–20 μL) are needed for the instrumental method, the time required for obtaining a clean sample by this method, assuming the components of interest are not retained, is the time it takes for an initial amount of mobile phase to pass through from top to bottom. Compared to the overnight time frame, such a cleanup time is quite minimal and does not diminish the speed of the instrumental methods.

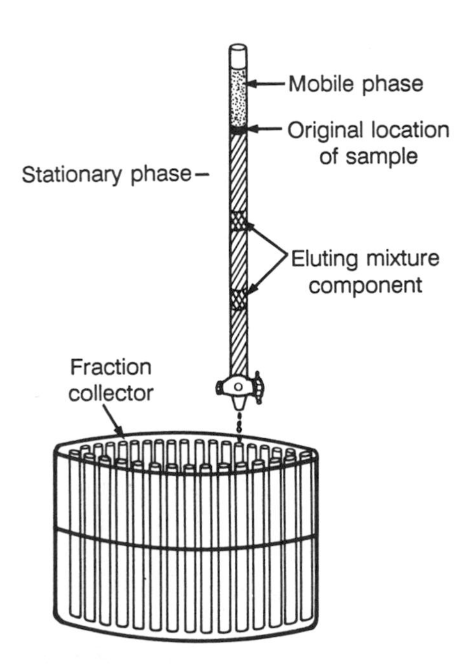

Figure 15.14. The classical open-column chromatography configuration with fraction collector. (From Kenkel, J., *Analytical Chemistry Refresher Manual*, Lewis Publishers, Chelsea, MI, 1992. With permission.)

Instrumental Chromatography. The concept of the finely divided stationary phase packed inside a column allowing the collection of the individual components as they elute, as discussed in the last section, presents a useful, more practical alternative. One can imagine such a column along with a continuous mobile-phase flow system, a device for introducing the mixture to the flowing mobile phase, and an electronic detection system at the end of the column – all of this incorporated into a single unit (instrument) used for repeated, routine laboratory applications. There are two such chromatography configurations which are in common use today, known as gas chromatography (GC) and high-performance liquid chromatography (HPLC). These techniques essentially can incorporate all types of column chromatography discussed thus far (HPLC) as well as those types in which the mobile phase can be a gas (GC). Both add a degree of efficiency and speed to the chromatography concept. HPLC, for example, is such a "high-performance" technique for liquid mobile-phase systems that a procedure that would normally take hours or days with open columns actually requires much less time. The full details of these instrumental techniques are discussed in Chapters 16 and 17.

15.7 ELECTROPHORESIS

Introduction

Another separation technique utilizes an electric field. An electric field is an electrically charged region of space, such as between a pair of electrodes connected to a power supply. The technique utilizes the varied rates and direction with which different dissolved ions migrate while under the influence of an electric field. This technique is called "electrophoresis". "Zone electrophoresis" refers to the common case in which a medium such as cellulose or gel is used to contain the solution. A schematic diagram of the electrophoresis apparatus resembles an electrochemical apparatus in many respects. A power supply is needed for connection to a pair of electrodes to create the electric field. The medium and sample to be separated are positioned between the electrodes. The basic concept of the technique and apparatus is illustrated in Figure 15.15.

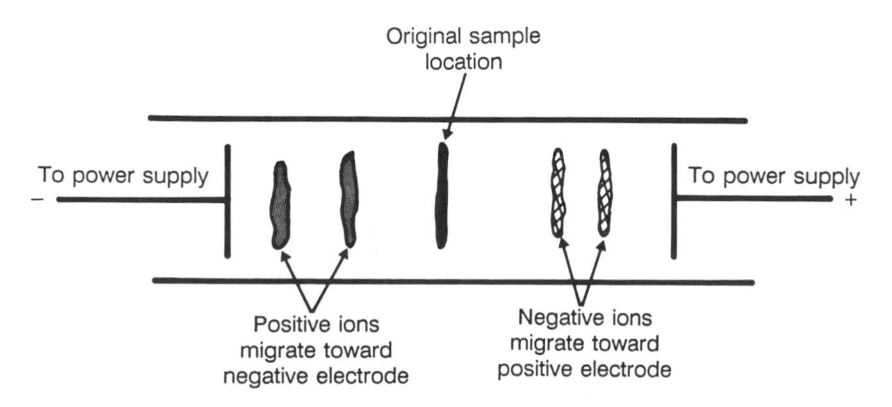

Figure 15.15. Illustration of the concept of electrophoresis. (From Kenkel, J., *Analytical Chemistry Refresher Manual*, Lewis Publishers, Chelsea, MI, 1992. With permission.)

Electrophoresis is for separating ions, since only ions will migrate under the influence of an electric field, negative ions to the positive electrode and positive ions to the negative electrode. Scientists have found electrophoresis especially useful in biochemistry experiments in which charged amino acid molecules and other biomolecules have to be separated. Thus, application to protein and nucleic acid analysis has been popular.

The principles of separation are (1) ions of opposite charge will migrate in different directions and become separated on that basis and (2) ions of like charge, while migrating in the same direction, become separated due to different migration rates. Factors influencing migration rate are charge values (i.e., −1 as opposed to −2, for example) and/or different mobilities. The mobility of an ion is dependent on the size and shape of the ion as well as the nature of the medium through which it must migrate. The biomolecules referred to above can vary considerably in size and shape, and thus electrophoresis is a powerful technique for separating them. As for the medium used, there are some options, including the use of an electrolyte-soaked cellulose sheet (paper electrophoresis), a thin gel slab (gel electrophoresis), or a capillary tube (capillary electrophoresis). The nature of the electrolyte solution used and its pH are also variable.

Paper Electrophoresis

Figure 15.16 represents a paper electrophoresis apparatus. The soaked cellulose sheet is sandwiched between two horizontal glass plates with the ends dipped into vessels containing more electrolyte solution. The electrodes are also dipped into these vessels as shown. The sample is spotted in the center of the sheet, and the oppositely charged ions then have room to migrate in opposite directions on the sheet. Qualitative analysis is performed much as with paper chromatography, by comparing the distances the individual components have migrated to those for standards spotted on the same sheet. (It may be necessary to render the spots visible prior to the analysis, as with paper chromatography.)

Problems associated with paper electrophoresis include the siphoning of electrolyte solution from one vessel through the paper to the other vessel when the levels of solution in the two vessels are different, causing the spots to possibly migrate in the wrong direction. The solution to this problem is to ensure that the levels in the two vessels are the same. Another problem stems from the fact that oxidation/reduction processes are occurring at the

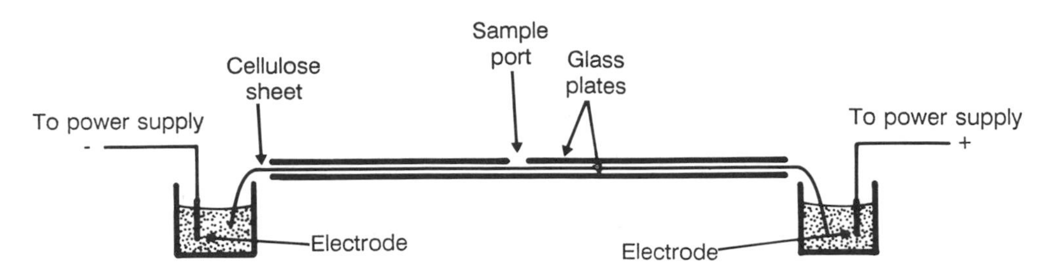

Figure 15.16. A paper electrophoresis apparatus. (From Kenkel, J., *Analytical Chemistry Refresher Manual*, Lewis Publishers, Chelsea, MI, 1992. With permission.)

surfaces of the electrodes. This may introduce undesirable contaminants to the electrolyte solution. These contaminants may in turn migrate onto the sheet. The solution to this problem is to isolate the electrodes while still allowing electrical contact, such as with the use of baffles, to keep the contaminants from diffusing from the vessels.

Gel Electrophoresis

A typical gel electrophoresis apparatus is shown in Figure 15.17. The thin gel slab referred to above is contained between two glass plates. The slab is held in a vertical position and has notches at the top where the samples to be separated are spotted or "streaked". In the configuration shown in the figure, only downward movement takes place; thus, only one type of ion, cation or anion, can be separated, since there is only one direction to go from the notch.

A tracking dye can be added to the sample so that the analyst can know when the experiment is completed (the leading edge of the sample solvent is visible via the tracking dye). Also, the slab can be removed from the glass plates and a staining dye can be applied which binds to the components, rendering them visible. The result is shown in Figure 15.18. Components with different mobilities through the gel show up as different bands or streaks on the gel. Qualitative analysis is performed as with paper electrophoresis – standards are applied alongside the samples and the components identified by their positions relative to the standards.

Finally, isoelectric focusing has been a useful extension of basic gel electrophoresis in protein analysis. In this technique, a series of ampholytes is placed on the slab via electrophoresis. An ampholyte is a substance whose molecule contains both acidic and basic functional groups. Solutions of different ampholytes have different pH values. Different ampholyte molecules

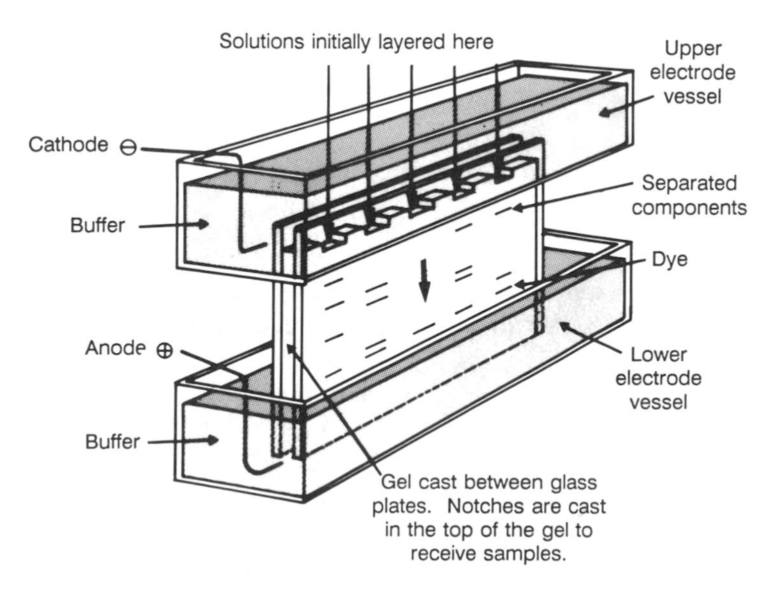

Figure 15.17. A gel electrophoresis apparatus. See text for description. (From Kenkel, J., *Analytical Chemistry Refresher Manual*, Lewis Publishers, Chelsea, MI, 1992. With permission.)

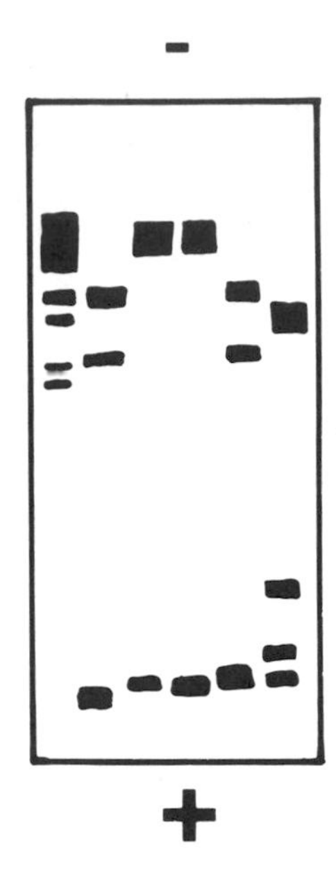

Figure 15.18. The gel slab after the electrophoresis experiment and after the components are rendered visible via staining. (From Kenkel, J., *Analytical Chemistry Refresher Manual,* Lewis Publishers, Chelsea, MI, 1992. With permission.)

differ in size and therefore will have varying mobilities in the electrophoresis experiment. Thus, these molecules migrate into the slab, take up different positions along the height of the slab, and create a pH gradient through the height of the slab. Amino acid molecules have different mobilities in different pH environments and also have their charges neutralized at particular pH values, rendering them immobile at some position in the gel. The pH at which the sample component is neutralized is called the "isoelectric point", and this technique is called "isoelectric focusing" since samples are separated according to their components' isoelectric points.

Capillary Electrophoresis

As the name implies, capillary electrophoresis is electrophoresis that is made to occur inside a piece (50–100 cm) of small-diameter capillary tubing, similar to the tubing used for capillary GC columns. The tubing contains the electrolyte medium, and the ends of the tube are dipped into solvent reservoirs as is the paper in paper electrophoresis. Electrodes in these reservoirs create the potential difference across the capillary tube. An electronic detector, such as those described for HPLC (Chapter 17), is "on line" and allows detection and quantitative analysis of mixture components.

The very small volume of sample (5–50 nL) is introduced, either by gravity flow or by force of the electrical field, on one end of the tube, typically the end dipped into the reservoir containing the positive electrode. When the experiment begins, the positive ions migrate quickly through the capillary tube toward the negative electrode in the opposite reservoir and separate on the basis of their mobilities, as in the other electrophoresis techniques. Ultimately, they pass individually through the detector and generate peaks on a recorder trace in a way very similar to HPLC.

Applications of this technique, like those described above, are very important in biochemistry. However, capillary electrophoresis offers some important advantages that point to a very bright future. These include the use of smaller quantities of samples, minimization of electrical heating during the separation, potential for automation, and the completion of qualitative and quantitative analysis in a much shorter period of time.

EXPERIMENTS

EXPERIMENT 40: TITRIMETRIC DETERMINATION OF THE EFFECTIVENESS OF AN EXTRACTION*

Discussion

A sample of an organic acid (mandelic acid) is weighed and dissolved in water. Portions of this solution are then tested for acid content (a) prior to any extraction, (b) after a single extraction using diethyl ether, and (c) after two extractions with diethyl ether. The objective is to determine the amount of mandelic acid remaining after one and two extractions.

SAFETY NOTE: Diethyl ether is highly flammable. It is very important not to have any open flames in the laboratory during this experiment. Safety glasses are required.

1. Weigh 5.0 g of mandelic acid into a 500-mL Erlenmeyer flask. Add about 250 mL of distilled water to the flask. Swirl or sonicate (see instructor) to dissolve. Swirl thoroughly after dissolving to make the solution homogeneous. The concentration of this solution need not be known accurately, but it must be homogeneous.
2. Prepare a 0.10 *M* NaOH solution. This solution need not be standardized.
3. Pipet 25 mL of the acid solution into each of three Erlenmeyer flasks. Titrate each with the NaOH to the phenolphthalein end point. All buret readings should agree to within 0.05 mL. If they do not, repeat until you have three that do or until you have a precision approved by your instructor. Average these readings and use this average as L_{NaOH} (before ext.) in the calculation in Step 7.
4. Obtain three clean 125-mL separatory funnels and place in a funnel rack. Pipet 25 mL of the acid solution into each of the three funnels. Then, with a dry pipet, pipet 25 mL of diethyl ether into each funnel. Proceed to extract the water by capping the separatory funnels and

*Adapted from Linstromberg, W. W. and Baumgarten, H. E., *Organic Experiments*, 6th ed., D. C. Heath and Company, Lexington, MA, 1987. With permission.

shaking (instructor will demonstrate extraction technique). After shaking, place the funnels back in the rack and allow the layers to separate.

5. Draw off the lower aqueous (water) layers through the stopcocks into 250-mL Erlenmeyer flasks. With a graduated cylinder, add 10 mL of distilled water to each separatory funnel as a rinse. Shake, allow to separate, and add the rinsings to the Erlenmeyer flasks. Repeat this rinsing step one more time with fresh distilled water.

6. Titrate the contents of each flask to the phenolphthalein end point with the 0.10 *M* NaOH.

7. Study the precision of the results from Step 6. If you think it can be improved, repeat Steps 4 through 6. Otherwise, calculate the average percentage of mandelic acid extracted as follows, using the average of your buret readings for L_{NaOH} (after ext.):

$$\% \text{ mandelic acid ext.} = \frac{L_{NaOH} \text{ (before ext.)} - L_{NaOH} \text{ (after ext.)}}{L_{NaOH} \text{ (before ext.)}} \times 100$$

8. If time permits, and if instructed to do so, perform the extraction procedure again, but this time do two extractions of the same 25-mL acid solution before titrating. This will show the effectiveness of two extractions as opposed to one.

EXPERIMENT 41: THE THIN-LAYER CHROMATOGRAPHY ANALYSIS OF COUGH SYRUPS FOR DYES

NOTE: Many cough medicine formulations exhibit the color they have because a food colorant has been added. Such colorants are standard FD&C dyes, typically blue #1, red #3, #33, and #40, or yellow #5, #6, and #10. In this experiment, identification of such dyes is accomplished by TLC. Safety glasses are required.

1. Weigh 10 mg of each dye standard into separate 50-mL volumetric flasks, dilute to the mark with distilled water, and shake. If the dye is not water soluble (Lake), dissolve it directly in the *n*-butanol in Step 3.

2. Pipet 5 mL of each standard and 5 mL of each sample into separate 50-mL Erlenmeyer flasks.

3. Add five drops of diluted HCl (prepared by diluting 23.6 mL of concentrated HCl to 100 mL with distilled water) and 5 mL of *n*-butanol to each flask, and shake on a shaker for 30 min.

4. Pour each into test tubes, and allow the layers to separate. Spot the butanol (top) layer onto a TLC plate (consult instructor).

5. Place the development solvent (1-propanol:ethyl acetate:concentrated NH_4OH, 1:1:1) in the developing tank, and develop the chromatograph so that the solvent migrates for 12 cm.
6. Allow the plate to dry. Observe and measure R_f values, if possible, and identify the dyes in the cough syrups.

EXPERIMENT 42: THE EXTRACTION AND ANALYSIS OF A PAIN RELIEF TABLET FOR CAFFEINE*

Introduction

A pain relief tablet containing both aspirin and caffeine is analyzed for caffeine. A solvent extraction procedure is used to separate the aspirin from the caffeine so that the caffeine can be analyzed spectrophotometrically free of interference from the aspirin.

Aspirin is a carboxylic acid, acetylsalicylic acid. If the tablet is first dissolved in chloroform, which is immiscible with water, and this solution extracted with an aqueous solution of a base, the aspirin will react with the base forming a water soluble salt. This salt, since it is insoluble in the organic solvent but soluble in the aqueous layer, will be extracted into the aqueous layer. Thus it is separated from the caffeine, which will remain in the organic layer. The chloroform layer can then be analyzed for caffeine by UV spectrophotometry. All glassware used to handle chloroform should be dry.

NOTE: Avoid contacting chloroform to your skin and work with it in a fume hood as much as possible.

Procedure

1. Accurately weigh one pain relief tablet containing both caffeine and aspirin (check label) on an analytical balance, then crush to a powder in a small beaker. Weigh an amount of this crushed tablet equal to one tenth of the tablet's weight on an analytical balance into a second small beaker (25 mL). Add 8 mL of chloroform and sonicate to dissolve. Any amount of powder that does not dissolve is probably tablet binding material and need not be dissolved. Quantitatively transfer the contents of the beaker to a 125-mL separatory funnel using small amounts of pure chloroform to rinse the beaker at least twice. Add the rinsings to the funnel.
2. Prepare 50 mL of an aqueous solution of sodium bicarbonate, 4% weight to volume, and chill in an ice bath. Add 10 mL of this solution to the separatory funnel and proceed to extract the aspirin from the chloroform using the shaking and venting technique described in this chapter, Section 15.4. Allow the lower chloroform layer to drain through filter paper into a 25-mL volumetric flask. Using a dropper, add a small amount (3–4 mL) of chloroform to the funnel to rinse the aqueous layer, shake and vent as before, and add this through the filter paper to the same 25-mL flask. Repeat one more time, then

*Adapted from Sawyer, D., Heineman, W., and Beebe, J., *Chemistry Experiments for Instrumental Methods*, Copyright © John Wiley & Sons, New York, 1984. With permission.

dilute the solution in the 25-mL flask to the mark with chloroform. This solution contains the caffeine from the original tablet portion weighed in step 1.

3. Prepare a dilution of the caffeine solution from step 2 by pipetting 0.5 mL (the pipet must be dry) into a second dry 25 mL volumetric flask and diluting to the mark with chloroform.
4. Prepare 25 mL of a standard solution of caffeine in chloroform with a concentration of 200 ppm. Prepare 25-mL of a second standard solution (chloroform solvent) with a concentration of 10 ppm by diluting the 200 ppm solution.
5. Measure the absorbances of the sample solution from step 3 and the 10 ppm standard solution from step 4 at 275 nm. Calculate the concentration in the sample solution using Equation 11.6 in Chapter 11.
6. Calculate the mg of caffeine in the tablet:

$$\text{mg in tablet} = \text{ppm found} \times .025 \times \frac{25}{0.5} \times \frac{\text{wt. of tablet}}{\text{wt. of sample}}$$

EXPERIMENT 43: THE EXTRACTION OF CAPSAICIN FROM CHILE PEPPERS*

Introduction

Capsaicin is an alkaloid with the formula $C_{18}H_{27}NO_3$. It is the chemical found in all varieties of peppers giving them their characteristic burning taste. In this experiment, a Soxhlet extraction (Section 15.5) is performed on a pepper sample for the purpose of extracting and concentrating the capsaicin in preparation for quantitative analysis using HPLC (Experiment 53).

1. Obtain a sample of a chile pepper, such as an Italian crushed pepper, a fresh Italian hot pepper, a cherry pepper, a jalapeno pepper, etc. If it is a fresh pepper, clean it, remove stems and seeds and chop it into fine pieces. If it is a dried pepper, chop it into fine pieces. If it is in ground form, use it without further preparation.
2. Prepare a Soxhlet extraction apparatus as depicted in Figure 15.5. Use a 125-mL flask with a ground glass top. Place approximately 5 g of the prepared pepper into the cellulose thimble and approximately 85 mL of methylene chloride solvent into the flask. Place the assembly on a hot plate, holding it in place with a ring stand and clamp. Be sure to circulate cold tap water through the condenser.
3. Adjust the hot plate setting to provide gentle, even boiling of the methylene chloride. Allow the filling/siphoning cycle to occur several times and observe the process to ensure smooth operation so that you

*This experiment, including the HPLC, was the subject of a poster displayed by Dr. Frank Torre, et al., of the Department of Chemistry, Springfield College, Springfield, Massachusetts, at the Fall 1993, American Chemical Society meeting in Chicago, IL. It is used here with his permission.

can confidently let it run unattended. Allow to run for a period of time determined by consulting your instructor. Several hours is recommended.

4. After the extraction, the capsaicin from the pepper sample is present in the methylene chloride "extract" in the flask. In order to prepare the extract for the HPLC analysis (Experiment 53), the methylene chloride must be evaporated and the residue dissolved with tetrahydrofuran solvent. To evaporate and reclaim the methylene chloride, a rotary evaporator (Section 15.4), operated at room temperature, may be used. Otherwise, evaporate the methylene chloride in a good fume hood, perhaps with the aid of a stream of nitrogen gas blowing over the surface of the liquid, until a resin is formed. If you decide to first transfer the extract to another container to facilitate the evaporation, it must be a quantitative transfer.
5. Pipet exactly 5.00 mL of tetrahydrofuran into the container with the resin. Dissolve the resin, then pour the solution into a large vial. Cap the vial and save for Experiment 53.

QUESTIONS AND PROBLEMS

1. Why is a study of modern separation science important in analytical chemistry?
2. Define recrystallization, distillation, fractional distillation, extraction, liquid-liquid extraction, solvent extraction, countercurrent distribution, liquid-solid extraction, chromatography.
3. In recrystallization, how is it that both soluble and insoluble impurities are removed from the solid being purified?
4. In recrystallization, why is it good to use the minimum amount of solvent at the elevated temperature that will completely dissolve the substance being purified?
5. What does vapor pressure have to do with purification by distillation?
6. How is it that dissolved solids, such as hardness minerals, are separated from water during distillation?
7. What special problems exist when trying to separate two organic liquids by distillation?
8. What is a theoretical plate and the height equivalent to a theoretical plate (HETP) as they pertain to distillation?
9. Describe what happens in a liquid-liquid extraction experiment.
10. What are the two important criteria for a successful separation by solvent extraction?
11. Describe the glassware article called the "separatory funnel" and tell for what purpose and how it is used.
12. If the distribution coefficient, K, for a given solvent extraction is 169,

(a) what is the molar concentration of the analyte found in the extracting solvent if the concentration in the original solvent *after* the extraction is 0.0270 *M*?

(b) what is the molar concentration of the analyte found in the extracting solvent if the concentration in the original solvent *before* the extraction was 0.0450 *M*?

13. How many moles of analyte are extracted if 50 mL of extracting solvent is brought into contact with 50 mL of original solvent, the concentration of analyte in the original solvent is 0.060 *M*, and the distribution coefficient is 238?

14. In an extraction experiment, it is found that 0.0376 g of an analyte is extracted into 50 mL of solvent from 150 mL of a water sample. If there was originally 0.192 g of analyte in this volume of the water sample, what is the distribution coefficient? What percent of analyte was extracted?

15. The distribution coefficient for a given extraction experiment is 527. If 0.037 g of analyte is found in 75 mL of the extracting solvent after extraction and 100 mL of original solvent were used, how many grams of analyte were in this volume of original solvent before extraction?

16. How does the distribution "coefficient" differ from the distribution "ratio"?

17. What is a Kuderna-Danish concentrator?

18. Is a separatory funnel appropriate for a liquid-solid extraction? Explain.

19. Give some examples of the types of samples and analytes to which liquid-solid extraction would be applicable.

20. Explain the use of the Soxhlet extraction apparatus.

21. Give a general definition of chromatography which would apply to all types and configurations. (To say that it is a separation technique is important but not sufficient.)

22. Name the four types of chromatography described in this chapter and give the details of the separation mechanism of each.

23. How does partition chromatography differ from adsorption chromatography?

24. Find, in a reference book, a description of the Craig countercurrent distribution apparatus and discuss its design as it relates to the description of countercurrent extraction presented in this chapter.

25. Consider a mixture of compound A, a somewhat non-polar liquid, and compound B, a somewhat polar liquid. Tell which liquid, A or B, would emerge from a chromatography column first under the following conditions and why.

    (a) a polar liquid mobile phase and a non-polar liquid stationary phase

    (b) a non-polar liquid mobile phase and a polar liquid stationary phase

26. We have studied four chromatography "types". One of these is partition chromatography. Answer the following questions concerning this type "yes" or "no":
    (a) Can the mobile phase be a solid?
    (b) Can the mobile phase be a liquid?
    (c) Can the mobile phase be a gas?
    (d) Can the stationary phase be a solid?
    (e) Can the stationary phase be a liquid?
    (f) Can the stationary phase be a gas?

27. Tell what each of the following refer to: GC, LC, GSC, LSC, GLC, LLC, BPC, IEC, SEC, GPC, GFC.

28. One type of chromatography separates small molecules from large ones. Name this type and tell how such a separation occurs.

29. Differentiate between the use of a cation exchange resin and an anion exchange resin in terms of whether the charged sites are positive or negative and whether cations or anions are exchanged.

30. List four chromatography methods which can be designated as different chromatography "configurations".

31. (a) Name four chromatography "configurations".
    (b) Choose one of your answers to (a) and tell how the stationary phase is configured relative to the mobile phase and what force is used to move the mobile phase through the stationary phase.

32. What does the abbreviation LSC stand for? Give two examples of chromatography "types" which can be abbreviated as LSC.

33. Match each item in the left column to a single item in the right column which most closely associates with it.

(a) paper chromatography	(i) thin-layer chromatography
(b) ion-exchange chromatography	(j) size exclusion chromatography
(c) electrophoresis	(k) stationary phase is water
(d) gas chromatography	(l) technically not chromatography
(e) adsorption chromatography	(m) column effluent is collected and analyzed
(f) gel permeation chromatography	(n) uses a stationary phase which trades ions with the mobile phase
(g) TLC	(o) an instrumental method
(h) open-column chromatography	(p) involves a mechanism in which the mixture components selectively "stick" to a solid surface

34. Fill in each blank with a term from the following list.

stationary phase	mobile phase
adsorption	partition
paper	thin layer
HPLC	ion exchange
size exclusion	electrophoresis
detector	

(a) In gas chromatography, the material packed within the column is usually a powdered solid that has a thin liquid film adsorbed on the surface. This thin liquid film is called the ____________. This type of gas chromatography falls into the general classification of ____________ chromatography.

(b) In one type of chromatography, the components of the mixture are separated on the basis of the relative sizes of the molecules. This is called ____________ chromatography.

(c) The technique in which separation of charged species is effected by the use of an electric field is called ____________.

(d) The fact that gravity flow of a liquid through a packed column is time-consuming led to the development of ____________.

(e) GSC is a type of ____________ chromatography.

(f) A type of chromatography in which a layer of adsorbent is spread on a glass or plastic plate is called ____________ chromatography.

35. Fill in the blanks in the following table:

Configuration	Type	Stationary Phase	To Separate Mixtures of
Paper	Partition	____________	____________
HPLC	____________	____________	Different size molecules
____________	Ion exchange	____________	____________
____________	Partition	____________	Gases and volatile liquids
____________	Adsorption	Layer of adsorbent spread on glass plate	____________

36. Answer the following questions either TRUE or FALSE.

(a) The stationary phase percolates through a "bed" of finely divided solid particles in adsorption chromatography.

(b) The mobile phase can be either a liquid or a gas.

(c) The mobile phase is a "moving" phase.

(d) Partition chromatography can only be used when the mobile phase is a liquid.

(e) Adsorption is a type of LSC chromatography.

(f) In partition chromatography, the mobile phase "partitions" or distributes itself between the sample solution and the stationary phase.
(g) If the stationary phase is a polar liquid substance, non-polar components will elute first.
(h) In GLC, the separation mechanism is partitioning.
(i) Size exclusion chromatography separates components on the basis of their charge.
(j) Gel permeation chromatography is another name for size exclusion chromatography.
(k) Ion-exchange chromatography is a technique for separating inorganic ions in a solution.
(l) Paper chromatography is a type of LLC.
(m) Thin-layer chromatography and open column chromatography are two completely different configurations of GSC.
(n) It is useful to measure R_f values in open column chromatography.
(o) R_f values are used for quantitative analysis.
(p) TLC refers to thin-layer chromatography.
(q) HPLC refers to high performance liquid chromatography.

37. Match each statement to one of the terms.

Partition chromatography
Ion-exchange chromatography
Paper chromatography
Open-column chromatography
High performance liquid chromatography
Electrophoresis
Adsorption chromatography
Size-exclusion chromatography
Thin-layer chromatography
Gas chromatography

(a) A chromatography configuration in which the stationary phase is spread across a glass or plastic plate.
(b) A chromatography type in which the stationary phase is a liquid.
(c) A chromatography type designed to separate dissolved ions.
(d) A chromatography configuration which utilizes a high pressure pump to move the mobile phase through the stationary phase.
(e) One of two chromatography types which have application in GC.
(f) A chromatography configuration which utilizes a "fraction collector".
(g) The only chromatography type described by the letters GLC.
(h) One of two chromatography configurations in which the mobile phase moves by capillary action opposing gravity.

38. Describe the following: electrophoresis, zone electrophoresis, paper electrophoresis, gel electrophoresis, isoelectric focusing, capillary electrophoresis.

39. Describe in detail the general mechanism of the separation of ions by electrophoresis.
40. What is "capillary electrophoresis" and what advantages does it have over other conventional electrophoresis techniques.

Chapter 16

Gas Chromatography

16.1 INTRODUCTION

One instrumental chromatography configuration mentioned in the preceding chapter is one in which the mobile phase is gaseous and the stationary phase is either liquid or solid. This configuration is called gas chromatography and is abbreviated simply GC. It may also be abbreviated either GLC or GSC in order to stipulate whether the stationary phase is a liquid (GLC) or a solid (GSC). Most gas chromatography procedures utilize a liquid stationary phase (GLC), and thus the chromatography "type" (see Chapter 15 for the distinction between "type" and "configuration") is partition chromatography most of the time. The only other possible "type" that is applicable here is adsorption chromatography. Thus, GSC refers to this latter type.

For the novice having just read Chapter 15, it may be difficult to visualize a chromatography procedure that utilizes a gas for a mobile phase. The gas, often called the "carrier gas", is typically purified helium or nitrogen. It flows from a compressed gas cylinder via the regulated pressure of the cylinder and a flow controller through a "column" containing the stationary phase where the separation takes place. There are two types of columns—two methods of holding the stationary phase in place—and these will be discussed in Section 16.4.

The fact that the mobile phase is a gas creates additional unique features, one of which has to do with the mechanism of the separation. When we discussed partition chromatography in Chapter 15, we were, for the most part, assuming a liquid mobile phase. The partitioning mechanism in that case involved only the relative solubilities of the mixture components in the two phases. With GLC, however, the mechanism, while still involving the solubilities of the mixture components in one phase, the liquid stationary phase, also involves the relative vapor pressures of the components, since we must have partitioning between one phase that is a liquid and another phase that is a gas. Further, since the mobile phase is a gas, the mixture components must also be gases, or at least liquids with relatively high vapor pressures (elevated temperatures are used), in order to be carried through the column as gases using a gaseous mobile phase.

Let us briefly review the concept of vapor pressure. Simply defined, vapor pressure is the pressure exerted by the vapors of a liquid above a liquid phase containing that liquid. It is a measure of the tendency of the molecules of a liquid substance to "escape" the liquid phase and become gaseous. If there is such a strong tendency, the vapor pressure is high (typical of non-polar, low molecular weight liquids). If, however, there is a weak tendency, then the vapor pressure is low (typical of highly polar and/or high-molecular-weight substances). Thus, examples of liquid substances with high vapor pressures are diethyl ether, acetone, and chloroform. Examples of liquid substances with low vapor pressures are water, high molecular weight alcohols, and aromatic halides. In gas chromatography, substances with high vapor pressures will be strongly influenced by the moving gaseous mobile phase and will emerge from the column quickly if their solubilities in the stationary phase are low. If their vapor pressures are high, but their solubilities in the stationary phase are also high, then they will emerge more slowly. If their vapor pressures are low, but they have high solubilities in the stationary phase, the time required for emergence from the column will be long. The time a given mixture component is retained by the stationary phase in the column from the time it is first introduced is called the "retention time". Figure 16.1 and Table 16.1 summarize the vapor pressure concepts discussed here.

16.2 INSTRUMENT DESIGN

Vapor pressures vary substantially with temperature. Because of this, it is useful to be able to use an elevated column temperature and also to be able to carefully control the temperature of the column. For this reason, the

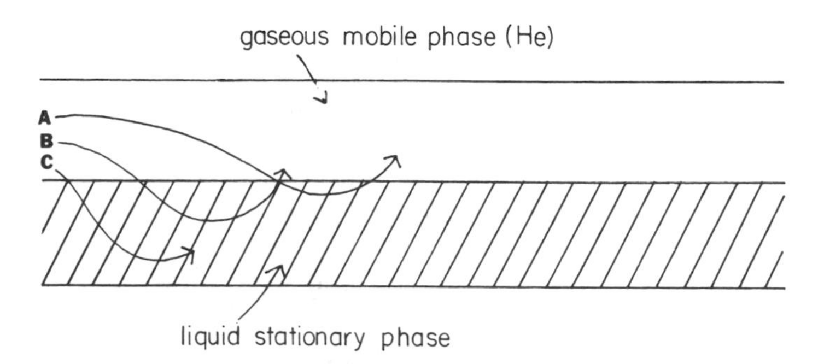

Figure 16.1. An illustration of the vapor pressure and solubility effects discussed in the text. Components A, B, and C have vapor pressures decreasing from A to B to C and have solubilities in the stationary phase increasing from A to B to C.

Table 16.1. A summary of retention concepts for GC.

Component's Vapor Pressure	Component's Solubility in Stationary Phase	Retention Time
High	Low	Short
High	High	Intermediate
Low	Low	Intermediate
Low	High	Long

column is always placed in a thermostatted oven inside the instrument. Additionally, a sample introduction system is needed at the head of the column which will allow either samples of gases to be introduced into the flowing carrier gas (gas sampling valves) or samples of volatile liquids to be introduced so that they are immediately vaporized and carried on to the column in the gas phase. This latter system utilizes an "injection" configuration which includes a high-temperature "injection port" equipped with a rubber septum. The sample to be separated is drawn into a microliter syringe with a sharp beveled tip which is then used to pierce the septum so that the sample is introduced into the flowing carrier gas, flash vaporized due to the high temperature, and then carried onto the column.

Finally, a detection system is required at the opposite end of the column which will detect when a substance other than the carrier gas elutes. This detector can consist of any one of a number of different designs, but the purpose is to generate the electronic signal to be recorded on a strip-chart recorder from which the qualitative and quantitative information is obtained.

The entire system is drawn in Figure 16.2. By way of summary, the helium or nitrogen flows from the compressed gas cylinder (pressure regulated by a gas bottle regulator) through the injection port, where the sample is introduced, onto the column and subsequently through the detector. The electronic signal generated by eluting components is recorded on the strip-chart recorder in the form of peaks, as indicated on the recorder portion in Figure 16.2. A sample "chromatogram", or strip-chart recording, for a four-

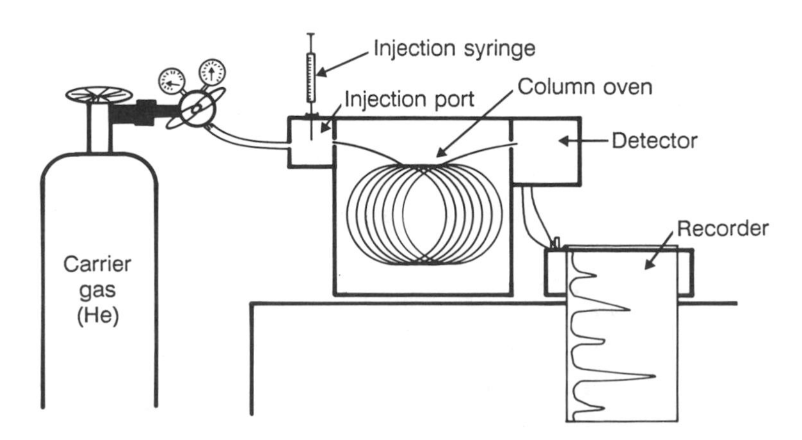

Figure 16.2. A drawing of a gas chromatograph. (From Kenkel, J., *Analytical Chemistry Refresher Manual*, Lewis Publishers, Chelsea, MI, 1992. With permission.)

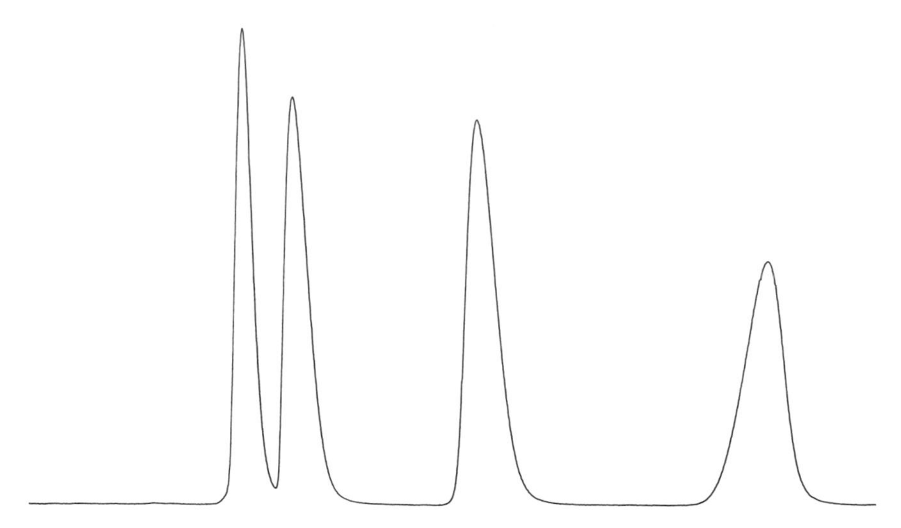

Figure 16.3. A sample gas chromatogram (strip-chart recording) of a four-component mixture.

component mixture is shown in Figure 16.3. When just pure carrier gas is passing through the detector, the recorder pen is zeroed. When something other than carrier gas emerges, the signal sent to the recorder changes, thus the reason for the peaks. Each component eluting then is recorded via its own individual peak.

Finally, it is important to recognize that there are several heated zones within the instrument. We've already mentioned the column oven. Temperatures here are typically in the 100°C–150°C range, but much higher temperatures are sometimes used. We've also mentioned the fact that the injection port area is heated (to cause vaporization of volatile liquid samples). The temperature in this zone depends on the volatility of the components but is typically in the 200°C–250°C range. Lastly, the detector must also be heated, mostly to prevent the condensation of the vapors as they pass through. GC detectors are designed to detect gases and not liquids. The temperature here is usually in the 200°C–250°C range also.

16.3 SAMPLE INJECTION

The injection port is designed to introduce samples quickly and efficiently. Most GC work involves the separation of volatile liquid mixtures. In this case, the injection port must be designed to flash vaporize small amounts of such samples so that the entire amount is immediately carried to the head of the column by the flowing helium. The most familiar design consists of a small glass-lined or metal chamber equipped with a rubber septum to accommodate injection with a syringe. As the helium "blows" through the chamber, a small volume of injected liquid (typically on the order of 0.1 to 3 μL) is thus flash vaporized and immediately carried onto the column. A variety of sizes of syringes and some additional features are available to make the injection easy and accurate. Syringes manufactured by the Hamilton Company, Reno, NV, are common. The mechanics for such an injection are shown in Figure 16.4. The rubber septum, after repeated

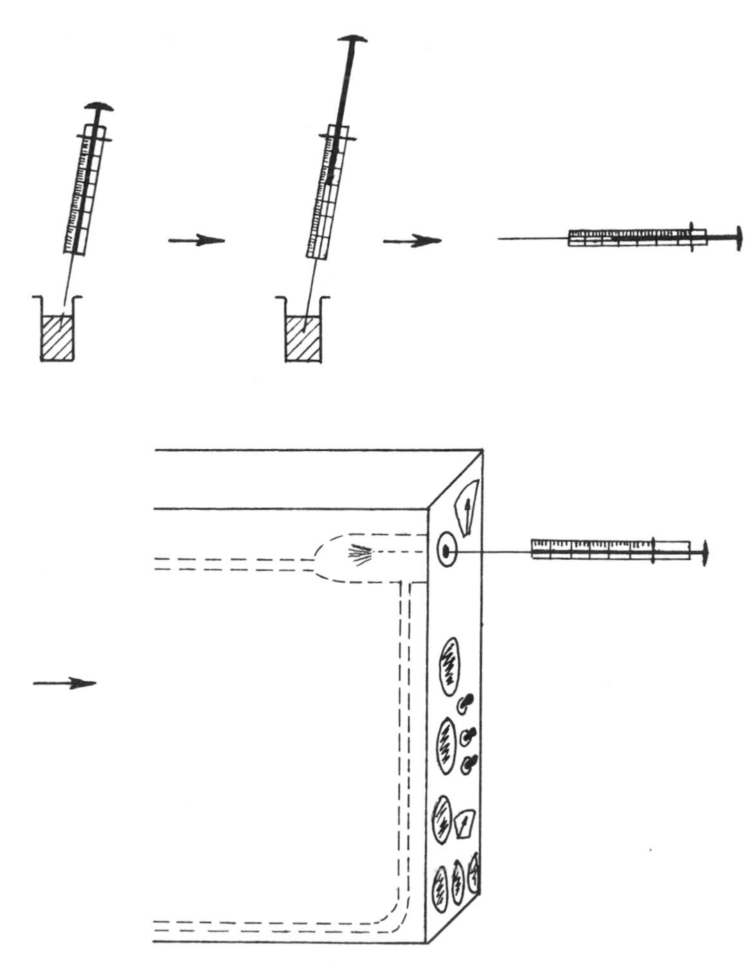

Figure 16.4. The mechanics of loading a gas chromatography syringe, showing the careful positioning of the plunger to the correct volume and the introduction of the sample to the injection port.

sample introductions, can be replaced easily. Sample introduction systems for gases (gas sampling valves) and solids are also available.

A volume of liquid as small as 0.1 to 3 μL (appearing as an extremely small drop) may seem to be extraordinarily small. When vaporized, however, such a volume is much larger and will occupy an appropriate volume in the column. Also, the detection system, as we will see later, is very sensitive and will detect very small concentrations even in such a small volume. As a matter of fact, too large a volume is a concern to the operator since columns, especially capillary columns, can become overloaded even with volumes that are very small. Overloading means that the entire vaporized sample will not fit onto the column all at once and will be introduced over a period of time. Obviously a good separation would be in jeopardy in such an instance. To guard against overloading of capillary columns, split injectors have been developed. In these injectors, only a fraction of the liquid from the syringe actually is passed to the column. The remaining portion is split from the sample and vented to the air.

As implied above, the appropriate range of sample injection volume depends on column diameter. As we will see in the next section, column diameters vary from capillary size (0.2 to 0.3 mm) to 1/8 in. and 1/4 in. Table 16.2 gives the typical injection volumes suggested for these column diameters. The capillary columns are those in which the overloading problem mentioned above is most relevant. Injectors preceding the 1/8 in. or larger columns are not split.

The accuracy of the injection volume measurement can be very important for quantitation, since the amount of analyte measured by the detector depends on the concentration of the analyte in the sample as well as the amount injected. In Section 16.9 a technique known as the internal standard technique will be discussed. (It is also discussed in Chapter 9.) Use of this technique negates the need for superior accuracy with the injection volume, as we will see. However, the internal standard isn't always used. Very careful measurement of the volume with the syringe in that case is paramount for accurate quantitation. An acceptable procedure for this is presented in Table 16.3. Of course, if a procedure calls only for identification (Section 16.8), then accuracy of injection volume is less important.

16.4 COLUMNS

Instrument Logistics

GC columns, unlike any other type of chromatography column, are typically very long. Lengths varying from 2 feet up to 300 feet or more are possible. Additionally, as mentioned previously, it is important for the column to be kept at an elevated temperature during the run in order to prevent condensation of the sample components. Indeed, maintaining an elevated temperature is very important for other reasons, as we shall see in

Table 16.2. Suggested typical injection volumes for various column diameters.

Column Diameters	Maximum Injection Volumes
1/4 in. (packed column)	100 μL
1/8 in. (packed column)	20 μL
Capillary (open tubular)	0.1 μL

Table 16.3. A syringe loading and injection method when accuracy of injection volume is important.

1. Flush the syringe throughly with clean sample solvent.
2. Expel the solvent from the syringe, then carefully retract the plunger (in air) to the 1.0 μL mark. A little less than 1 μL of solvent will be present (needle holdup).
3. Transfer the syringe to the sample container and slowly draw several microliters of sample into the syringe barrel. Remove the syringe needle from the sample container and expel sample until the plunger is at the 2 μL mark (for a 1-μL injection).
4. Retract the syringe plunger, pulling the needle load entirely into the barrel. Two liquid plugs will be seen—sample, and solvent without sample. Note the volume of the sample plug.
5. Insert the needle to its full length; inject the sample and quickly remove the syringe.

From operator's manual for Varian Model 3300/3400 gas chromatograph, Vol. 2, Varian Instruments, Walnut Creek, CA. With permission.

the next section. The obvious logistical problem is how to contain a column of such length and be able to simultaneously control its temperature.

Such a long column is wound into a coil and fits nicely into a small oven, perhaps 1 to 3 cubic feet in size. This oven probably constitutes about half of the total size of the instrument. (See Figure 16.2.) Connections are made through the oven wall to the injection port and the detector. The temperatures of column ovens typically vary from 50°C to 150°C, with higher temperatures possible in procedures that require them. A more thorough discussion of this subject is found in Section 16.5.

GC instruments are designed so that columns can be replaced easily by disconnecting a pair of brass fittings inside the oven. This not only facilitates changing to a different stationary phase altogether, but also allows the operator to replace a given column with a longer one containing the same stationary phase. The idea here is to allow more contact with the stationary phase, which in turn is bound to improve the separation. If a 6-ft column is useful for a partial separation, would not a 12-ft column be that much better?

Packed, Open-Tubular, and Preparative Columns

It was indicated previously that column lengths of up to 300 ft are not unusual. It should be mentioned here that the longer a column with stationary phase tightly packed (a so-called "packed" column), the greater the gas pressure required to sustain the flow of carrier gas. A 20-ft length is approximately the upper limit for the length of a packed column. It is more common now to use the "open-tubular" capillary column. Instead of tightly packed solid substrate particles holding the liquid stationary phase inside the column (see Section 15.6), the stationary phase is made to adsorb on the inside wall of a small-diameter capillary tube so that the tube remains open to gas flow in the center. A design such as this offers very little resistance to gas flow and can be made hundreds of feet long without having to utilize a large pressure. It is no exaggeration to say that such columns are so popular today that the packed column is fast becoming obsolete. (See Figures 16.5 and 16.6.)

In addition to the "analytical" columns (columns used mainly for analytical work), so-called "preparative" columns may also be encountered. Preparative columns are used when the purpose of the experiment is to prepare a pure sample of a particular substance (from a mixture containing the substance) by GC for use in other laboratory work. The procedure for this involves the individual condensation of the mixture components of interest in a cold trap as they pass from the detector and as their peak is being traced on the recorder. While analytical columns can be suitable for this, the amount of pure substance generated is typically very small since what is being collected is only a fraction of the extremely small volume injected. Thus, columns manufactured with very large diameters (on the order of inches) and capable of very large injection volumes (on the order of milliliters) are manufactured for the preparative work. Also, the detector used must not destroy the sample like the flame ionization detector (Section 16.7) does, for example. Thus, the thermal conductivity detector (Section 16.7) is used most often with preparative gas chromatography.

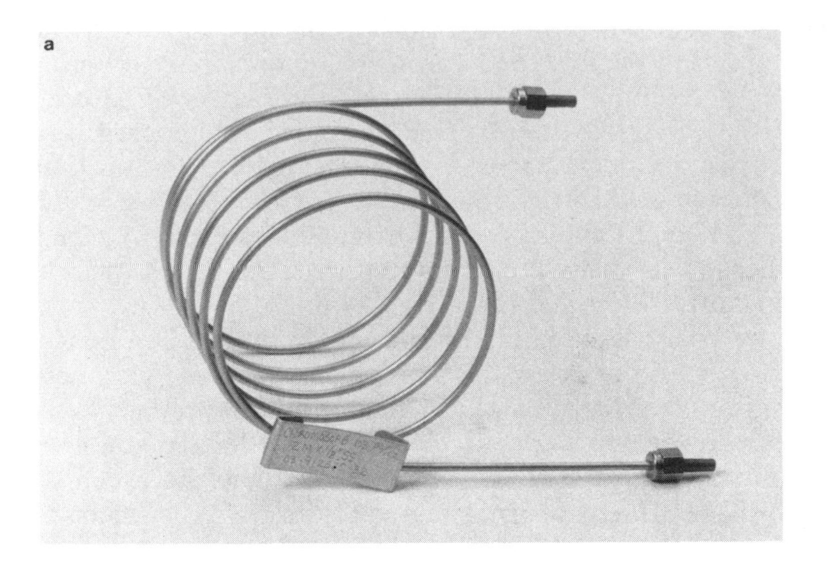

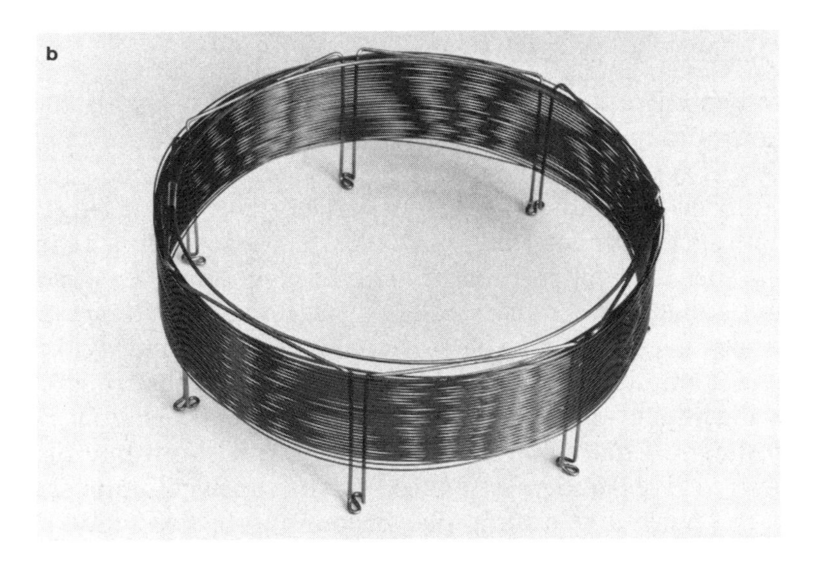

Figure 16.5. Examples of columns: (a) a 6-ft-long, $1/8$-in.-diameter packed column; (b) a 100-ft-long capillary column. (Courtesy of Varian Associates, Palo Alto, CA.)

The Nature and Selection of the Stationary Phase

The liquid stationary phase in a GLC packed column is adsorbed on the surface of a solid substrate (also called the "support"). This material must be inert and finely divided (powdered). The typical diameter of a substrate particle is 125–250 μm, creating a 60- to 100-mesh material. These particles are of two general types: Diatomaceous earth and Teflon®. Diatomaceous earth, the decayed silica skeletons of algae, is most commonly referred to by the manufacturer's (Johns Manville) trade name Chromosorb®. Various types of Chromosorb®, which have had different pretreatment procedures applied, are available, such as Chromosorb® P, Chromosorb® W, and Chromosorb® 101–104. The nature of the stationary phase as well as the nature

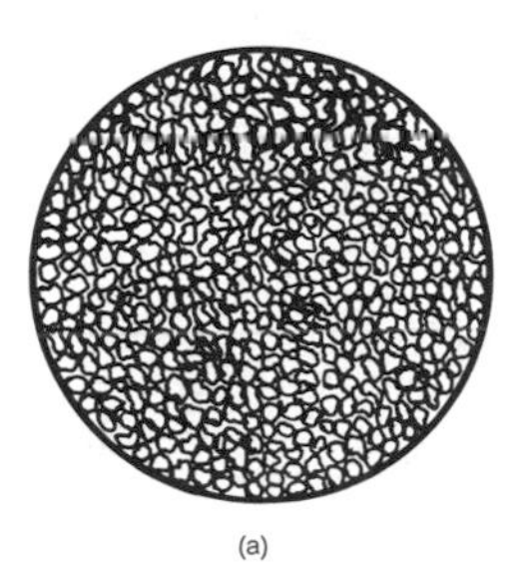

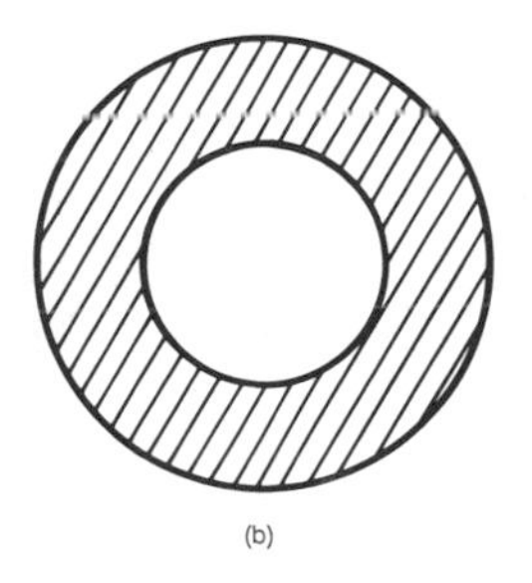

Figure 16.6. (a) Cross section of the packed column; (b) cross section of the open-tubular capillary column. (From Kenkel, J., *Analytical Chemistry Refresher Manual*, Lewis Publishers, Chelsea, MI, 1992. With permission.)

of the substrate material are both usually specified in a chromatography literature procedure, and columns are tagged to indicate each of these as well.

Since the interaction of the mixture components with the liquid stationary phase plays the key role in the separation process, the nature of the stationary phase is obviously important. Several hundred different liquids useful as stationary phases are known. This means that the analyst has an awesome choice when it comes to selecting a stationary phase for a given separation. It is true, however, that relatively few such liquids are in actual common use. Their composition is frequently not obvious to the analyst because a variety of common abbreviations have come to be popular for the names of some of them. Table 16.4 lists a number of common stationary phases, their abbreviated names, a description of their structures, and the classes of compounds (in terms of polarity) for which each is most useful.

The selection of a stationary phase depends largely on trial and error or experience, with consideration given to the polar nature of the mixture, as noted in Table 16.4 or a similar table. The usual procedure is to select a stationary phase, based on such literature information, and attempt the separation under the various conditions of column temperature, length, carrier gas flow rate, etc., to determine the optimum capability for separating the mixture in question. If this optimum resolution is not satisfactory (see Section 16.6), then an alternate selection is apparently required.

More experienced chromatographers may refer to the McReynolds constants for a given stationary phase as a measure of its resolving power. A complete discussion of this subject, however, is beyond the scope of this text.

16.5 OTHER VARIABLE PARAMETERS

Column Temperature

Both the vapor pressure of a substance and the solubility of a substance in another substance change with temperature. Figure 16.7 shows, for example, how the vapor pressure can change with temperature and Figure 16.8 shows how the solubility can change with temperature. It should not be surprising then that the precise control of the temperature of a GLC column is very important since, as we have indicated, the separation depends on

Table 16.4. Some stationary phases for GLC.

Abbreviated or Non-Descriptive Name	Structure, Descriptive Name, or Other Description	Useful for Mixtures of Compounds Which Are
FFAP	A Teflon®-based material	Highly polar
Casterwax	$CH_3(CH_2)_5-CH(OH)-CH_2-CH{=}CH-(CH_2)_7-C(=O)OH$	Highly polar
Carbowax® (variety of molecular weights)	$HO-[CH_2-CH_2-O]_n-H$	Polar
XE-60 (also XF1150, SF-1125	$Si(CH_3)_3-O-[Si(CH_3)(CH_2-CH_2-CH_2-C{\equiv}N)-O]_n-Si(CH_3)_3$	Polar
OV-17	Methyl, phenyl, silicone (a silicone oil)	Somewhat polar
OV-101	Liquid methyl silicone	Non-polar
OV-1	Methyl siloxane	Non-polar
SE-30	$-[Si(CH_3)_2-O-Si(CH_3)_2-O]_n-$	Non-polar
Apiezon (various types)	A grease	Non-polar
Squalane	High-molecular-weight hydrocarbon (C_{30})	Non-polar

From McNair, H. M. and Bonelli, E. J., Basic Gas Chromatography, Varian Analytical Instruments, Sunnyvale, CA, 1965. With permission.

both vapor pressure and solubility. Both isothermal (constant) and programmed (continuously changing) temperature experiments are possible. For simple separations, the isothermal mode may well be sufficient – there may be sufficient differences in the mixture components' vapor pressures and solubilities to effect a good separation at the chosen temperature. However, for more complicated mixtures, a complete separation is less likely in the isothermal mode.

For example, consider gasoline, which has a goodly number of highly volatile components as well as a significant number of less volatile compo-

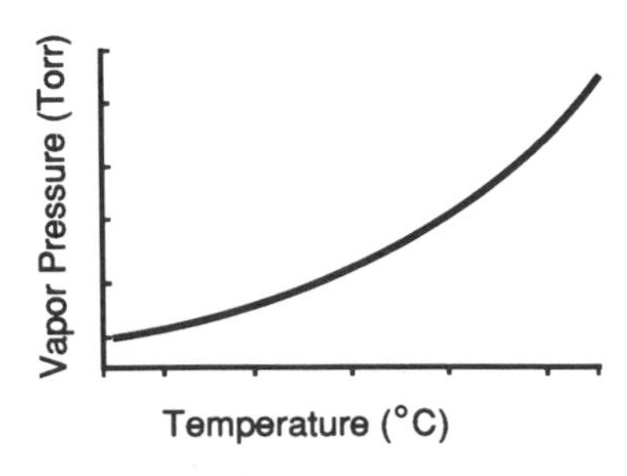

Figure 16.7. A graph showing a typical example of how the vapor pressure can change with temperature.

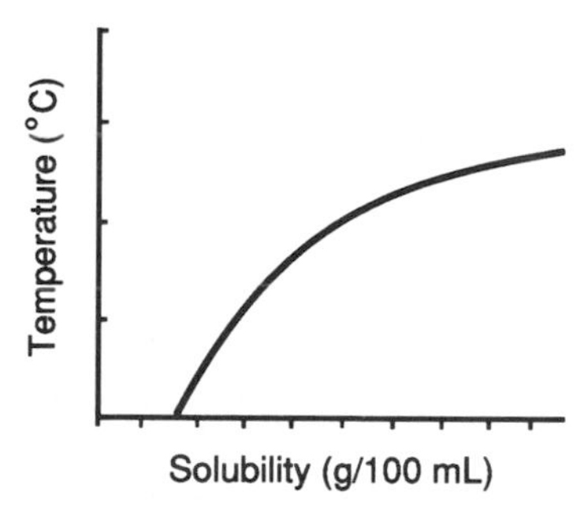

Figure 16.8. A graph showing how the solubility of a substance (in a hypothetical solvent) can change with temperature.

nents. It's possible that at a temperature of, say, 100°C some of the less volatile components will be resolved, but the more volatile ones will pass through unresolved and have very short retention times. A lower temperature of, say, 40°C may cause complete resolution of these more volatile components, but would result in unwanted long retention times for the less volatile components and perhaps also result in poorly shaped peaks for these. If we could increase the temperature from 40°C to 100°C or higher in the middle of the run, however, we could have the best of both worlds—complete resolution and reasonable retention times for all peaks. Thus, temperature-programmable ovens have been developed and are now commonplace on virtually all modern GC units. Temperature programming can consist of simple programs such as that suggested above—a single linear increase from a low temperature to a higher temperature—but it can also be more complex. For example, a chromatography researcher may find that several temperature increases, and perhaps even a decrease, must be used in some instances to effect an acceptable separation. Most modern GC units are capable of at least a slow temperature decrease in the middle of the run since they are equipped with venting fans that bring ambient air into the oven to cool it. Both a simple program and a more complex program are represented in Figure 16.9.

Carrier Gas Flow Rate

The rate of flow of the carrier gas affects resolution. A simple analogy here will make the point. Wet laundry hung out on a clothesline to dry will dry faster if it is a windy day. The components of the mixture will "blow" through the column more quickly (regardless of the degree of interaction with the stationary phase) if the carrier gas flow rate is increased. Thus, a minimum flow rate is needed for maximum resolution. It is well known, however, that at extremely slow flow rates resolution is dramatically reduced due to factors such as packing irregularities, particle size, column diameter, etc. The treatment of these factors and the quantitative determination of the optimum flow rate are beyond the scope of this text.

It is obvious that the flow rate must be precisely controlled. The pressure from the compressed gas cylinder of carrier gas, while sufficient to force the gas through a packed column, does not provide the needed flow control of

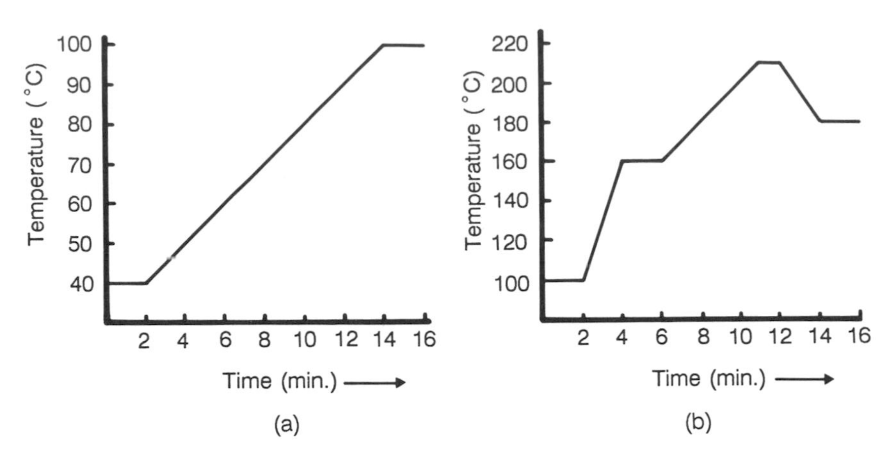

Figure 16.9. (a) A temperature program from 40°C to 100°C at 5°C/min; (b) a complex temperature program. (From Kenkel, J., *Analytical Chemistry Refresher Manual*, Lewis Publishers, Chelsea, MI, 1992. With permission.)

itself. Thus a flow controller, or needle valve, must be part of the GC system and is often incorporated into the face of the instrument. In addition, the flow rate must be able to be carefully measured so that one can know what the optimum flow rate is and be able to match it in subsequent experiments. Various flow meters are commercially available for this, and often the instrument manufacturer builds one into the instrument so that the flow rate is monitored continuously and is observable as one turns on the needle valve. In other cases, a simple soap bubble flow meter is often used and can be constructed easily from an old measuring pipet, a piece of glass tubing, and a pipet bulb. (See Figure 16.10.) With this apparatus, a stopwatch is used to measure the time it takes a soap bubble squeezed from the bulb to move between two graduation lines, such as the 0 and 10 mL lines. The flow rate in milliliters per minute can thus be calculated.

16.6 THE CHROMATOGRAM

The chart recording giving the written record of the resolved substances, or peaks, is called the chromatogram. All qualitative and quantitative information obtained from a GC experiment is found in the chromatogram. One piece of such information is the "retention time", symbolized as t_R. From the time a substance is injected into the injection port until it emerges from the column and passes through the detector, it is being retained by the column. This is the span of time referred to as the retention time. Since the chart paper is passing through the recorder at a constant rate (for example, 1 in./min), the recorder itself becomes a device for measuring retention time. A certain number of inches or centimeters of chart paper corresponds to a certain number of minutes. Figure 16.11 shows how this measurement is made on the chromatogram. Typically, retention times vary from a small fraction of a minute to about 20 min, although much longer retention times have been experienced.

Another parameter often measured is the adjusted retention time, t'_R. This is the difference between the retention time of a given component and the retention time of an unretained substance (t_M), which is often air. You

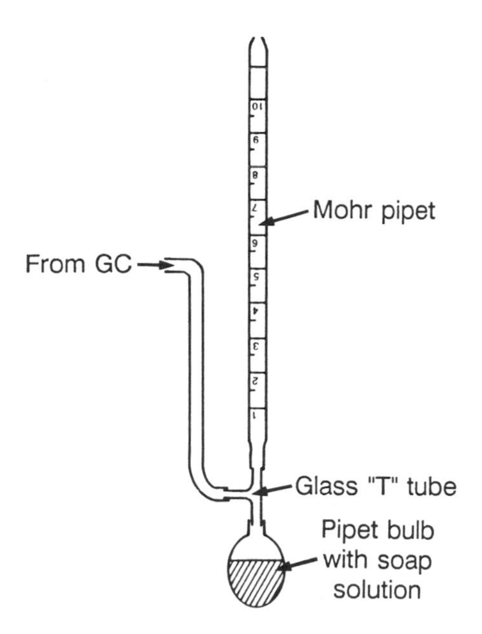

Figure 16.10. A homemade soap bubble flow meter constructed from an old Mohr pipet, a piece of glass tubing, and a pipet bulb. (From Kenkel, J., *Analytical Chemistry Refresher Manual*, Lewis Publishers, Chelsea, MI, 1992. With permission.)

will recall that the injection technique described in Table 16.3 involved the injection of some air with the sample. Air is usually completely unretained by a column, and thus the adjusted retention time becomes a measure of the exact time a mixture component spends in the stationary phase. Figure 16.12 shows how this measurement is made. The most important use of this

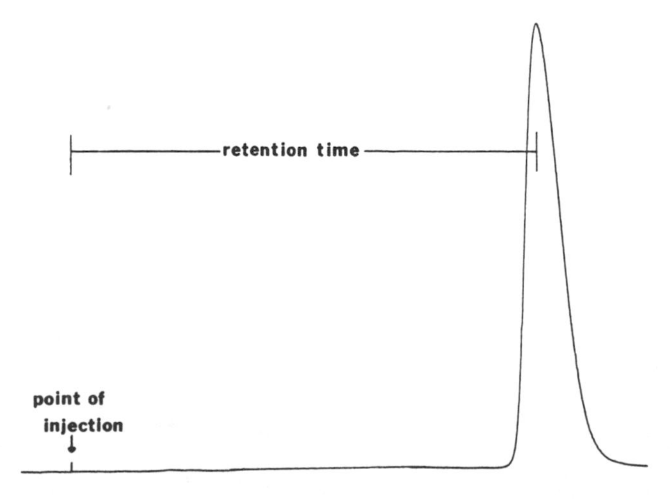

Figure 16.11. Retention time is the time corresponding to the length of chart paper measured from the point of injection to where the peak is at its apex.

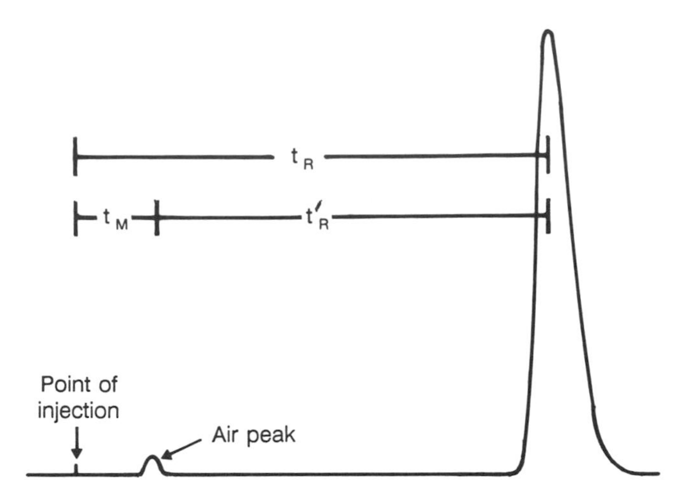

Figure 16.12. A chromatogram showing the definitions of t_R, t'_R, and t_M. (From Kenkel, J., *Analytical Chemistry Refresher Manual*, Lewis Publishers, Chelsea, MI, 1992. With permission.)

retention time information is in peak identification, or qualitative analysis. This subject will be discussed in more detail in Section 16.8.

Other parameters sometimes obtained from the chromatogram, which are mostly measures of the degree of separation and column efficiency, are "resolution", R; the number of "theoretical plates", N; and the "height equivalent to a theoretical plate", HETP, or H. These require the measurement of the width of a peak at the peak base. This measurement is made by first drawing the tangents to the sides of the peaks and extending these to below the baseline, as shown for the two peaks in Figure 16.13. The width at peak base, W_B, is then the distance between the intersections of the tangents with the baseline, as shown. Resolution is defined as the difference in the retention times of two closely spaced peaks divided by the average widths of these peaks, as shown mathematically in Figure 16.13. R values of 1.5 or more would indicate complete separation. Peaks that are not well resolved

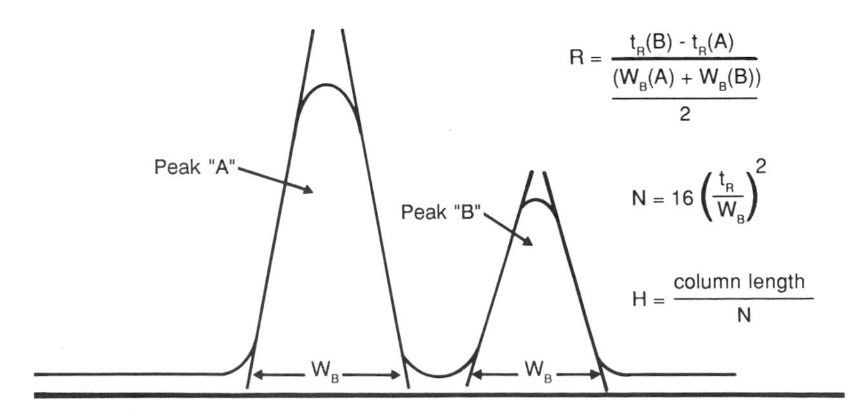

Figure 16.13. The measurement of the "width at base", which is needed for resolution and theoretical plate calculations. (From Kenkel, J., *Analytical Chemistry Refresher Manual*, Lewis Publishers, Chelsea, MI, 1992. With permission.)

would inhibit satisfactory qualitative and quantitative analysis (Sections 16.8 and 16.9). An example of a chromatogram showing unsatisfactory resolution of two peaks is shown in Figure 16.14.

The number of theoretical plates, N, is also mathematically defined in Figure 16.13. The concept of theoretical plates was discussed briefly in Section 15.3 for distillation. For distillation, one theoretical plate was defined as one evaporation/condensation step for the distilling liquid as it passes up a fractionating column. In chromatography, one theoretical plate is one "extraction" step along the path from injector to detector. You will recall that in Section 15.4 we spoke of chromatography as being analogous to a series of many extractions, but with one solvent (the mobile phase) constantly moving through the other solvent (the stationary phase) rather than being passed along through a series of separatory funnels. The equilibration that would occur in the fictitious separatory funnel is one theoretical plate in chromatography.

The height equivalent to a theoretical plate, also mathematically defined in Figure 16.13, is that length of column that represents one theoretical plate, or one equilibration step. Obviously, the smaller the value of this parameter the more efficient the column. The more theoretical plates packed into a length of column the better the resolution. Factors that influence the number of theoretical plates and the resolution are column length, column temperature, carrier gas flow rate, and other factors we have already discussed. Other parameters calculated from the chromatogram, including capacity factor and selectivity, are defined in Section 17.5.

16.7 DETECTORS

Detectors in gas chromatography are designed to generate an electronic signal when a gas other than the carrier gas elutes from the column. There have been a number of detectors invented to accomplish this. Not only do these detectors vary in design, but they also vary in sensitivity and selectivity. Sensitivity refers to the smallest quantity of mixture component for which it is able to generate an observable signal, and selectivity refers to the type of compound for which a signal can be generated. The flame ionization detector, for example, is a very sensitive detector, but does not detect everything; i.e., it is selective for only a certain class of compounds. The thermal conductivity detector, on the other hand, detects virtually everything; i.e., it is a "universal" detector, but is not very sensitive. What follows is a brief description of the designs of the detectors that are in common use, along with some indication of their sensitivity and selectivity.

Thermal Conductivity Detector (TCD)

The thermal conductivity detector (TCD) operates on the principle that gases eluting from the column have thermal conductivities different from that of the carrier gas, which is usually helium. Present in the flow channel at the end of the column is a hot filament, hot because it has an electrical current passing through it. This filament is cooled to an equilibrium temperature by the flowing helium, but is cooled differently by the mixture components as they elute, since their thermal conductivities are different from helium. This change in the cooling process causes the filament's electrical resistance to change and thus causes the current flowing through it and the

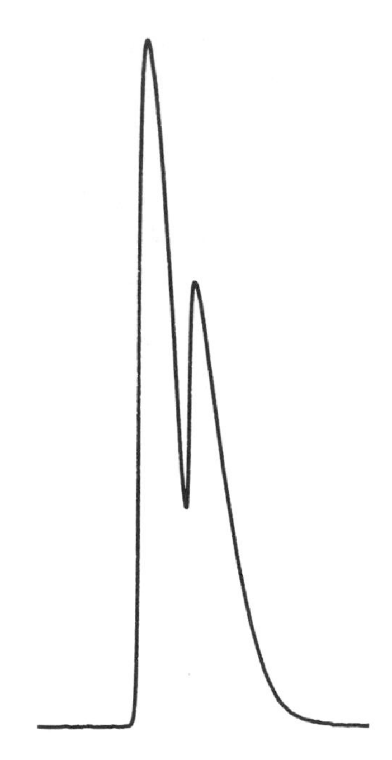

Figure 16.14. A chromatogram showing two peaks not satisfactorily resolved.

voltage drop across it to change each time a mixture component elutes. The recorder, which is constantly monitoring this voltage drop, thus records a peak.

The actual design includes a second filament within the same detector block. This filament is present in a different flow channel, however, one through which only pure helium flows. Both filaments are part of a Wheatstone bridge circuit (as shown in Figure 16.15), which allows a "comparison" between the two resistances and a voltage output to the recorder. Such a

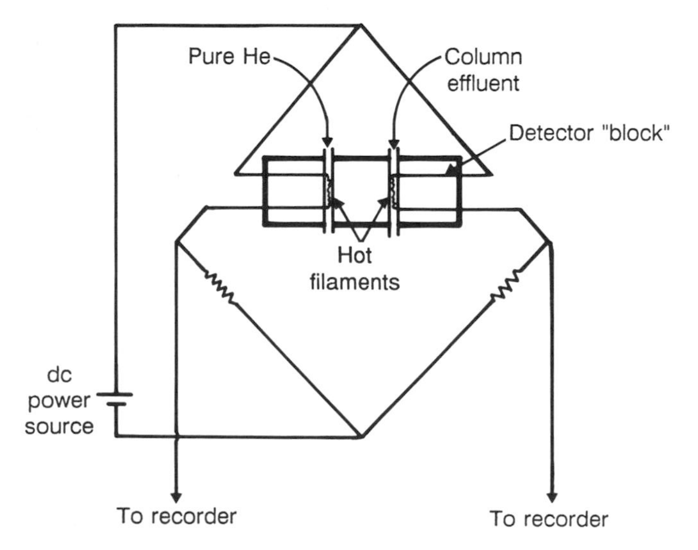

Figure 16.15. The thermal conductivity detector. (From Kenkel, J., *Analytical Chemistry Refresher Manual*, Lewis Publishers, Chelsea, MI, 1992. With permission.)

design is intended to minimize effects of flow rate, pressure, and line voltage variations.

Most recently, a flow modulated design has become popular. In this design, a single filament is used and the column effluent is alternated with the pure helium through the flow channel where the filament is located. This eliminates the need to use two matched filaments.

The thermal conductivity detector is universal (i.e., detects everything); it is non-destructive (i.e., can be used with preparative GC), but less sensitive than other detectors.

Flame Ionization Detector (FID)

Another very important GC detector design is the flame ionization detector, the FID. In this detector, the column effluent is swept into a hydrogen flame where the flammable components are burned. In the burning process, a very small fraction of the molecules become fragmented and the resulting positively charged ions are drawn to a "collector" (negatively charged) electrode, a metal cylinder above and encircling the flame, while electrons flow to the positively charged burner head. The negatively charged collector and the positively charged burner head are part of an electrical circuit in which the current changes when this process occurs and the change is amplified and seen as a peak on the recorder trace. Figure 16.16 shows a drawing of this detector. The design includes the hydrogen flame burner nozzle, the collector electrode, an inlet for air to surround the flame, and an igniter coil for igniting the hydrogen as it emerges from the nozzle.

Apparent on the exterior of the instrument, and located near the bench on which the GC unit sits, are pressure-regulated compressed gas cylinders of hydrogen and air as well as the helium. Metal tubing, typically of 1/8 in. diameter, connects the cylinders to the detector, with a needle valve for flow control in between. These valves are located in the instrument for easy access and control by the operator.

The FID is very sensitive, but is not universal, and it also destroys (burns) the sample. It only detects organic substances that burn and fragment in a hydrogen flame. These facts preclude its use for preparative GC or for

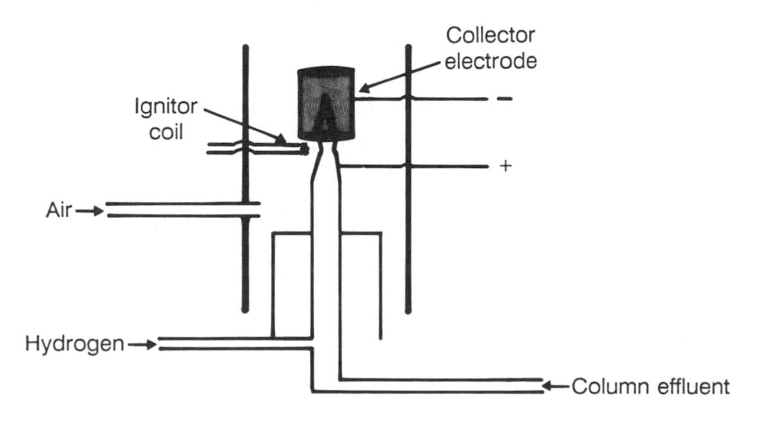

Figure 16.16. The flame ionization detector. (From Kenkel, J., *Analytical Chemistry Refresher Manual*, Lewis Publishers, Chelsea, MI, 1992. With permission.)

inorganic substances that do not burn, such as water, carbon dioxide, etc. Still, it is a very popular detector, given its sensitivity and given the fact that most analytical work involves flammable organic substances.

Electron Capture Detector (ECD)

A third type of detector, required for some environmental and biomedical applications, is the electron capture detector, the ECD. This detector is especially useful for large halogenated hydrocarbon molecules since it is the only one which has an acceptable sensitivity for such molecules. Thus, it finds special utility in the analysis of halogenated pesticide residues found in environmental and biomedical samples.

The electron capture detector is another type of ionization detector. Specifically, it utilizes the beta emissions of a radioactive source, often nickel-63, to cause the ionization of the carrier gas molecules, thus generating electrons which constitute an electrical current. As an electrophilic component, such as a pesticide, from the separated mixture enters this detector, the electrons from the carrier gas ionization are "captured", creating an alteration in the current flow in an external circuit. This alteration is the source of the electrical signal which is amplified and sent on to the recorder. A diagram of this detector is shown in Figure 16.17. The carrier gas for this detector is either pure nitrogen or a mixture of argon and methane.

An additional consideration regarding pesticides warrants mentioning here. Most of these compounds decompose on contact with hot metal surfaces. This problem has, however, been adequately solved for most pesticides by constructing the entire path of the sample out of glass or glass-lined materials. Thus, glass or glass-lined injection ports and all-glass columns are available.

In terms of advantages and disadvantages, the ECD is extremely sensitive, but only for a very select group of compounds – halogenated hydrocarbons. Other gases will not give a peak. It does not destroy the sample and thus may be used for preparative work.

The Nitrogen/Phosphorus Detector (NPD)

While the ECD is useful for chlorinated hydrocarbon pesticides, the NPD, also known as the "thermionic" detector, is useful for the phosphorus- and nitrogen-containing pesticides, the organophosphates and carbamates. The design, however, represents a slight alteration of the design of the FID. In the NPD, we basically have an FID with a bead of alkali metal salt posi-

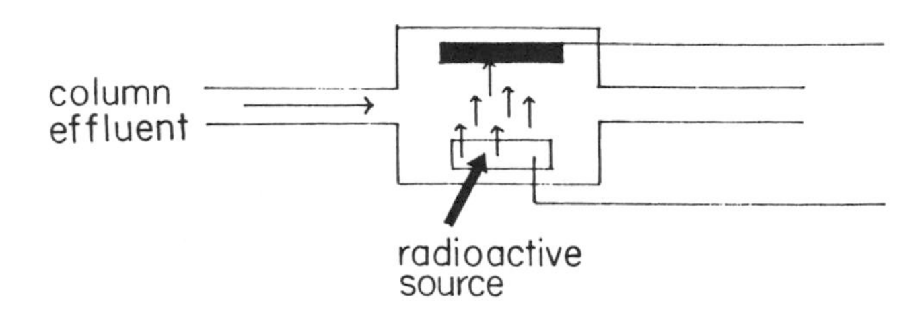

Figure 16.17. The electron capture detector.

tioned just above the flame. The hydrogen and air flow rates are lower than in the ordinary FID and this minimizes the fragmentation of other organic compounds. These changes result in a somewhat mysterious increase in both the selectivity and sensitivity for the pesticides.

Flame Photometric Detector (FPD)

A detector that is specific for organic compounds containing sulfur or phosphorus is the flame photometric detector (FPD). A flame photometer (Chapter 14) is an instrument in which a sample solution is aspirated into a flame and the resulting emissions from the flame are measured with a phototube detector. The FPD is a flame photometer positioned to accept the effluent from the column in place of the aspirated sample. The flame in this case is a hydrogen flame as in the FID. The basic operating principle is that the sulfur or phosphorus compounds burn in the hydrogen flame and produce light-emitting species. A monochromator, typically a glass filter, makes this detector specific for these compounds. The signal for the recorder is the signal proportional to light intensity that is produced by the phototube.

The advantages are that it is a very selective detector and also very sensitive. Disadvantages include the problems associated with the need to carefully control the flame conditions so that the correct species are produced (S=S for the sulfur compounds and HPO for the phosphorus compounds). Such conditions include the gas flow rates and the flame temperature. It is a destructive detector.

Electrolytic Conductivity (Hall®) Detector

The Hall® detector converts the eluting gaseous components into ions in liquid solution and then measures the electrolytic conductivity of the solution in a conductivity cell. The solvent is continuously flowing through the cell, and thus the conducting solution is in the cell for only a moment while the conductivity is measured and the peak recorded before it is swept away with fresh solvent. The conversion to ions is done by chemically oxidizing or reducing the components with a "reaction gas" in a small reaction chamber made of nickel positioned between the column and the cell. The nature of the reaction gas depends on what class of compounds is being determined. Organic halides, the most common application, use hydrogen gas at 850°C or higher as the reaction gas. The strong HX acids are produced, which give highly conductive liquid solutions.

The Hall® detector has excellent sensitivity and selectivity, giving a peak for only those components which produce ions in the reaction chamber. It is a destructive detector.

GC-MS and GC-IR

We discussed the fundamentals of mass spectrometry in Chapter 13 and those of infrared spectrometry in Chapter 12. The quadrupole mass spectrometer and the Fourier transform IR spectrometer have been adapted to and used with GC equipment as detectors with great success in recent years. Gas chromatography-mass spectrometry (GC-MS) and gas chromatography-infrared spectrometry (GC-IR) are very powerful tools for qualitative analysis in GC because they not only give retention time information, but due to their inherent speed they are able to measure and record the mass or IR spectra of the individual sample components as they elute from the GC column. It's like taking a photograph of each component as it elutes. (See Figure 16.18.) Coupled with the computer banks of mass and IR spectra, a component's identity is an easy chore for such a detector. Recently fabricated GC-MS units have become very compact, in contrast to older units which take up large amounts of space. It seems the only disadvantage remaining is the expense, although that also seems to be improving. The only other slight disadvantage is the fact the large amounts of computer memory space are required to hold the amount of spectral information required for a good qualitative analysis.

Both the GC-MS instrument and the GC-IR instrument obviously require that the column effluent be fed into the detection path. For the IR instrument, this means that the IR cell, often referred to as a "light pipe", must be situated just outside the interferometer (Chapter 12) in the path of the light, of course, but must also have a connection to the GC column and an exit tube where the sample may possibly be collected. The infrared detector is non-destructive. With the mass spectrometer detector, we have the problem of the low pressure of the MS unit coupled to the ambient pressure of the GC column outlet. A special device is needed as a "go-between".

Photoionization Detector (PID)

The photoionization detector, as the name implies, involves the ionization of eluting mixture components by light, specifically, UV light. The UV source emits a wavelength characteristic of the gas (either helium or argon) inside. This light passes into an "ionization chamber" through a metal fluoride window and into the path of the column effluent there. This is where the mixture components absorb the light and ionize. The resulting ions are detected through the use of a pair of electrodes in the ionization chamber,

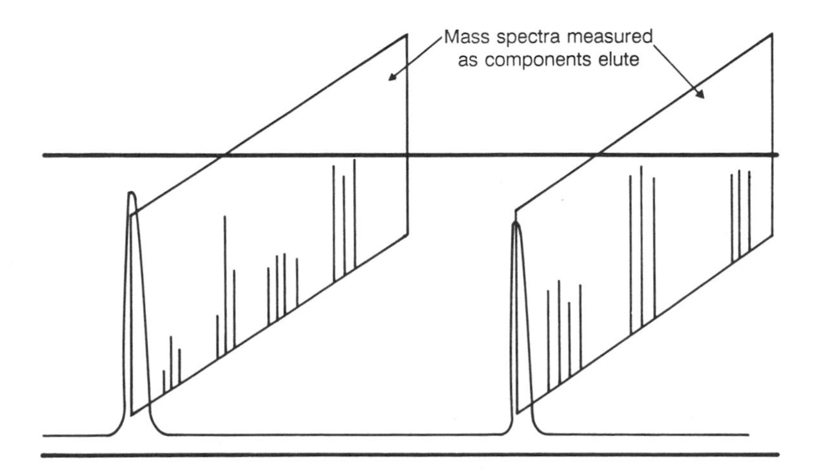

Figure 16.18. A "photograph" of individual mixture components is obtained with GC-MS and GC-IR instruments. (From Kenkel, J., *Analytical Chemistry Refresher Manual*, Lewis Publishers, Chelsea, MI, 1992. With permission.)

the current from which constitutes the signal to the recorder. The specific lamp and window are chosen according to the ionization energy needed for the compounds in the sample.

Since different lamps and windows are available, this detection method can often be selective for only some of the components present in the sample. Its sensitivity is especially good for aromatic hydrocarbons and inorganics. It is a very sensitive non-destructive detector.

16.8 QUALITATIVE ANALYSIS

As mentioned in Section 16.6, the parameters that are most important for a qualitative analysis using most GC detectors are the retention time, t_R, and the adjusted retention time, t'_R. Their definitions were graphically presented in Figure 16.11. Under a given set of conditions (the nature of the stationary phase, the column temperature, the carrier flow rate, the column length and diameter, and the instrument dead volume), the retention time is a particular value for each component. It changes only when one or more of the above parameters changes. Thus, when the temperature changes, retention time will change, when the carrier gas flow rate changes, the retention time will change, etc. Repeated injections into a given system under a given set of conditions should always yield a particular retention time for a given component, and qualitative analysis using this system only requires accurate measurement of t_R. When one of the parameters changes, such as when an analyst in another laboratory who is trying to duplicate a given qualitative analysis sets up with a different dead volume (the volume of the space between the injection port and the column packing and the space between the column packing and the detector) or perhaps a slightly different stationary phase composition, for example, then the retention time for that component will be slightly different. The adjusted retention time will correct for changes in the dead volume, but will not correct for any other change. A parameter defined as the "relative retention", however, will adjust for other changes. This parameter compares the retention of one component (1) with another (2) and is given the symbol alpha (α). It is defined as follows:

$$\alpha = \frac{t'_R(1)}{t'_R(2)} \qquad (16.1)$$

The relative retention is thus an important parameter for qualitative analysis if the work involves other setups with other instruments and columns which do not exactly match the original.

The usual qualitative analysis procedure, then, is to establish the conditions for the experiment, perhaps by trial and error in one's own laboratory or by matching conditions outlined in a given procedure, and then to match the retention time data, either ordinary retention time, t_R, or the relative retention, α, whichever is appropriate, for standards (pure samples) with that for the unknown. The analyst can then proceed to match the retention time data for the unknown to those of the pure samples to determine which substances are present (Experiment 44).

One caution is that there may be more than one component with the same retention (no separation) and thus further experimentation may be required. For example, when working with a complex mixture whose components are perhaps not all known, it may be necessary to change the experimental conditions to determine whether a given peak is due to one component (known) or more (e.g., one known and one unknown). Changing the stationary phase may prove useful. Such a change would produce a chromatogram with completely different retention times and probably a different order of elution. Thus, two components that co-eluted before may now be separated, evidence for which would be a different peak size for the known component.

16.9 QUANTITATIVE ANALYSIS

Peak Size Measurement

The physical size of the peak traced out by the strip-chart recorder is directly proportional to the amount of that particular component passing through the detector. Thus, it is imperative that we have an accurate method for determining this peak size if it is the quantity of a component in the mixture that is sought. There are a variety of methods that have been used for this over the years. Three manual methods are the peak height method, which simply measures the height from the baseline to the apex of the peak, the triangulation method, which measures the area of a triangle drawn to approximate the peak, and the half-width method, which measures the area of a rectangle drawn to approximate the peak. These three methods are illustrated in Figure 16.19.

While the first is a peak "height" method, the others are peak "area" methods. In the triangulation method, the height and base of the triangle are measured as shown and the area calculated (bh/2). In the half-width method, the height of the peak and the width of the peak at half-height are measured, and these represent the length and width of a rectangle; thus, the area is again easily calculated (hw). None of these methods are terribly accurate, but they are fast and do not require expensive equipment. The peak height method is especially useful (fast) when only a rough indication of quantity is desired.

The most popular method of measuring peak size is by integration. Integration is an area-measuring method in which a series of "heights" are measured from the moment the pen begins to deflect until the baseline is com-

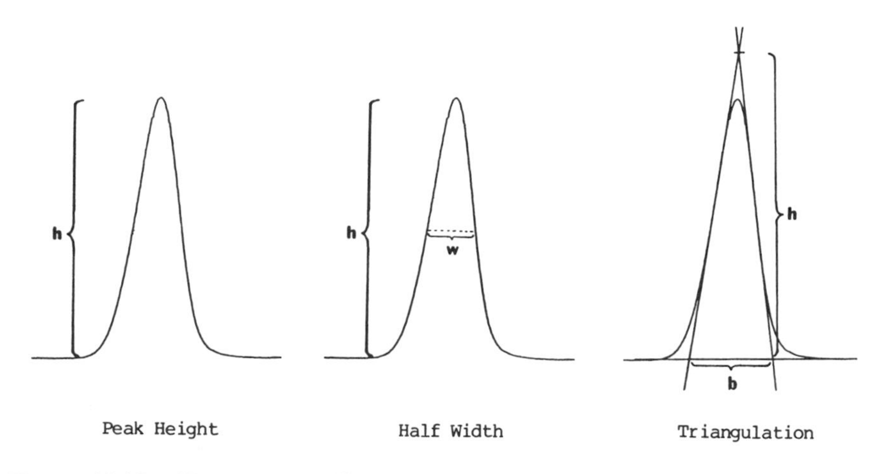

Figure 16.19. Three "manual" methods for measuring peak size.

pletely restored, as illustrated in Figure 16.20. This is conveniently done by computer, since a computer works with digital values derived from the analog data output by the instrument to the recorder (Chapter 9). The method is thus easy, fast, and accurate.

The computer hardware for integration can be one of a variety of designs, from a small unit designed only for measuring chromatography peaks (a "computing integrator") to a larger system, such as a microcomputer or other computer, programmed using independently prepared software. Figure 16.21 shows an example of a computing integrator (a) and the printed output (b). Such a device often replaces the ordinary recorder, since it prints the peaks as a recorder would. It also records the retention time next to each peak as the peak is recorded. Note the area values in the sample printout (under the "area" heading). These values represent the sum of the series of digital values represented by the heights illustrated in Figure 16.20.

Quantitation Methods

Several different approaches exist as to what peaks are measured and how the mixture component of interest is actually quantitated. We now discuss the more popular methods. (See also Chapter 9.)

The Response Factor Method. Consider a four-component mixture to be analyzed by GC. The chromatogram may look something like that shown in Figure 16.22. One might think it logical that in order to quantitate the mixture for, say, component B, all one would need to do is to measure the sizes of all four peaks and divide the size of the peak representing B by the total of all four:

$$\%B = \frac{Area_B}{Area_A + Area_B + Area_C + Area_D} \tag{16.2}$$

The problems with this approach are (1) without comparing the peaks to a standard or a set of standards, it is not known whether the result is a weight percent, volume percent, or mole percent; and (2) the instrument detector does not respond to all components equally. For example, not all components will have the same thermal conductivity, and thus the thermal con-

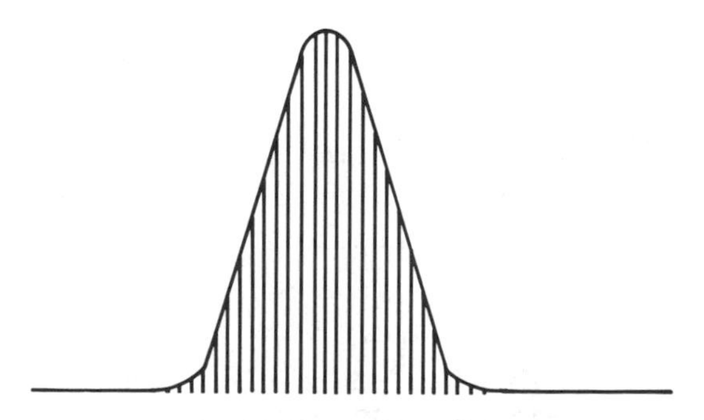

Figure 16.20. An illustration of the measurement of peak size by integration. The sum of a series of vertical "heights", such as those illustrated, represents the area.

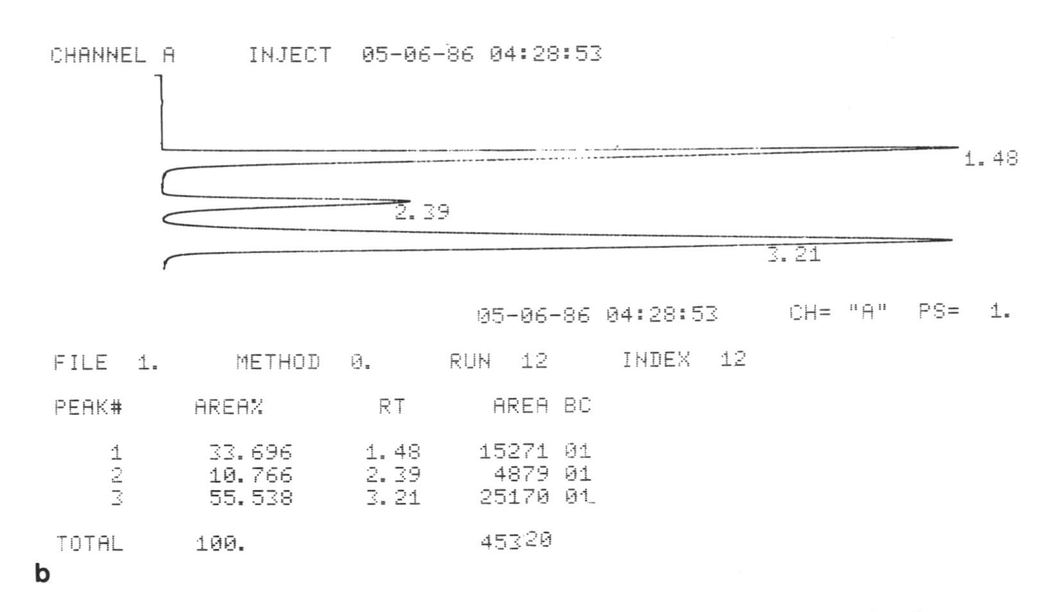

Figure 16.21. (a) A computing integrator; (b) the printout from a computing integrator.

ductivity detector will not give equal-sized peaks for equal concentrations of any two components. Thus, the sum of all four peaks would be a meaningless quantity, and the size of peak B by itself would not represent the correct fraction of the total.

It is possible, however, to measure a so-called response factor for the analyte, which is the area generated by a unit quantity injected, such as a microliter (μL) or microgram (μg). The procedure is to inject a known quantity of the analyte, measured by the position of the plunger in the syringe (μL) or by weighing the syringe before and after injection. The peak size that results is measured and divided by this quantity:

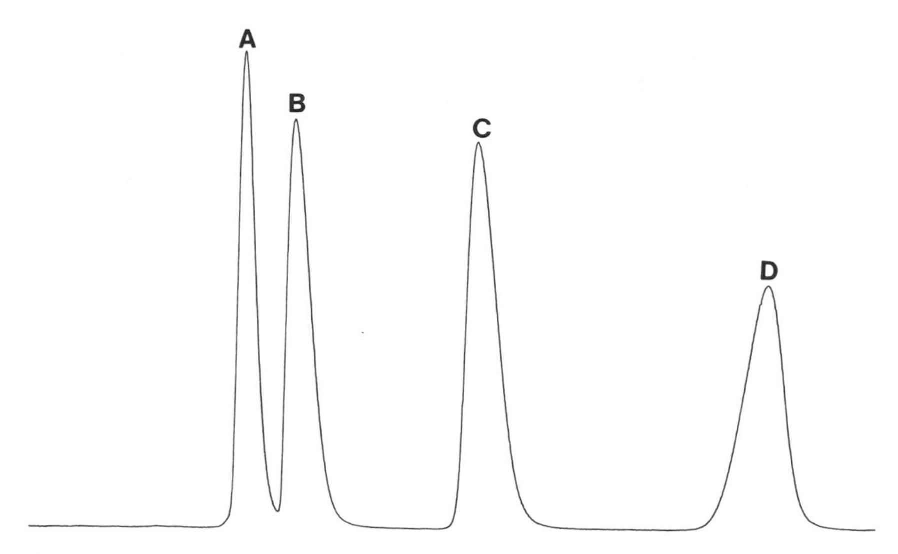

Figure 16.22. A chromatogram of a four-component mixture.

$$\text{response factor} = \frac{\text{size of peak}}{\text{quantity of pure sample injected}} \tag{16.3}$$

The quantity of analyte in an unknown sample is then determined by measuring the peak size of the analyte resulting from an injection of a known quantity of unknown sample and dividing by the analyte's response factor:

$$\text{quantity of analyte} = \frac{\text{peak size}}{\text{response factor}} \tag{16.4}$$

The percent of the analyte can then by calculated as follows:

$$\text{\% of analyte} = \frac{\text{quantity of analyte (from Equation 16.4)}}{\text{total quantity injected}} \tag{16.5}$$

In this method, only the peak of the analyte need be measured in the four-component mixture in order to quantitate this component.

Internal Standard Method. Since the peak size is directly proportional to concentration, one may think that one could prepare a series of standard solutions and obtain peak sizes for a plot of peak size vs. concentration, a method similar to Beer's law in spectrophotometry, for example. But since peak size also varies with amount injected, there can be considerable error due to the difficulty in injecting consistent volumes as discussed above and in Section 16.3. A method that does away with this problem is the internal standard method. (See Section 9.3.) In this method, all standards and samples are spiked with a constant known amount of a substance to act as what is called an internal standard. The purpose of the internal standard is to serve as a reference point for the peak size measurements so that slight variations in injection technique and volume injected are compensated for by the fact that the internal standard peak and the analyte peak are both affected by the slight variations, and thus the problem cancels out.

The procedure is to measure the peak sizes of both the internal standard peak and the analyte peak and then to divide the analyte peak size by the internal standard peak size. The "area ratio" thus determined is then plotted vs. concentration of the analyte. The result is a method in which the volume injected is not as important and, in fact, can vary substantially from one injection to the next.

Can just any substance serve as an internal standard? There are certain characteristics that the internal standard should have, and those are listed below:

1. Its peak, like the analyte's, must be completely resolved from all other peaks.
2. Its retention time should be close to that of the analyte.
3. It should be structurally similar to the analyte.

Standard Additions Method. This general analytical method was also discussed in Chapter 9. Increasing standard amounts of analyte are added to the sample and the resulting peak sizes, which should show an increase with concentration added (Figure 16.23), measured. This method is not as useful in GC as it would be in AA (see Chapter 14), since the sample matrix is not an issue in GC as it is in AA due to the fact that matrix components become separated. However, standard additions may be useful for convenience sake, particularly when the sample to be analyzed already contains a component capable of serving as an internal standard. Thus, standard additions could be used in conjunction with the internal standard method (see Experiment 47) and the internal standard would not have to be independently added to the sample and to the series of standards – it is already present, a convenient circumstance. Area ratio would then be plotted vs. concentration added and the unknown concentration determined by extrapolation to zero area ratio. Please refer to Chapter 9 for other details of the method of standard additions.

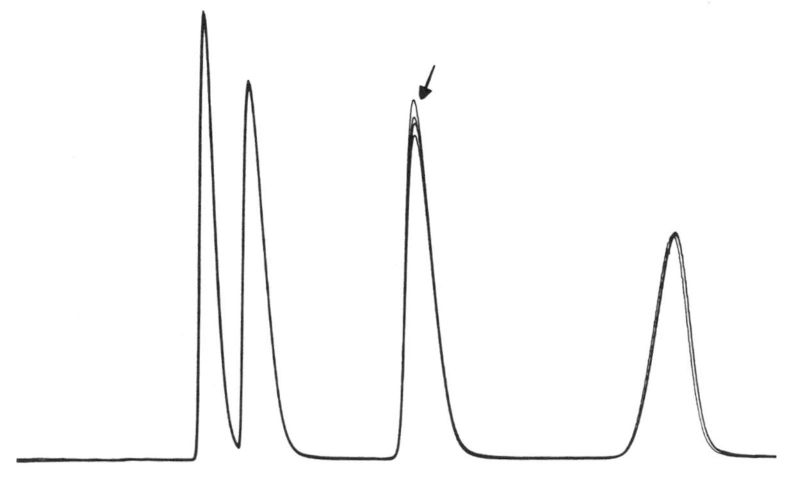

Figure 16.23. A series of chromatograms superimposed to illustrate the method of standard additions in GC. The peak due to the analyte grows as more analyte is added.

16.10 TROUBLESHOOTING

Problems that arise during a GC experiment usually manifest themselves on the chromatogram. Examples of such manifestations are peak shapes being distorted, peak sizes diminishing for reasons other than quantity of analyte, the baseline drifting, or the retention times changing for no apparent reason, etc. These kinds of problems can usually be traced to injection problems, problems with the column, or problems with the detector. There can, of course, be problems associated with the electronics of the instrument. However, we will not be concerned with those here because of the large number of different instrument designs that have been manufactured over the years. The operator can usually find assistance for these in a troubleshooting section of the manual that accompanies the instrument.*

In the following paragraphs we will address some of the most common problems encountered, pinpoint possible causes, and suggest methods of solving the problems.

Diminished Peak Size. We could also refer to this as reduced sensitivity. The peaks are smaller than expected based on previous observations when equal or greater quantities of a particular sample are injected. Such an observation usually means a problem with injection (less injected than assumed) or a problem with the detector such that a smaller electronic signal is sent to the recorder. One should check for a leaky or plugged syringe, a worn septum, a leak in the pre- and post-column connections, or a contaminated detector. Of course, detector attenuation, recorder sensitivity settings, electrical connections, and other associated hardware problems are potential causes.

Unsymmetrical Peak Shapes. Peak "fronting" or peak "tailing" (Figure 16.24) are typical examples of this problem. These could be indicators of poor injections, meaning too large an injection volume for the diameter of

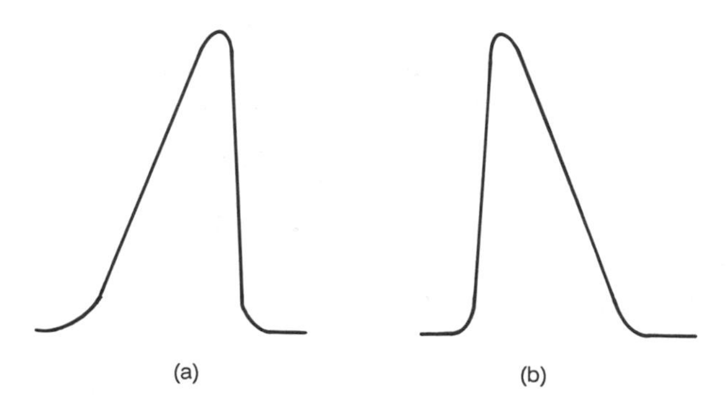

Figure 16.24. (a) A peak exhibiting fronting; (b) a peak exhibiting tailing. (From Kenkel, J., *Analytical Chemistry Refresher Manual*, Lewis Publishers, Chelsea, MI, 1992. With permission.)

*See also the "GC Troubleshooting" column published regularly in the monthly journal *LC·GC, The Magazine of Separation Science.*

the column in use, too slow with the syringe manipulation during injection, or not fully penetrating the septum. It may indicate a decomposition of thermolabile components in contact with the hot system components such as the metal walls of the injection port and column. It may also mean contamination of the injection port and/or column.

Altered Retention Time. This is usually caused by changes in the carrier gas flow rate or column temperature. Flow rate changes can be caused by leaks in the system upstream from the column inlet, such as in the injection port (e.g., the septum), by low pressure in the system due to an empty or nearly empty carrier supply, or by faulty hardware, such as the flow control valve or pressure regulator. Temperature changes can be caused by a faulty temperature controller, an improperly set temperature program, too short a cool-down period prior to the next injection in a temperature-programmed experiment, etc. Altered retention times could also be caused by overloading the column or by diminished effectiveness (decomposition?) of the stationary phase.

Baseline Drift. This occurs when a new column has not been sufficiently conditioned, when the detector temperature has not reached its equilibrium value, or when the detector is contaminated or otherwise faulty. New columns need to be conditioned, usually with an overnight "bakeout" at the highest recommended temperature for that column. Detector signals may very well change when the detector temperature changes. One should be sure that sufficient time has been given for the detector temperature to level off. The nature of detector problems depends of course on the detector. TCD filaments may become oxidized due to an air leak, ionization detectors may be leaking, or there may be a crack in the FID burner nozzle, etc.

Baseline Perturbations. If the perturbations are in the form of spikes of an irregular nature, the problem is likely to be detector contamination. Such spikes are especially observed when dust particles have settled into the FID flame orifice. Of course, the problem may also be due to interference from electrical pulses from some other source nearby. Regular spikes can be due to condensation in the flow lines causing the carrier, or hydrogen (FID), to "pulse", or they can be due to a bubble flow meter attached to the outlet of the TCD as well as the electrical pulses referred to above. These can also be caused by pulses in the carrier flow due to a faulty flow valve or pressure regulator.

Appearance of Unexpected Peaks. Unexpected peaks can arise from components from a previous injection that moved slowly through the column, contamination from either the reagents used to prepare the sample or standards, or from a contaminated septum, carrier, or column. The solution to these problems includes a rapid "bakeout" via temperature programming after the analyte peaks have eluted, using pure reagents, and replacing or cleaning septa, carrier, or column.

EXPERIMENTS

EXPERIMENT 44: A QUALITATIVE GAS CHROMATOGRAPHY ANALYSIS OF A PREPARED SAMPLE

NOTE: Safety glasses are required.

1. Obtain from your instructor an unknown contained in a vial equipped with a rubber system. It contains any number of the following organic liquids: benzene, toluene, chlorobenzene, bromobenzene, ethylbenzene, cyclohexane, and acetone.
2. The purpose of the experiment is to determine which of these liquids are in your unknown. To do this, obtain chromatograms and determine the retention times of the pure liquids. Your instructor has established the instrumental conditions and will demonstrate how to use the instrument to determine the retention times of each. The pure liquids should also be contained in vials equipped with rubber septa.
3. Obtain a chromatogram of your unknown. You may have to adjust the attenuator to get large, well-defined peaks (instructor will demonstrate). Measure the retention times for each peak and compare with your data for the pure liquids. Those whose retention times match your unknown are therefore contained in your unknown.
4. Report what organic liquids are in your unknown.

EXPERIMENT 45: A STUDY OF THE EFFECT OF THE CHANGING OF GC INSTRUMENT PARAMETERS ON RESOLUTION

NOTE: A mixture of two organic liquid is used for this study. The specific liquids will be selected by your instructor. They should be available in a vial equipped with a rubber septum. Your instructor has also selected the initial column packing and length. The two liquids should be in a ratio of approximately 1:1 by volume.

1. Set the carrier gas flow rate for your instrument to 20 mL/min. The method of measuring this flow rate will be demonstrated by your instructor. Set the column temperature to 70°C. Inject 1.0 μL of the mixture and observe the resolution. Now change the column temperature to 80°C, wait 5 min for the oven temperature to become stable, and inject 1.0 μL again, observing the resolution. Repeat at 90°C, 100°C, 110°C, and 120°C, observing the effect on resolution.
2. Set the column oven temperature to 100°C and wait for the temperature to become stable. Set the carrier gas flow rate to 10 mL/min. Inject 1.0 μL of the mixture and observe the resolution. Now increase the carrier gas flow rate by 5 mL/min so as to observe the resolution at 15, 20, 25, 30, and 35 mL/min.
3. Set the carrier gas flow rate to 20 mL/min and obtain a series of chromatograms, each at a different volume injected – 0.5, 1.0, 1.5, 2.0, and 2.5 μL. The attenuation should be set so that the peaks from the larger injection volumes will not be off scale.

4. If a longer column with the same stationary phase is available, change columns. Set the temperature and carrier gas flow rate to some combination of values that gave poor resolution in one of the previous steps. Inject 1.0 μL and observe the effect of a longer column on resolution.
5. Change columns to one with some other stationary phase, perhaps one suggested by your instructor. Set the column temperature to 100°C and the carrier gas flow rate to 20 mL/min. Inject 1.0 μL of the mixture. Assuming good resolution, observe the order of elution. Is it different from that observed with the former stationary phase? If so, explain how that could be. If not, compare the resolution here with that of a previous injection in which all the parameters were equal. Comment on the difference.
6. A numerical value for resolution can be calculated as follows:

$$R = \frac{2(t_2 - t_1)}{(w_1 + w_2)}$$

in which t_2 is the retention time for the component with the longest retention time (component 2), t_1 is that for the other component, and w_1 and w_2 represent the respective widths at the base bisected by the tangents to the sides (see discussion in Section 16.6). As directed by your instructor, calculate resolution for some or all of the above data and construct graphs of R vs. volume injected, R vs. flow rate, and R vs. temperature. Comment on the results.

EXPERIMENT 46: THE QUANTITATIVE GAS CHROMATOGRAPHY ANALYSIS OF A PREPARED SAMPLE FOR BENZENE BY THE INTERNAL STANDARD METHOD

NOTE: Refer to the text for a discussion of the internal standard method. Safety glasses are required. It is best to work with benzene in a good fume hood. Avoid contact between benzene and your skin. All flasks and pipets should be free of water and other solvents.

1. Prepare a series of standard solutions of benzene in toluene (0.5, 1, 2, 3% by volume) in 25-mL volumetric flasks. Add exactly 0.50 mL of ethylbenzene to each flask before diluting to mark with the toluene. The ethylbenzene is the "internal standard". Obtain an unknown, add the internal standard, and dilute with toluene to the mark.
2. Get a good chromatogram of each solution (the instrument settings have been determined by the instructor). You will have to adjust the attenuator setting to get the proper sized peaks. The toluene peak can be allowed to go off scale. If necessary, the instructor will demonstrate the use of a computer for data acquisition.
3. Determine the areas of the ethylbenzene and benzene peaks on each chromatogram and normalize to the same attenuation setting, if necessary, between chromatograms. (Consult instructor for the area measuring technique required.)
4. Plot the ratio of the benzene peak area to the ethylbenzene peak area vs. the benzene concentration and determine the benzene concentra-

tion in the unknown from the graph. A computer may be used for this.

EXPERIMENT 47: THE GAS CHROMATOGRAPHY DETERMINATION OF A GASOLINE COMPONENT BY THE METHOD OF STANDARD ADDITIONS AND THE INTERNAL STANDARD METHOD

NOTE: Refer to the text to refresh your memory concerning the method of standard additions and the internal standard method. Safety glasses are required. Use a good fume hood when preparing the standards. All flasks/pipets should be water-free.

1. Consult your instructor for proper separation conditions. Obtain a chromatogram of the gasoline sample and locate the peak to be identified and quantitated. Identify this peak by matching retention times with some pure samples, or consult your instructor.
2. Prepare three standard solutions in 25-mL volumetric flasks, using the gasoline to be tested as the diluent.
   (a) 0.25 mL of the chosen component (1% addition)
   (b) 0.50 mL of the chosen component (2% addition)
   (c) 0.75 mL of the chosen component (3% addition)
3. Obtain chromatograms of the three standards and also the pure gasoline (0% added). Give the column plenty of time to clear between injections—temperature programming may be useful. A computer may be used to acquire the data.
4. Select another peak that is well resolved from the others and use it as an internal standard. Obtain the peak sizes (consult instructor as to the method), calculate the peak size ratio (analyte to internal standard), and plot peak ratio vs. concentration added. This can also be done on a computer.
5. Extrapolate the line to zero peak ratio and obtain the concentration in pure gasoline.

EXPERIMENT 48: THE DETERMINATION OF DICHLOROMETHANE IN COMMERCIAL PAINT STRIPPERS BY GAS CHROMATOGRAPHY AND THE INTERNAL STANDARD METHOD

NOTE: Refer to the text for a discussion of the internal standard method. Safety glasses are required. CAUTION: Chloroform used in this experiment has been determined to be a carcinogen. It is necessary to avoid contact with the skin and the breathing of its vapor. A fume hood is required. Your instructor may suggest an alternate chemical. All glassware must be water-free.

1. Weigh a dry 25-mL volumetric flask, with a funnel in its mouth, on an analytical balance. Add about 5.0 g of the paint stripper to be tested and weigh the flask again. The weight gain is the sample weight. Rinse the funnel into the flask with toluene.

2. Pipet 2.5 mL of chloroform into the flask and dilute to the mark with toluene. Shake well. Be sure to keep the flask stoppered to discourage evaporation.
3. Prepare a series of standard solutions of dichloromethane in toluene by pipetting 1, 3, 5, and 7 mL of dichloromethane into 25-mL flasks. Before diluting to the mark with toluene, add 2.5 mL of chloroform (the internal standard) to each. Shake well and keep stoppered.
4. Your instructor will establish the instrumental conditions. Inject 1.0 μL of each solution into the chromatograph. Be careful to allow any insoluble residue in the sample flask to settle before filling the syringe. A computer may be used for data acquisition.
5. Obtain the sizes of the dichloromethane and the chloroform peaks in each chromatogram, divide the size of the dichloromethane peak by the size of the chloroform peak in each, and plot this ratio vs. concentration. A computer may be used for this.
6. Obtain the concentration of the dichloromethane in the unknown and calculate the percent dichloromethane in the paint stripper using the following calculation (refer to QUESTIONS AND PROBLEMS #45):

$$\text{Weight \%} = \frac{V_D \times D_D}{W} \times 100$$

in which V_D = volume of dichloromethane in unknown solution, D_D = density of dichloromethane (1.327), and W = weight of the paint remover.

EXPERIMENT 49: THE DETERMINATION OF ETHANOL IN WINE BY GAS CHROMATOGRAPHY AND THE INTERNAL STANDARD METHOD

NOTE: Safety glasses are required.

1. Prepare 4 to 6 standard solutions of ethanol in water such that the alcohol content of the wine (see the label) is in the "middle". For example, if the wine is 15% ethanol, standards of 5, 10, 20, and 25 are appropriate. Use 25-mL volumetric flasks and pipet the ethanol accurately. Also pipet 1.0 mL of acetone (the internal standard) into each flask after diluting to the mark with water. Also, fill a 25 mL flask with the wine and add 1 mL of acetone above the mark.
2. Your instructor will establish the instrumental conditions, including injection volume. Obtain chromatograms for all standards and all wine samples. A computer may be used for data acquisition and peak size measurement.
3. Obtain the areas of the ethanol peaks and acetone peaks, calculate the peak ratio, and plot peak ratio vs. ethanol concentration. A computer may be used for the plotting. Obtain the percent of ethanol in the sample from the graph.

QUESTIONS AND PROBLEMS

1. What do the abbreviations GC, GLC, and GSC refer to?
2. Which "types" of chromatography are applicable to GC?
3. Define carrier gas, column, vapor pressure.
4. What role does vapor pressure play in a GC separation?
5. In a GC experiment in which the liquid stationary phase is polar, which would have a shorter retention time – a nonpolar mixture component with a high vapor pressure or a polar mixture component with a low vapor pressure? Explain.
6. Why is the injection port in a gas chromatograph heated to a relatively high temperature?
7. What is the general role of the detector in GC?
8. There are three heated zones in a GC instrument. Which zones are these are why does each need to be heated?
9. Why must the size of a liquid sample injected into the GC be small? What would happen if it were too large?
10. Study the injection technique given in Table 16.3 and explain why such a procedure would give a reproducible injection volume.
11. What is meant by a GC open-tubular capillary column? Why has the development of such a column been useful?
12. Contrast the packed column and the open-tubular capillary column in terms of design, diameter, and length.
13. Contrast an open-tubular capillary column with a packed column in terms of design, length, separation ability, and amount of sample injected.
14. What is the difference between "analytical GC" and "preparative GC"?
15. What is meant by "temperature programming" in GC?
16. Tell how the temperature programming feature of most modern GCs can be useful in separating complex mixtures.
17. What is Chromosorb®? What is its use in GC?
18. Study Table 16.4 and tell what stationary-phase material would be useful for separating some low-molecular-weight alcohols.
19. Would higher carrier gas flow rates increase or decrease retention time?
20. What is the difference between retention time and adjusted retention time?
21. Consider the separation of components A and B on a GC column. The retention time for component A is 1.40 min and for component B is 2.10 min. The width at base for the peak due to A is 0.38 min and for the peak due to B is 0.53 min. Calculate the resolution, R, and tell whether the two peaks are considered to be resolved and why.

22. Calculate the number of theoretical plates indicated for the data for component A in #21 and also calculate the height equivalent to a theoretical plate given that the column length is 72 in. Comment on the significance of your answers.

23. List three ways to increase the number of theoretical plates of a GC experiment.

24. What does it mean to say that a GC detector is "universal"? What does it mean to say that one GC detector is "more sensitive" and "more selective" than another?

25. Which of the different types of GC detectors
    (a) requires the use of hydrogen gas?
    (b) is well-suited for pesticide residue analysis?
    (c) uses the abilities of the eluting gases to conduct heat?
    (d) is the least sensitive?
    (e) will not work for noncombustible mixture components?
    (f) breaks the eluting molecules into charged fragments, which are then analyzed in terms of charge and mass?
    (g) is part of an extremely powerful (and expensive) system abbreviated GC-MS?
    (h) uses a radioactive source?
    (i) uses an FID with a bead of alkali metal salt positioned just above the flame?
    (j) use an FTIR instrument so as to take an IR spectrum "photograph" of the mixture components as they elute?
    (k) is a detector that is specific for organic compounds of sulfur and phosphorus?
    (l) is a detector which utilizes a UV light beam to ionize the component molecules?
    (m) is a detector that converts eluting molecules into ions in solution so that they can then be detected by electrical conductivity measurements?

26. Which of the GC detectors we have studied
    (a) burn(s) the mixture components in a hydrogen flame?
    (b) is (are) not very sensitive?
    (c) is (are) "universal"?
    (d) is (are) good for pesticide residue analysis?
    (e) utilize(s) a hot filament in the flow stream?
    (f) utilize(s) a source of radioactivity?
    (g) would not detect water?
    (h) do (does) not destroy the mixture components?
    (i) convert(s) eluting molecules into ions?
    (j) make(s) a measurement like a photograph of the eluting molecules?

27. Match the detector with the items that follow.

    thermal conductivity detector   flame ionization detector
    electron capture detector   mass spectrometer detector

    (a) FID

(b) a GC detector good especially for pesticides
(c) a GC detector which utilizes hydrogen gas
(d) a GC detector which utilizes a radioactive source
(e) a GC detector which analyzes fragmented molecules according to their mass and charge
(f) a powerful GC detector which can perform qualitative analysis even without the use of retention times
(g) a GC detector which is based on the differences of the abilities of helium and the mixture components to conduct heat
(h) a GC detector in which noncombustible materials will not give a peak
(i) a GC detector especially good for electrophilic substances
(j) a GC detector which is part of the GC-MS assembly
(k) a universal but not very sensitive GC detector

28. Of these three detectors, thermal conductivity, flame ionization and electron capture, which
    (a) requires the use of hydrogen gas?
    (b) is universally applicable, but not very sensitive?
    (c) does not destroy the sample?
    (d) is good for pesticide analysis?
    (e) can only be used for samples that are able to be burned?
29. What advantages does
    (a) a thermal conductivity detector have over a flame ionization detector?
    (b) a flame ionization detector have over a thermal conductivity detector?
30. Explain the principles by which qualitative analysis can be performed in GC with the use of retention times.
31. A certain sample is a mixture of four organic liquids and these liquids exhibit the following retention times in a GC experiment:

Liquid	Retention Time
A	1.6 min
B	2.2 min
C	4.7 min
D	9.8 min

Some known liquids were injected into the chromatograph and the following data were determined:

Liquid	Retention Time
Benzene	0.5 min
Toluene	1.6 min
Ethylbenzene	3.4 min
n-Propylbenzene	4.7 min
Isopropylbenzene	5.8 min

(a) Can you tell what liquids from these five are definitely not present? If not, why not? If you can tell, which liquids are they?
(b) What liquids are possibly present?

(c) Why can one not tell with certainty what liquids are present, based on the information given here?

32. Fill in the blanks with terms chosen from the following list:

retention time
injection port
resolution
thermal conductivity detector
flame ionization detector
electron capture detector
temperature programming
open-tubular column
column
carrier gas
theoretical plates
preparative GC
peak size
column
mobile phase
stationary phase
detector

(a) Two terms which describe the helium used in gas chromatography are ______________ and ______________.
(b) The measurement that is made when doing a qualitative analysis of a mixture in gas chromatography is ______________.
(c) A type of detector in gas chromatography which is "universal" is the ______________.
(d) A type of detector in gas chromatography which is useful for pesticide analysis is the ______________.
(e) The three heated parts in a gas chromatography instrument are the ________, ________, and ________.
(f) In doing quantitative analysis with gas chromatography, the ________ is an important measurement.
(g) Increasing the length of a GC column is a way of increasing the number of ________ and therefore improving the ________.
(h) Sometimes a difficult separation in gas chromatography can be accomplished easily by changing the temperature of the column during the run. This is called ________.
(i) A gas chromatography instrument can be used to obtain pure samples of the components of a mixture to be used whenever pure samples are needed for some other experiment. This is called ________.
(j) In an experiment in which the ________ is used, one needs to have a source of hydrogen gas.
(k) The ________ has the stationary phase adsorbed directly onto the wall of the column.

33. On the chromatography peak in Figure 16.25, show clearly the manner in which the size is measured by triangulation and also by the half-width method. Label each measurement that is made and write the equations for calculating the area using these labels in your equation. (Use the labels a, b, c, . . . or h, w, ...)

34. An injection of 3.0 μL of methylene chloride (density = 1.327 g/mL) gave a peak size of 3.74 cm^2. The injection of 3.0 μL of an unknown sample (density = 1.174 g/mL) gave a methylene chloride peak size of 1.02 cm^2. Calculate a response factor for methylene chloride and the percent of methylene chloride in the sample.

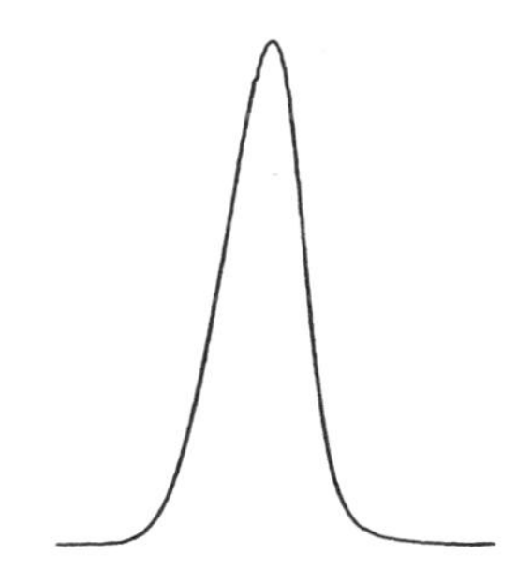

Figure 16.25. A chromatography peak for peak size measurement practice in Problem #33.

35. Compare and/or differentiate between the two items in each of the following (be complete, but concise):
    (a) computer integration vs. peak height methods of measuring peak size
    (b) the internal standard method and the standard additions method
36. Compare the internal standard method with the standard additions method in terms of
    (a) solution preparation.
    (b) what the chromatograms look like.
    (c) what is plotted.
37. Consider the quantitative gas chromatography analysis of alcohol-blended gasoline for ethyl alcohol by the internal standard method, using isopropyl alcohol as the internal standard. The peaks for these two substances are well resolved from each other and from other components. Assume there is no isopropyl alcohol in pure gasohol.
    (a) Describe how you would prepare a series of standard solutions for this analysis.
    (b) What would the chromatograms of these solutions look like?
    (c) What is plotted in an analysis of this type?
38. Select one of the quantitation procedures we have discussed (response factor method, series dilution method, internal standard method, or standard addition method) and describe
    (a) experimental details (solution preparation and/or sample preparation).
    (b) what the raw data are.
    (c) what is plotted.
39. What observation on a chromatogram would lead you to conclude that you injected too much sample for the diameter of column you are using?
40. If a chromatographer observes that a given mixture component gives a smaller peak than in previous work for the same sample and the same injection volume, what might be the problem?
41. What might be the cause of a drifting chromatogram baseline?

42. What can a GC analyst do to solve the problem of unexpected peaks on the chromatogram?

43. The following data were obtained for Experiment 46:

C	Benzene Peak Area	Ethylbenzene Peak Area
0.50	0.539	1.970
1.00	1.050	2.051
2.00	2.184	2.018
3.00	3.308	2.009
Unknown	1.382	2.015

Use the procedure outlined in this experiment to find the percent of benzene in the unknown.

44. In Experiments 47 and 49, why is there no extra calculation at the conclusion of the experiments to get the percent of the components in the samples?

45. Derive and discuss the equation in Step 6 of Experiment 48, especially demonstrating that the units cancel appropriately.

46. Consider the analysis of plant material for a pesticide residue by GC; 2 g of the material are chopped up and placed in a Soxhlet extractor (Chapter 15) and the pesticide quantitatively extracted into an appropriate solvent. Following this, the solvent is evaporated to near dryness; the residue is then diluted to volume in a 25-mL flask. 2.5 μL of this solution and standards are injected in a GC, with the following results:

C (ppm)	Peak Size
5.0	1168
10.0	2170
15.0	3214
20.0	4079
25.0	5392
Sample	3577

What is the ppm of pesticide in the original plant material?

Chapter 17

High-Performance Liquid Chromatography

17.1 INTRODUCTION

Basic Concepts

High-performance liquid chromatography (HPLC) is an instrumental chromatography method in which the mobile phase is a liquid. The principles of liquid chromatography (LC) in general, including "types" or separation mechanisms as well as a brief introduction to HPLC, are presented in Chapter 15. All types of liquid chromatography discussed in that chapter can be utilized in the HPLC configuration. Thus, we have partition (LLC), adsorption (LSC), bonded phase (BPC), ion exchange (IEC and IC), and size exclusion (SEC), including gel permeation (GPC) and gel filtration (GFC), all as commonly used types of HPLC. The instrumental design is based on concepts similar to gas chromatography (GC), especially in terms of the detection and recording schemes following the separation.

Simply stated, HPLC involves the high-pressure flow of a liquid mobile phase through a metal tube (column) containing the stationary phase, with electronic detection of mixture components occurring on the effluent end. The high pressure, often reaching 4000–6000 psi, is derived from a special pulsation-free pump, which will be described. The detection system can be any one of several designs, as with GC, and each of these will be discussed. In addition, special "solvent delivery" systems and injection systems are common and will also be described.

The rise in popularity of HPLC is due in large part to the advantages offered by this technique over the older, non-instrumental, "open column" method described in Chapter 15. The most obvious of these advantages is speed. Separation and quantitation procedures that require hours and sometimes days with the open-column method can be completed in a matter of minutes, or even seconds, with HPLC. Modern column technology and gradient solvent elution systems, which will be described, have contributed significantly to this advantage in that extremely complex samples can be resolved with ease in a very short time.

The basic HPLC system, diagrammed in Figure 17.1, consists of a solvent (mobile phase) reservoir, pump, injection device, column, and detector. The pump draws the mobile phase from the reservoir and pumps it through the column as shown. At the head of the column is the injection device which introduces the sample to the system. On the effluent end, a detector, pictured in Figure 17.1 as a UV absorption detector, detects the sample components and the resulting signal is displayed as peaks on a strip-chart recorder. Besides these basic components, an HPLC unit may be equipped with a gradient programmer (Section 17.2), an auto-sampler, a "guard column" and various in-line filters, and a computing integrator or other data handling system.

Comparisons with GC

There are many similarities between the HPLC configuration and the GC configuration. First, the stationary phase consists of small solid particles packed inside the column, which is similar to "packed" columns in GC. Second, there is an injection device at the head of the column through which the mixture to be separated is introduced into the flowing mobile phase. Third, there is a detection system on the effluent side of the column which generates an electrical signal when something other than the mobile phase elutes.

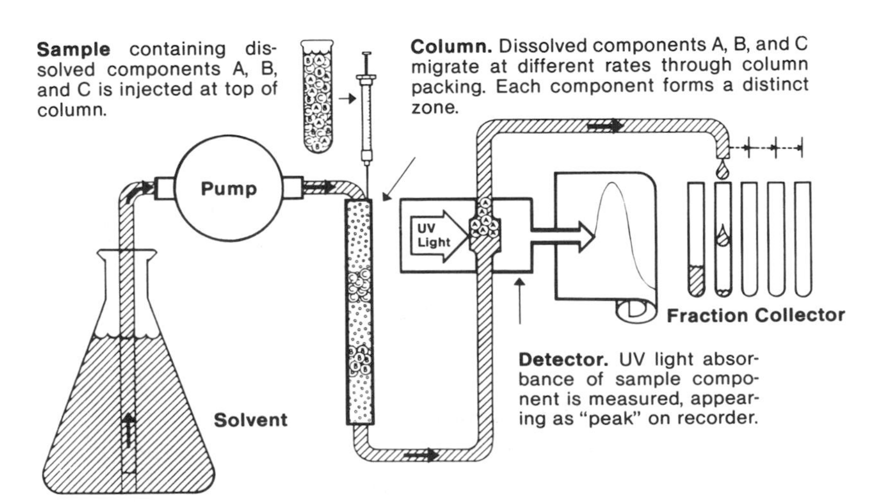

Figure 17.1. A diagram of an HPLC system in which mixture components A, B, and C are separated. (Courtesy of ISCO, Inc., Lincoln, NE.)

Fourth, the electronic signal is fed into a strip-chart recorder where peaks are recorded – a system identical to GC.

Because the mobile phase is a liquid, however, there are also some very obvious differences in the two configurations. First, the mechanism of separation in HPLC involves the specific interaction of the mixture components with a specific mobile phase composition, while in GC the vapor pressure of the components, and not their "interaction" with a specific carrier gas, is the most important consideration (see Chapter 16). Second, the force which sustains the flow of the mobile phase is that of a high-pressure pump rather than the regulated pressure from a compressed gas cylinder. Third, the injection device requires a totally different design due to the high pressure of the system and the possibility that a liquid mobile phase may chemically attack a rubber septum. Fourth, the detector requires a totally different design because the mobile phase is a liquid. Finally, the injector, column, and detector need not be heated as in GC, although the mode of separation occurring in the column can be affected by temperature changes, and thus sometimes elevated column temperatures are used.

Sample and Mobile-Phase Pretreatment

The packed bed of finely divided stationary-phase particles through which the mobile phase percolates is an excellent filter for the mobile phase and injected samples. Particles in the mobile phase as small as 5.0×10^{-6} cm in diameter can be filtered out by the stationary phase. The result of this is a decreased effectiveness of the column with time. In a reasonably short period of time the particles filtered out on the column will (1) mask the stationary phase, preventing the mixture from interacting with it and thus causing poor resolution; and (2) make it necessary to use extremely high pressures in order to get the mobile phase through the column at the prescribed flow rate. The result is a dramatic decrease in the life of the column, an item of significant expense.

The solution to this problem involves the pre-filtering of all mobile phases and samples before beginning the experiment. For mobile phases and large sample volumes, this involves utilizing a vacuum apparatus such as that pictured in Figure 17.2. There are many choices for filter materials. For non-aqueous solvents and their water solutions, paper is not a good choice due to the possibility of chemical attack which may cause contamination. For these, a Teflon®-based or other compatible material should be used. A common Teflon®-based designation is the polytetrafluoroethylene (PTFE) designation. When these are used, if the mobile phase contains some water the filter must be "wetted" first with some pure organic solvent in order to provide a reasonable filtration rate. Aqueous solutions are often impossible to filter unless the Teflon®-based filter is first wetted in this manner. In cases in which only very small amounts of sample (or standard solutions) are available, a small syringe-type filtering unit is used. Here again, the filter must be wetted first if the filter is a Teflon®-based material and the sample is partially or 100% non-aqueous.

In addition to these pre-filtering steps, a series of in-line filters and a "guard column" are often used. The first such filter is at the very beginning of the mobile-phase flow stream in the mobile-phase reservoir, while others

Figure 17.2. A vacuum filtration apparatus for mobile phases and large-volume samples. (Courtesy of Millipore Corporation, Bedford, MA.)

are at other strategic points, such as immediately following the injector. The guard column is usually placed just before the regular "analytical" column. Its function is often to remove not just particles but also other contaminating substances – substances that perhaps have long retention times on the analytical column and that eventually interfere with the detection in later experiments. Guard columns are inexpensive and disposable and are changed frequently.

Another problem is the appearance of air pockets in the HPLC system. If a sample or mobile phase has a significant amount of gas (air) dissolved in it, a pressure drop, which sometimes is experienced in an HPLC line, can cause these gases to withdraw from the solution. The air pockets that result are void spaces that can cause erratic readings from detectors, cause problems in pumps, and decrease the effectiveness of columns. The problem is alleviated by degassing the mobile phase and sample in advance of their entering the system. Some instruments are equipped with degassing units between the mobile-phase reservoirs and the pump. Usually, however, the analyst will degas the mobile phase and samples in a separate experiment in advance by creating a vacuum over the liquid with the use of a vacuum pump and/or agitating the liquid with the use of an ultrasonic bath. A time-saving technique is to filter and degas in one step, since both procedures can involve the use of a vacuum. To do this, the vacuum (receiving) flask can be placed in the ultrasonic bath as the filtration proceeds.

17.2 SOLVENT DELIVERY

Pumps

The pump that is used in HPLC cannot be just any pump. It must be a special pump that is capable of very high pressure in order to pump the mobile phase through the tightly packed stationary phase at a reasonable flow rate, usually between 0.5 and 4.0 mL/min. It also must be nearly free of pulsations so that the flow rate remains even and constant throughout. Only manufacturers of HPLC equipment manufacture such pumps.

There are several pump designs in common use. Probably the most common is the "reciprocating piston" pump shown in Figure 17.3. In this pump a small piston is driven rapidly back and forth, drawing liquid in through the inlet check valve during its backward stroke and expelling the liquid through the outlet check valve during the forward stroke. Check valves allow liquid flow only in one direction. This design is often a "twin piston" design in which a second piston is 180° out of phase with the first. This means that when one piston is in its forward stroke, the other is in its backward stroke. The result is a flow that is free of pulsations. With the single-piston design, a pulse-damping device following the pump is desirable.

The Gradient Programmer

There are two mobile-phase elution methods that are used to elute mixture components from the stationary phase. These are referred to as isocratic elution and gradient elution. Isocratic elution is a method in which a single mobile-phase composition is in use for the entire separation experiment. A different mobile-phase composition can be used, but the transfer to a new composition can only be done by stopping the flow, changing the mobile phase reservoir, and restarting the flow. Gradient elution is a method in which the mobile-phase composition is changed, often gradually, in the middle of the run, analogous to temperature programming in GC.

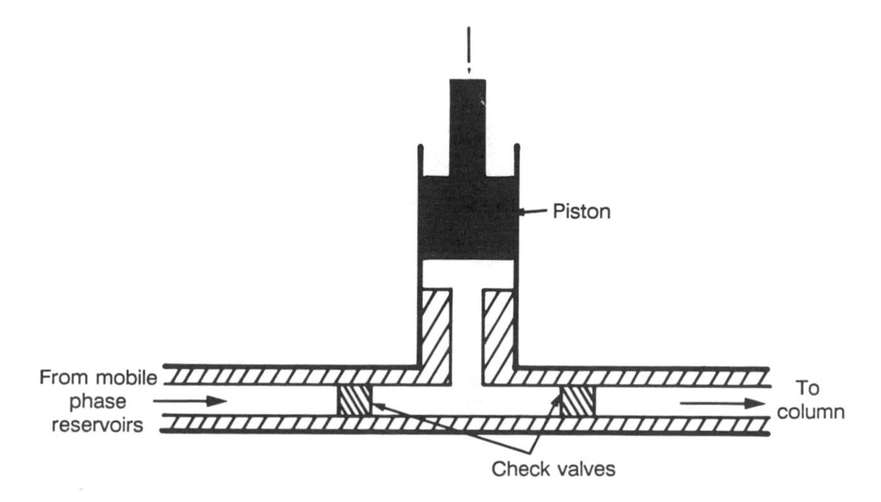

Figure 17.3. A diagram of a reciprocating piston HPLC pump. (From Kenkel, J., *Analytical Chemistry Refresher Manual*, Lewis Publishers, Chelsea, MI, 1992. With permission.)

In any liquid chromatography experiment, the composition of the mobile phase is very important in the entire separation scheme. In Chapter 15 we discussed the role of a liquid mobile phase in terms of the solubility of the mixture components in both phases. Rapidly eluting components are highly soluble in the mobile phase and insoluble in the stationary phase. Slowly eluting components are less soluble in the mobile phase and more soluble in the stationary phase. Retention times, and therefore resolution, can be altered dramatically by a change in the mobile-phase composition. The chromatographer can use this fact to his/her advantage by being able to change the mobile-phase composition in the middle of the run. This is the basis for the gradient elution method.

The gradient programmer is a hardware module used to achieve this goal. The gradient programmer is capable of drawing from at least two mobile-phase reservoirs at once and gradually, in a sequence programmed by the operator in advance, changing the composition of the mobile phase delivered to the HPLC pump. A schematic diagram of this system is shown in Figure 17.4a and a sample "program" is shown in Figure 17.4b.

Solvent "strength" is a designation of the ability of a solvent (mobile phase) to elute mixture components. The greater the solvent strength, the shorter the retention times.

17.3 SAMPLE INJECTION

As mentioned previously, introducing the sample to the flowing mobile phase at the head of the column is a special problem in HPLC due to the high pressure of the system and the fact that the liquid mobile phase may chemically attack the rubber septum. For these reasons, the use of the so-called "loop injector" is the most common method for sample introduction.

The loop injector is a two-position valve which directs the flow of the mobile phase along one of two different paths. One path is a sample loop, which when filled with the sample causes the sample to be swept into the column by the flowing mobile phase. The other path bypasses this loop

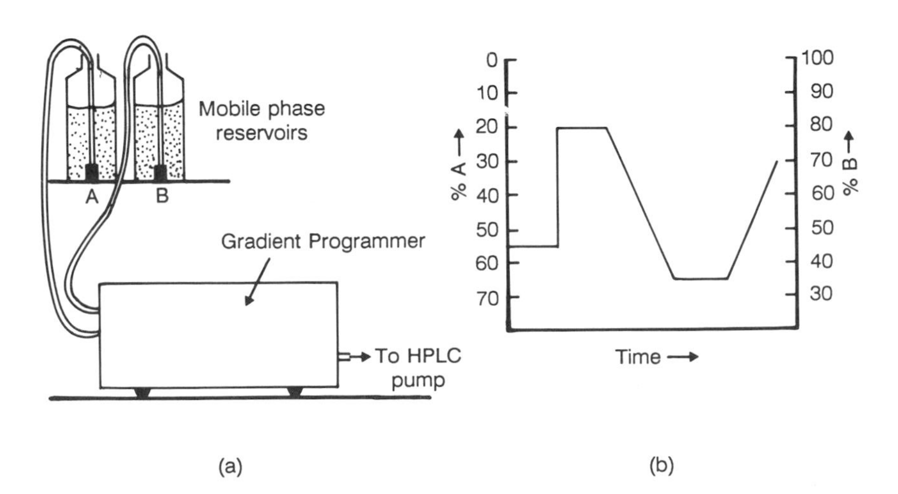

Figure 17.4. (a) A schematic diagram of a gradient programming system; (b) a sample program. (From Kenkel, J., *Analytical Chemistry Refresher Manual*, Lewis Publishers, Chelsea, MI, 1992. With permission.)

while continuing on to the column, leaving the loop vented to the atmosphere and able to be loaded with the sample free of a pressure differential. Figure 17.5 is a diagram of this injector, showing both the "LOAD" position and the "INJECT" position, as well as the flow of the mobile phase in both positions. The sample loop has a particular volume which is of such accuracy that measuring the sample volume with the syringe loader is unnecessary unless volumes smaller than this loop volume are required. This feature aids in the injection of an accurate, reproducible sample volume, which can increase the accuracy of a quantitation.

Automated injectors are often used when large numbers of samples are to be run. Most designs involve the use of the loop injector coupled to a robotic needle which draws the samples from vials arranged in a carousel-type autosampler. Some designs even allow sample preparation schemes such as extraction and derivatization (chemical reactions) to occur prior to injection.

17.4 COLUMN SELECTION

The stationary phases available for HPLC are as numerous as those available for GC. As mentioned previously, however, adsorption, partition, ion exchange, and size exclusion are all LC methods. We can therefore classify the stationary phases according to which of these four types of chromatography they represent. Additionally, partition HPLC, which is the most common, is further classified as "normal-phase" HPLC or "reverse-phase" HPLC. Both of these come under the broader heading of "bonded-phase" chromatography, which was described in Chapter 15. Let us begin with these.

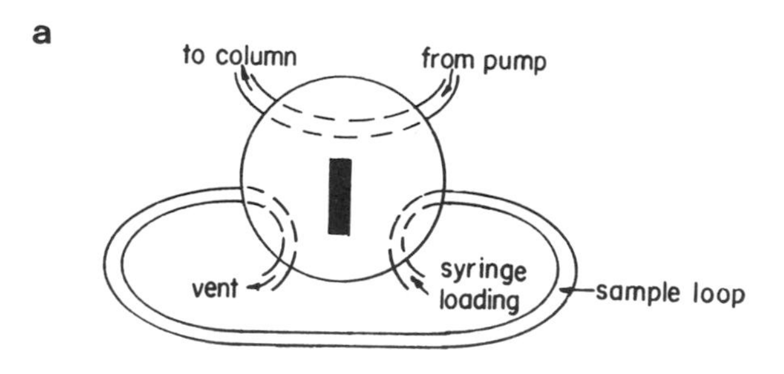

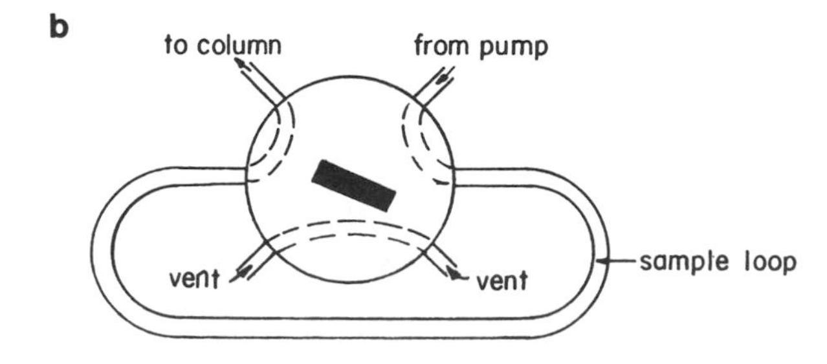

Figure 17.5. The loop injector for HPLC. (a) The "LOAD" position – the sample is loaded into the loop via a syringe at atmospheric pressure; (b) the "INJECT" position – the mobile phase sweeps the contents of the loop onto the column.

Normal-Phase Columns

Normal-phase HPLC consists of methods which utilize a non-polar mobile phase in combination with a polar stationary phase. Adsorption HPLC actually fits this description, too, since the adsorbing solid stationary-phase particles are very polar. (See discussion of adsorption columns in this section.) Normal-phase partition chromatography makes use of a polar liquid phase chemically bonded to these polar particles, which typically consist of silica (Si–O–) bonding sites. Typical examples of normal-phase bonded phases are those in which a cyano group (–CN), an amino group ($-NH_2$), or a diol group ($-CHOH-CH_2OH$) is part of the structure of the bonded phase. Designation of such structural features is often given in the manufacturer's names. Some examples of these are Chromegabond® DIOL, LiChrosorb® DIOL, MicroPak®-CN, μBondapak®-CN, Nucleosil®-NH_2, and Zorbax®-NH_2. Typical mobile phases for normal-phase HPLC are hexane, cyclohexane, carbon tetrachloride, chloroform, benzene, and toluene.

Reverse-Phase Columns

Reverse-phase HPLC describes methods which utilize a polar mobile phase in combination with a non-polar stationary phase. As stated above, the non-polar stationary-phase structure is a bonded phase – a structure that is chemically bonded to the silica particles. Here, typical column names often have the carbon number designation indicating the length of a carbon chain to which the non-polar nature is attributed. Typical designations are C_8 or C_{18} (or ODS, meaning "octadecyl"), etc. Some of these and other examples of reverse-phase stationary phases are Partisil ODS-2, μBondapak C_{18}, Spherisorb ODS, μBondapak Phenyl, Hyposil-SAS, and Nucleosil C-8. Common mobile-phase liquids are water, methanol, acetonitrile (CH_3CN), and acetic acid-buffered solutions.

Adsorption Columns

Adsorption HPLC is the classification in which the highly polar silica particles are exposed (no adsorbed or bonded liquid phase). Aluminum oxide particles fit this description, too, and are also readily available as the stationary phase. As mentioned earlier, this classification can also be thought of as normal-phase chromatography, but LSC rather than LLC. Typical normal-phase mobile phases (non-polar) are used here. The stationary-phase particles can be irregular, regular, or "pellicular", in which a solid core, such as a glass bead, is used to support a solid porous material. Examples of this classification are shown in Table 17.1.

Ion Exchange and Size Exclusion Columns

As discussed in Chapter 15, ion-exchange stationary phases consist of solid resin particles which have positive and/or negative ionic bonding sites on their surfaces at which ions are exchanged with the mobile phase (see Figure 15.10). Cation exchange resins have negative sites so that cations are exchanged, while anion exchange resins have positive sites at which anions are exchanged. Typical examples are given in Table 17.2. A popular modern name for HPLC ion exchange is simply "ion chromatography". Detection of ions eluting from the HPLC column has posed special problems which are

Table 17.1. Some adsorption HPLC stationary phases.

Names	Types
Partisil	Silica, irregular
Hypersil	Silica, regular
Chromasep PAA	Alumina, irregular
Spherisorb A-Y	Alumina, regular
Corasil I and II	Silica, pellicular
Pellumina HS	Alumina, pellicular
Zipax	Silica, pellicular

described in Section 17.6. The mobile phase for ion chromatography is always a pH-buffered water solution.

Size exclusion columns, as discussed in Chapter 15, separate mixture components on the basis of size by the interaction of the molecules with various pore sizes on the surfaces of porous polymeric particles. Size exclusion chromatography is subdivided into two classifications, gel permeation chromatography (GPC) and gel filtration chromatography (GFC). GPC utilizes nonpolar organic mobile phases, such as tetrahydrofuran (THF), trichlorobenzene, toluene, and chloroform, to analyze for organic polymers such as polystyrene. GFC utilizes mobile phases that are water-based solutions and is used to analyze for naturally occurring polymers, such as proteins and nucleic acids. GPC stationary phases are rigid gels, such as silica gel, whereas GFC stationary phases are soft gels, such as Sephadex®. Neither technique utilizes gradient elution because the stationary-phase pore sizes are sensitive to mobile-phase changes.

Column Selection

Since each type of HPLC just discussed utilizes a different separation mechanism, the selection of a specific column packing (stationary phase) depends on whether or not the planned separation is possible or logical with a given mechanism. For example, if a given mixture consists of different molecules all of approximately the same size, then size exclusion chromatography will not work. If a mixture consists only of ions, then ion chromatography is the logical choice. While the conclusions drawn from these examples are obvious, others are less obvious and require a study of the variables and the mechanisms in order to be able to logically choose a particular stationary phase.

Table 17.2. Some ion-exchange chromatography stationary phases.

Names	Type
Ion-X-SC	Cation
Partisil 10 SCX	Cation
Amberlite IR-120	Cation
Dowex 50W	Cation
Ion-X-SA	Anion
Partisil 10 SAX	Anion
Amberlite IRA-400	Anion
Dowex 1	Anion

Table 17.3 presents some guidelines about each choice which would be helpful in deciding which to use. While these guidelines may prove helpful as a starting point, additional facts about the planned separation need to be determined in order to select the most appropriate chromatographic system, including facts that can only be discovered through experimentation or by searching the chemical literature. Several different mobile phase/stationary phase systems may work. Comparing reverse phase with normal phase, for example, one can see that there would only be a reversal in the order of elution. Polar components would elute first with reverse phase, whereas nonpolar components would elute first with normal phase. Experimenting with various mobile-phase compositions, which may include a mixture of two or three solvents in various ratios, would be a logical starting point. Some considerations which would involve such experimentation are

1. The mixture components should have a relatively high affinity for the stationary phase compared to the mobile phase. This would mean longer retention times and thus probably better resolution.
2. The various separation parameters should be adjusted to provide optimum resolution. These include mobile-phase flow rate, stationary-phase particle size, gradient elution, and column temperature (using an optional column oven).
3. Use partition chromatography for highly polar mixtures and adsorption chromatography for very non-polar mixtures.

17.5 THE CHROMATOGRAM

As with GC, the chart recording which presents the written record of the separation is called the chromatogram. Please refer to Section 16.6 in the previous chapter for a brief related discussion and for definitions of retention time, t_R, adjusted retention time, t'_R, resolution, R, the number of theoretical plates, N, and the height equivalent to a theoretical plate, H. In the HPLC definition of t'_R, the reference substance is not air but the sample

Table 17.3. Summary of applications of the different types of HPLC.

Type	Useful for Components which
Normal and reverse phase	Have a low formula weight (<2000) Are non-ionic Are either polar or non-polar Are water or organic soluble
Adsorption	Have a low formula weight (<2000) Are non-polar Are organic soluble
Ion exchange	Have a low formula weight (<2000) Are ionic Are water soluble
Size exclusion	Have a high or low formula weight Are non-ionic Are water or organic soluble

From Kenkel, J., *Analytical Chemistry Refresher Manual*, Lewis Publishers, Chelsea, MI, 1992. With permission.

solvent, which usually gives a slight perturbation to the baseline at a very short retention time as it emerges from the column, depending on the type of detector used. Obviously, if t'_R is an important measurement in a given experiment, then the sample solvent must be different from the mobile phase so that this slight perturbation will be observed.

In addition to these parameters, liquid chromatographers are also concerned with the capacity factor, k′, and selectivity, α. The capacity factor is the adjusted retention time divided by the retention time of the solvent, t_M.

$$k' = t'_R/t_M \tag{17.1}$$

The capacity factor is a measure of the retention of a component per column volume. The greater the capacity factor, the longer that component is retained and the better the chances for good resolution. This, however, must be weighed against the high speed advantage of HPLC. While a large capacity factor is desirable, the experiment should be completed within about 15 min. An optimum range for k′ values is between 2 and 6.

Example 1

What is the capacity factor if the retention time for the component of interest is 3.2 min and the retention time of the sample solvent is 0.70 min?

Solution

$$k' = \frac{t'_R}{t_M} = \frac{(t_R - t_M)}{t_M} = \frac{(3.2 - 0.70)}{0.70} = 3.6$$

Selectivity, α, the retention of one component relative to another, also known as relative retention, is defined as the ratio of the adjusted retention time for one component to the adjusted retention time for another, the latter having a shorter retention time

$$\alpha = \frac{t'_R(A)}{t'_R(B)} \tag{17.2}$$

and is a measure of the "quality" of a separation. Selectivity values greater than about 1.2 are considered good. A selectivity equal to 1 would mean that the two retention times are equal, which means no separation has occurred at all.

Example 2

The retention time for a mixture component "A" is 5.4 min and the retention time for a mixture component "B" is 3.3 min. The retention time for the solvent is 1.1 min. What is the selectivity for component "A" relative to "B"?

Solution

$$\alpha = \frac{t'_R(A)}{t'_R(B)} = \frac{[t_R(A) - t_M]}{[t_R(B) - t_M]} = \frac{(5.4 - 1.1)}{(3.3 - 1.1)} = 2.0$$

17.6 DETECTORS

The function of the HPLC detector is, of course, to examine the solution that elutes from the column and output an electronic signal proportional to the concentrations of individual components present there. In Chapter 16 we discussed a number of detector designs that serve this same purpose for gas chromatography. The designs of the HPLC detectors, however, are more "conventional" in the sense that components present in a liquid solution can be determined with conventional instruments, such as a spectrophotometer. Thus, spectrophotometric and fluorometric detectors are common. Let us discuss some of the more popular HPLC detectors individually.

UV Absorption

The UV absorption HPLC detector is basically a UV spectrophotometer that is capable of measuring a flowing solution rather than a static solution. It has a light source, a wavelength selector, and a phototube as does an ordinary spectrophotometer. The sample compartment, however, is equipped with a flow cell, through which the column effluent flows, and the absorbance is monitored continuously. (See Figure 17.6.) The output of the phototube is sent to the recorder or integrator where the absorbance is continuously displayed with time. Peaks are recorded as the UV-absorbing components elute from the column.

The monochromator can be one of two different designs, either a simple light filter (a so-called fixed-wavelength detector) or a full-blown slit/dispersing element/slit monochromator (a variable wavelength detector), which has a control on the face of the detector for dialing in the wavelength as in a standard spectrophotometer. The fixed wavelength detector can be made to be variable in the sense that the light filter can be changed, but one cannot tune to the wavelength of maximum absorbance, and thus some sensitivity can be lost. The fixed wavelength version, however, is less expensive and is fine for many applications for which UV absorbance detection is appropriate. Filters for 254 nm and 280 nm are common.

While such a detector is fairly sensitive, it is not universally applicable. The mixture components being measured must absorb light in the UV region and, at least in the fixed wavelength design, they must absorb at the wavelength used for a particular experiment in order for a peak to appear on the recorder. Also, the mobile phase must not absorb an appreciable amount at the selected wavelength.

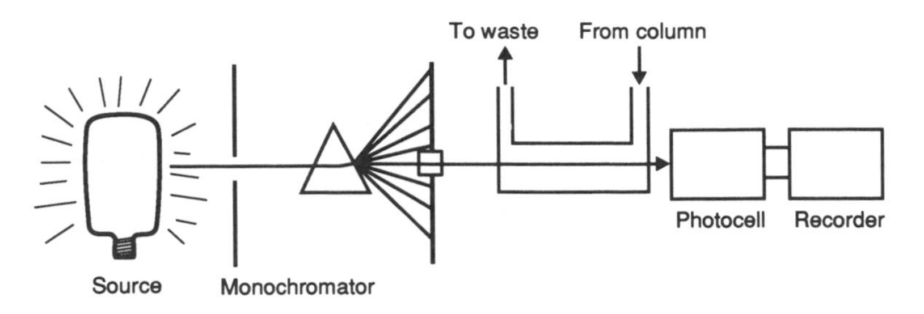

Figure 17.6. A drawing of a UV absorbance detector.

Diode Array

A diode array UV detector is a "multichannel" detector in which the light beam from the UV source is not dispersed into its component wavelengths until after it has passed through the flow cell. The dispersed light then sprays across an array of photodiodes, each of which detects only a narrow wavelength band. With the help of a computer, the entire UV absorption spectrum can be immediately measured as each individual component elutes. With computer banks containing a library of UV absorption spectral information, a rapid, definitive qualitative analysis is possible in a manner similar to GC-MS or GC-IR (Chapter 16). In addition, the peak displayed on the recorder/integrator can be the result of a rapid changeover of the wavelength by the computer. Thus, the peaks displayed can represent the maximum possible sensitivity for each component. Finally, a diode array detector can be used to "clean up" a chromatogram so as to only display the peaks of interest. This is possible since we can rapidly change the wavelength giving rise to the peaks.

Fluorescence

The basic theory, principles, sensitivity, and application of fluorescence spectrometry (fluorometry) were discussed in Chapter 11. Like the UV absorption detector described above, the HPLC fluorescence detector is based on the design and application of its parent instrument, in this case the fluorometer, and you should review Chapter 11 for more information about the fundamentals of the fluorescence technique.

In summary, the basic fluorometer, and thus the basic fluorescence detector, consists of a light source and a monochromator (usually a filter) for creating and isolating a desired wavelength, a sample "compartment", and a second monochromator (another filter) with a phototube detector for isolating and measuring the fluorescence wavelength. The second monochromator and detector are lined up perpendicular to the light beam from the source (the so-called "right angle" configuration).

As with the UV absorption detector, the sample compartment consists of a special cell for measuring a flowing, rather than static, solution. The fluorescence detector thus individually measures the fluorescence intensities of the mixture components as they elute from the column (see Figure 17.7). The electronic signal generated at the phototube is sent to the recorder or integrator where a peak is recorded each time a fluorescing species elutes.

The advantages and disadvantages of the fluorometry technique in general hold true here. The fluorescence detector is not universal (it will give a peak only for fluorescing species), but it is thus very selective (i.e., almost no possibility for interference) and very sensitive.

Refractive Index

The refractive index of a liquid or liquid solution is defined as the ratio of the speed of light in a vacuum to the speed of light in the liquid:

$$n = c_{vac} / c_{liq} \tag{17.3}$$

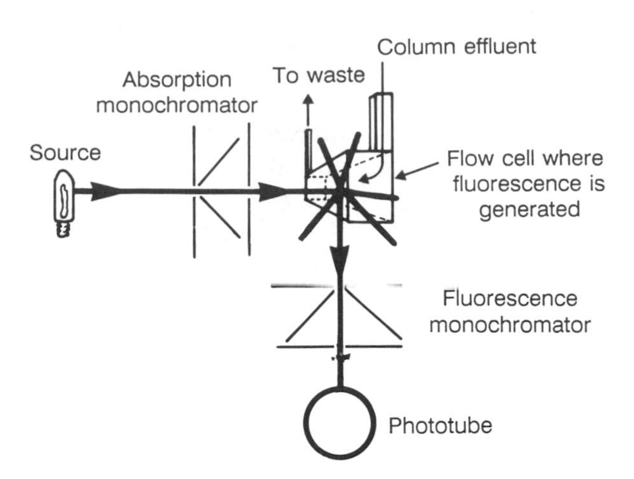

Figure 17.7. An illustration of a fluorescence detector. (From Kenkel, J., *Analytical Chemistry Refresher Manual*, Lewis Publishers, Chelsea, MI, 1992. With permission.)

Since the speed of light in any material medium is less than the speed of light in a vacuum, the numerical value of the refractive index for any liquid is greater than 1.

An instrument known as a refractometer has been invented and used for many years to measure the refractive index of liquids and liquid solutions for the purpose of both quantitative and qualitative analysis. A refractometer measures the degree of refraction (or "bending") of a light beam passing through a thin film of the liquid. This refraction occurs when the speed of light is different from a reference liquid or air. The refractometer measures the position of the light beam relative to the reference and is calibrated directly in refractive index values. It is rare for any two liquids to have the same refractive index, and thus this instrument has been used successfully for qualitative analyses.

The refractive index detector in HPLC is a modification of this basic instrument and actually can be purchased in two different designs, depending on the manufacturer. In probably the most popular design, both the column effluent and the pure mobile phase (acting as a reference) pass through adjacent flow cells in the detector. A light beam passing through both cells is focused onto a photosensitive surface, and the location of the beam when both cells contain pure mobile phase is taken as the reference point and the recorder pen is zeroed. When a mixture component elutes, the refractive index in one cell changes; the light beam is "bent" and becomes focused onto a different point on the photosensitive surface, causing the recorder pen to deflect and trace a peak. (See Figure 17.8.)

The major advantage of this detector is that it is almost universal. All substances have their own characteristic refractive index (it is a physical property of the substance). Thus, the only time that a mixture component would not give a peak is when it has a refractive index equal to that of the mobile phase, a rare occurrence. The disadvantages are that it is not very sensitive and the output to the recorder is subject to temperature effects. Also, it is difficult to use this detector with the gradient elution method because it is sensitive to changes in the mobile-phase composition.

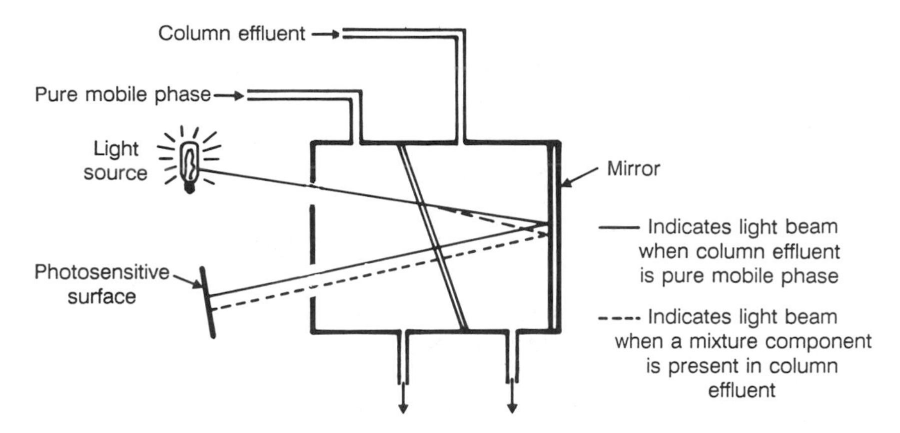

Figure 17.8. A representation of a refractive index detector. See text for description. (From Kenkel, J., *Analytical Chemistry Refresher Manual*, Lewis Publishers, Chelsea, MI, 1992. With permission.)

Electrochemical

Various detectors which utilize electrical current or conductivity measurements for detecting eluting-mixture components have been invented. These are called electrochemical detectors. Let us now examine some of the basic designs.

Conductivity. Perhaps the most important of all electrochemical detection schemes currently in use is the electrical conductivity detector. This detector is specifically useful for ion exchange, or ion, chromatography, in which the analyte is in ionic form. Such ions elute from the column and need to be detected as peaks on the recorder trace.

A well-known fact of fundamental solution science is that the presence of ions in any solution gives the solution a low electrical resistance and the ability to conduct an electrical current. The absence of ions means that the solution would not be conductive. Thus, solutions of ionic compounds and acids, especially strong acids, have a low electrical resistance and are conductive. This means that if a pair of conductive surfaces is immersed into the solution and connected to an electrical power source, such as a simple battery, a current can be detected flowing in the circuit. Alternatively, if the resistance of the solution between the electrodes is measured (with an ohmmeter), it would be low. Conductivity cells based on this simple design are in common use in non-chromatography applications to determine the quality of deionized water, for example. Deionized water should have no ions dissolved in it and thus should have a very low conductivity. The conductivity detector is based on this simple apparatus.

For many years the concept of the conductivity detector could not work, however. Ion chromatography experiments utilize solutions of high ion concentration as the mobile phase. Thus, changes in conductivity due to eluting ions were not detectable above the already high conductivity of the mobile phase. This was true until the invention of so-called ion "suppressors". Today, conductivity detectors are used extensively in HPLC ion chromatography instruments which also include suppressors.

A suppressor is a short column (tube) that is inserted into the flow stream just after the analytical column. It is packed with an ion exchange resin itself, a resin that removes mobile-phase ions from the effluent, much like a deionizing cartridge removes the ions in laboratory tap water, and replaces them with molecular species. A popular "mixed bed" ion exchange resin is used, for example, in deionizing cartridges, such that tap water ions are exchanged for H^+ and OH^- ions, which in turn react to form water. The resulting water is thus deionized. Of course, in the HPLC experiment the analyte ions must *not* be removed in this process, and thus suppressors must be selective only for the mobile-phase ions.

A typical design for a conductivity detector uses electrically isolated inlet and outlet tubes as the electrodes. This design is shown schematically in Figure 17.9.

Amperometric. A thorough discussion of electroanalytical techniques, including "polarography", "voltammetry", and "amperometry", is given in Chapter 18. An understanding of these would be useful for understanding the amperometric HPLC detector.

Electrochemical oxidation and/or reduction of eluting mixture components is the basis for amperometric electrochemical detectors. The three electrodes needed for the detection, the working ("indicator") electrode, reference electrode, and auxiliary electrode, are either inserted into the flow stream or imbedded in the wall of the flow stream. (See Figure 17.10.) The indicator electrode is typically glassy carbon, platinum, or gold, the reference electrode a silver/silver chloride electrode, and the auxiliary a stainless steel electrode. Most often, the indicator electrode is polarized so as to cause oxidation of the mixture components as they elute. The oxidation current is then measured and constitutes the signal sent to the recorder/integrator.

Advantages of this detector include broad applicability to both ionic mixture components as well as molecular components, as long as they are able to be oxidized (or reduced) at fairly small voltage polarizations. Selectivity can be improved by varying the potential. In addition, the sensitivity experienced with this detector is quite good – generally better than the UV detector, but not as good as the fluorescence detector. A disadvantage is that the indicator electrode can become fouled due to products of the electrochemical reaction coating the electrode surface. Thus, this detector must be

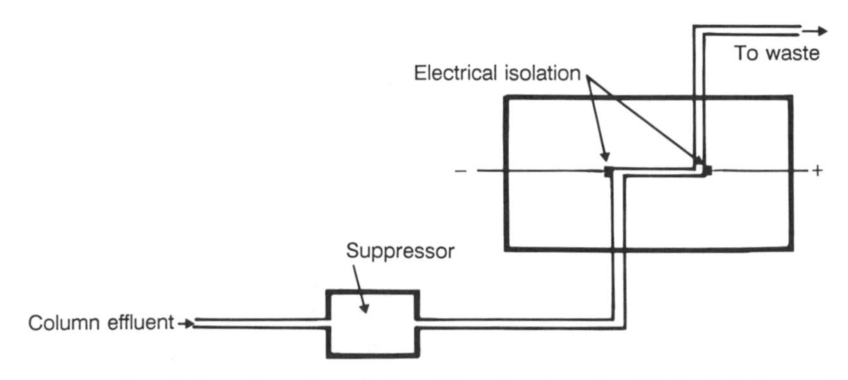

Figure 17.9. A drawing of a conductivity detector in which the inlet and outlet tubes are the electrodes. (From Kenkel, J., *Analytical Chemistry Refresher Manual*, Lewis Publishers, Chelsea, MI, 1992. With permission.)

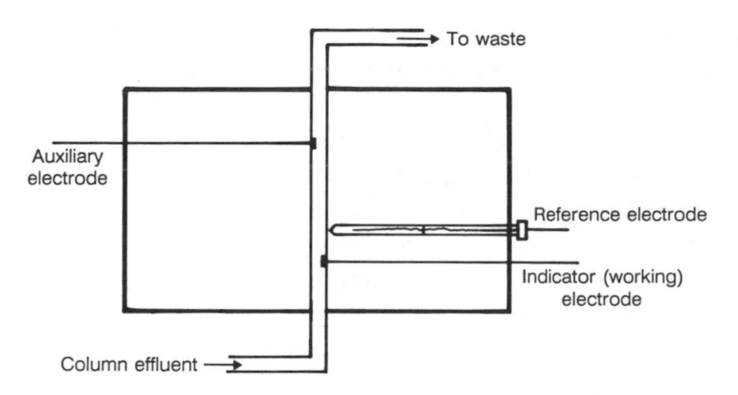

Figure 17.10. A drawing of an amperometric HPLC detector. (From Kenkel, J., *Analytical Chemistry Refresher Manual*, Lewis Publishers, Chelsea, MI, 1992. With permission.)

able to be disassembled and cleaned with relative ease, since this may need to be done frequently.

LC-MS and LC-IR

In Chapter 16, the use of mass spectrometry and FTIR for GC detection was discussed. Details of these techniques were individually given in Chapters 12 and 13. Much of the discussion presented in Chapter 16 is applicable here. Both mass spectrometry and infrared spectrometry have been adapted to HPLC detection in recent years.

FTIR is a "natural" for HPLC in that it (FTIR) is a technique that has been used mostly for liquids. The speed introduced by the Fourier transform technique allows, as was mentioned for GC, the recording of the complete IR spectrum of mixture components as they elute, thus allowing the IR "photograph" to be taken and interpreted for qualitative analysis. Of course, the mobile phase (and its accompanying absorptions) is ever-present in such a technique and water must be absent if the NaCl windows are used, but IR holds great potential, at least for non-aqueous systems, as a detector for HPLC in the future.

The mass spectrometer is also incompatible with the HPLC system, but for a different reason. The ordinary mass spectrometer operates under very low pressure (a high vacuum – see Chapter 13), and thus the liquid detection path must rapidly convert from a very high pressure and large liquid volume to a very low pressure and a gaseous state. Several approaches to this problem have been used, but probably the most popular is the "thermospray" (TS) technique. In this technique, the column effluent is converted to a fine mist (spray) as it passes through a small-diameter heated nozzle. The analyte molecules, which must be thermally stable, are preionized with the presence of a dissolved salt. A portion of the spray is introduced into the mass spectrometer. The analyte and mobile phase must be polar if the TS technique is used because the mobile phase must dissolve the required salt and the components must interact with the analyte molecule.

17.7 QUALITATIVE AND QUANTITATIVE ANALYSIS

Qualitative and quantitative analysis with HPLC are very similar to that with GC (Chapter 16, Sections 16.8 and 16.9). In the absence of diode array, mass spectrometric, and FTIR detectors which give additional identification information, qualitative analysis depends solely on retention time data, t_R and t'_R. (Remember that t'_R here is the time from when the solvent front is evident to the peak). Under a given set of HPLC conditions, namely the mobile- and stationary-phase compositions, the mobile-phase flow rate, the column length, temperature (when the optional column oven is used), and instrument dead volume, the retention time is a particular value for each component. It changes only when one of the above parameters changes. Refer to Section 16.8 for further discussion of qualitative analysis.

Peak size measurement and quantitation methods outlined for GC in Section 16.9 are also applicable here. The reproducibility of the amount injected is not nearly the problem with HPLC as it is with GC. Roughly 10 times more sample is typically injected (5–20 μL), and there is no loss during the injection since the sample is not loaded into a higher-pressure system through a septum. In addition, the sample loop is manufactured to have a particular volume and is often the means by which a consistent amount is injected, which means reproducibility is maximized through the consistent "overfilling" of the loop via the injection syringe. In this way, the loop is assured of being filled at each injection and a reproducible volume is always introduced. Sometimes, however, the analyst chooses to inject varying volumes of a single standard to generate the standard curve (Chapter 9) rather than equal volumes of a series of standard solutions. In this case, the injection syringe is used to measure the volumes – a less accurate method, but better than an identical method with GC since the sample volume is larger and there is less chance for sample loss. With this type of quantitation, the standard curve is a plot of peak size vs. amount injected rather than concentration.

The most popular quantitation "method", then, is the series of standard solutions method with no internal standard (i.e., "serial dilution" – see Chapter 9) or the variable injection of a single standard solution as outlined above.

17.8 TROUBLESHOOTING

Problems that arise with HPLC experiments are usually associated with abnormally high or low pressures, system leaks, worn injectors parts, air bubbles, and/or blocked in-line filters. Sometimes these manifest themselves on the chromatogram and sometimes they don't. In the following paragraphs we will address some of the most common problems encountered, pinpoint possible causes, and suggest methods of solving the problems.*

Unusually High Pressure. A common cause of unusually high pressure is a plugged in-line filter. In-line filters are found at the very beginning of the flow line in the mobile-phase reservoir, immediately before and/or after the injector and just ahead of the column. With time, they can become plugged

*See also the "LC Troubleshooting" column published regularly in the monthly journal *GC·LC, The Magazine of Separation Science.*

due to particles that are filtered out (particles can appear in the mobile phase and sample even if they were filtered ahead of time), and thus the pressure required to sustain a given flow rate can become quite high. The solution to this problem is to back-flush the filters with solvent and/or clean them with a nitric acid solution in an ultrasonic bath.

Other causes of unusually high pressure are an injector blockage, mismatched mobile and stationary phases, and a flow rate that is simply too high. An injector that is left in a position between "LOAD" and "INJECT" can also cause a high pressure, since the pump is pumping but there can be no flow.

Unusually Low Pressure. A sustained flow that is accompanied by low pressure may be indicative of a leak in the system. All joints should be checked for leaks (see below).

System Leaks. Leaks can occur within the pump, at the injector, at various fittings and joints, such as at the column, and in the detector. Leaks within the pump can be due to failure of pump seals and diaphragms, as well as loose fittings, such as at the check valves, etc. Leaky fittings should be checked for mismatched or stripped ferrules and threads, or perhaps they simply need tightening. Leaks in the injector can be due to a plugged internal line or other system blockage, gasket failure, loose connections, or use of the wrong size of syringe if the leak occurs as the sample is loaded. Detector leaks are most often due to a bad gasket seal or a broken flow cell. Of course, loose or damaged fittings and a blockage in the flow line beyond the detector are possible causes.

Air Bubbles. An air pocket in the pump can cause low or no pressure or flow, erratic pressure, and changes in retention time data. It may be necessary to bleed air from the pump or to prime the pump according to system start-up procedures. Air pockets in the column will mean decreased contact with the stationary phase and thus shorter retention times and decreased resolution. Tailing and peak splitting on the chromatogram may also occur due to air in the column. Air bubbles in the detector flow cell are usually manifested on the chromatogram as small spikes due to the periodic interruption of the light beam (e.g., in a UV absorbance detector). Increasing the flow rate, or restricting and then releasing the post-detector flow, so as to increase the pressure, should cause such bubbles to be "blown" out.

Column "Channeling". If the column packing becomes separated and a channel is formed in the stationary phase, the tailing and splitting of peaks will be observed on the chromatogram. In this case, the column needs to be replaced.

Recorder Zero vs. Detector Zero. Both the detector and the recorder have "zeroing" capability. The recorder zero control is used when there is zero recorder input, such as when the input terminals are shorted. The detector zero control is used to zero the recorder pen when only the mobile phase is eluting. The two can be "mismatched". This problem is obvious on the chromatogram when changing the attenuation setting on the detector. At a less sensitive attenuation the detector output may appear to be zero, but when the attenuation is changed to a more sensitive setting there may be a sudden jump in the baseline, indicating that the detector output is actually not zero since the more sensitive setting is able to show a small offset from zero. The solution to this problem is to initially zero the detector at the more sensitive setting.

Decreased Retention Time. When retention times of mixture components decrease, there may be problems with either the mobile or stationary phases. It may be that the mobile-phase composition was not restored after a gradient elution, or it may be that the stationary phase was altered due to irreversible adsorption of mixture components or simply chemical decomposition. Use of guard columns (see previous discussion) may avoid stationary-phase problems.

Baseline Drift. A common cause of baseline drift is a slow elution of adsorbed substances on the column. A column cleanup procedure may be in order, or the column may need to be replaced. This problem also may be caused by temperature effects in the detector. Refractive index detectors are especially vulnerable to this. In addition, a contaminated detector can cause drift. The solution here may be to disassemble and clean the detector.

You are also referred to the troubleshooting guide in Chapter 16 (GC) for possible solutions to problems.

EXPERIMENTS

EXPERIMENT 50: A STUDY OF THE EFFECTS OF MOBILE-PHASE COMPOSITION ON RESOLUTION IN REVERSE-PHASE HPLC

Introduction

The composition of the mobile phase in HPLC plays a key role in the success of the separation. In normal- and reverse-phase experiments, for example, the relative solubilities of the mixtures components in both the stationary and mobile phases dictate the extent of the separation. Mixture components that are highly soluble in the stationary phase but not highly soluble in the mobile phase will have a long retention time. In contrast, mixture components that are highly soluble in the mobile phase but not highly soluble in the stationary phase will have a short retention time. Because of the dependency of solubility on molecular polarities, the relative polarities of the mixture components as compared to the mobile and stationary phases are obviously important.

In this experiment, a mixture of methyl, propyl, and butyl paraben (structures shown in Figure 17.11) will be separated by reverse-phase HPLC. Mobile-phase compositions of varying polarities will be studied and their effects on resolution noted.

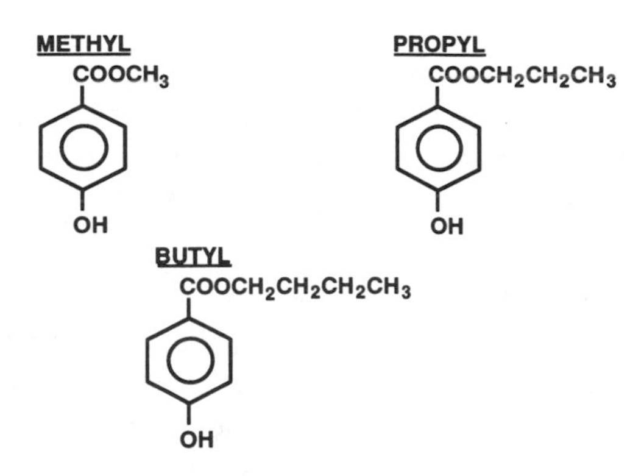

Figure 17.11. The structures of methyl, propyl, and butyl paraben.

Procedure

NOTE: Safety glasses are required.

1. Mobile-phase compositions for this experiment are polar methanol/water mixtures in the ratios 80/20, 70/30, and 60/40 by volume. The stationary phase is a reverse-phase nonpolar material. Prepare 200 mL of each mobile phase and then filter and degas each through 0.45-μm filters with the aid of a vacuum. (Instructor will demonstrate.)
2. Filter the sample with the aid of the syringe filter kit. (Instructor will demonstrate.) The sample is a solution of methyl, propyl, and butyl paraben in methyl alcohol.
3. Separate the mixture using the HPLC instrument repeating the run three times, each with a different mobile phase (beginning with the 80/20 ratio), and observe the effect of changing the mobile-phase composition on the resolution. Be sure to allow each new mobile phase to completely purge the system before injecting. The flow rate should be 1.5 mL/min. Your instructor will demonstrate and discuss the use of the instrument.
4. Optional – Calculate "R", "k′" and "α" values according to your instructor's directions and interpret the results.

EXPERIMENT 51: THE QUANTITATIVE DETERMINATION OF METHYL PARABEN IN A PREPARED SAMPLE BY HPLC

Introduction

In this experiment, the concentration of methyl paraben in the sample used in Experiment 50 will be determined. A series of standard solutions of methyl paraben in methyl alcohol will be prepared. The standard curve will be constructed by injecting equal volumes of each standard by "overfilling" the injector loop as discussed in the text.

Procedure

1. This experiment utilizes the same stationary phase as Experiment 50. Select the mobile phase from Experiment 50 which gave the best resolution and, if it is not already in the system, allow it to purge the system at 1.5 mL/min for 8 min.
2. The instructor has prepared a stock standard solution of methyl paraben that has a concentration of 2 mg/mL. From this stock solution prepare four standard solutions having concentrations of 0.05, 0.1, 0.15, and 0.2 mg/mL in 50-mL volumetric flasks. Filter these as you did the unknown when you performed Experiment 50.
3. Inject full sample loop volumes of each standard solution and the unknown and obtain a chromatogram for each. A computer may be used for data acquisition.
4. Repeat for the unknown sample from Experiment 50 and obtain the size of the methyl paraben peak on the resulting chromatogram.
5. Obtain the size of the lone peak found on the chromatogram for each standard, perhaps with the help of a computer or computing integrator, and then obtain a plot of peak size vs. concentration. From this plot obtain the concentration of the unknown.

EXPERIMENT 52: HPLC DETERMINATION OF CAFFEINE AND SODIUM BENZOATE IN SODA POP

NOTE: Safety glasses are required.

1. Prepare 500 mL of mobile phase (1.0 *M* acetic acid in 10% acetonitrile) as follows. Dilute 50 mL of acetonitrile and 28.5 mL of glacial acetic acid to 500 mL with water. Pour into a large beaker and place on a magnetic stirrer for pH adjustment. With a pH meter and a magnetic stirrer, adjust the pH to 3.0 by adding successive small amounts of a saturated solution of sodium acetate. Filter and degas as in Experiment 50 or as directed by your instructor. Prepare the instrument for use by flushing the system with this mobile phase for 8 min at 2.0 mL/min.
2. Prepare soda pop samples by filtering through paper 0.45-μm filters and degas as with the mobile phase.
3. Prepare 50 mL of a standard solution that is 0.20 mg/mL in caffeine and 0.50 mg/mL in sodium benzoate. Use distilled water as the diluent. Filter using the syringe filter kit.
4. Obtain chromatograms of the standard by injecting 5, 10, 15, and 20 μL. A computer may be used for data acquisition.
5. Obtain chromatograms of the samples, injecting a full 20 μL of each. Allow the column to completely clear before making another injection.
6. When finished, flush the system with a filtered and degassed neutral pH liquid, such as pure methanol or a methanol/water mixture.
7. Obtain peak sizes, graph the results, and obtain the concentrations of caffeine, and benzoate in the samples as follows:

$$C_u = C_s \times \frac{\text{amount injected (from graph)}}{20}$$

Refer to QUESTIONS AND PROBLEMS #53. A computer may be used for the graphing.

8. Calculate the mg of caffeine and benzoate present in one 12 oz. can of the soda pop. There are 0.02957 L/fl. oz.

EXPERIMENT 53: THE QUANTITATIVE ANALYSIS OF CHILE PEPPER EXTRACT FOR CAPSAICIN

NOTE: This experiment utilizes the extract solution from Experiment 43, so Experiment 43 must be performed first.

1. Filter part of the extract solution, using the syringe filter kit (0.45-μ filter cartridge), into a clean vial. This filtered extract is the sample to be injected in Step 4 below.
2. Accurately prepare 100 mL of a 500 ppm stock standard solution of capsaicin (8-methyl-*N*-vanillyl-6-noneamide) in tetrahydrofuran. Prepare 25 mL each of four standards in the range 0–100 ppm from this stock. Filter each into small vials as you did the sample in Step 1.
3. Prepare 300 mL of a 70:30 methanol:water mobile phase and filter and degas as in previous experiments. Flush the HPLC system (C_{18} reverse phase column) for 8 min at 2 mL/min with this mobile phase.
4. Obtain chromatograms for all standards and sample, injecting 20 μL of each. Your instructor may suggest that a microcomputer or another data station be used for data acquisition and analysis. Obtain the peak sizes for the capsaicin peak on each chromatogram and plot peak size vs. concentration. Obtain the concentration of the pepper extract from the graph.
5. Calculate the ppm capsaicin in the chile pepper.

EXPERIMENT 54: THE ANALYSIS OF MOUTHWASH BY HPLC – A RESEARCH EXPERIMENT

Introduction

Various brands of mouthwash have a variety of components that should potentially be able to be determined by HPLC. In this experiment, you are "on your own" to determine what component to analyze for, what mobile and stationary phases to use, what flow rate to use, what concentrations to use, etc.

Procedure

1. Visit your local pharmacy (outside of lab time) and examine the labels on various brands of mouthwash that are on the shelf. Select for laboratory analysis a brand that looks interesting and bring it into the laboratory.
2. Filter and degas a part of the sample. Prepare the instrument as you've done before, choosing a particular stationary- and mobile-phase system (such as a reverse-phase system using a methanol/water

mixture for the mobile phase and a nonpolar stationary phase) and flow rate that you will use as a first trial.

3. Inject the sample and observe the separation of peaks. If there is at least one peak that is resolved, proceed to Step 4. Otherwise, try changing the composition of the mobile phase and the flow rate in order to achieve good resolution of at least one peak. If you are still unsuccessful, change the stationary phase and try again.
4. Prepare and filter standard solutions of several of the components shown on the label. The concentrations should be reasonable guesses of what might match what is in the sample. Inject each individually and observe the retention time for each component. Check to see if one of these retention times matches the retention time for a resolved peak from the sample.
5. If you get a match of retention times, proceed to quantitate, comparing the peak size of the sample to that of the standard.
6. Optional: Just because the retention times match does not necessarily mean that you have identified the peak. To be sure, change the stationary phase and inject again. If a sample peak is again resolved from the others (and it is the same size), and if the retention time of the same standard peak matches the resolved peak, you can be sure that you have identified the peak and your quantitation in Step 5 is valid.

QUESTIONS AND PROBLEMS

1. "HPLC" stands for ________________________________.
2. Define LC, LLC, LSC, BPC, IEC, IC, SEC, GPC, and GFC.
3. Give a simple but total definition of HPLC and describe with some detail the basic HPLC system.
4. Why is HPLC an improvement over the "open column" technique?
5. Compare HPLC with GC in terms of
   (a) the force which moves the mobile phase through the stationary phase.
   (b) the nature of the mobile phase.
   (c) how the stationary phase is held in place.
   (d) what "types" of chromatography are applicable.
   (e) application of vapor pressure concepts.
   (f) sample injection.
   (g) mechanisms of separation.
   (h) detection systems.
   (i) recording systems.
   (j) data obtained.
6. Why must mobile phases and all samples and standards be finely filtered before an HPLC experiment?
7. Why is it important to know whether the mobile phase and samples contain an organic solvent when preparing to filter them?

8. For what types of samples and mobile phases is Teflon®-based filter material appropriate?
9. Explain the use of a "guard column" and "in-line filters".
10. Why must HPLC mobile phases and samples be degassed prior to use in the HPLC system?
11. Explain what "degassing" is and how it is accomplished.
12. Why is it that no ordinary liquid pump can be used as the pump in an HPLC system?
13. Describe what is meant by a reciprocating piston pump.
14. What is a gradient programmer?
15. Define the gradient elution method for HPLC, tell what instrument component is needed for it, and tell how this method is useful.
16. Distinguish between isocratic elution and gradient elution.
17. How is gradient elution HPLC similar to temperature programming in GC?
18. Why is it that a change in the mobile phase in the middle of the run will change the retention and resolution of the mixture components?
19. What is meant by "solvent strength"?
20. Give two reasons why an injection system similar to that in GC would not work for HPLC.
21. Describe in detail how the "loop injector" works and tell how it overcomes the problems that would be encountered with an injection port/septum system.
22. Show by means of a diagram the difference between the "LOAD" and "INJECT" positions of the HPLC injection system.
23. Distinguish normal-phase HPLC from reverse-phase HPLC.
24. If a given HPLC system is using a methanol/water mixture for the mobile phase and a C_{18} column for the stationary phase, what classification of chromatography would be in use? Explain.
25. List some typical mobile and stationary phases for (a) reverse-phase HPLC and (b) normal-phase HPLC.
26. What is meant by "bonded phase" chromatography? Would such a name describe normal phase, reverse phase, neither, or both? Explain.
27. Name two traditional stationary phases for adsorption chromatography. Are they polar or non-polar?
28. Distinguish between cation exchange resins and anion exchange resins in terms of the nature of the charged sites on the surface of the particles and in terms of application.
29. What is "ion chromatography" and what is a typical mobile-phase composition for ion chromatography?
30. Distinguish between gel permeation chromatography and gel filtration chromatography in terms of mobile phases that are used and application.

31. What type of HPLC should be chosen for each of the following separation applications?
    (a) All mixture components have formula weights less than 2000, are molecular and polar, and are soluble in non-polar organic solvents.
    (b) Mixture components have formula weights varying from very large to rather small and are non-ionic.
    (c) Mixture components have formula weights less than 2000, are molecular and polar, and are water soluble.
32. Explain why the order of elution of polar and non-polar mixture components would be reversed when switching from normal phase to reverse phase.
33. What are some HPLC system parameters that can be altered in an attempt to improve resolution?
34. Calculate the capacity factor when the retention time of a mixture component is 4.51 min and the retention time of the sample solvent is 0.95 min. Is it within the optimum range? Explain.
35. Calculate the selectivity for propyl paraben compared to methyl paraben given the following data:

    t_R (methyl paraben) = 3.23 min

    t_R (propyl paraben) = 5.16 min

    t_M = 0.87 min

    Is this a good separation? Explain.
36. Calculate the capacity factor for the two mixture components in #35.
37. Compare and/or differentiate between the two items in each of the following (be complete but concise):
    (a) the isocratic elution technique and the gradient elution technique
    (b) normal-phase chromatography and reverse-phase chromatography
    (c) advantages and disadvantages of the UV absorbance detector vs. the refractive index detector
38. Distinguish between the fixed-wavelength UV detector and the variable-wavelength UV detector in terms of design and use.
39. What is a diode array detector and what are its advantages?
40. In two words or less for each, give one advantage and one disadvantage for each of the following:
    (a) UV absorbance detector
    (b) refractive index detector
    (c) fluorescence detector
    (d) conductivity detector
    (e) amperometric detector
    (f) LC-MS
    (g) LC-IR

41. Why is it that a UV absorption detector in HPLC is not universal?
42. Discuss the advantages and disadvantages of any three HPLC detectors discussed in the text.
43. Which of the four HPLC detectors (UV absorbance, refractive index, fluorescence, or conductivity)
    (a) is the most sensitive?
    (b) is the most universal?
    (c) requires an ion suppressor to eliminate the ions present in the mobile phase?
    (d) is a popular detector because it is "almost" universal and very sensitive?
    (e) gives a signal based on the position of a light beam on the detector?
    (f) has a right angle configuration design?
    (g) has a single monochromator that can be either a glass filter or a slit/disperser/slit type?
    (h) is used frequently for "ion chromatography"?
44. In Chapter 16 we discussed the need to calculate response factors specifically when a TCD detector is used (Section 16.7). Would response factors need to be calculated in HPLC when a UV absorbance detector is used? Explain.
45. What is a "suppressor" and why is one needed in an ion-exchange HPLC experiment in which the mobile phase contains ions?
46. Why is FTIR considered a "natural" as an HPLC detector? What problem does the presence of the mobile phase pose, especially if water is present?
47. Discuss qualitative and quantitative analysis methods for HPLC and how they are different from those of GC.
48. What is the most common cause of an unusually high pressure in an HPLC flow stream and how is the problem solved?
49. What symptoms appear if air bubbles enter the HPLC flow line?
50. The splitting of a peak into two peaks is a symptom of what problem in the column? How is it solved?
51. Name some causes of baseline drift in HPLC.
52. Derive and discuss the calculation used to determine the concentration of capsaicin in the chile peppers in Step 5 of Experiment 53, especially demonstrating that the units cancel appropriately.
53. Refer to the calculation in Step 7 of Experiment 52. Derive and discuss this equation, especially demonstrating that the units cancel appropriately.

54. Consider the analysis of a soda pop sample for caffeine by the standard additions method. Construct a graph from the following data and report the milligrams of caffeine in one 12-oz. can of the soda pop.

C (ppm Added)	Peak Size
0	1368
5	1919
10	2431
15	2997

(Calculation hint – there are 0.02957 L/fl. oz.)

55. The concentration of furazolidone additive in livestock feed can be determined using HPLC. The furazolidone is extracted from the feed using a water/dimethylformamide solution. The sample is then filtered and injected into the chromatograph. If 9.7186 g of the feed were weighed and 55 mL of water/dimethylformamide were added for the extraction, what is the percent of furazolidone in the sample given the following?

C (ppm Added)	Peak Size
2.0	978
10.0	4621
30.0	14017
50.0	21071
70.0	28994

Chapter 18

Electroanalytical Methods

18.1 INTRODUCTION

The subject of electroanalytical chemistry encompasses all analytical techniques which are based on electrode potential and current measurements at the surfaces of electrodes immersed in the solution tested. Either an electrical current flowing between a pair of immersed electrodes or an electrical potential developed between a pair of immersed electrodes is measured and related to the concentration of some dissolved species.

Electroanalytical techniques are an extension of classical oxidation-reduction chemistry; indeed, oxidation and reduction processes occur at or within the two electrodes, oxidation at one and reduction at the other. Electrons are consumed by the reduction process at one electrode and generated by the oxidation process at the other. The complete system is often called a "cell", the individual electrodes "half-cells", and the individual oxidation and reduction reactions are the "half-reactions". Electrons flow on a conductor between the half-cells, and this flow constitutes the electrical current that is often measured. A "galvanic" cell is one in which this current flows spontaneously because of the strong tendency for the chemical species involved to give and take electrons. A battery that has its positive and negative poles externally connected is an example of a such a cell. An "electrolytic" cell is one in which the current is not a spontaneous current, but rather is the result of connecting an external power source, such as a battery, to the system. A rechargeable battery when it is positioned in the recharging unit would be an example of such a cell. Electroanalytical techniques utilize both general types of cells. Let us begin by studying the relationship between these concepts and oxidation-reduction chemistry.

18.2 TRANSFER OF ELECTRONS VIA ELECTRODES

The redox reactions we studied in Chapter 8 involve the direct transfer of electrons from one species to another, presumably as a result of collision between the two, or at least a close approach of one to the other. Representing an extremely important physicochemical phenomenon, however, is the fact that the electrons can be made to travel through another medium in order to get from one species to the other. As a result, not only is it possible for there to be no collision or close approach of the two species at all, but they can actually be placed in two completely separate containers. The applications of this are truly significant. First, an electrical current (the flow of electrons through a conductor), from which electrical energy can be derived, results. Second, it represents a way to convert chemical energy (the driving force of chemical reactions) to electrical energy, which in turn can be converted to mechanical energy for operating a wide variety of energy-consuming devices. The important aspect is that this is not a dream waiting to be realized. The conversion of chemical to mechanical energy in the manner just described has been around for some time. It occurs each time a battery is put to use. Inside a battery a redox process is occurring, and the electrons flow from one pole through the energy-consuming device to the other pole.

A battery is nothing more than a device which consists of two separate containers (known as "half-cells") in which two redox half-reactions occur: the oxidation half-reaction, which is the source of the electrons, and the reduction half-reaction, which is the ultimate destination of the electrons. A battery can be easily constructed in the laboratory. Consider for example the following oxidation-reduction reaction:

$$Cu + Ag^{+1} \rightarrow Ag + Cu^{+2}$$

The half-reactions are

$$\text{Oxidation: } Cu \rightarrow Cu^{+2} + 2e^-$$

$$\text{Reduction: } Ag^{+1} + 1e^- \rightarrow Ag$$

This reaction occurs readily by simply placing a strip of copper metal into a solution containing silver ions. After a short time, silver metal can be seen plating on the copper and copper ions can be found in the solution, the result of a direct collision of the Ag^{+1} ion with the Cu strip.

Let us now separate these two half-reactions into two half-cells, as shown in Figure 18.1, so that the chemical species found in the oxidation half-reaction are found in the beaker (half-cell) on the right and those for the reduction half-reaction are found in the half-cell on the left.

The pathway for the electrons to travel from one half-cell to the other is provided by connecting the two metal strips ("electrodes") with an electrical conductor, such as copper wire. In addition, there must be some form of contact between the two solutions to allow the diffusion of ions between the two half-cells. Such diffusion is required for the current to flow and represents the final step here in forming a completed circuit. The contact can be made using a so-called "salt bridge", which is a glass (or other) tube filled with a solution of some inert ions, such as K^{+1} and Cl^{-1}. The complete ar-

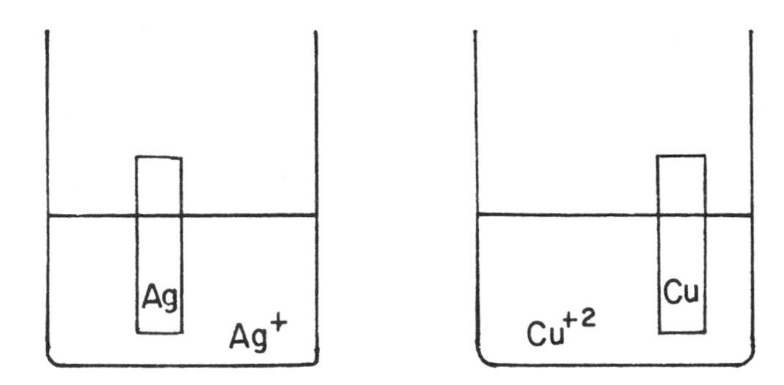

Figure 18.1. Two "half-cells".

rangement, including a switch and current-measuring device in the external circuit, is shown in Figure 18.2.

With the switch closed, electrons spontaneously flow from the copper strip to the silver strip as a result of copper atoms losing electrons to become copper ions and the silver ions accepting electrons to become silver atoms. Thus, we have a redox reaction occurring – not by the collision of two species in a solution, but through the transfer of electrons through an external conductor. The flowing electrons represent an electrical current which can be used to drive battery-driven devices.

18.3 TRANSFER TENDENCIES – STANDARD REDUCTION POTENTIALS

All of us are well aware of the need in everyday life for batteries of various voltages. For example, an ordinary flashlight requires "D"-size batteries, which are 1.5-volt batteries. Many calculators require the small 9-volt battery. Most automobiles require a relatively large 12-volt battery. An interesting question is, "What determines the voltage of a battery?"

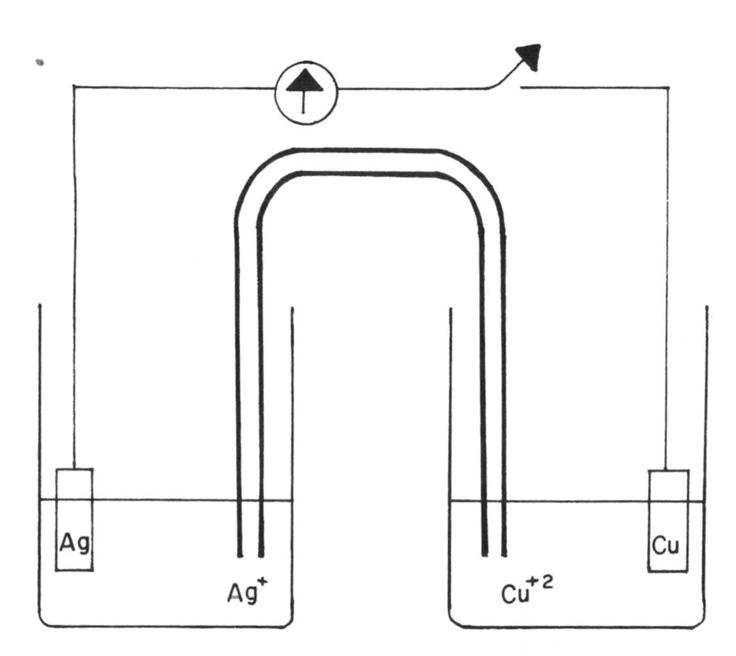

Figure 18.2. A laboratory "battery".

You will recall that in the previous section we referred to the "driving force" of a chemical reaction as a force representing the amount of chemical energy capable of being converted to electrical energy in a battery. It should be obvious that the amount of this driving force is based on the relative tendency for the reaction to proceed and specifically the relative tendency of the two half-reactions to proceed. What do we mean by "tendency"?

Consider the three metals sodium, magnesium, and aluminum. A sodium atom has a very strong tendency to give up one electron to become a sodium ion. The tendency is so strong, in fact, that the reaction of sodium with ordinary water borders on explosive! Magnesium also has a relatively strong tendency to give up its two outermost electrons, but much less than that of sodium. Fresh magnesium metal reacts with water, but only very slowly. Aluminum metal has an even lesser tendency to lose electrons, even though it does lose electrons to become the aluminum ion. Its reactivity with water, however, is almost nil. Thus, it has an even lesser tendency to lose its electrons. The same illustration can be made with the non-metals and their tendency to take on electrons. Obviously, the driving force of different redox half-reactions represents different reaction tendencies. As we shall see quantitatively in the next section, if an oxidation half-reaction of high tendency is coupled with a reduction half-reaction of high tendency, the result is a battery of relatively high voltage. In other words, the voltage of a battery is dependent upon the relative tendencies of the two half-reactions of which it is composed to proceed.

Half-reaction tendencies are very well catalogued. In the first place, the activity series of metals (Table 18.1) can be considered to be an ordered listing of some half-reaction tendencies. Those metals at the top (Na, K, etc.) dissolve in water, losing an electron. Metals further down the list (Zn, Fe, etc.) require weak acids for oxidation; HCl is required for oxidation with metals such as Ni, Sn, etc., and still further down the list (Cu, Mg, etc.) nitric acid is required for oxidation. Thus, those oxidation half-reactions of high tendency are at the top of the list, and the tendency decreases as one goes down the list (reread Section 2.4 for a related topic).

In a modern table of reaction tendencies, which can include all conceivable reduction half-reactions, numbers are assigned to all reduction half-reactions. These numbers, referred to as the standard reduction potentials, represent relative reaction tendencies. This means that one particular half-reaction is chosen as the basis for the table (much like the carbon-12 isotope is the basis for the table of relative atomic weights) and assigned a number, and all others then have a value representing the degree of tendency, lesser or greater than this standard.

Table 18.1. The activity series of metals.

Li > K > Ba > Sr > Ca > Na	Dissolve in water
Mg > Al > Mn > Zn > Cr > Fe > Cd	Dissolve in weak acids
Co > Ni > Sn > Pb	Dissolve in HCl
Sb > As > Bi > Cu > Ag > Hg	Dissolve in HNO_3
Pt > Au	Dissolve in "aqua regia"

Note: The ">" symbols refer to the levels of reactivity of the elements shown; the reactivity of lithium is greater than that of potassium, which is greater than that of barium, etc.

The standard in this case is the reduction of hydrogen ions to form hydrogen gas. It is assigned a standard reduction potential of zero. Table 18.2 is a table of some standard reduction potentials. The value of the standard reduction potential is given the symbol E^0 (read E-zero).

It is interesting to compare this table with the activity series of metals (Table 18.1). First, Table 18.2 seems to be inverted. The very active metals,

Table 18.2. Standard reduction potentials for selected half-reactions.

Half-Reactions	E^0 (volts)
$F_2 + 2e^- \rightarrow 2F^-$	+2.87
$O_3 + 4H^+ + 2e^- \rightarrow O_2 + 2H_2O$	+2.08
$H_2O_2 + 2H^+ + 2e^- \rightarrow 2H_2O$	+1.78
$Ce^{+4} + 1e^- \rightarrow Ce^{+3}$	+1.72
$2BrO_3^{-1} + 12H^+ + 10e^- \rightarrow Br_2 + 6H_2O$	+1.48
$MnO_4^- + 8H^+ + 5e^- \rightarrow Mn^{+2} + 4H_2O$	+1.51
$Cl_2 + 2e^- \rightarrow 2Cl^-$	+1.36
$Cr_2O_7^{-2} + 14\ H^+ + 6e^- \rightarrow 2Cr^{+3} + 7H_2O$	+1.23
$O_2 + 4H^+ + 4e^- \rightarrow 2H_2O$	+1.23
$Fe(phenanthroline)_3^{+3} + 1e^- \rightarrow Fe\ (phenanthroline)_3^{+2}$	+1.15
$Br_2 + 2e^- \rightarrow 2Br^-$	+1.09
$NO_3^{-1} + 3H^+ + 2e^- \rightarrow HNO_2 + H_2O$	+0.93
$Hg^{+2} + 2e^- \rightarrow Hg$	+0.85
$Ag^+ + 1e^- \rightarrow Ag$	+0.80
$Fe^{+3} + 1e^- \rightarrow Fe^{+2}$	+0.77
$MnO_4^{-1} + H_2O + 3e^- \rightarrow MnO_2 + OH^-$	+0.60
$I_2 + 2e^- \rightarrow 2I^-$	+0.54
$Cu^{+2} + 2e^- \rightarrow Cu$	+0.34
$Hg_2Cl_2 + 2e^- \rightarrow 2Hg + 2Cl^-$ (SCE)	+0.27
$AgCl + 1e^- \rightarrow Ag + Cl^-$ (Ag/AgCl Ref.)	+0.22
$Sn^{+4} + 2e^- \rightarrow Sn^{+2}$	+0.15
$S_4O_6^{-2} + 2e^- \rightarrow 2S_2O_3^{-2}$	+0.08
$2H^+ + 2e^- \rightarrow H_2$	0.00
$Fe^{+3} + 3e^- \rightarrow Fe$	-0.04
$Sn^{+2} + 2e^- \rightarrow Sn$	-0.14
$Ni^{+2} + 2e^- \rightarrow Ni$	-0.26
$Fe^{+2} + 2e^- \rightarrow Fe$	-0.45
$S + 2e^- \rightarrow S^{-2}$	-0.48
$Cr^{+3} + 3e^- \rightarrow Cr$	-0.74
$Zn^{+2} + 2e^- \rightarrow Zn$	-0.76
$Mn^{+2} + 2e^- \rightarrow Mn$	-1.19
$Be^{+2} + 2e^- \rightarrow Be$	-1.85
$Mg^{+2} + 2e^- \rightarrow Mg$	-2.37
$Na^+ + 1e^- \rightarrow Na$	-2.71
$K^+ + 1e^- \rightarrow K$	-2.93

From *Handbook of Chemistry and Physics*, 72nd ed., CRC Press, Boca Raton, FL, 1992. With permission.

Na, K, etc., are at the bottom in Table 18.2. The reason for this is the half-reactions are written in reverse—that is, as a reduction rather than an oxidation. Obviously, if these metals have a tendency to oxidize (lose electrons), their ions would not have a strong tendency to reduce. In other words, the ion is stable, the atom is not. Hence, if the tendency for the ion to take on an electron is low, the half-reaction is at the bottom of the table. The table is a listing of reduction half-reactions only. Thus, reduction half-reactions that have a strong tendency to occur are at the top of the list, while those that have a lesser tendency to occur are at the bottom of the list. In fact, those half-reactions in the bottom half of the table tend to go in the reverse direction. They all have a negative standard reduction potential. Another way to read the table is to say that if the reduced form in a half-reaction at the bottom of the table comes into contact with the oxidized form in a half-reaction above it, the reaction will occur. Stated in reverse, if the oxidized form in any half-reaction high in the table comes into contact with the reduced form in a half-reaction below it, a reaction will proceed. Mnemonic schemes (schemes which aid in memorizing) for this are the arrows shown in Figure 18.3a. Thus, MnO_4^{-1} would react with $H_2C_2O_4$ to form CO_2, and $H_2C_2O_4$ reacts with MnO_4^{-1} to form Mn^{+2}, as shown in Figure 18.3b.

The more the two half-reactions are separated, the greater is the tendency for this overall reaction to occur.

18.4 DETERMINATION OF OVERALL REDOX REACTION TENDENCY: E^0_{cell}

We have indicated that the voltage of a battery is determined from the standard reduction potentials, E^0 values, of the half-reactions involved, and the value of this voltage is an indication of the tendency of the overall redox reaction to occur. We will now present a scheme for determining this voltage, which is symbolized E^0_{cell}.

In the following scheme, it is assumed that there is a proposed redox system given so that the half-reactions and standard reduction potentials can be found in Table 18.2. An example follows.

Step 1: Write the equations representing the half-reactions as extracted from the overall reaction given and label as an "oxidation" and a "reduction".

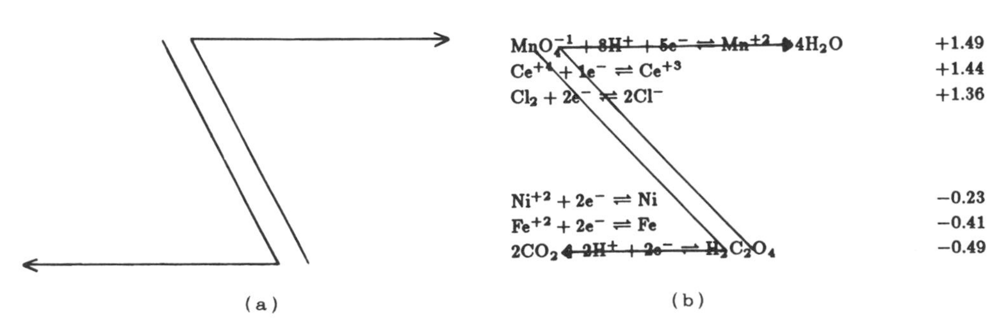

Figure 18.3. (a) Arrows that can be used as mnemonic devices, in conjunction with Table 18.2, to determine whether there will be a redox reaction between two chemical species; (b) the actual use of these arrows as discussed in the text.

Step 2: Locate the half-reactions in a table of standard reduction potentials, such as Table 18.2, and write the E^0 values adjacent to the respective half-reactions. For the oxidation half-reaction the sign of the E^0 must be changed since the reaction is written in reverse.

Step 3: Balance charges, equalize electrons in both half-reactions, and add the two equations together (as in the scheme for equation balancing—Chapter 8), and also add the E^0 values together. Do not multiply E^0 values by the multiplying coefficients. The resulting E represents the overall reaction tendency and also the voltage of the "battery".

Step 4: If E^0_{cell} is positive (+), the reaction will proceed spontaneously to the right. If it is negative (-), it will proceed to the left.

Example 1

What is the E^0_{cell} for the following redox system (previously discussed), and in which direction will the reaction spontaneously proceed?

Solution

$$Cu + Ag^{+1} \rightarrow Ag + Cu^{+2}$$

Step 1: Oxidation: $Cu \rightarrow Cu^{+2} + 2e^-$
Reduction: $Ag^{+1} + 1e^- \rightarrow Ag$

Step 2: Oxidation: $Cu \rightarrow Cu^{+2} + 2e^-$ $E^0 = -0.34$ V
Reduction: $Ag^{+1} + 1e^- \rightarrow Ag$ $E^0 = +0.80$ V

Step 3: Oxidation: $Cu \rightarrow Cu^{+2} + 2e^-$ $E^0 = -0.34$ V
Reduction: $2(Ag^{+1} + 1e^- \rightarrow Ag)$ $E^0 = +0.80$ V

$$Cu + 2Ag^{+1} \rightarrow Cu^{+2} + 2Ag \qquad E^0_{cell} = +0.46 \text{ V}$$

Step 4: E^0_{cell} is (+), reaction is spontaneous to the right.

It should be pointed out that if the reaction does not proceed in the direction desired, one can force it to do so by inserting a commercial battery into the external circuit. The cell thus becomes what is termed an "electrolytic" cell. Without the battery in the circuit, the cell is called a "galvanic" cell. In either case, the electrode at which reduction occurs is called the "cathode" and the electrode at which oxidation occurs is called the "anode".

18.5 THE NERNST EQUATION

Half and overall reaction tendencies change with temperature, pressure (if gases are involved), and concentrations of the ions involved. In the discussion thus far, standard conditions have been assumed. (Table 18.2 lists *standard* reduction potentials.) Standard conditions here are 25°C, 1.0 atm pressure, and 1.0 *M* ion concentrations. Thus, measured cell voltages are often not the same as predicted from the discussion thus far. How much are E values changed, and how is the scheme for determining the overall cell potential discussed in the previous section changed when conditions are not "standard conditions"? A physical chemist by the name of Nernst came up with the answers in the 1890s. His now famous "Nernst equation" is used to calculate the true E (cell voltage) from the E^0, temperature, pressure, and ion concentrations. For the half-reaction

$$qQ^{r} + ne^{-} \rightarrow qQ^{r-n} \quad (18.1)$$

the Nernst equation* is

$$E = E^{0} - \frac{0.059}{n} \log \frac{[Q^{r-n}]^{q}}{[Q^{r}]^{q}} \quad (18.2)$$

and for the general overall reaction

$$aA + bB \rightarrow cC + dD \quad (18.3)$$

the Nernst equation is

$$E_{cell} = E^{0}_{cell} - \frac{0.059}{n} \log \frac{[C]^{c}[D]^{d}}{[A]^{a}[B]^{b}} \quad (18.2)$$

If any species involved is a gas, the partial pressure of the gas is substituted for the concentration. If the temperature is different from 25°C, the "constant" 0.059 changes. The symbol "n" represents the number of electrons involved.

It is obvious from the foregoing discussion that the potential of electrodes and half-cells that one would actually measure is dependent on the concentration of the dissolved species involved. This is the basis for all quantitative potentiometry techniques and measurements to be discussed in this chapter.

Example 2

What is the E for the Fe^{+3}/Fe^{+2} half-cell if $[Fe^{+3}] = 10^{-4}$ *M* and $[Fe^{+2}]$ is 10^{-1} *M* at 25°C?

Solution

$$Fe^{+3} + 1e^{-} \rightarrow Fe^{+2} \qquad E^{0} = -0.77 \text{ V}$$

$$E = E^{0} - \frac{0.059}{1} \log \frac{[Fe^{+2}]}{[Fe^{+3}]}$$

$$= +0.77 - \frac{0.059}{1} \log \frac{10^{-1}}{10^{-4}}$$

$$= 0.77 - 0.059 \log 10^{3}$$

$$= 0.77 - 0.059 \times 3$$

$$= 0.77 - 0.177$$

$$= +0.59 \text{ V}$$

*Strictly speaking, the Nernst equation involves "activity" rather than concentration. Activity is directly proportional to concentration—the "activity coefficient" is the proportionality constant. For most applications, the activity coefficient is equal to "one" and thus the activity equals the concentration. Further discussion of activity and the activity coefficient is beyond the scope of this book.

Example 3

What is the new E for the $Cu^{+2}/Cu//Ag^{+1}/Ag$ cell if $[Cu^{+2}] = 0.1\ M$ and $[Ag^{+1}] = 10^{-2}\ M$ at 25°C?

Solution

From Example 1:

$$Cu + 2Ag^{+1} \rightarrow Cu^{+2} + 2Ag \qquad E^0 = +0.46\ V$$

$$E = E^0_{cell} - \frac{0.059}{2} \log \frac{[Cu^{+2}]}{[Ag^{+1}]^2}$$

NOTE: As with equilibrium constant problems, concentrations of pure undissolved solids (in this case Ag and Cu) do not appear in the expression.

$$E = +0.46 - \frac{0.059}{2} \log \frac{0.1}{(0.01)^2}$$

$$= +0.46 - \frac{0.059}{2} \log 10^3$$

$$= +0.46 - \frac{0.059}{2} (3)$$

$$= +0.46 - 0.0885 = +0.37\ V$$

18.6 POTENTIOMETRY

Electroanalytical techniques which measure or monitor electrode potential utilize the *galvanic* cell concept. Such techniques fall under the general heading of "potentiometry". Examples include the pH measurement, ion-selective electrode measurement, and potentiometric titrations. In these techniques, a pair of electrodes is immersed and the potential, or voltage, of one of the electrodes is measured relatively to the other, hence the name potentiometry. The application of the Nernst equation in chemical analysis can be inferred since, as the Nernst equation states, a voltage (or "potential", as it is scientifically termed) is proportional to ion concentrations. Thus, one can imagine that the measurement of the potential with an ordinary voltmeter or equivalent can be a measurement that could lead to the concentration of some dissolved species and thus be a tool in the chemical analysis of a sample. Thus, the measurement of potential in chemical analysis is what is called "potentiometry".

Reference Electrodes

The measurement of any voltage is a relative measurement and requires an unchanging reference point. For voltage measurements in most electronic circuitry, this reference is usually "ground"; it is often a wire that is connected to the frame of the electronic unit and the third prong in an electrical outlet, which in turn is connected to a rod that is pushed into the earth, hence the term "ground". Thus, an electronic technician measures voltages relative to ground.

In electroanalytical chemistry, the unchanging reference is a portable half-cell that is designed to develop a potential that is constant. There is more than one design for this half-cell, and we now proceed to describe several that have become popular over the years.

The Saturated Calomel Reference Electrode (*SCE*). The saturated calomel reference electrode is one such constant-potential electrode. A typical SCE available commercially is shown in Figure 18.4. It consists of two concentric glass tubes, which we will refer to here as an outer tube and an inner tube, each isolated from the other except for a small opening for electrical contact. The outer tube has a porous fiber plug in the tip which acts as the "salt bridge" to the analyte solution. A saturated solution of potassium chloride is in the outer tube. The saturation is evidenced by the fact that there is some undissolved KCl present. Within the inner tube is a paste-like material known as calomel. Calomel is made by thoroughly mixing mercury metal (Hg) with mercurous chloride (Hg_2Cl_2), a white solid. When in use, the following half-cell reaction occurs:

$$Hg_2Cl_2 + 2e^- \rightarrow 2Hg + 2Cl^-$$

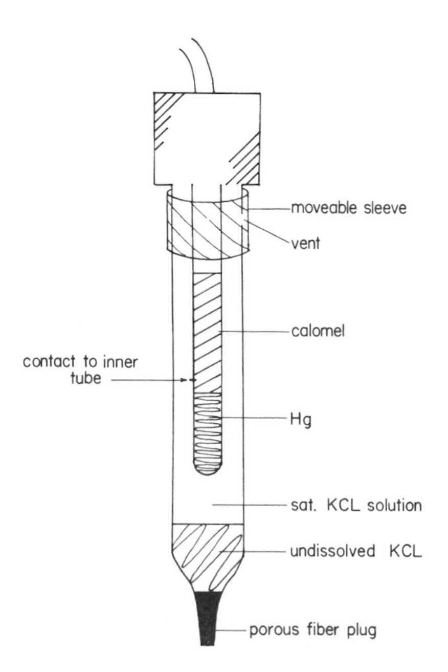

Figure 18.4. The saturated calomel reference electrode.

The Nernst equation for this reaction is

$$E = E^0 - \frac{0.059}{2} \log [Cl^-]^2$$

or

$$E = E^0 - 0.059 \log [Cl^-]$$

Obviously, the only parameter on which the potential depends is $[Cl^-]$. The saturated KCl present provides the $[Cl^-]$ for the reaction and, since it is a saturated solution, $[Cl^-]$ is a constant at a given temperature represented by the solubility of KCl at that temperature. If $[Cl^-]$ is constant, the potential of this half-cell, dependent only on the $[Cl^-]$, is also a constant. As long as the KCl is kept saturated and the temperature is kept constant, the SCE is useful as a reference against which all other potential measurements can be made. Its standard reduction potential at 25°C (see Table 18.2) is +0.27 volts. An advantage of this electrode is that, unlike its unsaturated counterparts, the $[Cl^-]$ doesn't change with evaporation of the water since the solution remains saturated. Unsaturated calomel electrodes, however, are not affected by temperature changes like the SCE.

The SCE is usually used by dipping it into the analyte solution which is part of a cell, and it includes an electrode typically referred to as an "indicator" electrode. A voltmeter is then externally connected across the leads to the two electrodes and the potential of the indicator electrode recorded vs. the constant reference. While the SCE is dipped into the solution, there will be a slight leakage of the potassium and chloride ions through the porous tip. In order for the SCE to be used accurately, the experiment must not be adversely affected by the slight contamination from these ions. It is a good idea to slide the movable sleeve (Figure 18.4) downward so that the outer tube is vented while the electrode is in use so that the ions do indeed freely diffuse through the porous tip. Also, the electrode, under these circumstances, should not be immersed into the solution so deep that the level of solution in the external tube is lower than the level of the solution tested. This would cause the solution to diffuse into the SCE rather than the reverse and thus would contaminate the KCl solution inside and possibly damage the SCE.

The vent hole may also be used prior to the experiment to refill the outer tube with more saturated KCl solution as this solution is lost with time. In addition, if the undissolved KCl disappears, more solid KCl can be added through the vent hole. The SCE should be stored in a beaker of distilled water (with the vent hole covered) for short-term storage and in a removable plastic cover usually provided with the electrode for long-term storage.

The Silver-Silver Chloride Electrode. The commercial silver-silver chloride electrode is similar to the SCE in that it is enclosed in glass, has nearly the same size and shape, and has a porous fiber tip for contact with the external solution. Internally, however, it is different. There is only one glass tube and a solution saturated in silver chloride and potassium chloride is inside. A silver wire coated at the end with a silver chloride "paste" extends into this solution from the external lead. (See Figure 18.5.) The half-reaction which occurs is

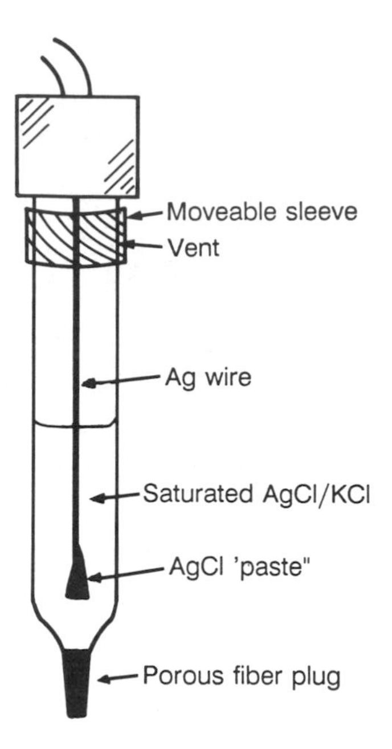

Figure 18.5. A typical silver-silver chloride reference electrode. (From Kenkel, J., *Analytical Chemistry Refresher Manual*, Lewis Publishers, Chelsea, MI, 1992. With permission).

$$AgCl(s) + 1e^- \rightarrow Ag(s) + Cl^-$$

and the Nernst equation for this is

$$E = E^0 - \frac{0.059}{1} \log [Cl^-]$$

The standard reduction potential for this half-reaction is +0.22 volts. The potential is only dependent on the $[Cl^-]$, as was the SCE, and once again the $[Cl^-]$ is constant because the solution is saturated. Thus, this electrode is also appropriate for use as a reference electrode. Refer to the segment above on the SCE for a discussion as to how it is physically handled while in use and in storage.

The Standard Hydrogen Electrode. The ultimate reference electrode utilizes the half-reaction on which Table 18.2 is based:

$$2H^+ + 2e^- \rightarrow H2 \qquad E^0 = 0.00 \text{ V}$$

This half-cell consists of a glass envelope housing a solution of constant pH in which a platinum wire of foil is immersed. (See Figure 18.6.) The platinum provides a surface for the exchange of electrons. Hydrogen gas bubbles through this solution at a constant pressure from some external source. Thus, since $[H^+]$ is a constant, and since the partial pressure of H_2 is a constant, the potential is a constant. The use of this cell as a reference electrode, however, is not very practical. It is cumbersome to use because it is not portable and requires a venting system for the hydrogen which otherwise is released into the laboratory.

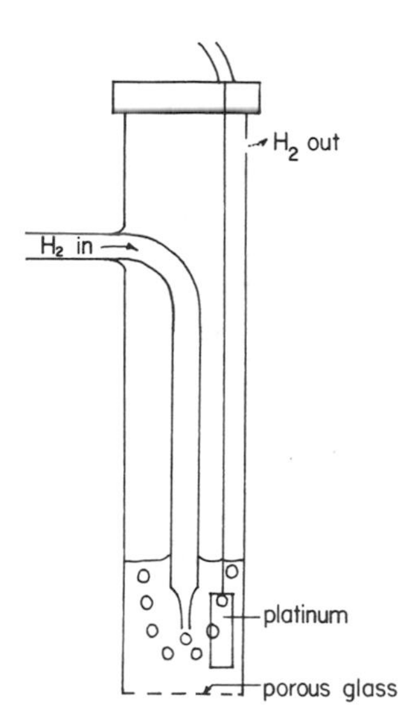

Figure 18.6. The standard hydrogen electrode.

Indicator Electrodes

As stated previously, the reference electrode represents half of the complete system for potentiometric measurements. The other half is the half at which the potential of analytical importance – the potential which is related to the concentration of the analyte – develops. There are a number of such "indicator" electrodes and analytical experiments which are of importance. Let us discuss these.

The pH Electrode. The measurement of pH is very important in many aspects of chemical analysis. Curiously, the measurement is based on the potential of a half-cell, the pH electrode.

The pH electrode, also called the glass electrode, consists of a closed-end glass tube that has a very thin, fragile glass membrane at the tip. Inside the tube is a saturated solution of silver chloride that has a particular pH. It is typically a 1 *M* solution of HCl. A silver wire coated with silver chloride is dipped into this solution to just inside the thin membrane. While this is almost the same design as the silver-silver chloride reference electrode, the presence of the HCl and the fact that the tip is fragile glass and does not have a porous fiber plug point out the difference. (See Figure 18.7.)

The purpose of the silver-silver chloride combination is to prevent the potential that develops due to possible changes in the interior of the electrode. The potential that develops is a "membrane" potential. Since the glass membrane at the tip is thin, a potential develops due to the fact that the chemical composition inside is different from the chemical composition outside. Specifically, it is the difference in the concentration of the hydrogen ions on opposite sides of the membrane that causes the potential – the membrane potential – to develop. There is no half-cell reaction involved. The Nernst equation is

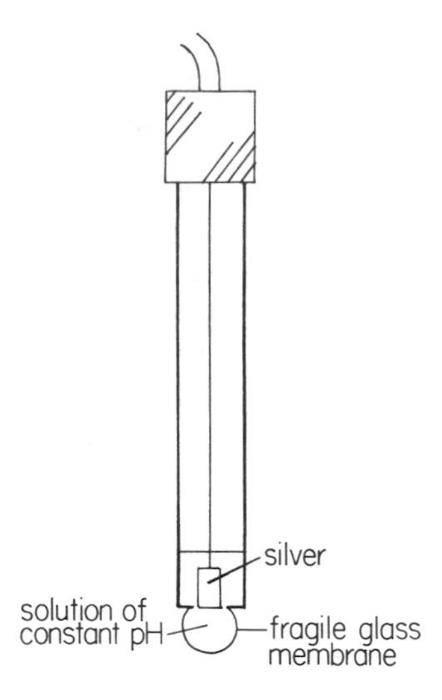

Figure 18.7. The pH or "glass" electrode.

$$E = E^0 - 0.059 \log \frac{[H^+](\text{internal})}{[H^+](\text{external})}$$

or, since the internal $[H^+]$ is a constant, it can be lumped into E^0, which is also a constant, giving a modified E^0, E^*, and eliminating $[H^+]$(internal):

$$E = E^* + 0.059 \log [H^+](\text{external})$$

In addition, we can recognize that pH = –log $[H^+]$ and substitute this into the above equation:

$$E = E^* - 0.059 \text{ pH}$$

The beauty of this electrode is that the measured potential (measured against a reference electrode) is thus directly proportional to the pH of the solution into which it is dipped. A specially designed voltmeter, called a pH meter, is used. In a pH meter, the meter readout is typically calibrated in both pH units and volts, making the pH meter a rather versatile device.

The pH meter is standardized (calibrated) with the use of buffer solutions. Usually, two buffer solutions are used for maximum accuracy. The pH values for these solutions "bracket" the pH value expected for the sample. For example, if the pH of a sample to be measured is expected to be 9.0, buffers of pH = 7.0 and pH = 10.0 are used. Buffers with pH values of 4.0, 7.0, and 10.0 are available commercially specifically for pH meter standardization. Alternatively, of course, homemade buffer solutions (see Chapter 4) may be used. In either case, the meter is adjusted, using the appropriate standardization method (refer to manufacturer's literature), to read the pH of the buffer into which the electrodes (the pH electrode and a reference electrode) are dipped. The solution of unknown pH is then determined. The pH electrode should be stored in a solution of distilled water or with a protective plastic sleeve over the tip.

The Combination pH Electrode. In order to use the pH electrode described above, two half-cells (sometimes called "probes") are needed – the pH electrode itself and a reference electrode, either the SCE or the silver-silver chloride electrode – and two connections are made to the pH meter. A modern development in this area is the invention of the "combination" pH electrode. This electrode incorporates both the reference probe and pH probe into a single probe. The reference portion is a silver-silver chloride reference. A drawing of the combination pH electrode is given in Figure 18.8.

The pH portion of this electrode is found in the center of the probe, as shown. It is identical to the pH electrode described above – a silver wire coated with silver chloride immersed in a solution saturated with silver chloride and having a $[H^+]$ of 1.0 *M*. This solution is in contact with a thin glass membrane at the tip. The reference portion is in an outer tube concentric with the inner pH portion. It has a silver wire coated with silver chloride in contact with a solution saturated with silver chloride and potassium chloride. A porous fiber plug in the wall of the outer tube connects the outer tube with the solution tested, as shown. This electrode either has two wires protruding from the lead (for connection to older style pH meters) or makes both connections using the "bnc" type or similar connector.

Ion-Selective Electrodes. The concept of the pH electrode has been extended to include other ions as well. Considerable research has gone into the development of these "ion-selective" electrodes over the years, especially in studying the composition of the membrane that separates the internal solution from the analyte solution. The internal solution must contain a constant concentration of the ion analyzed for, as with the pH electrode. Today we utilize electrodes with (1) glass membranes of varying compositions,

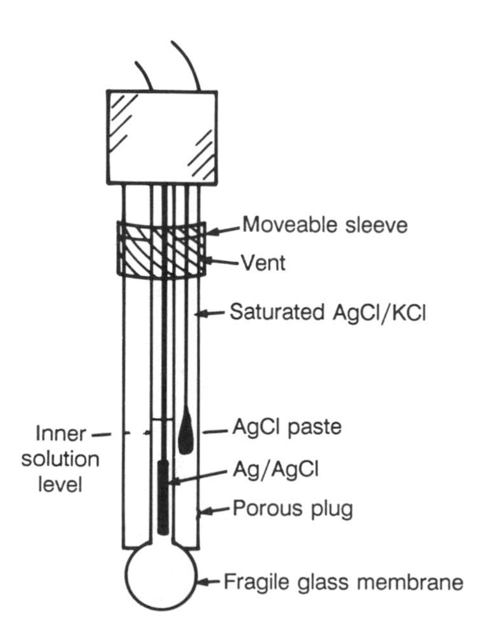

Figure 18.8. The combination pH electrode. (From Kenkel, J., *Analytical Chemistry Refresher Manual*, Lewis Publishers, 1992. With permission.)

(2) crystalline membranes, (3) liquid membranes, and (4) gas-permeable membranes. In each case, the interior of the electrode has a silver-silver chloride wire immersed in a solution of the analyte ion.

Examples of electrodes which utilize a glass membrane are those for lithium ions, sodium ions, potassium ions, and silver ions. Varying percentages of Al_2O_3, SiO_2, along with oxides of the metal analyte are often found in the membrane as well as other metal oxides. Selectivity and sensitivity of these electrodes vary.

With crystalline membranes, the membrane material is most often an insoluble ionic crystal cut to a round, flat shape and having a thickness of 1 or 2 mm and a diameter of about 10 mm. This flat "disc" is mounted into the end of a Teflon® or PVC tube. The most important of the electrodes with crystalline membranes is the fluoride electrode. The membrane material for this electrode is lanthanum fluoride. The fluoride electrode is capable of accurately sensing fluoride ion concentrations over a broad range and to levels as low as 10^{-6} *M*. Other electrodes that utilize a crystalline membrane but with less impressive success records are chloride, bromide, iodide, cyanide, and sulfide electrodes. The main difficulty with these is problems with interferences.

Liquid membrane electrodes utilize porous polymer materials such as PVC or other plastics. An organic liquid ion-exchanger immiscible with water contacts and saturates the membrane from a reservoir around the outside of the tube containing the water solution of the analyte and the silver-silver chloride wire. (See Figure 18.9.) Important electrodes with this design are the calcium and nitrate ion-selective electrodes.

Finally, gas-permeable membranes are used in electrodes which are useful for dissolved gases such as ammonia, carbon dioxide, and hydrogen cyanide.

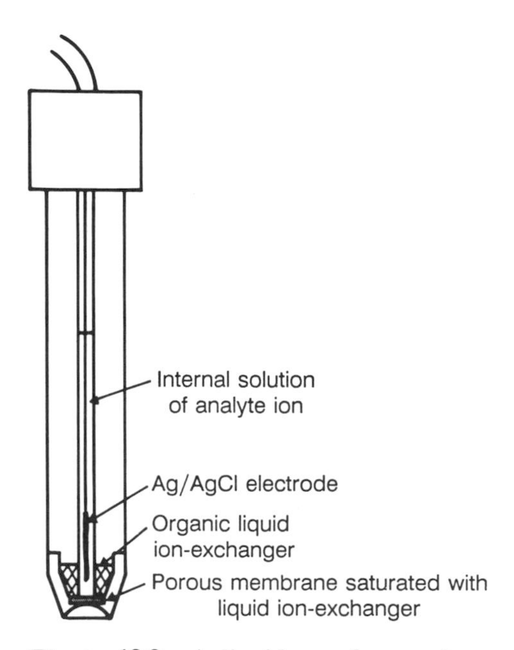

Figure 18.9. A liquid membrane electrode. (From Kenkel, J., *Analytical Chemistry Refresher Manual*, Lewis Publishers, 1992. With permission.)

These membranes are permeated by the dissolved gases, but not by solvents or ionic solutes. Inside the electrode is a solution containing the reference wire as well as a pH probe, the latter positioned so as to create a thin liquid film between the glass membrane of the pH probe and the gas-permeable membrane. As the gases diffuse in, the pH of the solution constituting the thin film changes, and thus the response of the pH electrode changes proportionally to the amount of gas diffusing in.

Calibration of ion-selective electrodes for use in quantitative analysis is usually done by preparing a series of standards as in most other instrumental analysis methods (see Chapter 9), since the measured potential is proportional to the logarithm of the concentration. The relationship is

$$E = E^* + \frac{0.059}{z} \log [\text{ion}]$$

in which z is the signed charge on the ion. The analyst can measure the potential of the electrode immersed in each of the standards and the sample (vs. the SCE or silver-silver chloride reference), plot E vs. log [ion], and find the unknown concentration from the graph. A typical plot and unknown determination is shown in Figure 18.10 for some real data using a fluoride ion-selective electrode. Alternatively, the standard additions technique (Chapter 9) may be used.

Potentiometric Titrations

It is possible to monitor the course of a titration using potentiometric measurements. The pH electrode, for example, is appropriate for monitoring an acid-base titration and determining an end point in lieu of an indicator. The procedure has been called a "potentiometric titration" and the experimental setup is shown in Figure 18.11. The end point occurs when the

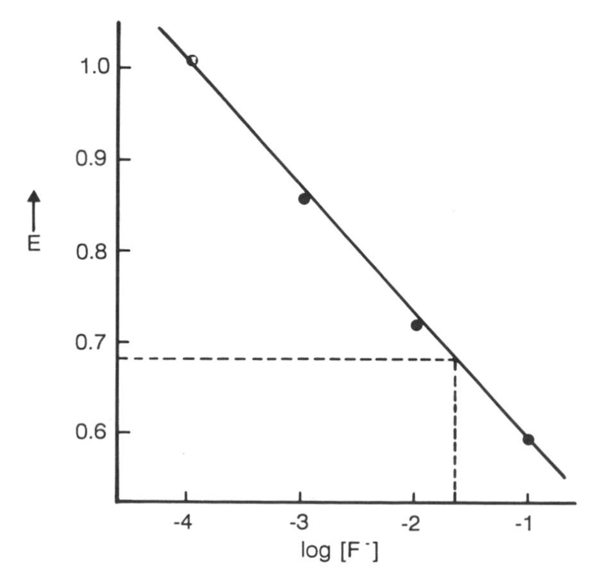

Figure 18.10. A calibration curve and unknown determination for an experiment using an ion-selective electrode.

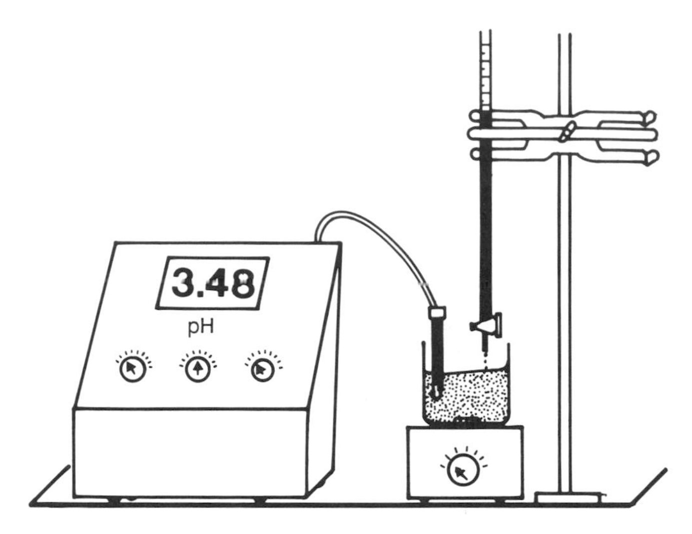

Figure 18.11. A potentiometric acid-base titration using a combination pH electrode. (From Kenkel, J., *Analytical Chemistry Refresher Manual*, Lewis Publishers, 1992. With permission.)

measured pH undergoes a sharp change – when all the acid or base in the titration vessel is reacted. The same procedure can be used for any ion for which an ion-selective electrode has been fabricated and for which there exists an appropriate titrant.

In addition, potentiometric titration methods exist in which an electrode other than an ion-selective electrode is used. A simple platinum wire surface can be used as the indicator electrode when an oxidation-reduction reaction occurs in the titration vessel. An example is the reaction of Ce(IV) with Fe(II):

$$Ce^{+4} + Fe^{+2} \rightarrow Ce^{+3} + Fe^{+3}$$

If this reaction were to set be up as a titration, with Ce^{+4} as the titrant and the Fe^{+2} in the titration vessel, and the potential of a platinum electrode dipped into the solution monitored (vs. a reference electrode) as the titrant is added, the potential would change with the volume of titrant added. This is because as the titrant is added, the measured E would change as the $[Fe^{+2}]$ is decreased, the $[Fe^{+3}]$ is increased, and the $[Ce^{+3}]$ is increased. At the end point and beyond, all the Fe^{+2} is consumed and the $[Fe^{+3}]$ and $[Ce^{+3}]$ change only by dilution, and thus the E is dependent mostly on the change in the $[Ce^{+4}]$. At the end point there would be a sharp change in the measured E.

Automatic titrators have been invented which are based on these principles. A sharp change in a measured potential can be used as an electrical signal to activate a solenoid and stop a titration. (See also Chapter 19.)

18.7 POLAROGRAPHY AND VOLTAMMETRY

Introduction

Techniques which utilize the electrolytic cell concept, rather than the galvanic cell concept as in potentiometry, are techniques which, as stated in Section 18.1, utilize an external power source to drive the cell reaction one way or the other. In these techniques, the current that results from this, and not the potential, is measured and related to the concentration of the analyte. Techniques in this category fall under the general heading of "amperometry" and include "polarography", which utilizes a special non-stationary mercury electrode to be described shortly, and "voltammetry", which utilizes only stationary electrodes.

Electrolytic cells can involve significantly larger currents than galvanic cells. The two-electrode system and the solution composition discussed for potentiometry usually must be modified in order to obtain desirable results. A reference electrode, such as the SCE or the silver-silver chloride electrode, cannot handle larger currents. In addition, the solution into which the electrodes are dipped must be able to sustain such a current. This wasn't true with potentiometric methods. Thus, for amperometric methods, a three-electrode system is usually used and a "supporting electrolyte" must be added to the solution.

The Three-Electrode System. The current measured is usually the current due to a specific oxidation or reduction process involving the analyte species at the surface of one of the electrodes. A three-electrode system includes a "working" electrode, at which the oxidation or reduction process of interest occurs, a reference electrode, such as the SCE or the silver-silver chloride electrode, and an "auxiliary" or "counter" electrode, which carries the bulk of the current (instead of the reference electrode) and "counters" the process that occurs at the working electrode. The three electrodes are connected to the power source, which is a specially designed circuit for precise control of the potential applied to the working electrode and is often called a "potentiostat" or "polarograph". The working electrode can be made positive or negative with the flip of a polarity switch on the instrument. The use of the reference electrode is crucial for the precise control of the potential of the working electrode. The process occurring at the auxiliary electrode is unimportant analytically. However, it is important not to allow the products of the reaction occurring there to interfere with the process occurring at the working electrode. For this reason, the auxiliary electrode is often placed in a separate chamber with a fritted glass disc allowing electrical contact with the rest of the cell but not allowing diffusion of undesirable chemical species to the working electrode. (See Figure 18.12.)

The typical experiment consists of scanning a potential range within which the process of interest is forced to occur at the surface of the working electrode. The current that flows as a result is measured by the instrument and displayed, usually on a recorder. This visual display has characteristics which depend on the particular potential application used, as we will see, and also on whether the electrode is stationary or non-stationary. The current due to the process of interest is determined from this display and related to the concentration of the analyte species.

The Supporting Electrolyte. The need for the supporting electrolyte mentioned above is obvious from the standpoint that the solution tested must be an electrolyte solution if it is to sustain a measurable current. Real-world sample solutions and standards which have very low analyte concentrations often do not have a sufficient electrolyte content. In addition, the presence

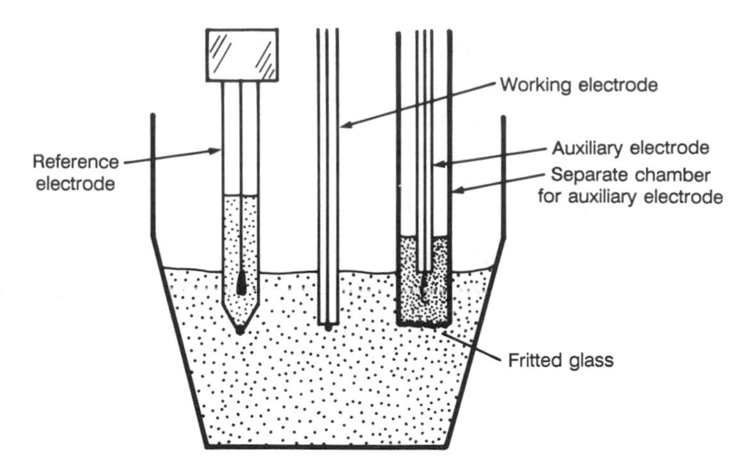

Figure 18.12. A drawing of a three-electrode system typical of most amperometric experiments. (From Kenkel, J., *Analytical Chemistry Refresher Manual*, Lewis Publishers, 1992. With permission.)

of a supporting electrolyte can help bring the potential of the electrode process into the desired range and free it from interferences due to the possibility of complexation and other reactions between the analyte species and the supporting electrolyte species. Such a reaction can create a complex ion, for example, which could shift the potential required to an interference-free range. The possibility of the supporting electrolyte species itself interfering with the oxidation or reduction of the analyte is eliminated by choosing an electrolyte species that is difficult to oxidize or reduce compared to the analyte species or the complex ion. The instrument is capable of precise control of the potential and, thus, capable of differentiating between the different species present in the solution. Thus, the analyte species can be "electroactive" (oxidized or reduced at the electrode surface) when the supporting electrolyte is not.

Polarography

Introduction. We have previously defined polarography as an amperometric technique which utilizes a special non-stationary electrode for the working electrode. We now proceed to describe this electrode in detail and continue with a discussion of the modern techniques associated with it.

The electrode is a "dropping mercury electrode", or DME. It consists of liquid mercury flowing through a very narrow-bore capillary tube, as shown in Figure 18.13. The mercury flows through the capillary by gravity from a reservoir of mercury. Drops of mercury form at the tip of the capillary, grow, fall, and re-form in fairly rapid intervals (0.5–10 sec). This sequence of events is depicted in Figure 18.13. Figure 18.14 shows the complete apparatus with connections made from the electrodes to the instrument.

The DME was invented in the early 1900s by a Nobel Prize-winning chemist by the name of Heyerovsky. Its fluid nature eliminates two problems normally associated with amperometric measurements: (1) contamination of the electrode surface resulting from electrode reactions (the mercury surface is continuously renewed in the case of the DME), and (2) the decay of the

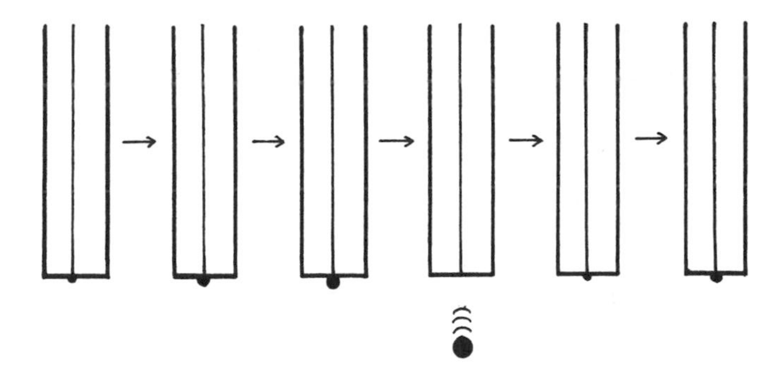

Figure 18.13. The dropping mercury electrode (DME) showing the growth and fall, growth and fall, etc. of the mercury drop.

current to small values with time due to the diffusion-limited transportation of analyte to the electrode surface when the solution is not stirred (the solution *is* stirred in the case of the DME each time the drop falls).

As we will soon see in the sections to come, it is desirable for the instrument to know the precise moment at which the drop will fall. Mechanical drop timers have been invented which "knock" the capillary at regular intervals such that the drop falls when required. Modern drop timers utilize a larger-bore capillary and a spring-loaded polyurethane "plug" at the top of

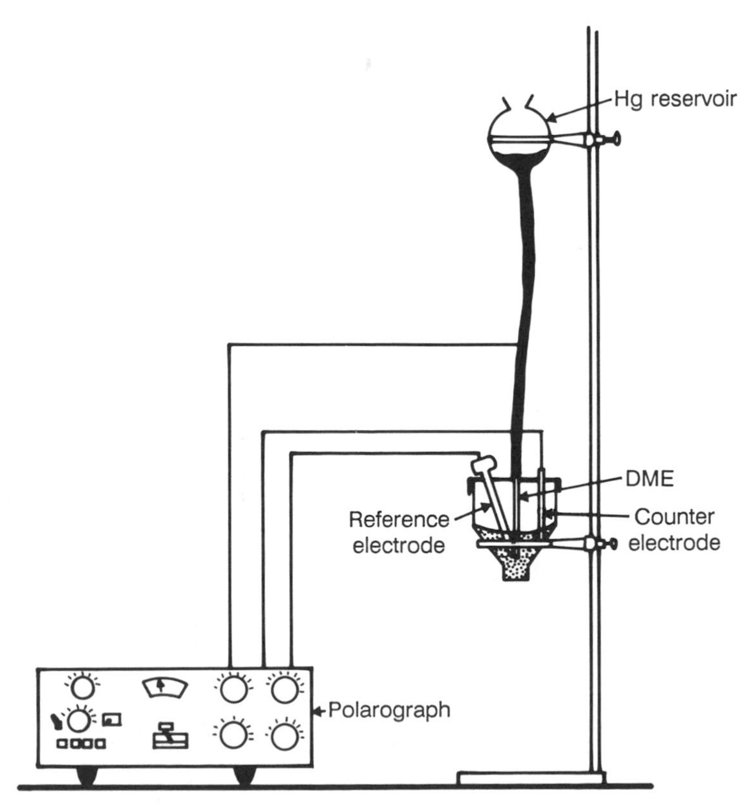

Figure 18.14. The "complete" polarography apparatus. DME = dropping mercury electrode. (From Kenkel, J., *Analytical Chemistry Refresher Manual*, Lewis Publishers, 1992. With permission.)

the capillary for controlling the drop. Figure 18.15 shows this more recently developed unit. In either case, the precise time the drop is to fall is controlled electrically such that the instrument makes its measurement at that exact moment, usually just before the drop falls.

Classical Polarography. The "classical" or "normal" polarography experiment consists of scanning the potential range in which the analyte becomes electroactive and monitoring the current continuously at the same time.

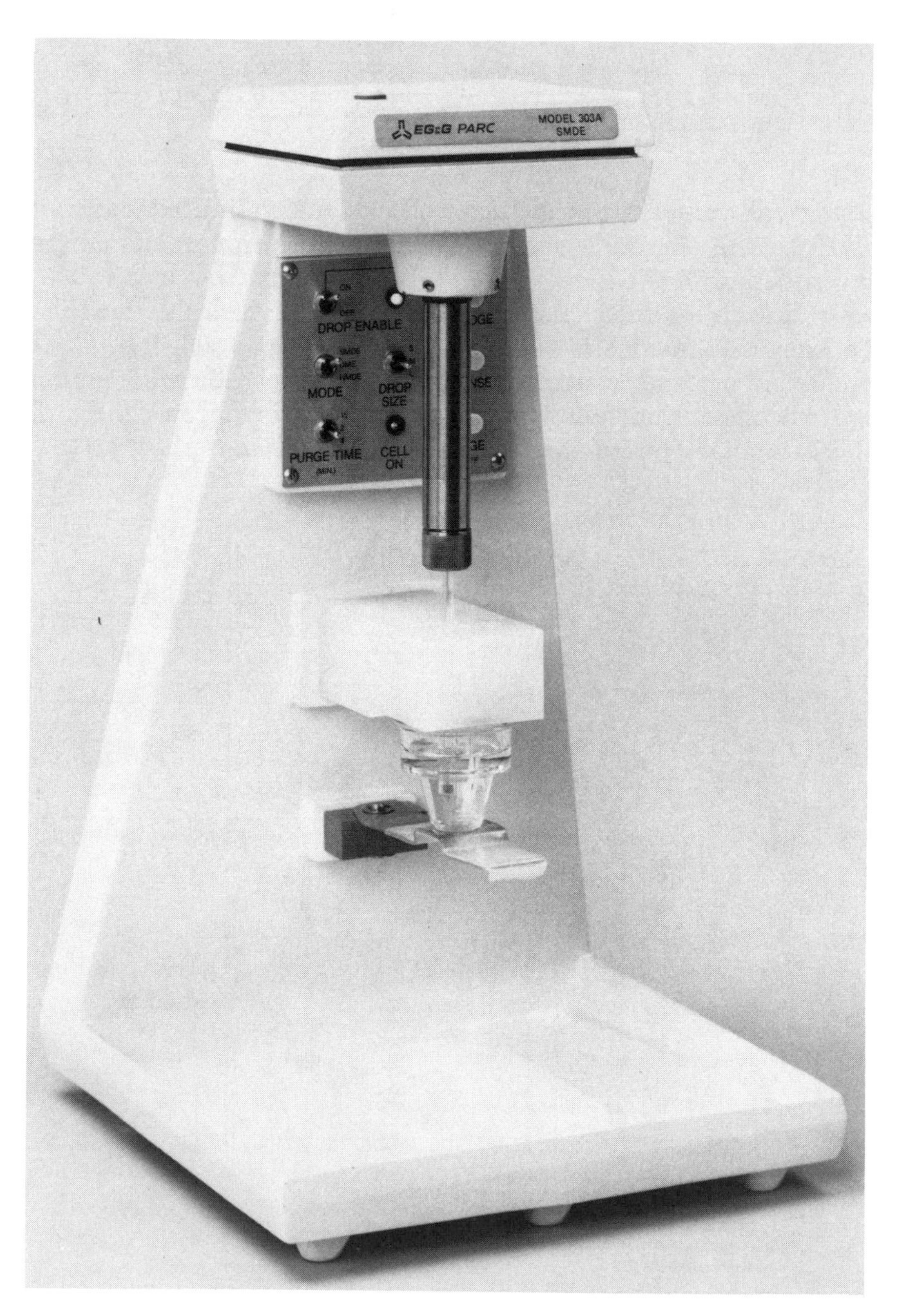

Figure 18.15. A modern apparatus for the mechanical control of the drop time and size. (Courtesy of EG&G Princeton Applied Research, Princeton, NJ).

Since the current that will flow at any electrode surface is proportional to the area of the electrode surface exposed to the solution (in addition to the analyte concentration), the current vs. potential display on the recorder shows a considerably unstable (but uniform) current. As the drop grows the current increases, but when the drop falls the current drops. The continuous growing and falling of the drop is thus accompanied by a continuous up and down motion of the recorder pen. The potential-time characteristic and the resulting current display (polarogram) for the reduction of cadmium ion in 0.1 *M* $NaNO_3$ are shown in Figure 18.16. The potential-time characteristic is often called a "ramp" because of the linear increase apparent in the figure.

Parameters associated with the polarogram are the "diffusion current", i_d, the "half-wave potential", $E_{1/2}$, and the "background current". The diffusion current is the current which is due directly to the presence of the analyte species in the solution. It is the increase in the measured current that occurs when the potential for the reduction or oxidation of the analyte species is reached as the electrode potential is linearly increased. The half-wave potential is the potential at which this increase in the current is half of its full value. The background current is the current that would be observed if the analyte species were not present in the solution. Sometimes this background current includes the current due to other oxidation or reduction processes occurring in the event there are other species in solution that can be reduced or oxidized at lower potentials. In the absence of these other species, this current would be very small and is called the "residual" or "charging" current. The diffusion current, half-wave potential, and residual current are all graphically defined in Figure 18.17.

The half-wave potential is often taken to be the potential at which reduction or oxidation of a given species occurs. It can be thought of as the "barrier" to be crossed in order for oxidation or reduction to occur. This barrier can be crossed from either direction, but must be crossed in order to observe the process desired. Usually we cross it from one direction only and observe the magnitude of the current that results. In a few techniques, however, it is desirable to cross it while proceeding in one direction, then

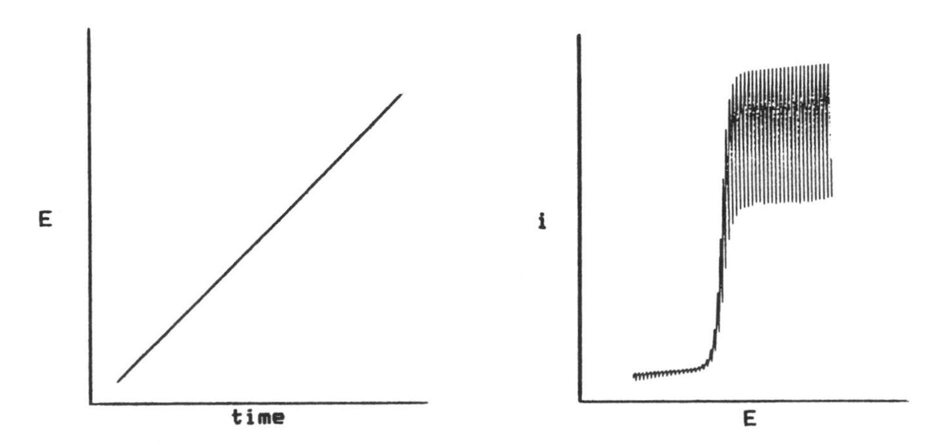

Figure 18.16. The time characteristic (left) and the recorder trace (right) for the classical polarography experiment described in the text. The "up and down" pattern on the right is due to the continuous growth and fall of the mercury drop.

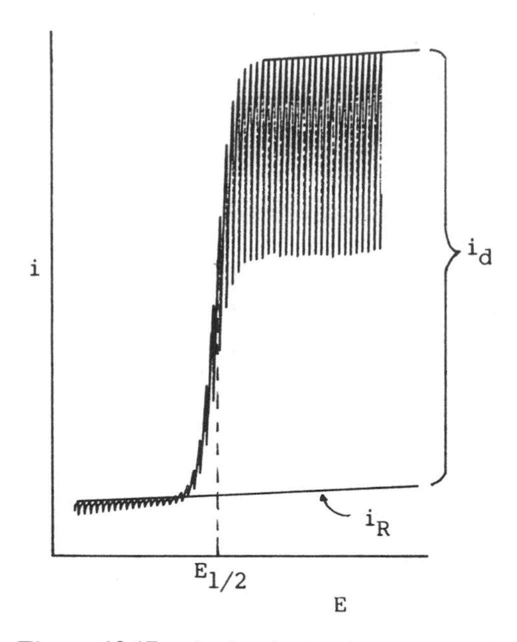

Figure 18.17. A classical polarogram and the definitions of diffusion current (i_d), half-wave potential ($E_{1/2}$), and residual current (i_R).

reverse the scan and cross it from the other direction too. We will describe these techniques in the next several subsections. Probably the most notable fact about the half-wave potential is that it is the one parameter which identifies a current increase to be due to a particular dissolved chemical species. One might think of it then as a parameter for qualitative analysis. However, the greatest usefulness of this lies in the fact that one can use it to identify the precise potential at which the current due to the analyte is expected and found and thus allow the correct diffusion current to be measured. This is important because it is the magnitude of the diffusion current that is related to concentration, and one must know which diffusion current to measure in the event that there are several current increases in a given polarogram. Figure 18.18 shows an example of one such polarogram.

One additional note about the half-wave potential is that its value is affected by the nature of the supporting electrolyte present in the solution. This is because the ions constituting the supporting electrolyte can react chemically with the analyte species and change its chemical makeup, thus causing a different chemical species, with a different tendency to give or take electrons, to be present in the solution. For example, a given metal ion forms different complex ions with different ligands supplied by the supporting electrolyte. Thus, Cd^{+2} in 0.1 *M* KCl solution has a half-wave potential of -0.60 volts. In a solution of 1 *M* NH_3 and 1 *M* NH_4Cl, it has a half-wave potential of -0.81 volts. In 1 *M* KCN, its half-wave potential is -1.18 v.

Notice that the residual and background currents in Figure 18.18 are steadily increasing slightly as the potential is increased. This is due to the fact that the mercury drop is becoming more and more charged as the potential is increased. The instrument sees this as a slight current flow and displays it on the recorder. This is a "non-Faradaic" type of current. A "Faradaic" current is one that is due to the oxidation or reduction of a

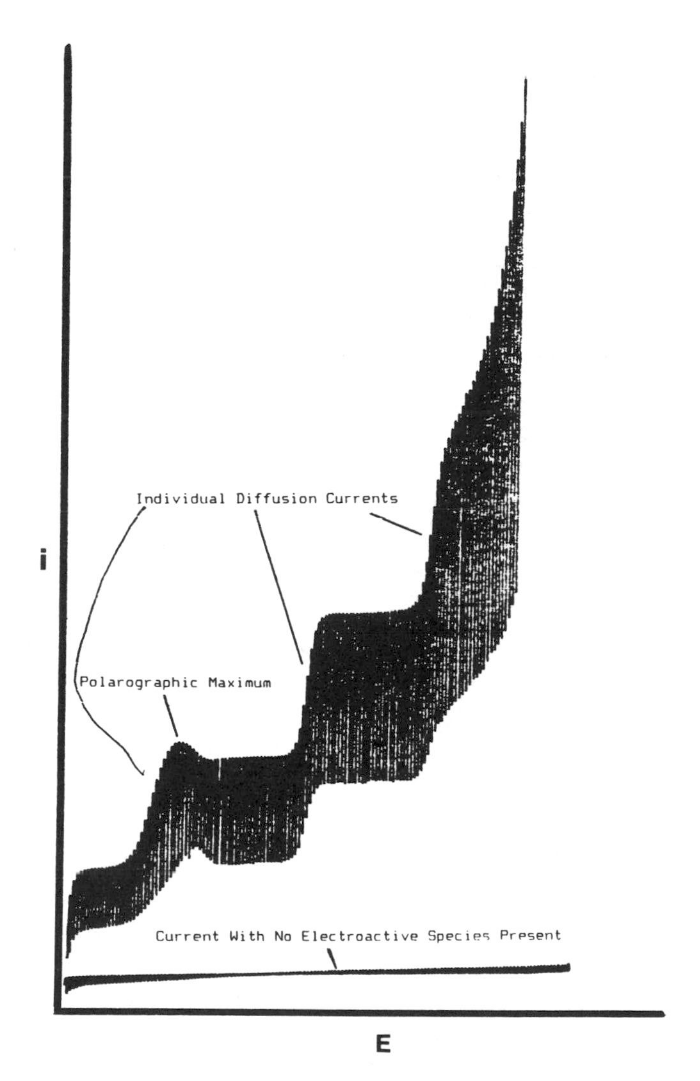

Figure 18.18. A polarogram showing several diffusion currents due to the presence of several electroactive species in the solution. The analyst may identify the analyte species' diffusion current by the magnitude of the associated half-wave potential.

dissolved chemical species. It is due to an electron transfer process. A non-Faradaic current is one that is not due to an electron transfer process.

Polarographic "maxima" are sometimes encountered in the current/potential display, and these can inhibit the accurate measurement of the diffusion current. Polarographic maxima are distortions of the current manifested by a large "peak" of current immediately beyond the half-wave potential (Figure 18.18). The cause is not well-understood; however, the problem can usually be solved by the addition of small amounts of gelatin or a surfactant, such as Triton® X-100, to the solution.

Modern Polarographic Techniques. Modern polarography instruments are capable of several modifications of the classical procedure just described. These are "sampled dc polarography", "pulse polarography", and "differential pulse polarography". Each differs from classical polarography because of either the nature of the potential-time characteristic or the mode of current measurement, or both. In most of these, the sensitivity is dramatically increased over the classical mode. In one of them, the resolution of two closely spaced half-wave potentials is improved. The techniques are summarized in terms of their potential-time characteristics and their current-potential recorder trace in Figure 18.19.

Sampled dc polarography is identical to the classical technique except that the current is only "sampled" and displayed just before the drop falls. The advantage is that it makes the diffusion current easier to measure – such current is free of the continuous up and down pattern.

Pulse polarography is like sampled in that the current is sampled just before the drop falls. However, the potential-time characteristic is very different – a series of increasing potential pulses is applied to the DME, as shown, each time just before the drop falls. The current sampling is done while the pulse is being applied, which is for less than 0.1 sec. Thus, the following sequence of events occurs in rapid order: (1) the pulse is applied, (2) the current is measured and displayed, (3) the pulse drops back, and (4) the drop falls. After a delay (on the order of seconds), this sequence begins again, but with a higher pulse. Once the pulse is high enough to cross the half-wave potential, the current begins to rise due to the reduction or oxidation of the analyte. The advantage is a much higher current for the same concentration and thus a better sensitivity.

With differential pulse polarography, both the potential-time characteristic and the current sampling method are different from the others. The potential-time characteristic is one in which pulses of constant magnitude are superimposed on the ramp as shown. The current is sampled both before and during the application of the pulse. These two current values are subtracted from each other and divided by the potential difference, creating a signal which represents the rate of change or the "derivative" of the original polarogram. It is this signal that is then displayed on the recorder as the sampled currents were for the others. The trace resembles a peak because it is the rate of change of the original polarogram; in the original there is very little change before and after the rise in current, while a sharp change occurs where there is a rise. Thus, a peak occurs at or near $E_{1/2}$. The advantage of this technique is not only that it is very sensitive (typically more sensitive than any of the others); also, two species that have similar $E_{1/2}$ values, while not distinguished before, can be distinguished with this technique.

Quantitative Analysis. In classical (and sampled dc) polarography, the diffusion current, the current increase that occurs when the half-wave potential of the analyte is reached and surpassed, is directly and linearly related to the concentration of the species involved. Quantitative analysis is then based on the measurement of the diffusion current. The usual procedure is to prepare a series of standard solutions and measure the diffusion current for each. A plot of i_d vs. concentration then is useful in determining the concentration in an unknown. In pulse polarography, the current increase observed for each of a series of standard solutions is plotted vs. concentration. We have already stated that this current is typically much higher than that for classical and sampled dc, and thus quantitative analy-

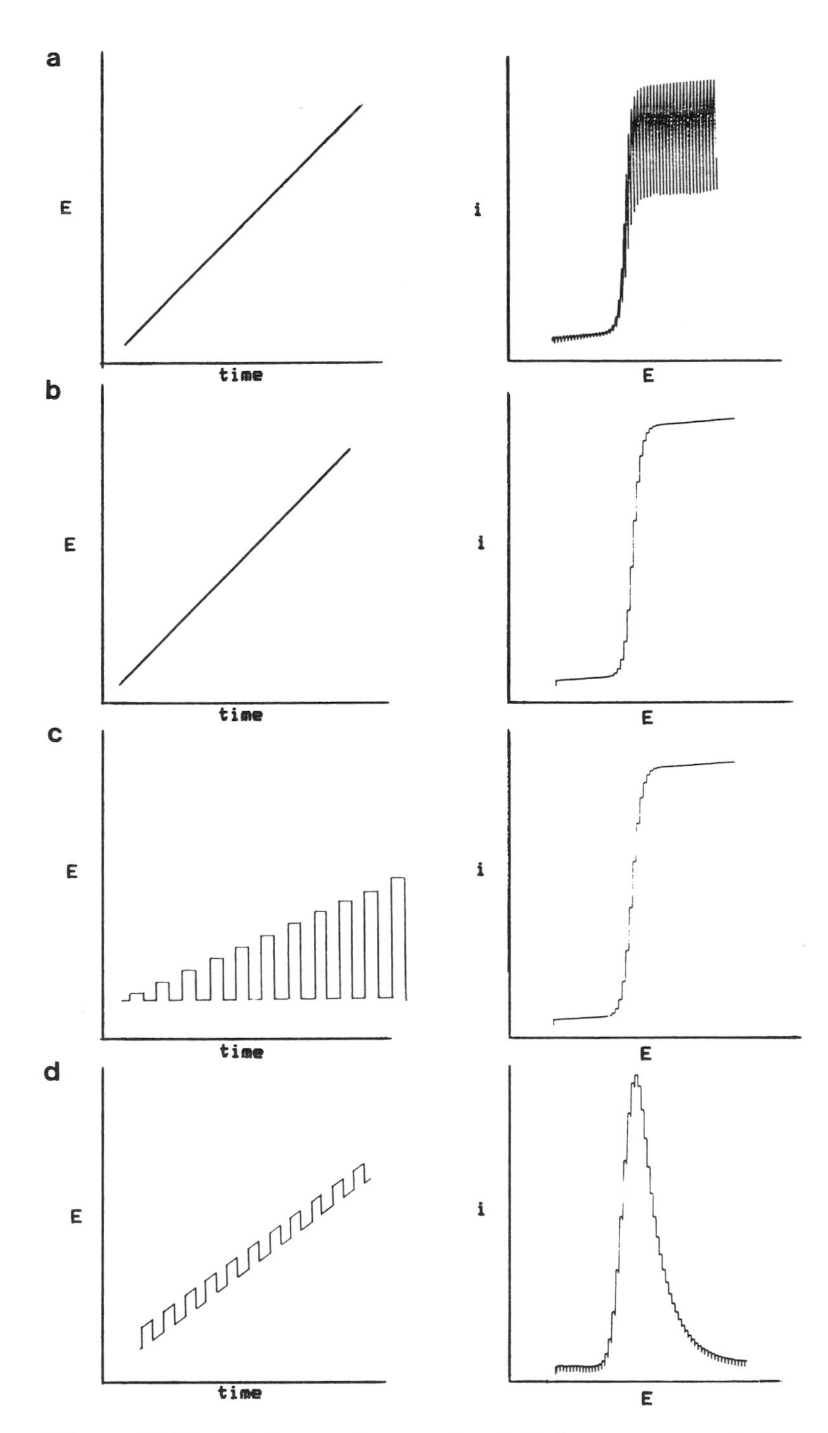

Figure 18.19. Modern polarography techniques compared in terms of potential-time characteristic and current-potential trace. (a) classical dc; (b) sampled dc; (c) pulse polarography; (d) differential pulse polarography.

sis is much more sensitive. See Figure 18.20 for a comparison between sampled dc and pulse polarography.

The height of the peak in differential pulse polarography is proportional to concentration. Thus, in this technique, the peak height for each of a series of standard solutions is measured and plotted vs. concentration.

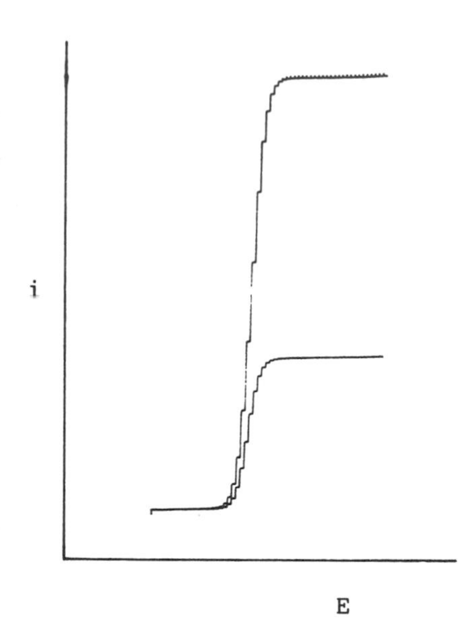

Figure 18.20. A comparison of a pulse polarogram with a sampled dc polarogram for the same concentration of electroactive species.

Comparison to Other Techniques. Polarographic techniques utilize a hazardous substance: mercury. The details of the hazards are presented in Chapter 19. In this subsection, however, the selection of polarography as an analytical procedure is discussed. Obviously, one would likely decide not to use it if a non-polarography technique would do as well.

The key words are "if it would do as well". Polarography is useful for all types of dissolved chemical species, molecules and ions, organic or inorganic. Thus, a decision whether to use it does not usually hinge on its applicability. What it does hinge on is its sensitivity weighed against the hazards. For example, if atomic absorption techniques would provide better sensitivity for dissolved iron, then atomic absorption would likely be chosen for iron analysis. If polarography would provide better sensitivity for vitamin C analysis than UV-visible spectrophotometry, then polarography may be chosen if the better sensitivity is necessary. Thus, polarography is used if necessary to attain a sensitivity that may be lacking with other techniques.

Voltammetry

Introduction. As stated previously, "voltammetry" is an amperometric technique in which the working electrode is some stationary electrode and not the DME. Typical electrodes here include small platinum (or gold) strips, wire coils, or beads sealed into the tip of a glass tube so that a small area or cross-section is exposed and polished. They may also consist of a small bead of graphitic carbon sealed into the tip of a glass tube and also finely polished as the others. This latter electrode is referred to as a "glassy carbon" electrode. A third type is the so-called hanging mercury drop electrode (the HME). This electrode is like the DME, except the mercury is not flowing but is held stationary.

The nature of the voltammetry techniques we will now describe dictates that the electrode be stationary. In one case, the electrode serves as a collection point for the analyte, thus precluding the need to have a continuously renewed surface. In another case, the electrode monitors the chemical reactivity near the electrode surface, thus precluding the need to have the solution stirred.

Stripping Voltammetry. This technique is most useful for metals that are able to be electroplated onto the surface of a working electrode that has been polarized to some negative potential so that reduction of the metal ion to the metal atom will take place, i.e., at a potential more negative than the half-wave potential for such a reduction. This reduction occurs over a period of time, perhaps even on the order of 1 or 2 hr, while the solution is stirred (with the use of a stirrer) to aid in the transport of the metal ion to the electrode surface. Over this period of time a fairly large quantity of metal ions are reduced, and the metal either deposits onto the surface of the electrode or, in the case of the HME, a very common electrode for this technique, dissolves into the mercury drop. After the preset period of time has elapsed, a "reverse" scan is performed so that the applied potential crosses the $E_{1/2}$ "barrier" and all the metal is oxidized back to the metal ion, in effect "stripping" it from the surface. The resulting oxidation current (usually the *total* current, measured in coulombs) is compared to that of a standard that has undergone the same experiment and the concentration determined. The potential-time characteristic and the resulting current pattern are shown in Figure 18.21. The advantage is that a solution of a very low concentration can be analyzed because the analyte is preconcentrated in or on the electrode, over a period of time, making the stripping current easily measurable.

Cyclic Voltammetry. This is a technique that is seldom used for quantitative analysis, but is a popular technique for other reasons, such as the determination of electrode reactions mechanisms. The potential-time characteristic is shown in Figure 18.22.

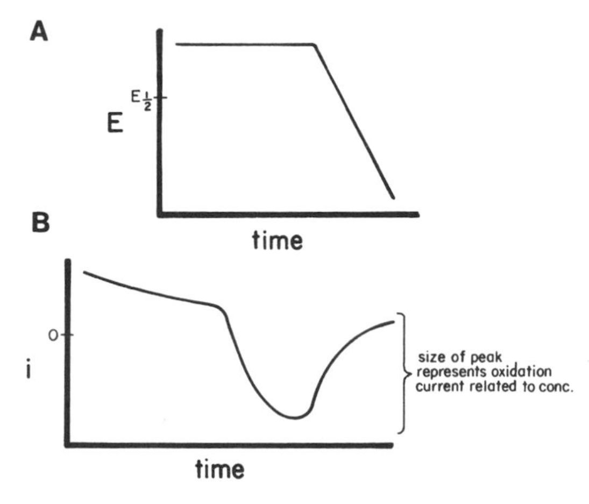

Figure 18.21. The potential (A) and current (B) characteristics of a stripping voltammetry experiment.

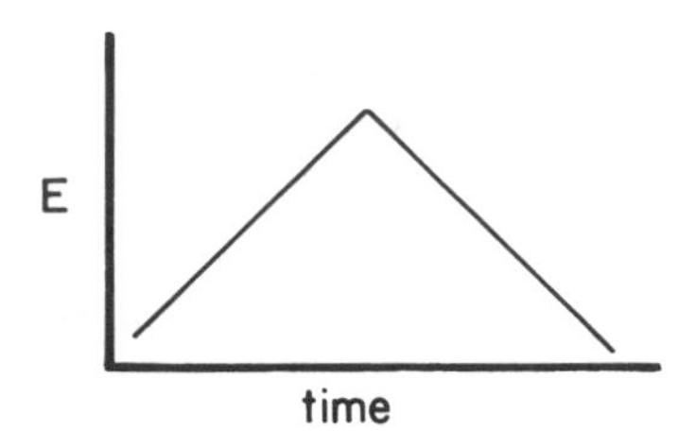

Figure 18.22. The potential-time characteristic in cyclic voltammetry.

As the potential is increased past $E_{1/2}$ the current is monitored continuously. The potential scan is then reversed and returned at an equal rate back to the starting point. The current/potential curve, if the electrode reaction is reversible, is shown in Figure 18.23a. If it is not reversible – that is, if the product of the electrode reaction has been reacted to form another species – the result is as shown in Figure 18.23b. The oxidation current evident in (a) shows that the product of the reduction is still available at the electrode surface to be oxidized back. In (b) it is not available.

Amperometric Titration. The concept of current measurement in voltammetry can be applied to a titration experiment much like potential measurements were in Section 18.6 (potentiometric titration). Such an experiment is called an "amperometric titration", a titration in which the end point is detected through the measurement of the current flowing at an electrode.

The potential of the measuring (working) electrode, which is typically a rotating platinum disk embedded in a Teflon® sheath, is held constant at some value beyond the half-wave potential of the analyte. The solution is stirred due to the rotation of the electrode. The resulting current (the diffusion current) is then measured as the titrant is added. The titrant reacts with the electroactive species, removing it from the solution and thus de-

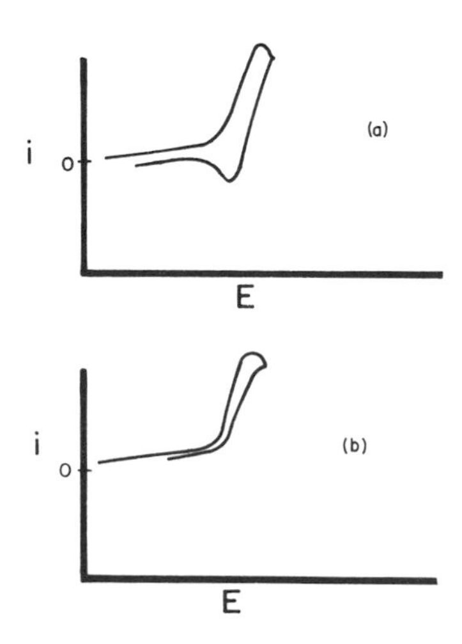

Figure 18.23. Cyclic voltammograms of (a) a reversible electrode process and (b) an irreversible process.

creasing its concentration. The measured current therefore also decreases. When all of the analyte has reacted with the titrant, the decrease will stop; this signals the end point.

EXPERIMENTS

EXPERIMENT 55: DETERMINATION OF THE pH OF SOIL SAMPLES

NOTE: Safety glasses are required.

1. Obtain soil samples as directed by your instructor. These samples should be thoroughly dried and crushed to very small particles using a mortar and pestle.
2. Weigh 5 g of each soil sample into separate 50-mL beakers. An analytical balance is not necessary for this.
3. Prepare a pH meter and combination pH electrode with a pH = 7 buffer.
4. Perform this step one at a time for each sample. Pipet 5 mL of distilled water into one of the beakers. Swirl vigorously by hand for 5 sec. Let stand 10 min. Swirl lightly, dip the pH electrode in, and measure the pH.
5. Repeat Step 3 for each sample. During the 10-min waiting period, the next sample(s) may be prepared, but be sure to measure each sample immediately after the 10-min period.

EXPERIMENT 56: DETERMINATION OF FLUORIDE IN MUNICIPAL WATER (OR OTHER SAMPLE) WITH THE USE OF AN ION-SELECTIVE ELECTRODE

NOTE: Safety glasses are required.

1. Prepare 100 mL of a stock standard solution of sodium fluoride, NaF, that is 1000 ppm F.

   $$\text{Calculation: } \frac{1000 \text{ mg F} \times \text{FW (NaF)}}{\text{atomic weight of F}} \times 0.100 = \text{mg NaF to weigh}$$

   (Review Section 7.5 for the explanation of this calculation.)
2. Prepare 100 mL (use volumetric flasks) each of four additional standards that are 100 ppm F, 10 ppm F, 1 ppm F, and 0.1 ppm F. Prepare these by serial dilution. That is, dilute the 1000 ppm to make the 100, dilute the 100 ppm to make the 10, etc. Mix thoroughly.
3. Pipet 25 mL of each standard solution and 25 mL of TISAB (total ionic strength adjustment buffer) into clean, dry beakers. Also pipet 25 mL of each of the samples and 25 mL of TISAB into clean, dry beakers. These mixtures may be prepared and measured according to Step 4, one at a time or together.
4. Immerse the tips of the fluoride electrode and the SCE (rinse the tips with distilled water and dab dry first) into each beaker and measure the potentials with the millivolt function on a pH meter.

5. Plot the E vs. log [F^-] and obtain the concentration of the unknown.

EXPERIMENT 57: FAMILIARIZATION WITH POLAROGRAPHIC TECHNIQUES

NOTE: Safety glasses are required. If this experiment is to be done in conjunction with Experiment 58, preparing the solutions required for Experiment 58 will automatically make available the solutions needed for Experiment 57.

1. Prepare the following solutions:
   (a) 100 mL of 0.1 *M* $NaNO_3$ (doesn't have to be accurate)
   (b) 100 mL of 0.1 *M* $NaNO_3$ that is also 10 ppm Cd (can dilute a stock 1000 ppm Cd)
2. With solution (a) obtain a normal dc polarogram of the oxygen dissolved in the solution. (Your instructor will demonstrate use of instrument.)
3. Remove the oxygen from this solution by "degassing" with some inert gas, such as nitrogen or helium. Obtain a second polarogram of the solution and observe that the oxygen has indeed been removed.
4. Obtain classical, sampled, pulse, and differential pulse polarograms of solution (b). Observe the current/voltage characteristics obtained in each of the modes and discuss the results with your instructor in light of the information presented in this chapter.

EXPERIMENT 58: POLAROGRAPHIC ANALYSIS OF A PREPARED SAMPLE FOR CADMIUM

NOTE: Safety glasses are required.

1. Prepare 1 L of 0.1 *M* $NaNO_3$ for use as the supporting electrolyte.
2. Prepare 100 mL of a 100-ppm Cd solution from the available 1000 ppm solution. (Use a 100-mL volumetric flask and use the 0.1 *M* $NaNO_3$ for the dilution.)
3. Prepare four standards, 5, 10, 15, and 20 ppm Cd, from the 100 ppm solution prepared in Step #2. Again use the 0.1 *M* $NaNO_3$ for the dilution. Use 100-mL volumetric flasks.
4. Prepare the sample by diluting to the mark with 0.1 *M* $NaNO_3$.
5. Prepare the instrument and cell as directed by your instructor and obtain differential pulse polarograms of each standard, all on the same recorder paper. Now run the sample using a different pen color. A computer may be used for data acquisition.
6. Plot peak height vs. concentration and determine the concentration of your unknown. A computer may be used for this.

QUESTIONS AND PROBLEMS

1. Differentiate between the two major classifications of electroanalytical methods.
2. How are oxidation-reduction chemistry and electroanalytical chemistry related?
3. Define cell, half-cell, electrolytic cell, galvanic cell.
4. Distinguish between an electrolytic cell and a galvanic cell.
5. Why does the transfer of electrons via electrodes translate into a way of converting chemical energy into electrical energy?
6. Why is it correct to say that a battery represents a device for converting chemical energy to electrical energy?
7. What is a "salt bridge" and why is it needed in a laboratory "battery"?
8. Explain how or why the power available from a battery is derived from the tendency for an oxidation-reduction reaction to proceed.
9. Explain how the activity series of metals is a partial listing of redox reaction tendencies.
10. Explain the relationship between the activity series of metals and a table of standard reduction potentials.
11. Define standard reduction potential, E^0.
12. Without doing any calculations, look at the table of standard reduction potentials and indicate whether or not there would be a reaction between
    (a) Cu and Cr^{+3}.
    (b) Hg and Ce^{+4}.
    (c) Fe and Ag^{+1}.
13. What is the E^0_{cell} for a cell in which the following overall reaction occurs?

$$Ce^{+3} + Cu^{+2} \rightarrow Ce^{+4} + Cu$$

14. Will the reaction in #13 proceed spontaneously to the right? Why or why not?
15. A certain voltaic cell is composed of a Ce^{+4}/Ce^{+3} half-cell and a Sn^{+4}/Sn^{+2} half-cell. The overall cell reaction is

$$Ce^{+4} + Sn^{+2} \rightarrow Sn^{+4} + Ce^{+3}$$

    (a) What is E^0_{cell}?
    (b) Is the reaction spontaneous as written, left to right? Why or why not?
16. What is the E^0_{cell} for the following redox system and in which direction will the reaction spontaneously proceed? (Assume standard conditions.)

$$Zn + Cu^{+2} \rightarrow Zn^{+2} + Cu$$

17. What is the E^0_{cell} for a cell in which the following overall reaction occurs?

$$Cu^{+2} + Ni \rightarrow Cu + Ni^{+2}$$

18. What is the E^0_{cell} for a cell in which the following overall reaction occurs?

$$Ce^{+3} + Sn^{+4} \rightarrow Ce^{+4} + Sn^{+2}$$

19. Compare Question #18 with Question #15. How are the answers different and why?

20. A voltaic cell is composed of a copper electrode (Cu) dipping into a solution of copper ions (Cu^{+2}) and a mercury electrode in a solution of Hg^{+2} ions. The cell reaction is

$$Cu + 2Hg^{+2} \rightarrow Cu^{+2} + 2Hg$$

What is the E^0 for this cell? Is the reaction spontaneous? Why or why not?

21. Under standard conditions, what is the E for a cell in which nickel ions react with magnesium metal to form nickel metal and magnesium ions?

22. Assuming standard conditions, what is the E for a cell in which iron metal reacts with manganous ion to form ferrous ion and manganese metal?

23. Define anode and cathode.

24. The E^0_{cell} for the following reaction is +0.45 V.

$$2Ag^{+} + Cu \rightarrow 2Ag + Cu^{+2}$$

If $[Ag^{+}] = 0.010\ M$ and $[Cu^{+2}] = 0.0010\ M$ initially, what is the true E for this cell at 25°C?

25. The concentration of Ag^{+} ions in a cell is 0.010 *M*. The concentration of Mg^{+2} ions is 0.0010 *M*. The cell reaction is

$$Mg + Ag^{+} \rightarrow Mg^{+2} + Ag$$

The E^0 for the cell is +3.18 V. What is the E under the above concentration conditions?

26. What is the E for the Sn^{+4}/Sn^{+2} half-cell if $[Sn^{+4}] = 0.10\ M$ and $[Sn^{+2}] = 0.0010\ M$?

27. What is the E for a Ce^{+4}/Ce^{+3} half-cell if $[Ce^{+4}] = 0.01\ M$ and $[Ce^{+3}] = 0.001\ M$?

28. If the E (under standard conditions) for the reaction of zinc ions with iron metal (to give ferric ions) is -0.72 V, what is the E if the zinc ion concentration is 0.01 *M* and the ferric ion concentration is 0.001 *M*?

29. What is the E for a cell in which tin(II) ions react with magnesium metal to give tin metal when the concentrations are magnesium ions = 0.01 *M*, tin(II) ions = 0.001 *M*?

30. What is the E for a $Cr^{+3}/Cr//Ni^{+2}/Ni$ cell if $[Cr^{+3}] = 0.000010\ M$ and $[Ni^{+2}] = 0.010\ M$?

31. What is the E for a $Cu^{+2}/Cu//Zn^{+2}/Zn$ cell if $[Cu^{+2}] = 0.010\ M$ and $[Zn^{+2}] = 0.0010\ M$?

$$Cu + Zn^{+2} \rightarrow Cu^{+2} + Zn$$

32. What is the E for a $Sn^{+2}/Sn^{+4}//Zn^{+2}/Zn$ cell if $[Sn^{+2}] = 0.010$ *M*, $[Sn^{+4}] = 0.00010$ *M*, and $[Zn^{+2}] = 0.0010$ *M*?

$$Sn^{+2} + Zn^{+2} \rightarrow Sn^{+4} + Zn$$

33. Define potentiometry, reference electrode, ground, indicator electrode.
34. Tell how the SCE electrode works (in terms of the Nernst equation) as a reference electrode.
35. Tell how the concept of the salt bridge is put into practice with reference electrodes.
36. In terms of the Nernst equation, tell how the silver-silver chloride electrode works as a reference electrode.
37. Tell how the pH electrode works (in terms of the Nernst equation) as an electrode for determining the pH of a solution.
38. Concerning the pH electrode, the following defines the potential that develops when it is immersed into a solution.

$$E = E^0 + 0.059 \log \frac{[H^+](\text{external})}{[H^+](\text{internal})}$$

Given this, explain how we can then say that this potential is directly proportional to the pH of the solution into which it is immersed.

39. Briefly describe how a pH meter is standardized.
40. What is a combination pH electrode? Describe its construction.
41. What is an ion-selective electrode? Identify and describe four different types.
42. The following data were obtained using a nitrate electrode for a series of standard solutions of nitrate:

$[NO_3^-]$ (*M*)	E (mV)
10^{-1}	85
10^{-2}	150
10^{-3}	209
10^{-4}	262

(a) Plot the calibration curve for this analysis.
(b) What is the nitrate ion concentration in a solution for which E = 184 mV?

43. Why is a platinum electrode needed in some half-cells?
44. What is a potentiometric titration?
45. Does potentiometry utilize galvanic cells or electrolytic cells? Does amperometry utilize galvanic cells or electrolytic cells?
46. Define working electrode, reference electrode, counter electrode, auxiliary electrode, potentiostat, polarograph.
47. What is the "supporting electrolyte" and why does it not interfere with the measurement of the analyte in a polarography experiment?
48. What are the two advantages of the DME over some stationary electrodes?

49. Sketch a sampled dc polarogram and indicate on your sketch what is meant by half-wave potential, diffusion current, and residual current.
50. Draw a rough sketch of a sampled dc polarogram showing the reduction of a species having a half-wave potential of −1.20 V and a diffusion current of 7.5 μamp. Make it clear in your sketch what is meant by the half-wave potential and the diffusion current.
51. Define half-wave potential and discuss its usefulness in a quantitative analysis.
52. Explain the reference to half-wave potential as an electrochemical "barrier".
53. Define Faradaic current, non-Faradaic current, polarographic maximum.
54. In each of the following, tell what potential/time characteristic is applied to the DME and what is the current potential/potential output results.
    (a) classical polarography
    (b) sampled polarography
    (c) pulse polarography
    (d) differential pulse polarography
55. In an analysis of a solution for Cd^{+2} as in Experiment 58, the following data were obtained:

C (ppm)	Peak Height
5	86.0
10	171.0
15	261.0
20	349.0
Unknown	162.0

    What is the concentration of Cd^{+2} in the unknown?
56. Under what circumstances would one want to choose polarography over, say, atomic absorption? (State two such circumstances.)
57. Differentiate between polarography and voltammetry.
58. Explain what is meant by (a) stripping voltammetry, (b) cyclic voltammetry, (c) amperometric titration.
59. Answer the following questions TRUE or FALSE.
    (a) The SCE reference electrode is not required for many polarography applications.
    (b) In polarography the working electrode is also the auxiliary electrode.
    (c) The supporting electrolyte is necessary in order to pass the current, of whatever magnitude, to be measured.
    (d) The supporting electrolyte's concentration must be kept low so as not to interfere with the analysis.
    (e) "DME" stands for "dropping mercury clectrode".
    (f) One advantage of the DME is that metals do not contaminate its surface by building up over a period of time.
    (g) With the DME the analyte species must be transported to the surface by diffusion.

(h) Voltammetry and polarography are the same thing.
(i) The half-wave potential is half the diffusion current.
(j) The diffusion current, i_d, is the current due to the species of interest in a given polarography experiment.
(k) The supporting electrolyte may give a sudden and large increase in current on a polarogram.
(l) The "up and down" motion of the recorder pen evident in a classical dc polarogram is due to the fact that the analyte must diffuse to the surface of the mercury.
(m) In sampled dc polarography, the current is measured and displayed on the recorder in the split second before the drop falls off.
(n) Pulse polarography is less sensitive than sampled.
(o) In pulse polarography the standard potential "ramp" is applied to the DME.
(p) In differential pulse polarography, pulses are superimposed on a ramp.
(q) In pulse polarography the "rate of change" of the normal polarogram is displayed on the recorder.
(r) In differential pulse polarography the peak appears at the half-wave potential.

60. Match the term in the list with the phrases that follow. The terms may be used more than once.

(a) classical polarography
(b) sampled dc polarography
(c) pulse polarography
(d) differential pulse polarography
(e) stripping voltammetry
(f) cyclic voltammetry
(g) amperometric titration

1. a technique in which a series of increasing potential pulses are applied to the DME
2. a technique that is not intended for quantitative analysis
3. a technique characterized by the repeated "up and down" motion of the recorder pen
4. a technique in which a series of pulses superimposed on a "ramp" is applied to the DME
5. a technique in which a current is measured in order to tell when a separate reaction is complete
6. a technique in which a peak is recorded on the recorder
7. a technique other than classical polarography which is used for quantitative analysis but is not more sensitive than classical polarography
8. a technique in which the analyte is reduced onto or into an electrode over a period of time and then oxidized again
9. a technique in which the sensitivity is controlled according to a time factor in the beginning of the experiment

10. a technique which measures the "rate of change" of a standard polarogram
11. a technique which records a smooth polarogram that coincides with the top of the recorder trace of the classical polarogram
12. a technique in which the current is sampled twice during the growth of the mercury drop

61. A sample of pond water is analyzed for lead content by classical dc polarography and the standard additions method as follows: 25.00 mL of the sample are measured into a beaker, enough concentrated HNO_3 is added to make the sample 0.1 *M* in HNO_3, and the polarogram is obtained. The lead ion reduction occurs at $E_{1/2}$ = 0.37 volts and the diffusion current is determined to be 7.5 μA. Next, 0.01 mL of 100 ppm Pb^{+4} is added and the diffusion current rises to 10.8 μA. What is the concentration of lead in the sample in ppm? (Hint—consider the dilution effect of the concentrated HNO_3 to be negligible.)

Chapter 19

Applications Summary

19.1 INTRODUCTION

The purpose of this chapter is to give you some idea of the variety of applications that the methods of chemical analysis introduced in this text have in the modern analytical laboratory. Specific applications are as diverse and as numerous as the different kinds of laboratories that exist. First, we have quality assurance laboratories, whose purpose is to ensure that the specifications of manufactured products and the raw materials that go into these products are met. Examples here would include food, pharmaceuticals, and other consumables. Next, we have private testing laboratories, which contract with government agencies and private parties for chemical analysis. A good example is with environmental analysis, such as soil, air, and water. Third, we have the governmental agencies themselves, which often have their own laboratories. Examples of these would include water and wastewater treatment plants (city government); state health departments, which use chemical analysis to analyze samples impacting human health; and federal agencies, such as the Soil Conservation Service and the Environmental Protection Agency. Fourth would be hospital and other clinical laboratories that employ medical laboratory technicians and medical technologists. Laboratory work in this area includes the analysis of urine, blood, plasma, mucous, and tissue samples in order to assist medical doctors in the diagnosis and treatment of diseases. Finally, research labs exist in almost all the situations noted above. Industries are continuously researching new analytical methods, for example, and academic researchers at universities rely heavily on analytical chemistry as an aid to their research.

This list of examples is not intended to be comprehensive, but only to give you an idea of the tremendous number and variety of chemical analysis applications that may be encountered.

19.2 WET METHODS

As stated earlier, classical wet methods (gravimetric and titrimetric) have largely been replaced by instrumental methods. However, both gravimetric and titrimetric methods continue to be used in situations in which (1) they are more convenient, (2) analyte concentrations are so high that instrumental methods are not appropriate, and (3) they are cost-effective.

If a gravimetric procedure requires a simple physical separation of the analyte from the sample, as opposed to requiring a chemical reaction to form a weighable product (precipitate), then such a procedure is likely to be more convenient than any instrumental method. An example would be in a wastewater treatment plant laboratory in which total solids and suspended solids are determined. The procedures for such analytes are purely gravimetric and involve simple physical separation schemes. In the case of total solids, a wastewater sample is added to a preweighed evaporating dish, the water evaporated in an oven, and the dish is weighed again. The difference in the two weights is the total solids in the sample. For suspended solids, a preweighed filtering crucible is used to filter the undissolved solids. After drying, this crucible is weighed again, and the difference in the two weights is the suspended solid content of that sample.

Similarly, moisture or any volatile component in various types of samples can be determined by evaporation and weighing the container before and after. Any gravimetric procedures involving a chemical reaction to form a weighable precipitate, such as in Experiments 2 and 3 in this textbook, have become obsolete and are virtually non-existent in today's laboratory.

An example of a titrimetric procedure that is useful because of its convenience is total alkalinity. Alkalinity is the capacity of dissolved solutes to neutralize an added strong acid. The strong acid is conveniently added from a buret and the neutralization point (end point) conveniently signaled with the use of an indicator or pH meter. This procedure is in common use in laboratories in which total alkalinity of water-based samples needs to be determined.

We have mentioned the Kjeldahl procedure earlier (Chapter 6). The procedure as outlined in Chapter 6 continues to be very popular for nitrogen-containing samples such as grain, flour, soil, cereals, animal feed supplements, etc. There are some modern developments in this area, however, which speed up the process and can be considered at least partially instrumental.

We have mentioned water hardness and discussed the details of the analysis at length (Chapter 7). Examples of oxidation-reduction applications have already been given (Chapter 8).

An example of an analyte concentration being too high for instrumental methods is in the analysis of plating solutions. Some, but not all, of the ingredients in such solutions are present at such high concentrations that the instrumental methods cannot be used without considerable dilution. It is not unusual to encounter titrimetric procedures in use in plating industry laboratories for such analytes as chloride (the Mohr method—Chapter 8), cyanide, sulfates, etc. The instrumental methods are usually most useful for low concentrations (ppm level) as you have seen in many of the experiments in this text.

19.3 INSTRUMENTAL METHODS

We have already noted how some instrumental techniques are better suited for some analytes than others. We will now summarize the specific applications of each.

UV-vis Molecular Spectrophotometry

This is an analysis of samples for specific dissolved molecular or ionic species that are capable of absorbing UV or visible light in that form. The samples must be able to be measured in the form of a solution held in a glass or plastic cuvette. Most analytes are dissolved molecular or ionic species and are capable of light absorption, and thus this technique has very wide application in analytical laboratories. It is frequently necessary, however, for certain chemical conditions to be met first, and thus considerable sample preparation may be involved before the solutions can be measured accurately. These chemical conditions can mean a pH adjustment, the addition of a ligand to produce the required absorbing species, the addition of an oxidizing or reducing agent to give the required oxidation state (see Experiments 22 and 26), an extraction or other separation procedure (Experiment 43), a chemical reaction to remove an interfering species, etc. Interfering species may be compensated for in other ways, such as what was done in Experiment 24. It is the presence of interfering species, however, that often is the cause for the inapplicability of this technique.

This latter point is especially evident with applications in the UV region. Nearly all chemicals absorb UV light to some degree at almost all wavelengths; thus, if more than one substance is present in a solution to be measured, then an interference is virtually guaranteed. The solution to this problem would be to cleanly separate the analyte from the interference, such as by extraction or column chromatography. The word "cleanly" is important, since even small quantities would cause interference given the sensitivity of the technique.

A modern solution to this problem, of course, is HPLC, which is a technique specifically developed to deal with complex mixtures in liquid solvents. The chromatography accomplishes the required "clean" separation, and this is immediately followed by an on-line detection system which frequently is a UV absorbance spectrophotometer (Chapter 17). Organic species, such as drugs, pesticides, herbicides, and vitamins, are especially important applications for such a system.

With visible spectrophotometry the list of applications is virtually endless. Entire volumes have been written that are devoted solely to these so-called "colorimetric" methods. Some examples, in addition to the experiments in this text, which include iron and nitrate in water, metals in alloys, ozone in air, and phosphorus in soil and water, are nitrite in meats, lead in air or other samples, phenols in water, and glucose in clinical samples.

Infrared Spectrometry

The two most important applications of infrared spectrophotometry can be summarized as characterization and identification of molecular (nonionic) compounds. Substances are characterized in terms of the nature of their covalent bonds and in terms of what organic functional groups are present. Identification results from the matching of the unknown's IR spectrum to that of a known from a catalog of spectra, usually following the characterization step. In other words, once a substance has been characterized in terms of what bonds and functional groups are present, one can examine known spectra that contain those bonds and identify the substance by matching the spectra.

Identification of pollutants in water, air, and soil, identification of contaminants in industrial raw materials and products, and characterization and identification of new products are just a few examples of applications of IR spectrometry. It must be stressed that a solution of the analyte is generally not useful here. The substances must be pure in order to have accurate and reliable identification. Chromatography, distillation, recrystallization (see Chapter 15), and other methods of organic purification may be required.

Fluorometry

The selectivity, sensitivity, and applicability of fluorometry were discussed in Chapter 11. In summary, fused benzene ring systems have the greatest applications potential. Thus, riboflavin, thiamine and other vitamins, many drugs, and some pesticides and other carcinogens are analyzed with this technique. In addition, as was mentioned in Chapter 11, metal ions (e.g., Al) that can be complexed with fluorescent ligands and inorganic ions that quench fluorescence (e.g., nitrate) can be analyzed by fluorescence. In the case of complex mixtures which contain interfering substances, HPLC instruments incorporating fluorescence detectors are very useful.

Atomic Absorption and Emission

In the case of atomic techniques, the key word is *metals*. Metal ions capable of atomization in a flame, graphite furnace, or ICP source are the analytes with this technique. Occasionally, nonmetal species are measured, but only indirectly as a result of the measurement of some metal.

A broad range of sample classifications contain metals. Agricultural applications include soils, fertilizers, plant tissue, and feeds. Environmental applications include water (surface, ground, and sea) and air. Food product applications include meat, fish, oils and fats, milk, and beverages. In addition, there are geochemical applications, including rocks, minerals, and ores; biochemical applications, including serum, urine, blood, and tissue; and industrial applications, including plating solutions, metallurgical samples, petroleum products, and pharmaceuticals. It's almost an endless list. Atomic absorption and emission instruments have become very commonplace in analytical testing labs today.

Problems with interferences are not as evident here as with the molecular procedures. The reason is the use of line spectra as opposed to continuous spectra. As was stated previously, particularly for the UV region, all molecular species absorb all wavelengths to some degree, and thus you only rarely can eliminate an interference by changing the wavelength. With atomic techniques, however, only very narrow wavelengths bands (lines) get absorbed, and thus there is seldom an overlap; when there *is* an overlap, the effect an be eliminated by zeroing in on a secondary line.

Thus, the separation procedures that often must be executed in the molecular case often need not be considered here. In fact, it is often possible to aspirate a liquid sample (such as environmental water) directly into the atomizer without any pretreatment. Samples such as soils, rocks, feeds, and other solids must, of course, undergo a dissolution or extraction procedure before they can be measured, just as in the molecular case.

NMR and Mass Spectrometry

NMR and mass spectrometry both are designed mostly for structure elucidation of organic analytes. Mass spectrometry has found a substantial application in conjunction with gas chromatography, and potentially with HPLC, as a technique for identifying components of complex mixtures.

Gas Chromatography

As indicated in Chapter 16, applications using GC are restricted to samples that are already gaseous or samples of liquids and solids that can easily be made gaseous. The vast majority of examples of applications are for volatile liquids. Thus, most applications are for organic liquid mixtures or substances that are extractable into organic liquids. While this restriction may appear to limit application potential, the actual application range is quite broad. We have, for example, environmental analysis, to include air pollutants, such as from automobile exhausts and industrial manufacturing; water pollutants, such as pesticides and herbicides from agricultural applications; and soil pollution, such as from underground storage tanks for organic chemicals and hazardous wastes. We also have food additives and natural food ingredients, such as carbohydrates and other organics. We have industrial manufacturing quality control, such as pharmaceuticals, pesticide formulations, and paint products, and we also have biomedical applications, such as steroids, amino acids, and proteins.

HPLC

Applications with HPLC are probably greater in number than with GC for two reasons: (1) the samples need not be gaseous, and (2) in addition to most of the organic applications noted with GC, we also have inorganic applications – dissolved inorganic ions (ion-exchange chromatography and ion chromatography). In addition, as you saw in our discussion of molecular absorption and fluorescence, HPLC is a tool that can quickly and conveniently separate an analyte prior to UV or fluorescence detection.

Electroanalytical Techniques

Potentiometry techniques have found widespread use considering the fact that the use of the pH meter, pH electrode, and reference electrode to measure pH is so commonplace in virtually all analytical (and other) laboratories. Not only is the measurement of the pH of analytical laboratory samples often important; the adjustment of the pH of extraction solutions, mobile phases, and the like has become equally important. Despite potential difficulties with interferences, ion-selective electrodes are also in common use. An example is the analysis for fluoride in municipal drinking water. An example of an application of potentiometric titration is the "bipotentiometric" titration of various samples for water—the so-called Karl Fischer titration.

While the potential application of polarographic techniques is quite good, we find relatively few such procedures in actual use. Probably the major reason for this is that working with mercury is potentially unsafe. The hazard with mercury is its vapor. Like all liquids, mercury has a vapor pressure, which means that if mercury is left exposed to laboratory air, mercury vapor will be in the air and the laboratory workers will be breathing it. Short-term exposure to such vapor is not a big problem, but long-term exposure is serious since mercury is a cumulative poison.

It is a simple matter to keep all mercury containers stoppered so that it is not exposed. If some of the mercury should happen to be spilled on the floor, however, a through cleaning procedure is very important, since it can creep into cracks and crevices and never be seen. Special mercury clean-up kits are manufactured for the purpose of cleaning up such spills.

The hazards aside, however, polarographic techniques are quite good analytically and have application in a large number of sample classifications. These include water, wastewater, plating solutions, vitamins, clinical samples, foods, and pharmaceuticals. The requirement is that the analyte be capable of oxidation or reduction at an electrode surface at relatively small voltage values. If this requirement is met, the choice of this technique over the others listed then depends on sensitivity differences and applicability.

19.4 REFERENCE SOURCES

Table 19.1 summarizes the application information just presented. In addition, it may be important for a technician to be able to glean information from reference sources. The nature of such information may be additional theory about individual techniques (other than what has been presented in this text), specific procedural information relating to a given analysis, or perhaps entire procedures for a given analysis. Some excellent reference sources for these are given in Appendix 10.

19.5 INTRODUCTION TO AUTOMATION

There is a continuing effort to streamline analytical laboratory procedures. Over the years, such procedures have developed from the time-consuming classical wet methods of analysis to the more sophisticated and faster instrumental methods. In turn, the instruments have undergone continuous improvement and streamlining to the point where large samples and standards are able to be tested in a very short period of time through what are called "automated" procedures. Today, automatic titrators, automatic analyzers which incorporate instrumental designs and concepts, and automatic samplers which are attached to traditional instruments are commonplace. Even complex sample preparation schemes involving sample dissolution and extraction utilize robotics to help streamline the procedures. In

Table 19.1. A summary of the applications of the major analytical techniques.

Technique	Current General Application
Gravimetric analysis	Analytes separated by physical means and easily weighed
Titrimetric analysis	Analytes present at high concentrations, or for which there are convenient, well-established methods.
UV-vis spectrophotometry	Molecular and ionic analytes capable of absorbing UV or visible wavelengths while in dilute solution
Infrared spectrometry	Molecular analytes only, most often for qualitative analysis
Fluorometry	Molecular or ionic analytes capable of fluorescence or fluorescence quenching while in dilute solution.
Nuclear magnetic resonance	Characterization of the structure and identification of organic and inorganic analytes
Mass spectrometry	Characterization of the structure and identification of organic and inorganic analytes
Atomic absorption and emission	Metals in dilute solution, natural liquids, and extracts and solutions of solids
Gas chromatography	Mixtures of volatile organics, organic solvent extracts and gases
HPLC	Complex mixtures (solutions) of analytes, including liquids and solids, organic and inorganic
Potentiometry	Ionic or molecular analytes capable of altering the potential of an electrode dipped into the solution

this chapter we discuss the design and theory of automatic titrators, the Technicon AutoAnalyzer®*, flow injection analyzers (FIA), and robotics.

Automatic Titrations

When a large number of titrations is part of a laboratory's daily workload, it is possible for the laboratory to employ a partially automated or even a fully automated system. This system may be something simple, such as a unit that automatically refills the buret or refills the buret with the use of a vacuum bulb, after a titration. In this case, a turn of the stopcock causes the buret to refill either by gravity or by vacuum from a large reservoir of titrant. It may also be an automatic dispensing buret with a digital display of titrant volume. Such a unit fits into a large reservoir of titrant solution and dispenses the solution into the titration flask. The operator starts and stops the titration according to his/her observations of the indicator, but the titrant volume is automatically measured and displayed, thus eliminating meniscus reading errors.

A fully automated system involves what is called an automatic titrator or autotitrator. Such a unit not only utilizes an automatic dispenser, such as described above, but also has automatic potentiometric, amperometric, or coulometric end-point detection. The unit can even employ a computer-controlled carousel in which samples rotate one after the other automatically through the unit. Such a system automatically activates reagent addi-

*Registered trademark of Bran & Lubbe, Inc., Buffalo Grove, IL.

tion, stirring, measuring, titrating, aspirating, rinsing, and reconditioning of electrodes.

Segmented Flow Methods

The Technicon AutoAnalyzer®, developed in the late 1950s and early 1960s, represents the widely used analysis system that is based on "segmented" flow. It is an automation system in which the sample solutions, and various reagents with which the sample solutions are to be mixed, flow from their original containers through a maze of plastic tubes, the diameters of which define the rate of flow and thus the proportions to be mixed. At appropriate and strategic locations along the flow route, the reagents and samples come together and mix, creating the solution to be measured. In addition, air is drawn into the system, also through one or more plastic tubes, and thus air bubbles are introduced as a result which help to divide the flowing liquid into segments, hence the "segmented flow" designation. (See Figure 19.1.) The complete system consists of a series of specific modular units assembled for a particular analysis. Modular units common to all analyses using this system are an automatic sampler (an "auto-sampler"), a peristaltic proportioning pump, a colorimeter, and a recorder. The proportioning pump is "peristaltic", which means the pumping action is performed through the repeated squeezing of the tubes through the action of metal rollers. (See Figure 19.2.) The system of tubes positioned in the pump along with the interconnecting tubes, mixing coils (see below), and glass and plastic pieces is called the "manifold". The auto-sampler consists of a rotating carousel which positions vials containing the sample solutions in such a way that they are drawn out, one after another, through one of the tubes leading

Figure 19.1. An illustration of a plastic tube through which an air-segmented solution is flowing.

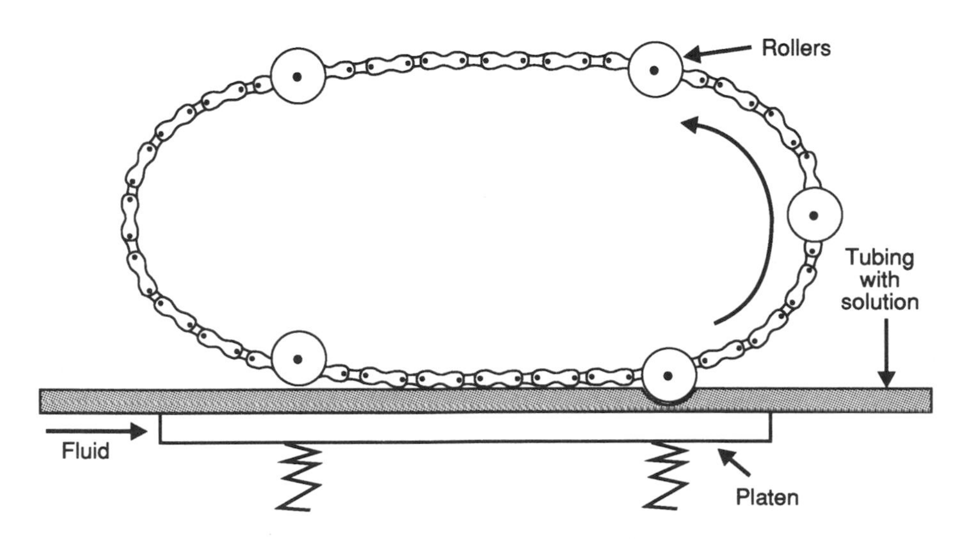

Figure 19.2. An illustration of a peristaltic proportioning pump.

to the proportioning pump. The length of time during which the sample is drawn out is determined by the operator through an interchangeable cam in the carousel.

At the same time, the pump also delivers the air bubbles, a diluent if desired, and one or more reagents needed for color development, such that all of these ultimately meet and move through a single tube. Thorough mixing then takes place by channeling this mixture, with the air bubbles, through coiled glass tubing for mixing. The air bubbles are of special assistance to the mixing process. Without air bubbles, a layer of a liquid adheres to the wall of the tubing, causing contamination of the flow stream to follow. After mixing and color development, the air bubbles are drawn off (via a "debubbler"), the solution flows through a detector, usually a colorimeter equipped with a flow-through cuvette, and the transmittance level is traced on the recorder and/or fed into a computer. The entire system, assembled for an analysis such as just described, is illustrated in Figure 19.3.

Specific methods and analytes require specific tubing diameters, junction tubes, mixing coils, etc. Transforming a system designed for phosphate analysis, for example, into one for chloride analysis requires a complete dismantling and re-assembly of the manifold, unless the manifold systems are stored so that they can be re-installed later. Technicon manufactures and sells all tubing and parts needed for a particular manifold. The tubing is color-coded, and tubing junctions and mixing coils are given part numbers which are specified in AutoAnalyzer® system flow diagrams available from Technicon or found in method books, such as in *Standard Methods for the Examination of Water and Wastewater*. Figure 19.4 gives two such system flow diagrams, one for chloride in water and wastewater (a) and one for phosphate in water and seawater (b). The color codes, such as "BLK/BLK" and "ORN/GRN", symbolize the colors of plastic clips that are found on the opposite ends of a given tube, and their associated flow rates are given within the parentheses next to the stated color code as shown. The clips are used to stretch and install the tubes in the pump. Part numbers for the tubing junctions and mixing coils are also given in these diagrams. You will notice, for example, that in the chloride system (Figure 19.4a) a 14-turn mixing coil has part number 116–0152–02.

You should also notice in Figure 19.4b that one of the mixing coils, the one positioned just before the colorimeter, requires an elevated temperature

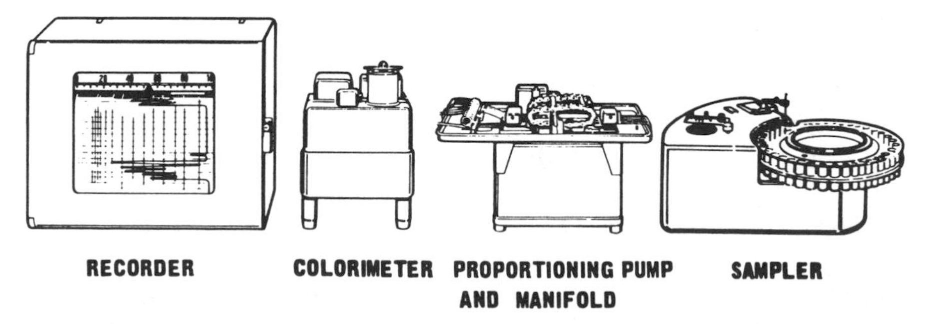

Figure 19.3. A drawing of the hardware components of the Technicon AutoAnalyzer® system. (Adapted from Publication No. TN0-0169-10, Technicon Instrument Corporation, Tarrytown, NY; Bran & Luebbe Analyzing Technologies, Buffalo Grove, IL. With permission.)

a

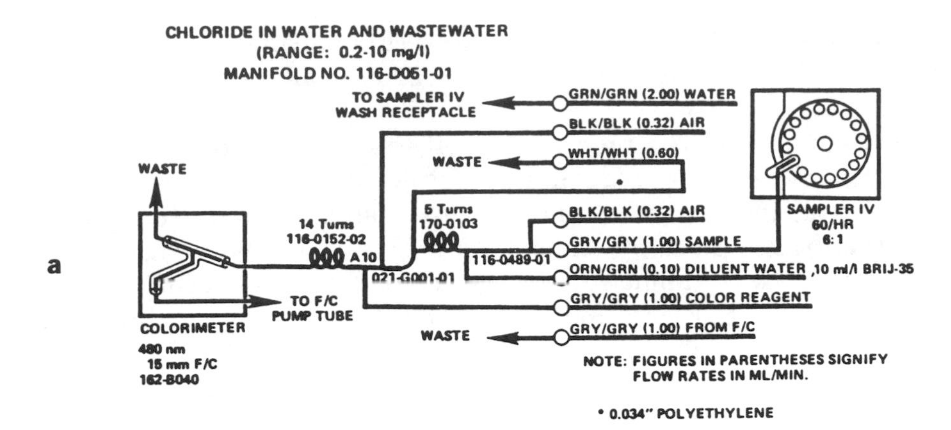

b

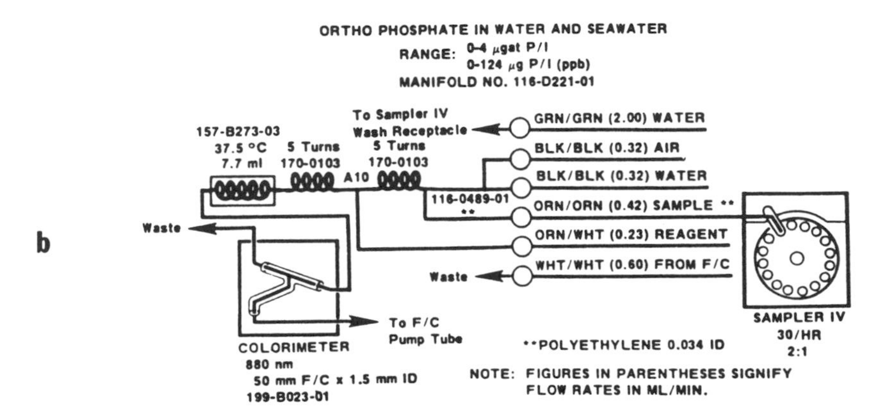

Figure 19.4. (a) The AutoAnalyzer® system for chloride; (b) the AutoAnalyzer® system for phosphate. (Courtesy of Technicon Instrument Corporation, Tarrytown, NY; Bran & Luebbe Analyzing Technologies, Buffalo Grove, IL.)

(37.5°C). A high-temperature bath is another module that can be added to the system. Yet another module is a dialyzer which is needed for glucose analysis. The colorimeter (see Chapter 11) consists of a visible light source, glass filters for monochromators, a flow cell (cuvette), a photocell, and the associated electronics required to produce a %T signal to be sent to the recorder. The air bubbles must be removed prior to the solution entering the colorimeter. A "debubbler" is used for this. Figure 19.5 is a diagram of the debubbler in combination with the flow cell. The air bubbles float and pass upward and out while the solution is drawn down into the flow cell by the vacuum created due to a connection to the pump.

The recorder records the %T signal over time. (It is a strip-chart recorder.) It is calibrated in %T units. As sample after sample (separated by distilled water volumes) drawn from the auto-sampler pass through the system, a recorder signal for each is generated, and these signals are then recorded one after the other on the chart paper. Figure 19.6 shows a sample of such a recorder trace. A popular modern alternative to the recorder is data acquisition with a computer.

A wealth of information, including additional details of the system set-up and details of specific methods and manifolds for various analytes,

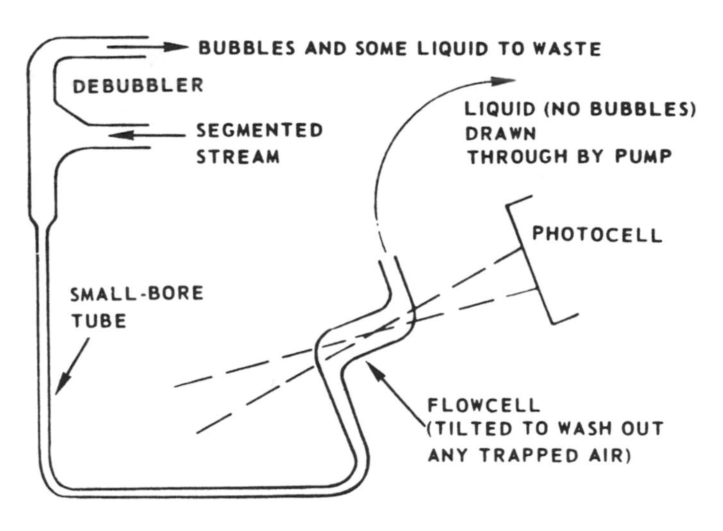

Figure 19.5. Flow cell and debubbler. (Reprinted from Publication No. TN1-0169-00, Technicon Instrument Corporation, Tarrytown, NY; Bran & Luebbe Analyzing Technologies, Buffalo Grove, IL. With permission.)

is available from the Bran & Luebbe Analyzing Technologies, Buffalo Grove, IL.

Flow Injection Methods

More recently, an automated flow system in which the various liquid flow streams are *not* segmented has been developed. To say that they are not segmented means that air bubbles, which are characteristic of the Technicon

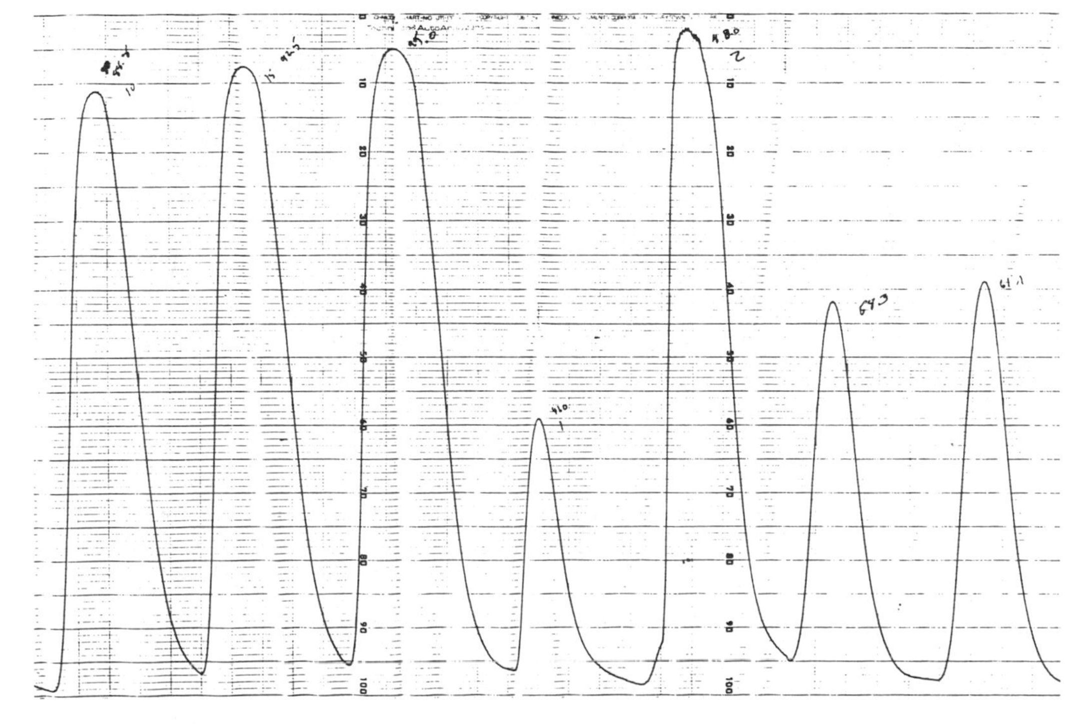

Figure 19.6. A typical AutoAnalyzer® strip chart recording of a series of solutions.

system described in the previous section, are not introduced at any time and, thus, the various flow streams are not characterized by segments separated by the bubbles. This non-segmented type of system is known as flow injection analysis, or FIA.

In the segmented flow method, the air bubbles serve to help mix the various flow streams after they come together in the system. The obvious question with regard to FIA then is, "How do the various flow streams adequately mix prior to measurement?" With FIA the tubing diameters are much smaller (0.5-mm i.d.), and a much smaller amount of sample is "injected" with a valve similar to HPLC systems described in Chapter 17. Thus, the need for thorough mixing is greatly reduced, as adequate mixing of such small quantities occurs more readily without the need for air bubbles. A small "plug" of sample is thus carried through the flow system and sharp spikes are observed on the recorder trace. The advantages are (1) greater sample throughput, meaning a more rapid analysis of large numbers of samples per unit time; (2) shorter time from injection to detection; (3) faster start-up, shut-down, and changeover; and (4) less sophisticated equipment. For an excellent review of FIA, please refer to a paper which appeared in *Analytical Chemistry* in 1978* and again in Volume 2 of *Instrumentation in Analytical Chemistry* in 1982**.

Robotics

Given the large volume of samples to be analyzed in many analytical laboratories, total automation of laboratory procedures has been a goal of analytical chemists and instrument manufacturers for years. With the advent of automation equipment, such as that described in this chapter, and the ever more sophisticated computer and computer software, discussed in Chapter 9, this dream is coming closer to reality. It is now commonplace for sample solutions and standards to be rapidly diluted and mixed with appropriate reagents, measured one after another quickly as they pass through a flow channel. Along with this, the resulting data can be acquired, manipulated, statistically adjusted, and stored by a computer, all with only a press of a button or a computer key, in an operation that is not only automated but may even be unattended for a substantial period of time. All that remains is for the chemist to take the raw samples and perform the dissolution and/or extraction techniques (described in Chapter 15) manually prior to executing the automated procedure.

The interesting news is that even these latter operations can be automated, since robots are available that will perform sample preparation tasks. Such robots are not androids or some other two-legged variety that may immediately come to mind in light of some futuristic TV show or movie. A robot typically consists of a robotic "arm" that is capable of 360° rotation around a table equipped with the necessary auxiliaries, such as tube racks, balances, extraction and dissolution solvent reservoirs, vortex mixers, shakers, dispensers, and centrifuges. (See Figure 19.7.) The robotic arm is programmed to move samples from one treatment station to another

*Betteridge, D., in *Analytical Chemistry*, 50(4), 832A (1978).

**Borman, Stuart A., Ed., *Instrumentation in Analytical Chemistry*, Volume 2, American Chemical Society, Washington, D.C., 1982.

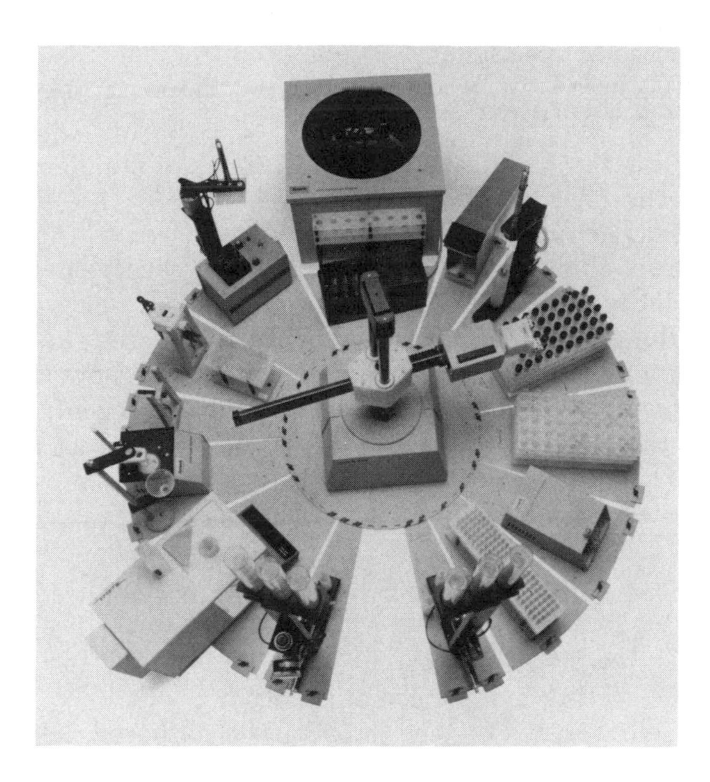

Figure 19.7. An example of a laboratory robotics system. (Courtesy of Zymark Corporation, Hopkinton, MA.)

in a sample preparation scheme designed to emulate the manual procedures. Besides having the ability to perform the routine preparation schemes without becoming bored, requiring a lunch or coffee break, or being limited to an 8-hr day, the robot is able to work in hazardous environments, which is a significant advantage given the increasing publicity of health risks associated with handling the chemicals needed for such schemes. Analytical chemists can fully expect further developments in this area as time goes on.

QUESTIONS AND PROBLEMS

1. Using library reference materials find a "real world" analysis application for each of six instrumental techniques – atomic absorption, gas chromatography, polarography, HPLC, UV or vis spectrophotometry, and fluorometry – and give the details of the procedure according to the following scheme:

   A. Title

   B. General information
      1. Type of sample to analyze
      2. Substance(s) to be determined
      3. Technique to use (AA, GC, polarography, etc.)
      4. How will the sample be collected?
      5. What will sample preparation involve? (Be concise but complete.)

C. Specifics of the technique to be used

Atomic absorption (AA)

1. Standard additions technique, series of standard solutions, or other?
2. Wavelength to be used
3. Fuel and oxidant to be used (e.g., air-acetylene)
4. Slit setting
5. Tell exactly how the data will be gathered, keeping in mind your response to (1) (in other words, tell how you would proceed with the technique chosen). Give concentration levels of standard solutions, state how you will make them up, etc.
6. Any possible problems?

Polarography

1. Standard additions, series of standard solutions, or other?
2. Supporting electrolyte
3. Half-wave potential
4. Operating mode (normal dc, etc.)
5. Tell exactly how the data will be gathered, keeping in mind your response to (1) (in other words, tell how you would proceed with the technique chosen). Give concentration levels of standard solutions, how you will make them up, etc.
6. Any possible problems?

Gas chromatography

1. Internal standard technique or other
2. Stationary-phase and solid support; length of column (if known)
3. Column temperature and the flow rate (if known)
4. Type of detector
5. Tell exactly how the data will be gathered, keeping in mind your response to (1) (in other words, tell how you would proceed with the technique chosen). Give concentration levels of standard solutions, how you will make them up, etc.
6. Any possible problems?

Liquid chromatography (HPLC)

1. What quantitation technique?
2. Type of LC (normal phase, reverse phase, etc.) and specific column packing used
3. Solvent characteristics (also tell whether gradient or isocratic and, if gradient, give details)
4. Type of detector
5. Tell exactly how the data will be gathered, keeping in mind your response to (1) (in other words, tell how you would proceed with the technique chosen). Give concentration levels of standard solutions, how you will make them up, etc.
6. Any possible problems?

UV or vis spectrophotometry

1. Beer's law plot, or other
2. Color-developing reagent(s) (for visible) or equivalent for UV.
3. Wavelength
4. Type of instrument needed (include light source, sample cell, and detector)
5. Tell exactly how the data will be gathered, keeping in mind your response to (1) (in other words, tell how you would proceed with the technique chosen). Give concentration levels of standard solutions, how you will make them up, etc.
6. Any possible problems?

Fluorometry

1. Is the substance self-fluorescing, or must you add something to make it fluoresce? If so, what?
2. Excitation wavelength
3. Fluorescence wavelength
4. pH required and how to adjust it
5. Tell exactly how the data will be gathered, keeping in mind your response to (1) (in other words, tell how you would proceed with the technique chosen). Give concentration levels of standard solutions, how you will make them up, etc.
6. Any possible problems?

D. Explain fully how your data are worked up. Include the kind of graphs required (example?) and how the information is extracted from the graph. Also include the calculation, etc. for the determination of the final desired result.

E. References

Appendix 1

Significant Figures

In the analytical chemistry laboratory, many measurements are made and the accuracy of these measurements obviously is a very important consideration. Different measuring devices give us different degrees of accuracy. A measurement of 0.1427 g is more accurate than a measurement of 0.14 g simply because it contains more digits. The former (0.1427 g) was made on an analytical balance, while the latter was make on an ordinary balance. A measurement recorded in a notebook should always reflect the accuracy of the measuring device. It doesn't make sense to use a very accurate measuring device and then record a number that is less accurate. For example, suppose a weight on an analytical balance was found to be 0.1400 g. It would be a mistake to record the weight as 0.14 g, even if you know personally that the weight is 0.1400 g. Presumably there are other people in the laboratory using the notebook, and your entry will be construed as to contain only two digits. The following example further illustrates this point.

Figure A1.1 shows a meter, such as a pH meter, or a readout meter on the face of some other laboratory instrument. The correct reading on this meter is 56.5. The temptation may be to write down 56 or 57. These latter readings are not correct in the sense that they do not reflect the accuracy of this measuring device. Measuring devices should always be used to their optimum capability, and this means recording all the digits that are possible from the device. The general rule of thumb for a device such as the meter in Figure A1.1 is to write down all the digits you know with certainty and then estimate one more. Obviously, the meter in Figure A1.1 shows a reading between 56 and 57, or "fifty-six point something". This "something" is the

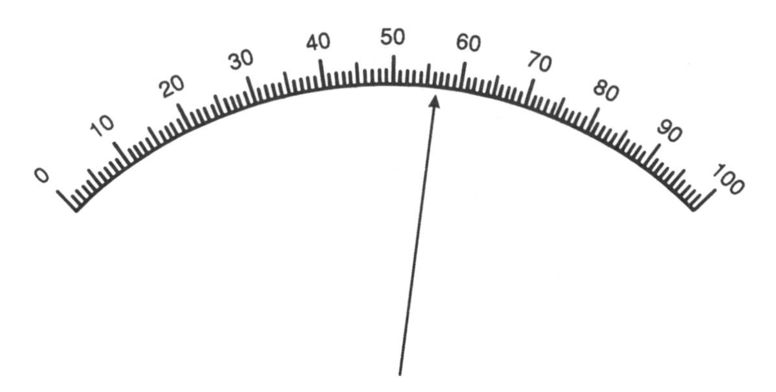

Figure A1.1. A measuring device registering a reading of 56.5 and not 56 or 57.

estimated digit and is estimated to be a 5. The correct reading is 56.5. For digital readouts, such as with an analytical balance, this estimation is done for you by the device.

The digits that are actually part of analytical measurements have come to be known as "significant figures". A knowledge of the subject of significant figures is important from the standpoint that (1) one needs to know the accuracy of a measurement from just seeing it in a notebook (and not necessarily from actually seeing it displayed on the measuring device) and (2) calculations are usually performed using the measurement, and the correct number of significant figures must be shown in the result of the calculation.

To cover point number 1 above, the following rules apply:

1. Any non-zero digit is significant. Example: 1.27 (three significant figures)
2. Any zero located between non-zero digits is significant regardless of the position of the decimal point. Example: 1.027 (four significant figures)
3. Any zero to the left of non-zero digits is not significant unless it is covered also by Rule #2. Such zeros are shown only to locate the decimal point. Example: 0.0127 (three significant figures)
4. Any zero to the right of non-zero digits and to the right of a decimal point is significant. Example: 1.270
5. Any zero to the right of non-zero digits and to the left of a decimal point may or may not be significant. Such zeros may be shown only to locate the decimal point, or they may be part of the measurement; one doesn't know unless he/she personally made the measurement. Such numbers are actually incorrectly recorded. They should be expressed in scientific notation to show the significance of the zero, because then Rule #4 would apply.

Example:

1270. (incorrect)

1.270×10^3 (four significant figures) or 1.27×10^3 (three significant figures)

To cover point number 2, the following rules apply:

1. The correct answer to a multiplication or division calculation must have the same number of significant figures as in the number with the least significant figures used in the calculation. Example: $1.27 \times 4.6 = 5.842 = 5.8$. Hence, 5.842 is the calculator answer, and 5.8 is the correctly rounded answer.
2. The correct answer to an addition or subtraction has the same number of digits to the right of the decimal point as in the number with the least such digits in the calculation.

   Example:

   $4.271 + 6.96 = 11.231 = 11.23$

   11.231 is the calculator answer; 11.23 is the correctly rounded answer.
3. When several steps are required in a calculation, no rounding would take place until the final step.
4. When both Rule #1 and Rule #2 apply in the same calculation, the number of significant figures in the answer is determined by following both rules in the order in which they are needed, keeping in mind that Rule #3 also applies.

   Example:

   $(7.27 - 4.8) \times 56.27 = 138.9869 = 1.4 \times 10^2$

   138.9869 is the calculator answer; 1.4×10^2 is the correctly expressed answer.
5. In cases in which a conversion factor that is an exact number is used in a calculation, the number of significant figures in the answer depends on all other numbers used in the calculation and not this conversion factor. To say that a number is exact means that it has an infinite number of significant figures and as such would never limit the number of significant figures in the answer.

   Example:

   $1.247\ \text{m} \times 100\ \text{cm/m} = 124.7\ \text{cm}$ (four significant figures)
6. In cases in which the logarithm of a number needs to be determined, such as in converting $[H^+]$ to pH or in the conversion of percent transmittance to absorbance, the number of digits in the mantissa of the logarithm (the series of digits to the right of the decimal point) must equal the number of significant figures in the original number.

   Example:

   $[H^+] = 4.9 \times 10^{-6}\ M$

   $\text{pH} = -\log[H^+] = 5.31$

This would appear to be an increase in the number of significant figures compared to the original number (three in 5.31, and only two in 4.9×10^{-6}), but the characteristic of the logarithm, the digit(s) to the left of the decimal point, represents the exponent of 10, which serves to designate only the position of the decimal point in the original number and, as such, is not significant.

Appendix 2

Table of Some K_{sp} Values

Substance	Solubility Product
Aluminum hydroxide	3.7×10^{-15}
Barium carbonate	2.58×10^{-9}
Barium chromate	1.17×10^{-10}
Barium fluoride	1.84×10^{-7}
Barium sulfate	1.07×10^{-10}
Cadmium sulfide	1.40×10^{-29}
Calcium fluoride	1.46×10^{-10}
Calcium sulfate	7.10×10^{-5}
Cupric sulfide	1.27×10^{-46}
Ferric hydroxide	2.64×10^{-39}
Lead carbonate	1.46×10^{-14}
Lead iodide	8.49×10^{-9}
Lead sulfate	1.82×10^{-8}
Magnesium hydroxide	5.61×10^{-12}
Manganese sulfide	4.65×10^{-14}
Nickel sulfide	1.07×10^{-21}
Silver bromide	5.35×10^{-13}
Silver carbonate	8.45×10^{-12}
Silver chloride	1.77×10^{-10}
Silver chromate	1.12×10^{-12}
Silver iodide	8.51×10^{-17}
Zinc sulfide	2.93×10^{-25}

From *Handbook of Chemistry and Physics*, 72nd ed., CRC Press, Boca Raton, FL, 1992. With permission.

Appendix 3

Table of K_a and K_b Values

Acids	Step	K_a
Acetic		1.76×10^{-5}
Ascorbic	1	7.94×10^{-5}
Ascorbic	2	1.62×10^{-12}
Benzoic		6.46×10^{-5}
Boric	1	7.3×10^{-10}
Boric	2	1.8×10^{-13}
Boric	3	1.6×10^{-14}
Bromoacetic		2.05×10^{-3}
n-Butyric		1.54×10^{-5}
Carbonic	1	4.30×10^{-7}
Carbonic	2	5.61×10^{-11}
Chloroacetic		1.40×10^{-3}
p-Chlorobenzoic		1.04×10^{-4}
trans-Cinnamic		3.65×10^{-5}
Citric	1	7.10×10^{-4}
Citric	2	1.68×10^{-5}
Citric	3	4.1×10^{-7}
Dichloroacetic		3.32×10^{-2}
Formic		1.77×10^{-4}
Iodoacetic		7.5×10^{-4}
Maleic	1	1.42×10^{-2}
Maleic	2	8.57×10^{-7}
Malonic	1	1.49×10^{-3}
Malonic	2	2.03×10^{-6}
Oxalic	1	5.90×10^{-2}
Oxalic	2	6.40×10^{-5}
Phenol		1.28×10^{-10}

Table of K_a and K_b values (cont.).

Acids	Step	K_a
Phenylacetic		5.2×10^{-5}
Phosphoric	1	7.52×10^{-3}
Phosphoric	2	6.23×10^{-8}
Phosphoric	3	2.2×10^{-13}
o-Phthalic	1	1.3×10^{-3}
o-Phthalic	2	3.9×10^{-5}
Sulfuric	2	1.20×10^{-2}
Sulfurous	1	1.54×10^{-2}
Sulfurous	2	1.02×10^{-7}
Trichloroacetic		2×10^{-1}

Bases	Step	K_b
Acetamide		2.34×10^{-1}
Ammonium hydroxide		1.79×10^{-5}
Aniline		2.34×10^{-5}
Arsenous oxide		1.1×10^{-4}
Beryllium hydroxide	2	5×10^{-11}
n-Butylamine		1.69×10^{-11}
t-Butylamine		1.48×10^{-11}
Calcium hydroxide	1	3.74×10^{-3}
Calcium hydroxide	2	4.0×10^{-2}
Hydrazine		1.7×10^{-6}
Hydroxylamine		1.07×10^{-8}
Lead hydroxide		9.6×10^{-4}
Pyridine		5.62×10^{-6}
Silver hydroxide		1.1×10^{-4}
Urea		7.94×10^{-1}
Zinc hydroxide		9.6×10^{-4}

From *Handbook of Chemistry and Physics*, 72nd ed., CRC Press, Boca Raton, FL, 1992. With permission.

Appendix 4

Compositions of Some Concentrated Commercial Acids and Bases

Acid or Base	Molarity	Density	% Composition (w/w)
Acetic acid ($HC_2H_3O_2$)	17	1.05	99.5
Ammonium hydroxide (NH_4OH)	15	0.90	58
Hydrobromic acid (HBr)	9	1.52	48
Hydrochloric acid (HCl)	12	1.18	36
Hydrofluoric acid (HF)	26	1.14	45
Nitric acid (HNO_3)	16	1.42	72
Perchloric Acid ($HC10_4$)	12	1.67	70
Phosphoric Acid (H_3PO_4)	15	1.69	85
Sulfuric Acid (H_2SO_4)	18	1.84	96

Appendix 5

Summary of Calculations for Titrimetric Analysis

Some chemical analysts prefer to work with formula weights, moles and molarity (what we have called the "FMM" system in Chapter 6), while others prefer to work with equivalent weights, equivalents and normality (the "EEN" system). This appendix summarizes the solution preparation, standardization, and percent constituent calculations for both systems. As discussed in Section 6.2, a molar ratio to convert moles of substance titrated to moles of titrant (or vice-versa) is needed for some calculations in the case of the FMM system (refer to Equations 6.18 and 6.19 and accompanying discussion).

1. Preparation of a solution from a pure, solid chemical:

   FMM: $L_D \times M_D \times FW_{Sol}$ = grams to be weighed

   EEN: $L_D \times N_D \times EW_{Sol}$ = grams to be weighed

   where L = liters, D = Desired, M = molarity, FW = formula weight, Sol = solute, N = normality, EW = equivalent weight.

2. Preparation of solutions by dilution:

FMM: $$L_B \times M_B = L_A \times M_A$$

EEN: $$L_B \times N_B = L_A \times N_A$$

where B = before dilution, A = after dilution.

3. Standardization with a primary standard:

FMM: $$L_T \times M_T = \frac{\text{grams}_{ST}}{FW_{ST}} \times \text{molar ratio}$$

EEN: $$L_T \times N_T = \frac{\text{grams}_{ST}}{EW_{ST}}$$

where T = titrant, ST = substance titrated.

4. Standardization with another standard solution:

FMM: $$L_T \times M_T = L_{ST} \times M_{ST} \times \text{molar ratio}$$

EEN: $$L_T \times N_T = L_{ST} \times N_{ST}$$

5. Percent constituent, direct or indirect titration:

FMM: $$\%\ \text{Const} = \frac{L_T \times M_T \times FW_{Const} \times \text{molar ratio}}{\text{Sample Weight}} \times 100$$

EEN: $$\%\ \text{Const} = \frac{L_T \times N_T \times EW_{Const}}{\text{Sample Weight}} \times 100$$

where Const = constituent.

6. Percent constituent, back titration:

FMM:
$$\%\ \text{Const} = \frac{\left(L_T \times M_T - L_{BT} \times M_{BT} \times \text{molar ratio}\right) \times FW_{Const} \times \text{molar ratio}}{\text{Sample Weight}} \times 100$$

EEN: $$\%\ \text{Const} = \frac{(L_T \times N_T - L_{BT} \times N_{BT}) \times EW_{Const}}{\text{Sample Weight}} \times 100$$

where BT = back titrant.

Appendix 6

Recipes for Selected Acid-Base Indicator Solutions

Methyl violet	0.01–0.05% in water
Cresol red	0.1 g in 26.2 mL of 0.01 N NaOH + 223.8 mL water
Thymol blue	0.1 g in 21.5 mL of 0.01 N NaOH + 229.5 mL water
Methyl orange	0.01% in water
Bromcresol green	0.1 g in 14.3 mL of 0.01 N NaOH + 235.7 mL water
Methyl red	0.02 g in 60 mL of ethanol + 40 mL water
Bromthymol blue	0.1 g in 16 mL of 0.01 N NaOH + 234 mL water
Phenolphthalein	0.05 g in 50 mL of ethanol + 50 mL water
Thymolphthalein	0.04 g in 50 mL of ethanol + 50 mL water
Clayton yellow	0.1% in water

From *Handbook of Chemistry and Physics*, 72nd ed., CRC Press, Boca Raton, FL, 1992. With permission.

Appendix 7

Basic Electrical Terminology

Because electronic instrumentation is in such heavy use in analytical laboratories today, an understanding of basic electrical terminology is quite appropriate. At various times throughout this text, reference is made to such electrical terms as "charge", "voltage", "potential", "current", "resistance", and "electric circuit". This appendix gives the definitions of these terms.

Charge. Electrical charge is a property of matter characterized by the buildup of electrons (negative charge) or the depletion of electrons (positive charge) on a body of matter. Examples are the negative and positive electrodes described in Chapter 18 and positive and negative ions. Like charges repel, while unlike charges attract. The existence of charge is also evidenced in life outside the laboratory. "Static cling" and the charge that builds up in your body when you slide your feet across a carpet on a dry day are also examples. Charged matter has a tendency to "discharge" given a sufficient opportunity for electrons to flow or to be transferred. Examples of this are lightning and the "snap" you feel when you touch a grounded conductor after sliding your feet across a carpet. We see an example of this in Chapter 14, in the discussion of "arc or spark emission spectrography".

Current. The movement of charge through a conductor constitutes current. Most often, this movement involves the flow of electrons, such as through a metal wire, but it can also be the movement of positive and negative ions, such as through a conducting solution. An analogy is often made that the flow of electrons through a metal wire is like the flow of water through a pipe. Some materials conduct a current better than others, much like some pipes conduct water better than others, depending on their diameters. Current can be precisely measured by devices known as ammeters in units of amperes, milliamperes, microamperes, etc. These are often abbreviated "amps" (A), "milliamps" (mA), "microamps" (μA), etc.

Resistance. All materials, except the so-called "superconductors", offer resistance to current flow. The amount of resistance depends on the material (the diameter of the pipe). With some materials, a high resistance is accompanied by energy loss as heat, such as in the burner on an electric stove, the heating element in an electric distiller or water heater, or the light source in an infrared spectrophotometer (Chapter 12). Resistance is measured and expressed in the unit "ohm", which is symbolized by the Greek letter omega (Ω). In addition, small electronic components called "resistors" are available. These typically consist of a material placed between two conducting wires that gives a segment of a circuit in a particular number of ohms of resistance. This concept is important when you want to cause a voltage drop (see the next definition) at a particular point in a circuit for some particular reason, such as when measuring current as in several of the GC detectors in Chapter 16. The pictorial symbol for resistance is -WWWW- .

Voltage. Water flowing through a pipe will flow at a rate depending not only on the diameter of the pipe, but also on the pressure of the water. This pressure is controlled either by a pump or by gravity. The parameter analogous to this pressure in electricity is voltage. The higher the voltage (measured in volts), the higher the pressure electrons feel to flow through the conductor. Current therefore depends on both voltage and resistance. A basic law of electricity is the law that relates voltage (V) to current (I) and resistance (R). It is called "Ohm's law":

$$V = IR$$

Thus, a given current flowing through a given resistance gives rise to a particular voltage drop (the water pressure on one side of a small diameter pipe inserted between two larger ones is different from that on the other side.) It is this voltage drop that constitutes the signal sent from a GC detector (Chapter 16) to the strip-chart recorder. In the flame ionization detector, for example, the current flow through a resistor in the circuit changes when a mixture component elutes from the column. The voltage drop thus changes and that is what the recorder measures.

Potential. In physics, potential is defined as the ability of a charge to do work, the work of interacting with another charge nearby. In the previous definition we referred to the fact that the voltage (pressure) on the opposite sides of a resistor is different, and we spoke of a voltage drop (analogous to a pressure drop). We can also describe this difference as a "potential difference". Thus, voltage and potential are two terms which describe the same phenomenon, and both use the unit "volt" (V) or "millivolt" (mV). Potential and voltage are both symbolized by "E", and a voltage source, such as a battery, is symbolized as (—⊣ ⊢+ —).

Electric Circuit. In order for a current to flow, there must be an uninterrupted pathway from the source of the electrons to their destination, such as from one pole of a battery to the other (see Chapter 18). This pathway is called a circuit. We often speak of a "completed" circuit. This term is associated with a circuit in which some electrical component, such as a switch, is used to achieve the pathway required at a certain instant. Thus, the switch (—⁄ —) in Figure 18.2, for example, completes the circuit in that figure.

Appendix 8

Common Metric Prefixes

Prefix (abbreviation)	Value	Example
Mega (M)	10^6	1 megaohm is 10^6 or 1 million ohms (1 MΩ)
Kilo (k)	10^3	1 kilometer is 10^3 or 1 thousand meters (1 km)
Deci (d)	10^{-1}	1 decigram is 10^{-1} or 1 tenth of a gram (dg)
Centi (c)	10^{-2}	1 centimeter is 10^{-2} or 1 hundredth of a meter (1 cm)
Milli (m)	10^{-3}	1 millivolt is 10^{-3} or 1 thousandths of a volt (mv)
Micro (μ)	10^{-6}	1 microliter is 10^{-6} or 1 millionth of a liter (1 μL)
Nano (n)	10^{-9}	1 nanometer is 10^{-9} or 1 billionth of a meter (1 nm)
Pico (p)	10^{-12}	1 picogram is 10^{-12} or 1 trillionth of a gram (1 pg)

Common units in analytical chemistry:

Length: meter (m), centimeter (cm), millimeter (mm), kilometer (km), millimeter (mm), nanometer (nm)
Mass: gram (g), kilogram (kg), centigram (cg), milligram (mg), microgram (μg), nanogram (ng), picogram (pg)
Volume: liter (L), milliliter (mL), microliter (μL)
Voltage: volt (V), millivolt (mV)
Current: ampere (A), milliampere (mA), microampere (μA)
Amount of substance: mole (mol), millimole (mmol)

Appendix 9

Answers to Selected Questions and Problems

CHAPTER 1

1. Qualitative analysis is identification—analysis for "what" is in a sample. Quantitative analysis is the analysis for quantity, or "how much" is in a sample.
2. Wet methods involve only reactions occurring in solution and the stoichiometry involved. Instrumental methods involve electronic instrumentation for measuring some parameter related to quantity or quality.
3. Mass is a measure of the amount of a sample. Weight is the measure of the gravitational pull on this amount of sample. On the surface of the earth, a weight measurement is taken to be the same as a mass measurement, utilizing the same units, although technically they are not the same.
4. A weighing device is called a balance because a weight is often determined by balancing the object to be weighed with a series of known weights across a fulcrum. Modern "torsion" balances, however, do not operate this way, but are still called "balances".
5. A single-pan balance is a balance with a single pan, which is used only for the object to be weighed. In the older styles, a constant counterbalancing weight is used while a number of removable weights on the same side as the object are added or removed in order to obtain the

weight of the object. The more modern single-pan balance are of the torsion variety.

6. To say a balance has the "tare" feature means that the balance can be zeroed with an object, such as a piece of weighing paper, on the pan. This helps to obtain the weight of a chemical directly without having to obtain and subtract the weight of the weighing paper.
7. Analytical balances are balances that measure to 0.1 or 0.01 mg.
8. A number of considerations are important. For example, the analytical balance must be level, the pan must be protected from air currents, the object to be weighed must be at room temperature, the object must be protected from fingerprints, etc. Refer to Section 1.7 for more details.
9. It depends on how heavy the sample must be. If it is less than 1 g, then an analytical balance must be used because only such a balance would give four significant figures. However, if the sample is greater than 10 g, then an ordinary balance is satisfactory since it would give four significant figures.

10. 50.594% 11. 27.239% 12. 48.301%

13. 55.992% 14. 22.009% 15. sand, 78.565%; salt, 21.435%

16. (a) 0.57445 (b) 0.3430 (c) 0.80847 (d) 1.4732 (e) 0.7071 (f) 0.6065 (g) 0.5501 (h) 1.103

17. (a) 1.8894 (b) 1.0447

18. (a) 0.8882

19. (b) 0.76950 (c) 0.83533

21. (a) 0.3622 (b) 0.9309 (c) 0.96618

22. 0.2693 g

25. 49.18%

26. 0.1840 g 27. 59.12% 28. 34.39%

31. 93.91% 33. 26.91% 34. 26.31%

35. The stirring rod is wet with your sample solution, and rinsing it back into the beaker ensures that all of your sample is present in the beaker for subsequent reactions and/or measurement.
36. It is a desiccant that changes color when saturated with adsorbed water.
37. There is no chance for the weighed sample to be anywhere but in your beaker. Also, the chance of contamination is eliminated.
38. This is done to get the precipitate particles to "clump" together to make them more filterable.
39. Avoid getting fingerprints on an item from the time you weigh it the first time until you weigh it the second time. It is during that time that the weight of a sample would be inaccurate due to fingerprints, since such added weight would alter the calculated weight of the sample.

40. This precipitate is AgCl, not $BaSO_4$. Its presence only signals the presence of chloride in the rinsings, thus indicating that more rinsing is needed.

41. Even though they are heat stable, the weight of such a label will change during heating. If there is a weight change for this reason, then the weight of the crucible contents calculated later will be incorrect.

42. The precipitate and filter paper are sopping wet at this point and would saturate the desiccant. Desiccators are mostly used to keep things dry once they are already dry, not to dry things that are wet.

43. A filter paper is "ashless" if it is completely volatilized when burned, leaving no ash residue. In Experiment 2, if the filter paper were not ashless, a residue would remain in the crucible and add weight to the precipitate. This in turn would increase the calculated percent and the results reported would be high.

44. The final results would be lower than the correct answer because some of the wax from the pencil would volatilize during the second heating, causing the crucible with the precipitate to weigh less than it should, thus causing the analyst to report a lower precipitate weight than he/she would otherwise.

45. The percent would be lower than the true percent because the weight of the precipitate (the numerator in the percent calculation) would be lower.

46. An ordinary balance would suffice because this solution is used only for a qualitative test – the identification of chloride in the rinsings.

CHAPTER 2

1. A representative sample must be obtained and transported to the laboratory safely without alteration.

3. For all of these (a–f), it's a case of sampling all parts of the systems that might be different so as to come up with a sample that represents the whole system. The final sample would then consist of the combination (well mixed) of all the individual samples taken. The exception would be if there is obviously a portion of the system, such as a corner of your yard (part "c"), that is different from the rest. In that case, it may be better to take two or more samples and run them separately.

4. Glass containers can leach trace levels of metals and contaminate the sample.

5. It can be refrigerated to slow down the bacterial action.

7. (a) water (b) HNO_3 (c) HF, $HClO_4$, or aqua regia
   (d) HCl (e) aqua regia (f) HF
   (g) H_2SO_4

8. (a), (g), (l) – sulfuric acid
   (b), (h), (i) – hydrofluoric acid
   (c), (e), (f) – nitric acid
   (d), (j) – hydrochloric acid
   (k) – perchloric acid
9. Aqua regia is a solution of HNO_3 and HCl in the ratio 1:3 by volume.
10. Refer to Table 2.2 and accompanying discussion.
11. (a) nitric acid (b) hydrochloric acid (c) hydrochloride acid
    (d) sulfuric acid (e) hydrofluoric acid (f) sulfuric acid
    (g) sulfuric acid
12. HCl – (b)
    HNO_3 – (e)
13. (a) acetonitrile, acetone, methanol
    (b) *n*-hexane, methylene chloride, benzene, toluene, diethyl ether, chloroform
14. Of those listed under 13b above, *n*-hexane, benzene, toluene, and diethyl ether are less dense than water and thus would be the top layer in an extraction experiment. Liquids with a higher density than water would sink to the bottom of the separatory funnel.
15. (a) All the organic liquids mentioned are toxic to a certain extent, and certainly they should never be ingested. However, some are more toxic than others and require special handling. These are acetonitrile, benzene, methanol, diethyl ether, and chloroform.
    (b) *n*-hexane, acetonitrile, acetone, benzene, toluene, methanol, and diethyl ether
    (d) *n*-hexane, methylene chloride, benzene, toluene, diethyl ether, and chloroform
18. Benzene is highly flammable and toxic. It is especially noted as a carcinogen. Breathing of vapors and skin contact should be avoided.
20. (1) analysis of water for pesticide residue; (2) analysis of soil for metals
21. Refer to Chapter 15, Section 15.4, for a thorough discussion of the separatory funnel.
22. "Extraction" is the selective dissolution, as opposed to total sample dissolution, of the analyte from a sample. The sample can be either a solid material, or a liquid, such as a water solution.
23. (a) Chloroform, ether, and benzene are common.
    (b) This is used anytime one wants to keep solid matter out of the extracting solvent, or when you want to sample to be contacted only by fresh solvent. See Chapter 15.
    (c) A "grab" sampler is used for sampling water and wastewater samples at various depths. It typically consists of a container that is lowered into the water and opened after it has been immersed using a remote device held by the operator.
24. Fusion is the dissolving of a sample with the use of a molten inorganic salt called a "flux".

26. Such material must be insoluble in the flux. Examples are platinum, gold, nickel, and porcelain.
27. The quality of the reagent must be assured in order for the analyst to have confidence that the sample and prepared reagents will not be contaminated and give inaccurate results.
28. "Reagent grade" refers to a rather high grade of purity, the limits for which have been established by the American Chemical Society or by the manufacturer.
29. No, they cannot be used because they generally are not pure enough.
30. "Spectro grade"
31. "HPLC grade"
32. It is very important for a sample or reagent to be properly labeled so that all laboratory personnel will have the knowledge of exactly what is in each container. This is essential for both safety and sample and reagent integrity. The label must include, as a minimum, the name of the sample or chemical and the date the sample was gathered or the solution prepared.

CHAPTER 3

1. There is a certain amount of error associated with all measurements regardless of the care with which the device was calibrated. This error is the uncertainty inherent in the measurement and the human error, either determinate or indeterminate, that can creep into an experiment. If the analyst and his/her client are to rely on the results, this error must be taken into account.
4. This is a determinate error because it is an error that is known to have occurred. Indeterminate errors are either errors inherent in a measurement or human errors that are not known to have occurred.
6. Yes, the results can be trusted if all determinate errors are eliminated and if indeterminate errors are taken into account by statistics.
8. They are very precise because they are within the uncertainty in the last significant figure expressed. It is not known with certainty if the results are accurate, although good precision usually indicates good accuracy.
10. The results are still quite precise (see the answer to #8 above) but, given the additional information, we now know that they are not accurate.
12. The mean is 38.51 and the median 37.23. The mode is not defined, since no measurement appears more than once and there is not a large number of measurements.
15. In numerical order, the relative ppt deviations are 0.03, 0.04, 0.04, 0.16, 0.00, 0.08, 0.05, 0.02.
16. average deviation = 0.0009, standard deviation = 0.001, relative standard deviation = 0.00007

17. The percent standard deviation = 1.3%. The stated precision is exceeded.
18. This is because it relates the value of the standard deviation to the value of the mean.
20. Using all eight measurements:

    s_R (ppt) = 0.075 → data not acceptable

    After rejecting #4:

    s_R = 0.042 → data acceptable
22. Range = 0.11
23. No, it is not rejectable because Q = 0.35, which is less than the 0.41 in Table 3.1 for the 90% confidence level for 10 measurements.
24. The results of the Q test are such that 0.1025 can be rejected. Thus, the average of five measurements is to be reported. It is 0.1032.
25. (a) 54.32% (b) 54.33% (c) 54.33% (d) in the order given: 0.11, 0.28, 0.25, 0.01, 0.21, 0.19, 0.01, 0.15, 0.03 (e) in the order given: 2.0, 5.2, 4.6, 0.18, 3.9, 3.5, 0.18, 2.8, 0.55 (f) 0.14 (g) 0.18 (h) 18% (j) 0.74
26. Q = 0.17. No value can be rejected.
27. 0.492 ± 0.011
28. 14.28 ± 0.16
34. Yes – The relative ppt deviation is 2.4; anything greater than 2.0 is not acceptable. (See Experiment 7.)
35. Yes; 0.09982 *N*

CHAPTER 4

3. $NaCl \rightleftharpoons Na^{+1} + Cl^{-1}$
4. $PbBr_2 \rightleftharpoons Pb^{+2} + 2Br^{-1}$
5. 8.49×10^{-9}
6. 5.08×10^{-5} *M*
7. 4.77×10^{-5} *M*
8. $[Ag^{+1}]$ = 2s because every time one Ag_2SO_4 unit dissolves, 2 Ag^{+1}'s result. The quantity "2s" is squared because the definition of K_{sp} calls for it to be squared, since the balancing coefficient for Ag^{+1} in the balanced equation is "2".
11. 8.39×10^{-12}
12. $2.14 \times 10^{-10} = K_{sp}$
13. 9.2×10^{-6}
14. $2.65 \times 10^{-2} = [H^{+1}]$; pH = 1.58
15. 4.87 = pH
16. $6 \times 10^{-9} = K_a$
17. 1.82 = pH

22. 1.57 = pH

23. 11.70 = pH

25. (a) correct (b) incorrect (c) correct
(d) correct (c) incorrect (f) correct
(g) incorrect (h) correct (i) incorrect
(j) approximation

26. (a) T (b) O (c) T
(d) A (e) T (f) T
(g) A (h) T (i) T
(j) O

27. (a) 1.30×10^{-2} *M*
(b) (i) entirely valid, since it is a true statement
(ii) valid in the sense that $[H^{+1}]$ is small compared to C_{HL}

28. (a) $PbSO_4 \leftrightarrow Pb^{+2} + SO_4^{-2}$ $K_{sp} = [Pb^{+2}][SO_4^{-2}]$

$K_{sp} = s^2$ $[Pb^{+2}] = [SO_4^{-2}] = s$

$$s = \sqrt{K_{sp}} = \sqrt{1.6 \times 10^{-8}}$$

$$= 1.3 \times 10^{-4}$$

(b) (i) entirely valid, since it is a true statement
(ii) same as i

29. (b) 30. (d) 31. (c)

32. (d) 33. (b) 34. (a)

CHAPTER 5

3. (a) 2.8 g of KOH are weighed, placed in the 500-mL flask, dissolved, and diluted to the mark.
(b) 2.2 g of NaCl in 250-mL flask, dissolved and diluted
(c) 36 g of glucose in 100-mL flask, dissolved and diluted
(d) 4.2 mL of 12 *M* HCl are diluted to 500 mL with water.
(e) 12.5 mL of the 2.0 *M* solution are diluted to 100 mL with water.
(f) 56 mL of the 18 *M* H_2SO_4 are diluted to 2.0 L with water.

4. (a) 3.5 g of NaOH are dissolved and then diluted to 250 mL.
(b) 7.3 mL of the 12.0 *M* solution are diluted to 250 mL water.

5. 4.7 g 6. 34 mg 8. 4.7 mL 9. 41 mL

10. 29 mL 11. 9.5 mL

12. (a) Weigh out 18 g of KNO_3, place in a container with a 500-mL calibration line, add water to dissolve, then dilute to the mark and shake.

(b) Measure out 39 mL of the 4.5 *M* solution, place it in a container with a 500-mL calibration line, then dilute to the mark and shake.

15. Weigh out 11 g of anhydrous sodium acetate and place it in a container with a 500-mL calibration line. Then, measure out 5.0 mL of concentrated acetic acid and place it in the same container. Add water to dissolve and then dilute to the 500 mL line and shake.

17. (a) false (b) true (c) true
(d) false (e) false (f) true
(g) true (h) false (i) false
(j) false (k) false (l) false
(m) false

18. (a) H_2SO_4 and $K_2Cr_2O_7$
(b) duopet and lambda (c) measuring
(d) the last bit of solution that remains in the pipet after gravity flow remains in the pipet and is not "blown out" into the receiving vessel. The calibration line is affixed accordingly.
(e) It is hygroscopic and cannot be weighed accurately.
(f) the thin liquid film that is present will dilute the solution to be transferred, unless they have been rinsed with this solution.

19. A volumetric pipet is more accurate, because the graduation line is located on a narrower diameter glass tube.

20. No, it should not, because the graduation line is affixed for the purpose of containment and not delivery.

21. It is calibrated for blow-out.

22. It is a substance that can be weighed accurately for the purpose of preparing a solution accurately, or for whatever purpose an accurate weight measurement is needed. See text for other desirable qualities.

23. A volumetric flask has a large base and a narrow neck where the calibration line is affixed. It is very accurate for measuring a contained volume. An Erlenmeyer flask is a "conical" flask that may or may not have calibration lines affixed. It is not very accurate for volume measurement.

24. The equivalence point is the point in a titration at which a titration reaction is complete. The end point is the point at which the reaction is complete as indicated by the indicator.

25. The thin liquid film present on the inside wall of the pipet will dilute the solution to be pipetted unless the pipet is first rinsed thoroughly.

26. The serological pipet is calibrated through the tip while the Mohr pipet is not. Thus, with a Mohr pipet, the meniscus may not drop below the last calibration line during a transfer.

27. (a) The volumetric flask is calibrated "to contain", not "to deliver". Thus, a small amount of the 50 mL contained in the flask would remain behind, wetting the inside walls of the flask, and not be delivered as desired. (b) For accurate transfer, the student should use a piece of glassware calibrated for transfer, such as a pipet or buret.

28. concentrated sulfuric acid and potassium dichromate
29. Sodium hydroxide is hygroscopic and cannot be weighed accurately.
30. See #25 above.
33. (a) Use an analytical balance for the weight of the chemical and a volumetric flask to measure and contain the volume of solution.
40. A graduated cylinder is good enough. Accuracy for this is not important because it is needed only to dissolve the solid.

CHAPTER 6

1. 0.1121 *M* 2. 0.09511 *M* 3. 0.1150 *M* 4. 0.1076 *M*
5. 0.09433 *M* 6. 0.1047 *M* 7. 0.1082 *M* 8. 0.09418 *M*
9. formula weight divided by

	Acid	Base
(a)	1	1
(b)	2	1
(c)	1	2
(d)	3	1
(e)	1	2
(f)	2	1
(g)	1	1
(h)	1	1
(i)	1	2

10. (a) NaOH – 40.00; H_3PO_4 – 98.00
(b) KH_2PO_4 – 87.09; $Ba(OH)_2$ – 85.68
14. 6.8 g
15. Dilute 1.5 mL of the 18 *M* solution to 500 mL.
16. Dissolve 7.1 g and dilute to 750 mL.
18. 15 g 20. 20 g 22. 1.0 mL
25. 0.08037 *M* 26. 0.1117 *N* 27. 0.08903 *N*
29. 0.2250 *N* 31. 0.1423 *N* 33. 0.3742 *N*
35. No. The THAM accepts only one hydrogen either way, so the calculations are identical.
37. 11.85% 39. 37.03%
41. 36.94% 42. 4.832%
44. A titration in which an excess of titrant is added with the excess being titrated with a second, or "back", titrant.
45. (a) conc. H_2SO_4 (b) conc. NaOH (c) NH_3 (g) indirect
(d) acid (e) acid (f) base
47. The quantity "$L_{BT} - N_{BT}$" represents the equivalents of titrant added in excess of the end point. Thus, it must be subtracted from the total equivalents added to represent the equivalents that actually reacted with the substance titrated.
50. No. Phenolphthalein's color change range is pH 8–10. An indicator that changes color in the pH 3–5 range is needed.

51. No. Acetic acid would require a color change in the pH range 7–11. Bromcresol green's is 3.5–5.5.

CHAPTER 7

1. Monodentate, bidentate, hexadentate – these are adjectives which describe a ligand in terms of the number of bonding sites (pairs of electrons) available for bonding to the metal ion. Mono = 1, bi = 2, hexa = 6.

   Ligand – the charged or uncharged chemical species which reacts with a metal ion forming a complex ion.

   Complex ion – a charged aggregate consisting of a metal ion in combination with one or more ligands.

   Coordinate covalent bond – a covalent bond in which the two shared electrons are contributed to the bond by only one of the two atoms involved.

   Water hardness – a term used to denote the Ca^{+2}, Mg^{+2}, and Fe^{+3} content of water both quantitatively and qualitatively.

   Aliquot – a portion of a larger volume of a solution, usually a transferred or pipetted volume.

3. $CoCl_4^{-2}$ = complex ion; Cl^{-1} = ligand
5. It is bidentate, because there are two sites (hence "bi" . . .) at which a metal ion will bond.
7. (a) 6 (b) 1 (c) hexadentate
10. A basic pH is needed to drive the titration reaction to completion. It cannot be too basic, however, or Mg^{+2} ion will precipitate as the hydroxide ($Mg(OH)_2$). pH = 10 represents a happy medium.
11. 6.3 mg of Mg are dissolved and diluted to 250 mL.
12. 23 mg
14. 7.5 mg of Mg are dissolved and diluted to 500 mL.
15. 7.5 mg = 0.0075 g
19. 9 mL of the 1000 ppm solution are diluted to 600 mL.
20. 13 mL of the 500 ppm solution are diluted to 250 mL.
22. 13 mg $CuSO_4$ are dissolved and diluted to 100 mL.
23. 45.6 mg
24. (a) 13 mL (b) 0.013 g (c) 0.022 g
26. 0.025 g
29. 4.7 g of H_2Na_2Y are dissolved and diluted to 500 mL.
31. 0.0103 *M*
32. 0.0196 *M*
33. 0.03711 *M*
36. 0.00775 *M*
37. 0.006418 *M*

39. 0.008402 *M*
41. 404.2 ppm $CaCO_3$
42. 536.4 ppm $CaCO_3$
44. 160.7 ppm
46. 314.5 ppm

CHAPTER 8

1. Oxidation – the loss of electrons, or the increase in oxidation number.

   Reduction – the gain of electrons, or the decrease in oxidation number.

   Oxidation number – a number indicating the state an element is in with respect to bonding.

   Oxidizing agent – a chemical that causes something to be oxidized while being reduced itself.

   Reducing agent – a chemical that causes something to be reduced while being oxidized itself.

2. (a) +5 (b) +3 (c) +6 (d) +5 (e) +7
   (f) +1 (g) +4 (h) +6 (i) +3 (j) +3

3. (a) +1 (b) -1 (c) +5 (d) 0 (e) +3 (f) +7

8. (a) CuO (or Cu) has been reduced; NH_3 (or N) has been oxidized.
   (b) oxidizing agent – Cl_2; reducing agent – KBr

10. (c) Yes – both Na and H have changed oxidation number.
    (e) No – no oxidation number changes have occurred.
    (f) Yes – both K and Br have changed oxidation number.
    (g) Yes – both Cl and O have changed oxidation number.
    (h) No – this is neutralization.

11. (b) is redox; Cu is oxidized, HNO_3 (N) is reduced. Cu is the reducing agent, HNO_3 the oxidizing agent.

12. (a) (2) is redox
    (b) HCl (or H)
    (c) Zn
    (d) lose

13. (s) $16H^+ + 2\,MnO_4^{-1} + 5H_2C_2O_4 \rightarrow 2Mn^{+2} + 8H_2O + 10CO_2 + 10H^+$
    (t) $6I^- + 14H^+ + Cr_2O_7^{-2} \rightarrow 3I_2 + 2Cr^{+3} + 7H_2O$
    (u) $Cl_2 + H_2O + NO_2^- \rightarrow 2Cl^- + NO_3^- + 2H^+$
    (v) $3S^{-2} + 8H^+ + 2NO_3^- \rightarrow 3S + 2NO + 4H_2O$
    (w) balanced as is

14. (a) $\frac{K_2Cr_2O_7}{6}$ (b) $\frac{Cl_2}{2}$

15. (a) $\frac{Na_2SO_3}{4}$ (b) $\frac{KI}{1}$ (c) $\frac{Cu}{2}$ $\frac{HNO_3}{3}$

16. 17.319 g/mol
17. 1.92 g
21. Dissolve 1.6 g $KMnO_4$ and dilute to 500 mL.
22. 1.7 g are dissolved and diluted to 500 mL.
24. (a) $6S_2O_3^{-2} + 14H^+ + Cr_2O_7^{-2} \rightarrow 3S_4O_6^{-2} + 2Cr^{+3} + 7H_2O$
    (b) 6.2 g are dissolved and diluted to 500 mL.
    (c) 0.4475 *N*
27. (a) 3.2 g (b) 0.1412 *N*
29. 2.957 *N*
31. 0.4341 *N*
33. 9.879%
35. 38.21%
39. The intensely purple colored MnO_4^- solution becomes easily visible when there is no longer any "ST" available to react.
40. (a) It takes on electrons readily.
41. An indirect titration is one in which the "ST" is determined indirectly by titrating a second species which is proportional to the "ST". A back titration is one in which the end point is intentionally overshot and the excess back titrated.
42. Iodometry is titrimetric method involving iodine. It is an indirect method because the product of the reaction of "ST" with I^- (I_2) is titrated.
43. KI – the "titrant" from which the I_2 is liberated
    $Na_2S_2O_3$ – the titrant for the liberated I_2
    $K_2Cr_2O_7$ – the primary standard for the $Na_2S_2O_3$

CHAPTER 9

4. 0.0170
6. 7.3 ppm
9. The linearity (or lack of linearity) of the readings is only known for the range of concentrations of the standards prepared. Without testing other standards, it is not known if the linearity extends beyond this range, and so the answer to the unknown cannot be reliably determined.
10. 1.85
12. 0.140
13. 4.84
14. The interpolation of the unknown results relies on the establishment of the linear relationship between the readout and the concentration.
18. slope, y-intercept, correlation coefficient, and concentrations of samples

19. Perfectly linear data refers to data in which the instrument readout is exactly the same multiple of the concentration at all concentrations measured and all the points lie exactly on the line.

21. Serial dilution is the preparation of a series of solutions by always diluting the solution just prepared to make the next one. For example, to make solutions with concentrations of 10, 20, 30, 40, and 50 ppm, serial dilution would mean to prepare the 50 first, then to prepare the 40 from the 50, the 30 from the 40, the 20 from the 30, etc.

22. The series of standard solutions does not work well when the instrument readout is dependent on some other variable factor in addition to the concentration, such as variable injection volume in gas chromatography or variable solution viscosity in atomic absorption.

23.

Concentration	mL of 1000 ppm Needed
1	0.05
2	0.10
3	0.15
4	0.20
5	0.25

The mL of 1000 ppm needed is pipetted into separate 50-mL volumetric flasks and water is added to each to the 50-mL mark. Each flask is then shaken to make the solutions homogeneous. The pipet needed would be a small serological pipet, perhaps 0.50-mL capacity. Alternatively, a pipetter, such as described in Chapter 5, can be used.

28. First, select the size of volumetric flask to use, perhaps 25 mL. Then calculate the amounts of benzene and ethylbenzene needed:

Benzene Concentration	mL of Benzene	mL of Ethylbenzene
0.5%	0.125	0.25
1.0%	0.258	0.25
2.0%	0.508	0.25
3.0%	0.758	0.25

Each solution is diluted to the mark with toluene and shaken to make homogeneous. A 1-mL serological pipet would be satisfactory.

30. Matrix matching refers to the attempt to match the composition of all standard solutions and blanks to that of the unknown sample solution.

34. Most of the time the unknown sample requires some pretreatment, such as dilution, extraction, or, if it is a solid, dissolving. The analytical concentration of the analyte in the untreated sample usually must be reported rather than the concentration in the sample solution. Thus, a calculation is usually required to obtain the final answer.

35. 0.0131 mg

37. 1.87×10^3 ppm

39. 535 ppm

41. 750 ppm

43. 0.65 ppm

44. Data acquisition by computer refers to the use of a computer to obtain data (instrument readout values) directly by interfacing to the instrument. These data are then found in the computer's memory, or on disk, and are not necessarily recorded independently on a recorder or in a notebook.

CHAPTER 10

4. IR light has a longer wavelength and lower frequency and wave number than UV light.

5. Energy and frequency: radio waves < infrared light < visible light < UV light < X-rays
   Wavelength: X-rays < UV light < visible light < infrared light < radio waves.

7. (a) 8.31×10^{-5} (c) 4.297×10^{-5} cm

8. (a) 0.317 nm (b) 5.11×10^{3} nm

9. (a) 4.79×10^{-2} cm

10. (a) 7.04×10^{13} sec^{-1} (b) 4.13×10^{14} sec^{-1}

11. (c) 5.06×10^{-16} erg

12. (a) 1.03×10^{4} sec^{-1}

13. (a) 4.30×10^{-19} erg

14. (a) 4.63×10^{7} cm

15. (a) 176 cm^{-1}

16. (a) 1.70×10^{15} sec^{-1}

17. (a) 1.26×10^{4} cm^{-1}

18. (a) 2.00×10^{-8} cm

19. (a) 2.54×10^{-5} cm (b) 1.18×10^{15} sec^{-1}
    (c) 7.81×10^{-12} erg (d) 3.94×10^{4} cm^{-1}

20. 627 Å

22. 591 nm

25. (a) A (b) B (c) A

26. (a) decreased (b) decreased (c) decreased

27. (a) UV has more energy than IR.
    (b) UV causes electronic transitions; IR causes vibrational transitions.

29. If the energy of light striking the atoms or molecules exactly matches an energy transition possible within the atoms or molecules, the transition will occur and the energy that once was light is now possessed by the atoms or molecules, i.e., absorption.

31. Electronic transitions require the most energy – the energy of visible or UV light. Rotational transitions require the least energy – longer-wavelength IR light.

34. A molecular absorption spectrum is of the "continuous" variety, while an atomic absorption spectrum is a "line" spectrum. Only specific wavelengths get absorbed by atoms because only specific energy transitions are possible (no vibrational transitions – only electronic). Both vibrational and electronic transitions are possible with molecules, and thus all wavelengths get absorbed to some degree.
38. Polychromatic ("many colors") light is light consisting of a broad wavelength band. Monochromatic ("one color") light is light consisting of a very narrow wavelength band.

CHAPTER 11

1. Because no two chemical substances display identical spectra.
3. A monochromator is a wavelength selector. It consists of two slits and a dispersing element in combination. The dispersing element splits the light from a source into the wavelengths of which it is composed. Upon rotating this element, different wavelengths pass through the "exit slit", and thus the position of the dispersing element dictates what wavelength is selected.
6. UV – quartz; visible – colorless glass or plastic; IR – inorganic salt crystals.
8. $T = I/I_0$ I_0 = intensity of light striking the detector with the blank in the path of the light. I = intensity of light striking the detector with a sample in the path of the light.
10. The intensity of light from the light source can vary dramatically when the wavelength changes. When measuring transmittance, we do not want the intensity to change for any reason other than the concentration of the sample, thus the need for recalibration with the blank.
11. (a) 0.0857
12. (a) 0.331
13. (a) 0.239
14. (a) 40.6%
15. 1. You don't have to continually replace sample with blank.
    2. Errors due to light source fluctuations are minimized.
    3. Rapid scanning of wavelengths is possible.
17. An absorption spectrum is a plot of absorbance vs. wavelength. A transmittance spectrum is a plot of transmittance vs. wavelength. Since $A = -\log T$, one appears to be the inversion of the other.
19. Beer's law is "A = abc". "A" is absorbance, "a" is absorptivity, "b" is pathlength, and "c" is concentration.
21. 3.72×10^{-5} *M*
22. 9.27×10^{-6} *M*
23. (a) 0.118 (b) 0.981 L/mol cm (c) 7.87×10^{-6} *M*
25. 1.46×10^{5} L/mol cm
27. 1.83×10^{-5} *M*

28. 0.404 cm

32. (a) 0.259 (b) 8.38×10^3 L/mol cm (c) 0.505
(d) 0.428 (e) 90.2%

34. 3.79 ppm

35. 3.31 ppm

36. d-T, f-A, a-I, g-I_0, h-a, b-b, e-Beer's law,
i-%T, c-ϵ

37. 522 nm, because this is the wavelength of maximum absorbance.

42. The wavelength giving the most absorbance is the wavelength giving the best sensitivity.

45. You need one to isolate the wavelength of absorption and one to isolate the wavelength of fluorescence.

46. The fluorescence is measured (with a monochromator and phototube/readout) at right angles to the incoming light beam.

49. fluorescence intensity

50. 0.12 ppm

52. To say fluorometry is more selective means that there are fewer interferences. To say that it is more sensitive means that it can detect smaller concentrations.

54. It is more sensitive and more selective.

55. Advantage – virtually no interferences exist.
Disadvantage – it is not useful for very many things.

56. (b) Fluorometry is more sensitive.
(c) Absorption spectrophotometry is more highly applicable.

58. C_u = 0.47 ppm, mg riboflavin = 16 mg.

CHAPTER 12

1. Infrared absorption patterns can be more directly assigned to more specific structural features.

2. Inorganic compounds consist of ionic bonds which do not absorb infrared light and therefore would present no interfering absorption bands.

4. (1) sealed cell (2) demountable cell (3) sealed demountable cell

6. the spacer between the NaCl windows.

8. (1) The windows may fog or become disfigured since water dissolves them.
(2) Water will interfere with the detection of alcohols in the spectrum.

10. The solvent will exhibit absorption bands which may interfere with those of the analyte. Carbon tetrachloride has only one kind of bond, the C–Cl bond, and this bond will not absorb IR light at wavelengths that are usually important.

13. (1) The best pellets are made from dry KBr. (2) The presence of water will result in absorption patterns that may cause the analyst to make erroneous conclusions.
16. Neither utilize solvents or other materials that would present possibly interfering absorption bands in the spectrum.
19. Fourier transform infrared spectrometry
21. (a) FTIR
    (b) FTIR
    (c) double-beam dispersive
    (d) FTIR
    (e) double-beam dispersive
    (f) FTIR
22. An interferometer is a device that utilizes a movable and a fixed mirror to manipulate the wave patterns of a split light beam so as to create constructive and destructive interference in this beam.
24. (a) FTIR
    (b) FTIR
    (c) double-beam dispersive
    (d) FTIR
    (e) double-beam dispersive
27. The infrared instrument is double beam in space while the UV/vis instrument is double beam in time. The infrared light is dispersed after it passes through the sample rather than before. The detector for the infrared instrument is a thermocouple, while the detector for UV/vis is a photomultiplier tube.
29. 12.15 (a) benzaldehyde (b) 1-propanol (c) ethylbenzene (d) 2-butanone
30. 12.16 (a) benzophenone (b) thymol (c) *n*-pentane

CHAPTER 13

1. nuclear magnetic resonance
2. The nuclear energy transitions that occur with the absorption of radio frequency light occur only in a strong magnetic field.
3. Radio frequency wavelengths are needed, because the energy required to cause the nuclear energy transitions in a magnetic field are on the order of radio frequency energy.
4. The two states are: When the small magnetic field due to the spinning nucleus (1) aligns with the applied magnetic field, and (2) when it opposes the applied magnetic field. The latter is a slightly higher energy state.
6. Most of the time, the nucleus measured is the hydrogen nucleus, which is essentially a proton. Thus, the "P" in "PMR" stand for "proton".
9. y-axis: absorption; x-axis: magnetic field strength
10. The "gauss" is a unit of magnetic field strength.

11. hertz: cycles per second; megahertz: million cycles per second.

14. The "chemical shift" is the effect that the environment of a nucleus (more specifically, the small magnetic fields of nearby electrons) has on the position of an absorption peak in a spectrum. Such peaks are "shifted" to an extent governed by this environment.

17. There are two kinds of hydrogen, the methyl group hydrogen and the hydrogen of the alcohol functional group, thus two peaks.

18. Since there are three separate peaks, there are probably three different kinds of hydrogens in the structure.

19. The integrator trace tells us that the number of hydrogens represented by the peak at 5 ppm is one half the number at 4.2 ppm and one third the number at 1.8 ppm.

20. The existence of three peaks means that there are three different kinds of hydrogens. There are three different kinds of hydrogens in ethyl alcohol, but only two different kinds in diethyl ether.

21. The integrator trace – see the answer to #19 above.

24. The high-energy electron beam is used to fragment the molecules of the analyte into particles of mass to charge ratios characteristic of that molecule.

27. A high-vacuum system is needed so that the components of air will not be fragmented and cause an interference.

28. y-axis: fragment count; x-axis: mass to charge ratio

30. The compound is *o*-dibromobenzene. The mass to charge ratio 154.9 is due to the loss of a bromine. Others:

mass to charge		
	233.9	$C_6H_4{}^{79}Br_2$
	234.9	$C_5{}^{13}CH_4{}^{79}Br_2$
	235.9	$C_6H_4{}^{79}Br^{81}Br$
	236.9	$C_5{}^{13}CH_4{}^{79}Br^{81}Br$
	237.9	$C_6H_4{}^{81}Br_2$
	238.9	$C_6{}^{13}CH_4{}^{79}Br_2$

Bromine consists of two isotopes of mass 79 and 81 which are essentially equal in abundance. Therefore, bromine-containing fragments appear as doublets separated by two mass units.

31. The compound is benzophenone. Mass to charge ratio of 182 is the molecular fragment. Mass to charge ratio of 105 is the molecular fragment with loss of a benzene ring. Mass to charge ratio of 77 is 105 with loss of C=O. Mass to charge of 77 is most likely residual nitrogen in the vacuum system.

CHAPTER 14

1. It converts ions to atoms.
2. (a) An atomizer is a device that forms atoms from ions.
   (b) graphite furnace, ICP source, vapor generators, etc.
   (c) Solution evaporates, 2. ions atomize, 3. atoms are raised to excited states, 4. excited atoms drop back to ground state and emit light.
4. An oxygen/acetylene flame has a high burning velocity which decreases the completeness of atomization and thus lowers the sensitivity.
6. Since the solution, air, and fuel are premixed, some solution droplets will not make it all the way to the flame. These collect in the bottom of the mixing chamber unless they are allowed to drain out.
9. Potassium and sodium atoms have their own unique sets of energy states above the ground state. The energy lost when excited atoms return to lower states therefore represents unique energy values representing specific and characteristic colors.
11. Flame AA is therefore useful because there is a relatively large quantity of unexcited atoms present that can absorb light from the light source.
13. The element to be analyzed is contained in the cathode, since its atoms become excited and emit light and this light is what is needed for absorption in the flame. No monochromator is needed because the wavelength is already specific for the atoms in the flame.
15. Many hollow cathode lamps are needed because each element, since it must be contained in the cathode, requires a different lamp. Some lamps, however, are "multielement". The element analyzed must be contained in the cathode so that its line spectrum will be generated and absorbed by the same in the flame.
18. The reference beam in the AA instrument does not pass through the blank, but merely bypasses the flame. Thus, the fluctuations in light intensity are accounted for, but the blank adjustment must be made at a separate time.
20. With AA, absorbance is measured. With FP, no absorption takes place – the intensity of emitted light is measured.
22.

	AA	FP
(a)	yes	yes
(b)	yes	no
(c)	no	yes
(d)	yes	no
(e)	no	yes
(f)	yes	no

23. It is used to release the calcium from the sample matrix so that it can be atomized in the flame.

24. It is useful because the chemical makeup of the sample solution doesn't need to be known.
27. Safety problems relating to the use of a compressed gas cylinder, the use of a flammable gas, the fact that toxic combustion products may form, and the fact that flashbacks are possible are examples. Compressed gas cylinders must be appropriately secured, independently vented fume hoods must be in place, etc. See subheading "Safety and Maintenance" under "Flame Atomic Absorption" in this chapter.
30. The absorbance signal originates from a very small volume of solution placed in the furnace, and since the furnace is continuously flushed with an inert gas, the vapors from this volume are swept out of the furnace after a short time.
32. Advantages: (1) high sensitivity, (2) only small volumes of sample are needed. Disadvantages: (1) matrix effects, (2) poor precision.
34. arsenic, bismuth, germanium, lead, antimony, selenium, tin, and tellurium
36. ICP = Inductively coupled plasma. It is more closely related to FP because emission is measured and not absorption.
38. Advantages: (1) more sensitive, (2) broader concentration range measurable, (3) multielement analysis possible. Disadvantages: (1) cost.
40.

	FP	AA	SE	AF
(a)	Drop back to ground state of atoms excited by flame	Elevation to excited state by absorption of light	Drop back to ground state by atoms excited by arc or spark source	Drop from higher state to intermediate lower state of atoms excited by light source
(b)	No light source; only flame, monochromator, and detector/readout	Light source, flame, monochromator, and detector/readout	No light source except for arc or spark; no exit slit on monochromator; photographic film detection	Light source, flame, monochromator, and detector (at right angles to incident light)
(c)	I vs. C	A vs. C	Qualitative–location of lines.	F vs. C

42.

(a) T	(b) T	(c) F	(d) T	(e) F
(f) F	(g) T	(h) F	(i) T	(j) T
(k) T	(l) F	(m) T	(n) F	(o) F
(p) T	(q) F	(r) T	(s) T	(t) T

CHAPTER 15

3. Insoluble impurities are removed by filtering at the elevated temperature–they are filtered out while all other mixture components pass through the filter. Soluble impurities are removed after cooling again by filtration–the purified crystals are captured by the filter while the soluble impurities pass through with the filtrate.

5. Vapor pressure is a measure of the tendency of a substance to be in the gas phase at a given temperature. At the boiling point of a liquid mixture, the component with the higher vapor pressure will have a higher concentration in the condensing vapors and thus will be purer than it was before.
7. Two liquids in a mixture typically have a significant vapor pressure and similar boiling points. Thus, a clean separation does not occur with only a simple distillation. A fractional distillation is usually required.
10. The two liquid phases must be immiscible and the analyte must be more soluble in the extracting solvent.
12. (a) 4.6 M (b) 0.0447 M
14. Distribution coefficient = 0.731, percent extracted = 19.6%.
16. These two quantities are the same, except that distribution ratio takes into account all dissolved forms of the analyte while the distribution coefficient takes into account only one form. If only one form exists, then the two are identical.
18. No. Besides the convenient use of the separatory funnel for the actual extraction, they are also designed for easy separation of two immiscible liquids through the stopcock after the extraction. It would not be easy to separate a liquid from a solid through the stopcock.
21. Chromatography is a separation technique in which mixture components are separated based on the differences in the extent of their interaction with two phases, a mobile phase and a stationary phase.
23. Partition chromatography utilizes the varying solubilities of the mixtures components in a liquid stationary phase. Absorption chromatography utilizes the varying tendencies for mixture components to adhere to the surface of a solid stationary phase.
25. (a) "B" would emerge first because polar mixture components tend to dissolve more in the polar mobile phase and thus will come through the column with the mobile phase and emerge first. "A", being non-polar, will tend to remain behind in the stationary phase.

    (b) "A", since it is non-polar, will emerge first since it will tend to dissolve more in the non-polar mobile phase.
28. Size exclusion chromatography. The separation occurs because the stationary phase particles are porous; the small molecules enter the pores and are slowed from passing through the column, while the large molecules pass through more quickly since they do not enter the pores.
30. paper chromatography, thin-layer chromatography, open-column chromatography, gas chromatography, and high-performance liquid chromatography
33. a-k, b-n, c-l, d-o, e-p, f-j, g-i, h-m
34. (a) stationary phase; partition

    (b) size exclusion

(c) electrophoresis
(d) HPLC
(e) adsorption
(f) thin layer

35. Blank lines, left to right, starting upper left: water, liquids or dissolved solids, size exclusion, porous polymer beds, any liquid type, polymer beads with ionic "site", ions, gas, thin liquid film, thin layer, liquids or dissolved solids.

36. (a) false (b) true (c) true (d) false (e) true
(f) false (g) true (h) true (i) false (j) true
(k) true (l) true (m) false (n) false (o) false
(p) true (q) true

37. (a) thin-layer chromatography
(b) partition chromatography
(c) ion-exchange chromatography
(d) high-performance liquid chromatography
(e) partition chromatography and adsorption chromatography
(f) open-column chromatography
(g) partition chromatography
(h) paper chromatography and thin-layer chromatography

40. Capillary electrophoresis is an electrophoresis technique in which the mixture components are separated in a capillary tube and detected with an "on-line" detector after the separation occurs. The advantages include smaller quantity of sample and qualitative and quantitative analysis in a much shorter time.

CHAPTER 16

2. partition and adsorption

4. Vapor pressure is a measure of the tendency of a substance to be in the gas phase at a given temperature. Since different mixture components will have different vapor pressures, the separation occurs in part due to the different tendencies of the components to be in the mobile gas phase.

6. to flash vaporize the liquid samples

8. The injection port, to flash vaporize the sample; the column, since the mixture components must remain gaseous and since vapor pressure depends on temperature; and the detector, in order to keep the mixture components from condensing.

11. An open-tubular capillary column is a very long (30–300 ft), narrow-diameter tube in which the stationary phase is held in place by adsorption on the inside wall. Such a column is useful because it allows the use of a very long column (for better resolution) with minimal gas pressure required.

14. "Analytical GC" is the use of GC solely for analysis, qualitative or quantitative. "Preparative GC" is the use of GC for preparing pure samples for use in another experiment.

15. Temperature programming involves changing the column temperature in the middle of the run.
18. Low-molecular-weight alcohols are highly polar; thus, FFAP or Castorwax® would be useful in their separation.
21. R = 1.5. The two peaks are considered to be satisfactorily resolved since the resolution is 1.5 or greater.
23. (1) lengthen column
    (2) lower column temperature
    (3) decrease helium flow rate
    (4) change stationary phase
25. (a) FID
    (b) ECD
    (c) thermal conductivity
    (d) thermal conductivity
    (e) FID
    (f) mass spectrometer (GC-MS)
    (g) mass spectrometer (GC-MS)
    (h) ECD
    (i) NPD
    (j) IR (GCIR)
    (k) flame photometric
    (l) photoionization
    (m) Hall
28. (a) flame ionization
    (b) thermal conductivity
    (c) thermal conductivity
    (d) electron capture
    (e) flame ionization
29. (a) more universal, does not destroy sample, safer to use.
    (b) more sensitive
30. Peaks for unknown are recorded. Peaks for a number of known liquids are recorded. If retention times match, it is likely that these unknown components have been identified. However, there may be other untested liquids that will give the same retention time, so further experimentation, possibly with a different stationary phase, may be necessary.
31. (a) yes—benzene, ethylbenzene, and isopropylbenzene
    (b) toluene, *n*-propylbenzene
    (c) There may be other compounds whose retention times have not been measured which may exhibit the same retention times as toluene and *n*-propylbenzene. Also, there is no match for compounds B and D among the known liquids measured.
32. (a) carrier gas, mobile phase
    (b) retention time
    (c) thermal conductivity
    (d) electron capture
    (e) injection port, column, detector
    (f) peak size
    (g) theoretical plates, resolution
    (h) temperature programming
    (i) preparative GC
    (j) FID
    (k) open-tubular column

36. (a) Internal standard method (IS) – a series of standard solutions is prepared in which the analyte is present at increasing known concentrations and a second substance, the internal standard, is present at a constant concentration. This amount of internal standard is also added to the unknown.

    Standard additions method (SA) – standards are prepared by adding a constant known amount of the analyte to the unknown sample, with the chromatogram measured after each addition. Alternatively, a series of standards could be prepared with the unknown as the diluent.

    (b) IS – the analyte peak and the internal standard peak are well-resolved from each other and the solvent peak(s).
    SA – see Figure 21.

    (c) IS – area ratio vs. concentration
    SA – peak size vs. concentration added (extrapolation required)

37. (a) Pipet increasing amounts of ethyl alcohol into the flasks. Pipet constant amount of isopropyl alcohol into the flasks. Dilute standards with an appropriate solvent. Unknown must have isopropyl added.

    (b) Ethyl alcohol and isopropyl alcohols peaks are well resolved. All other peaks need not be resolved (there would be many in the gasohol sample.)

    (c) area ratio vs. concentration

38. Response factor method:

    (a) No sample or solution preparation is required. One chromatogram of unknown sample is required. Considerable experimentation is required to determine response factors, however (see text).

    (b) One chromatogram of unknown is needed. Also, one chromatogram of each compound in the unknown in the pure state is required in order to determine response factors.

    (c) No plotting required

    Serial dilution method:

    (a) Prepare a series of standards with analyte in increasing concentration. Unknown is measured as is.

    (b) Only the analyte peak need be resolved.

    (c) peak size vs. concentration

    Internal standard method and standard additions method – see Question 36.

40. a leaky or plugged syringe, a worn septum, a leak in the pre- or post-column connections, or a contaminated detector

42. Perform a rapid "bakeout" via temperature programming after the analyte peaks have eluted, use pure reagents, or replace or clean septa, carrier, or column.

44. The sample was injected directly without any prior preparation procedure that would require a calculation.

CHAPTER 17

1. high-performance (or pressure) liquid chromatography
2. LC – liquid chromatography
   LLC – liquid-liquid chromatography
   LSC – liquid-solid chromatography
   BPC – bonded-phase chromatography
   IEC – ion-exchange chromatography
   IC – ion chromatography
   SEC – size exclusion chromatography
   GPC – gel permeation chromatography
   GFC – gel filtration chromatography
4. speed and overall performance
6. Mobile phases and samples contain small particulates that can damage the column unless they are removed by filtration.
7. Whether the samples or mobile phase contain an organic solvent dictates what filter material to use.
8. It is appropriate for those samples and mobile phases that contain an organic solvent.
10. It is possible for the mobile-phase system to undergo a sharp pressure drop if the pump fails, for example. When that happens, dissolved gases can withdraw from solution and create void spaces in the system, which can be damaging to equipment and analytical results.
12. The HPLC pump must be capable of providing extremely high pressure and pulsation-free flow.
14. A gradient programmer is that part of an HPLC instrument that provides for the gradual changing of the mobile composition in the middle of the run.
16. Isocratic elution – one mobile phase composition is used for entire run. Gradient elution – mobile-phase composition is altered during the run in some preprogrammed manner.
17. A parameter that alters retention time (and hence, resolution) is altered in the middle of the run in each case.
20. (1) The septum material may not be compatible with all mobile phases, thus creating the possibility for contamination. (2) The system is under too high a pressure to make the septum-piercing method a viable possibility.
22. See Figure 17.5.
23. Normal phase – stationary phase is polar; mobile phase is nonpolar. Reverse phase is just the opposite.
26. Bonded-phase chromatography is a type of liquid-liquid chromatography in which the liquid stationary phase is chemically bonded to the support material (as opposed to being simply adsorbed). The stationary phase can be either polar or nonpolar and thus both normal and reverse phase are possible.

28. Cation exchange resins have negatively charged bonding sites for exchanging cations, while anion exchange resins have positively charged sites for exchanging anions.

31. (a) normal- and reverse-phase HPLC
    (b) size-exclusion HPLC
    (c) normal- and reverse-phase HPLC

33. mobile-phase composition, stationary-phase composition, flow rate, temperature of stationary phase

35. Selectivity = 1.82. This is a good separation based on the criterion that a value of 1.2 or better is considered good.

38. The fixed-wavelength detector utilizes a glass filter as the monochromator. The variable-wavelength utilizes a slit-dispersing element-slit monochromator. The latter design is used in order to maximize sensitivity by setting the monochromator to the wavelength of maximum absorbance. If this is not important, then the fixed-wavelength design may be preferred because it is less expensive.

40.

	Detector	Advantage	Disadvantage
(a)	UV	Sensitive	Not universal
(b)	RI	Universal	Not sensitive
(c)	F	Very sensitive	Not universal nor highly applicable

41. Not all potential mixture components will absorb UV light.

42. See #40 for some examples.

44. Yes, because the absorptivity of mixture components varies.

46. FTIR, like UV absorbance, refractive index, etc., is a technique for liquid solutions but has an advantage in that it is fast, allowing a complete spectrum to be obtained as a given mixture component elutes, making it an extremely powerful tool for qualitative analysis.

48. A plugged in-line filter is a typical cause of unusually high pressure. The solution is to backflush the filter or otherwise clean it.

50. This is a symptom of "channeling" in the column. The problem is solved by replacing the column.

CHAPTER 18

12. (a) no (b) yes (c) yes

13. -1.38 V

14. It will proceed to the left because E^0_{cell} is negative.

15. (a) +1.57 V
    (b) Yes, because E^0_{cell} is positive.

17. +0.60 V. Reaction proceeds to the right because E^0_{cell} is positive.

20. E^0_{cell} = +0.51 V; yes—E^0_{cell} is positive.

21. +2.11 V

24. +0.42 V

25. +3.15 V

27. +1.78 V

30. −0.52 V (for $2Cr^{+3} + 3Ni \rightarrow 2\ Cr + 3\ Ni^{+2}$)

35. The salt bridge for reference electrodes consists of porous fiber tips that provide for the diffusion of ions in and out of this half-cell.

40. A combination pH electrode consists of a pH electrode and a reference electrode in a single probe.

41. An ion-selective electrode is a half-cell that is sensitive to a particular ion like the pH electrode is sensitive to the hydrogen ion.

43. A platinum electrode is needed in some half-cells to provide a surface at which electrons can be exchanged. Such a surface is lacking where there is no solid metal as part of the half-reaction, such as with the ferrous/ferric half-reaction.

45. Potentiometry utilizes galvanic cells, while amperometry uses electrolytic cells.

47. The supporting electrolyte is a substance which supplies the ions needed to make the analyte solution conducting. It doesn't interfere because its reduction (or oxidation) potential is well out of the way of that of the analyte.

48. (1) Its surface is continuously renewed, thus avoiding problems with surface contamination.
    (2) The solution is automatically stirred when the drop falls.

49. See Figure 18.17.

52. When the applied potential is more negative than the half-wave potential, the reduced form of the analyte is stable. When the applied potential is more positive than the half-wave potential, the oxidized form is stable. Thus, crossing the half-wave potential from either direction would cause a reversal of direction of the electrochemical reaction taking place.

54. Refer to Figure 18.19.

56. (1) When it is more sensitive and (2) when a molecular species is the analyte, rather than a metal.

58. (a) A technique in which a metal is electroplated onto the surface of a stationary electrode and then "stripped" off via a positive potential scan. The current measured for the "stripping" action is related to the concentration of this metal's ion in the original solution.

59. (a) F (b) F (c) T (d) F (e) T (f) T
(g) F (h) F (i) F (j) T (k) T (l) F
(m) T (n) T (o) F (p) F (q) T (r) F
(s) T

60. (1) c (2) f (3) a (4) d (5) g (6) d
(7) b (8) e (9) e (10) d (11) b (12) d

Appendix 10

Reference Sources

ANALYTICAL CHEMISTRY BOOKS, TECHNICIAN LEVEL

1. Kenkel, J., *Analytical Chemistry Refresher Manual*, Lewis Publishers, Chelsea, MI, 1992.
2. Hajian, H., and Pecsok, R., *Modern Chemical Technology*, Vol. I and II, Prentice-Hall, Englewood Cliffs, NJ, 1988.
3. Sarner, S., *Introduction to Analytical Chemistry*, Vol. 1 and 2, 4th ed., Pumpkin Press, Wilmington, DE, 1990.
4. Shugar, A., and Ballinger, J., *Chemical Technicians Ready Reference Handbook*, 2nd ed., McGraw-Hill, New York, 1990.

ANALYTICAL CHEMISTRY BOOKS, BACCALAUREATE LEVEL

1. Christian, G., *Analytical Chemistry*, 5th Ed., John Wiley & Sons, New York, 1994.
2. Harris, D., *Quantitative Chemical Analysis*, 3rd ed., W.H. Freeman and Company, San Francisco, 1991.
3. Day, R., and Underwood, A., *Quantitative Analysis*, 6th ed., Prentice-Hall, Englewood Cliffs, NJ, 1991.
4. Kennedy, J., *Analytical Chemistry: Principles* and *Analytical Chemistry: Practice*, 2nd ed., W.B. Saunders, Philadelphia, 1990.
5. Skoog, D., Holler, F., and West, D., *Analytical Chemistry, An Introduction*, 6th ed., 1994; and *Fundamentals of Analytical Chemistry*, 6th ed., W.B. Saunders, Philadelphia, 1992.
6. Skoog, D., and Leary, J., *Principles of Instrumental Analysis*, 4th ed., W.B. Saunders, Philadelphia, 1992.
7. Hargis, L., *Analytical Chemistry, Principles and Techniques*, Prentice-Hall, Englewood Cliffs, NJ, 1988.

8. Manahan, S., *Quantitative Chemical Analysis*, Brooks/Cole Publishing Co., Monterey, CA, 1986.

ATOMIC ABSORPTION SPECTROSCOPY

1. Beaty, R., *Concepts, Instrumentation and Techniques in Atomic Absorption Spectrophotometry*, The Perkin-Elmer Corporation, Norwalk, CT, 1988.

INFRARED SPECTROMETRY

1. Chia, L., and Ricketts, S., *Basic Techniques and Experiments in Infrared and FTIR Spectroscopy*, The Perkin-Elmer Corporation, Norwalk, CT 1988.

NMR AND MASS SPECTROSCOPY

1. Abraham, R., Fisher, J., and Loftus, P., *Introduction to NMR Spectroscopy*, John Wiley & Sons, New York, 1988.
2. Davis, R., Frearson, M., and Prichard, P. E., *Mass Spectrometry*, John Wiley & Sons, New York, 1987.

SOLVENT EXTRACTION AND CHROMATOGRAPHY

1. Touchstone, J., *Practice of Thin-Layer Chromatography*, John Wiley & Sons, New York, 1992.
2. Rydberg, J., Musikas, C., and Choppin, G., *Principles and Practices of Solvent Extraction*, Marcel Dekker, New York, 1992.

GAS CHROMATOGRAPHY

1. Grob, R., *Modern Practice of Gas Chromatography*, John Wiley & Sons, New York, 1985.
2. McNair, H., and Bonelli, E., *Basic Gas Chromatography*, Varian Associates, Palo Alto, CA, 1969.
3. Buffington, R., and Wilson, M., *Detectors for Gas Chromatography, A Practical Primer*, Hewlett-Packard Co., Avondale, PA, 1987.
4. Ettre, L., *Basic Relationships of Gas Chromatography*, The Perkin-Elmer Corporation, Norwalk, CT, 1979.
5. Hammarstrand, K., *Gas Chromatographic Analysis of Pesticides*, Varian Associates, Palo Alto, CA, 1976.

LIQUID CHROMATOGRAPHY

1. Lindsay, S., *High Performance Liquid Chromatography*, 2nd ed., John Wiley & Sons, New York, 1992.
2. Yost, R., Ettre, L., and Conlon, R., *Practical Liquid Chromatography, An Introduction,* The Perkin-Elmer Corporation, Norwalk, CT, 1980.
3. Schram, S., *The LDC Basic Book of Liquid Chromatography*, The Milton Roy Co., St. Petersburg, FL, 1980.

ORGANIC CHEMISTRY, INCLUDING INFRARED, UV/VIS, MASS, AND NMR SPECTROSCOPY

1. Morrison, R., and Boyd, R., *Organic Chemistry*, 6th ed., Prentice-Hall, Englewood Cliffs, NJ, 1992.
2. Wade, L., *Organic Chemistry*, 2nd ed., Prentice-Hall, Englewood Cliffs, NJ, 1991.
3. Silverstein, R., *Spectrometric Identification of Organic Compounds*, 5th ed., John Wiley & Sons, New York, 1991.
4. Shriner, R., et al., *The Systematic Identification of Organic Compounds, A Laboratory Manual*, 6th ed., John Wiley & Sons, New York, 1980.

DATA HANDLING

1. Dux, J., *Handbook of Quality Assurance for the Analytical Chemistry Laboratory*, Van Nostrand Reinhold, 1986.
2. Taylor, J. K., *Quality Assurance of Chemical Measurements*, Lewis Publishers, Chelsea, MI, 1987.

ANALYTICAL METHODS

1. *Standard Methods for the Examination of Water and Wastewater*, 18th ed., WEF, AWWA, APHA, Washington, D.C., 1992.
2. *Official Methods of Analysis of the Association of Official Analytical Chemists (AOAC)*, 15th ed., AOAC, McLean, VA, 1990.
3. *Methods of Soil Analysis*, 2nd ed., American Society of Agronomy, 1982.
4. Mueller, W., and Smith, D, *Compilation of EPA's Sampling and Analysis Methods*, Lewis Publishers, Chelsea, MI, 1991.
5. Lodge, J. P., Jr., *Methods of Air Sampling and Analysis*, 3rd ed., Lewis Publishers, Chelsea, MI, 1989.
6. Schirmer, R, *Modern Methods of Pharmaceutical Analysis*, 2nd ed., CRC Press, Boca Raton, FL, 1991.
7. *NIOSH Manual of Analytical Methods*, Vol. 1 & 2, 3rd ed., National Institute for Occupational Safety and Health, 1984.

Index

AA, see Atomic absorption
A to D converter, 190–191
Absorbance, in UV/vis absorption spectrophotometry, 221, 225–226
Absorption of light, 207–211
Absorptivity, 228
  molar, 228–229
Accuracy
  precision vs., 48, 49
  when important and when not, 10–11, 98–99
Acetic acid, concentrated
  in buffer solutions, 72–74
  density, molarity, percent composition, 483
  in HPLC mobile phases, 400
Acetone, 38
Acetonitrile, 37, 400
Acid/base dissociation constants, K_a
  general discussion, 68–71
  tables of, 481–482
  titration curves and, 123–126
Acids and bases
  composition of some concentrated, 483
  densities of some concentrated, 483
  equivalent weights of, 108–110
  monoprotic, 68
  polyprotic, 70, 126
  reactions, 105–106
  for sample preparation, 35–36
Activity, 63, 428
Activity coefficient, 63, 428
Activity Series of Metals, 424
Adjusted retention time
  in gas chromatography, 366–368
  in HPLC, 402–403
Adsorption chromatography, 334, 393, 400, 402
Affinity chromatography, see Adsorption chromatography
Air, sampling of, 33, 237
Alconox, 96
Aliquot, 151
Alkalinity, 460
Ammonia, in Kjeldahl method, 117–119
Ammonium hydroxide
  in buffer solutions, 72–75, 144
  properties of concentrated, 483
Amperometric detectors, 408–409
Amperometric titration, 450–451
Amperometry, 408, 439

Ampholyte, isoelectric focusing and, 343–344
Analog vs. digital signals, 190
Analytical balances, see Balances
Analytical columns, 361
Analytical separations, 321–345, see also Chromatography, Distillation, Electrophoresis, Extraction, Recrystallization
Anion exchange resin, 334, 400–401
Anode, 427
Aqua Regia, 36
Arc emission spectrography, 307–308, 310
Ascending chromatography, 336
Atomic absorption (AA), 209–212, see also Atomic spectroscopy; Flame atomic absorption
  applications of, 305, 462–463, 465
  atomization for, 289–292, see also specific methods
  detection limits of, 303, 304
  fuels and oxidants in, 290
  graphite furnace method, 304–306, 310
  instrumentation for, 298–301
  interferences in, 302–303
  matrix matching in, 302
  nebulizer in, 291
  safety, 303
  sensitivity, 303, 304
Atomic emission spectrography, see Arc emission spectrography
Atomic fluorescence, see Fluorescence
Atomic spectroscopy, see also Atomic absorption (AA); Atomic emission spectrography; Flame photometry; Inductively coupled plasma (ICP)
  burners for, 290–292
  excitation of atoms in, 292–293
  general discussion, 289–309
Atomization, see Atomic absorption
AutoAnalyzer, Technicon, 465, 466–469
Automatic samplers, 464
Automatic titrators, 438, 465–466
Automation, 464–471
  flow injection methods, 469–470
  robotics, 470–471
  segmented flow methods, 466–469
  titrations, 465–466
Auxiliary balances, see Balances
Auxiliary electrode, 439
Auxochrome, 228
Average deviation, 51
Background current, 443
Back titration, see Titration
Balances, 4–10
  analytical, 7–10
  auxiliary, 7
  basic concept of, 5–6
  instructions for use of, 9–10
  less accurate, 7
  multiple-beam, 7
  single pan, 5
  top-loading, 7, 8
Balancing oxidation-reduction equations, see Oxidation-reduction reactions
Band pass, 219
Band spectrum, see Continuous spectrum
Barium chloride hydrate, determination of water of hydration in, 15–18
Baseline drift
  in gas chromatography, 382
  in HPLC, 412
Bases, see Acids and bases
Bathochromic shift, 226
Beam splitter, 258, see also Chopper
Beer-Lambert Law, see Beer's Law
Beer's Law, 228–231
  deviations from, 230–231
  flame atomic absorption and, 298
Beer's Law plot, 229
Benzene, 38, 400
Bipotentiometric titrations, 464
Blank solutions, 186–188, 222
Blow-out pipet, see Pipets
Bonded-phase chromatography (BPC), 333, 393, 400
Boric acid, in Kjeldahl experiment, 117–119
Bronsted-Lowry base, 106
Buffer solutions
  general discussion, 72–75
  preparation of, 74–75, 150
Burets, 82–84, 94–95
Burners, atomic spectroscopy, see specific type

Cadmium, determination of, by polarography, 452
Caffeine
  determination of, in pain relief tablet, 347–348
  HPLC determination of in soda pop, 414

Calcium, atomic absorption determination of, 302
Calibration, spectrophotometer, 222
Capacity factor
 in gas chromatography, 369
 in HPLC, 403
Capillary columns, in gas chromatography, 361, 362
Capsaicin
 extraction of from chile peppers, 348–349
 determination of in chile pepper extract, 415
Carbon tetrachloride, 252, 400
 IR spectrum of, 253
Carrier gas, 355, 365–366
Catalyst, Kjeldahl, 117
Cathode, 427
Cation exchange resin, 334, 400–401
Cells, in electroanalytical methods, 421
Cerium, 166, 438
Charge, electrical, 489
Charging current, 443
Chelate, 141
Chelating agents, 141
Chemical analysis, defined, 1, 2
Chemical equilibrium, 63–75
Chemical shifts, in NMR spectroscopy, see NMR spectroscopy
Chemically pure (CP) grade of chemical, 40
Chloride, determination of, by Mohr method, 170–171
Chlorine in wastewater plant effluent, determination of, 169
Chloroform, 38, 400
Chopper
 in an IR double-beam spectrometer, 257
 in flame atomic absorption, 300
 in a UV/vis double-beam spectrophotometer, 224
Chromatogram
 in gas chromatography, 358, 366–369
 in HPLC, 402–403
Chromatography, see also specific types and configurations
 adsorption, see Adsorption chromatography
 applications of, 463
 ascending, 336
 bonded-phase (BPC), see Bonded-phase chromatography
 column methods of, 336
 configurations, 336–341
 descending, 336
 gas, see Gas chromatography (GC)
 HPLC, see High-performance liquid chromatography (HPLC)
 instrumental, 341
 ion, see Ion-exchange chromatography
 ion-exchange, see Ion-exchange chromatography
 mobile phase of, 331
 open column, 339–340
 paper, 336–339
 partition, see Partition chromatography
 planar methods of, 336
 radial, 336
 size-exclusion, see Size-exclusion chromatography
 stationary phase of, 331
 thin-layer, 336–339
 types of, 332–336
Chromic acid, 86
Chromosorb, 362
Chromophore, 226–228
Class A glassware, 88, 90–91
Cleaning solutions,
 general discussion, 96
 recipes for, 96
 safety and, 96
Colorimeter, 212
 segmented flow and, 466, 468
Colorimetry, 212
Column chromatography, see Chromatography
Columns, see also specific type
 gas chromatography, 360–363
  temperature of, 363–365
 HPLC, 399–402
Combination pH electrode, 435
Complex ions, 139, see also Water hardness
 examples of, 140–142, 143
Computers, use of, 190–193
Computing integrators
 in instrumental chromatography, 377
Concentrated acids and bases, see Acids and bases
Concentrators, 328–329
Conductivity detectors in HPLC, 407–408
Confidence interval, 53, 55
Confidence limits, 53, 55–56
Continuous spectrum, 211–212

Control samples, 188
Conversion factors, metric system, 206
Coordinate covalent bond, 140
Correlation chart, see Infrared spectrometry
Correlation coefficient, 183–184
Cough syrups, analysis of, by TLC, 346–347
Countercurrent distribution, 328
Counter electrode, see Auxiliary electrode
CP, see Chemically pure
Current, 489–490
  background, 443
  diffusion, 443
  Faradaic, 444
  non-Faradaic, 444
Cuvette, 220–221
  matched, 231–232
Cyclic voltammetry, 449–450

Data
  acquisition by computer, 190–191
  display and analysis, 192
  from instruments, 177–184
  handling, 192–193
  management and reporting of, 193
  stations, 192
Debubbler, in segmented flow methods, 468
Degassing, in HPLC
Degrees of freedom, 56
Deionized water, 408
Delves cup, 307
Demountable cell, 250–252
Desiccant, 17
Desiccator, 17
Density of concentrated acids and bases, 483
Descending chromatography, 336
Detectors,
  in spectroscopy, 221
  in gas chromatography, 369–375
  in liquid chromatography, 404–409
Determinate errors, 47–48
Deviation, 49
  average, 51
  relative, 50–51
  relative standard, 51–52
  standard, 51
Dichloromethane, determination in paint stripper, 385–386, see also Methylene chloride
Diethyl ether, 38
Differential pulse polarography, 445, 446, 447
Diffraction grating, see Dispersing element
Diffuse reflectance, see Infrared spectrometry, solid sampling
Diffusion current, see Current
Digestion
  definition, 22
  in Kjeldahl analysis, 117, 132
  of precipitates, 22
Digital signals, 190
Dilution, solution preparation by, 85, 147
Dilution factor, 188, 189
Diode array detector, 405
Dispersing element, 218, 219–220
Disposable pipets, see Pipets
Distillation, 322–324
  in Kjeldahl method, 117, 118, 132
Distribution coefficient, 326–327
DME, see Dropping mercury electrode
Double-beam instruments
  atomic absorption, 300–301
  dispersive IR, 256–258
  in space, 257
  in time, 257
  UV/vis, 218, 223–224
Drierite, 17
Dropping mercury electrode, 440, 441
Dual nature of light, see Light
Duopettes, see Pipets
Dynodes, 221

$E_{1/2}$, see Half-wave potential
ECD, see Electron capture detector
EDTA, see Ethylenediaminetetraacetic acid
Electroanalytical methods, 421–451, see also Polarography, Potentiometry, Voltammetry
  applications of, 448, 463–464
Electrochemical detectors, 407–409
Electrodeless discharge lamp, 300
Electrodes, see also specific types (Auxiliary, Combination pH, Dropping mercury, Hanging mercury drop, Indicator, Ion-selective, pH, Reference, Saturated calomel, Silver-silver chloride, Standard hydrogen, Working)
  in amperometric detector, 408–409
  in polarography, 439, 440–442

in three-electrode system, 439
in voltammetry, 439, 448–450
Electrolytic cell, 421, 438
Electromagnetic radiation, see Light
Electromagnetic spectrum, 202, 204–205
Electron capture detector, 372
Electronic analytical balance, see Balances
Electronic energy transitions, 207–212
Electrophoresis, 341–345
Electroplating, 449, 460
Elution, in HPLC
Emission of light, see Light
Emission spectrography, see Arc emission spectrography
Emission spectroscopy, see specific methods (Arc emission spectrography, Flame photometry, Inductively coupled plasma
End point, in titrations, 98, 437, 438, 450, 451
Energy of light, see Light
Energy transitions, see specific types (Electronic, Rotational, Vibrational)
Equilibrium, chemical, see Chemical equilibrium
Equivalence point, in titrations, 98
Equivalent, concept of, 108–109
Equivalent weight
of acids/bases, 109–110
of oxidizing and reducing agents, 162–163
Eriochrome black T
preparation of indicator solution of, 150
in water hardness analysis, 145
Error analysis, 47–48
Ethanol, determination of, in wine, 386
Ethylenediaminetetraacetic acid
in complexation reaction, 142–144
in water hardness analysis, 144–145
Evaporator, rotary, see Concentrators
Extinction coefficient, see Absorptivity
Extraction
liquid-liquid, 37
liquid-solid, 37
sample preparation by, 37
Soxhlet, 37, 330, 331
Extraction theory, 326–328

Faradaic current, see Current
Ferrous ammonium sulfate (Mohr's salt), 166
equivalent weight of, 162
as primary standard for potassium permanganate solutions, 162, 164
FIA, see Flow injection analysis
FID, see Flame ionization detector
Filtering, in HPLC sample preparation, 395–396
Fingerprint region, IR spectra, see Infrared spectra
Flame atomic absorption, 292, 297–304, 310
application of, 301–304
instrumentation for, 298–301
optimization of instrument parameters, 309–312
Flame emission, 294–297
Flame ionization detector, 371–372
Flame photometer, 296, 373
Flame photometric detector, 373
Flame photometry, 292, 295–296, 310
applications of, 296–297
Flame test, 294
Flashback, 292, 303
Flask, volumetric, see Volumetric flask
Flow injection analysis, 469–470
Fluorescence
molecular, see Fluorometry
atomic, 308–309, 310
Fluorescence detector, 405
Fluoride in drinking water, determination of, 451–452
Fluorometry, 232–235
applications of, 234–235, 462
comparison to molecular absorption spectrophotometry, 234–235
Formation constant, 141–142
Fourier Transform Infrared Spectrometry (FTIR), 258–259
Fractional distillation, 323–324
Fraction collector, 340
Frequency of light, see Light
Fronting, of gas chromatography peaks, 381–382
FTIR, see Fourier Transform Infrared Spectrometry
Fuels, for atomic spectroscopy flames, 290
Fusion, 39

Galvanic cell, 421, 429, 438
Gas chromatography (GC), 330, 341, 355–382, 463

applications of, 463
baseline drift in, 382
baseline perturbations in, 382
carrier gas flow rate in, 365–366
chromatogram parameters in, 366–369
columns in, 360–363
temperature of, 363–365
detectors in, 369–375
selectivity of, 369
sensitivity of, 369
diminished peak size in, 381
HPLC vs., 394–395
instrument design for, 356–358
qualitative analysis in, 375–376
quantitative analysis in, 377–380
internal standard method, 379–380
peak size measurement, 376–377
response factor method, 377–379
standard additions method, 380
sample injection, 358–360
stationary phases of, 362–363
troubleshooting, 381–382
unexpected peaks appearing in, 382
unsymmetrical peak shapes in, 381–382
Gas chromatography-infrared spectrometry, 373–374
Gas chromatography-mass spectrometry, 373–374
Gas-liquid chromatography (GLC), 330, 331, 355–356, 362–363
Gasoline, analysis of, by gas chromatography, 385
Gas-solid chromatography (GSC), 330, 331, 355–356, 362–363
Gaussian distribution, 53
GC, See Gas chromatography
GC-IR, See Gas chromatography-infrared spectrometry
GC-MS, See Gas chromatography-mass spectrometry
Gel electrophoresis, 343–344
Gel filtration chromatography, see Size-exclusion chromatography
Gel permeation chromatography, see Size-exclusion chromatography
Glass electrode, see pH electrode
Glassware
burets, 94–95
Class A, 88, 91
cleaning, 95–96
pipets, 90–93
pipetting devices, 93–94
storage of, 97
volumetric flask, 86–90
GLC, see Gas-liquid chromatography
GFC, see Size-exclusion chromatography
GPC, see Size-exclusion chromatography
Gradient elution in HPLC, 397–398
Gradient programmer, in HPLC, 397–398
Graphite furnace atomizer, 305
Graphs
linear relationships, 181–183
plotting of, 191
Grating, see Dispersing element
Gravimetric analysis, 4
applications of, 460
Gravimetric factors, 13–15, 16
GSC, see Gas-solid chromatography
Guard column, in HPLC, 395–396

Half-cells, in electroanalytical methods, 421, 422, 423
Half reactions, 161
Half-wave potential, 443, 444
Half-width method, in gas chromatography, 376
Hall detector, 373
Hanging mercury drop electrode (HME), 448, 449
Height equivalent to a theoretical plate (HETP)
in distillation, 324
in gas chromatography, 368, 369
in HPLC, 402
HETP, see Height equivalent to a theoretical plate
Henderson-Hasselbalch equation, 73
n-Hexane, 37, 400
Heyerovsky, 440
High-performance liquid chromatography (HPLC), 393–412
adsorption, 393, 400
air bubbles in, 411
applications of, 461, 463
baseline drift in, 412
basic concepts of, 393–394
chromatogram parameters in, 402–403
column “channeling” in, 411
column selection for, 401–402
decreased retention time in, 412
detectors in, 404–410
filtering and degassing in, 395–396
gas chromatography vs., 394–395

ion-exchange, 393, 400–401
normal-phase, 400
qualitative analysis with, 410
quantitative analysis with, 410
recorder zero vs. detector zero in, 411–412
reverse-phase, 400
sample and mobile phase pretreatment in, 395–396
sample injection in, 398–399
size-exclusion, 393, 400–401
solvent delivery in, 397–398
system leaks in, 411
troubleshooting in, 410–412
unusually high/low pressure in, 410–411
High-performance liquid chromatography-mass spectrometry (LC-MS), 409
HME, see Hanging mercury drop electrode
Hollow cathode lamp, 299–300
HPLC, see High-performance liquid chromatography
HPLC grade of chemical, 40
Hydrobromic acid, concentrated, 483
Hydrochloric acid, concentrated
properties of, 35, 36, 483
sample preparation with, 35, 36
standardization of solution of, 128–130
Hydrofluoric acid, concentrated
properties of, 35, 36, 483
sample preparation with, 35, 36
Hydrogen electrode, standard, see Standard hydrogen electrode
Hyperchromic effect, 228
Hypsochromic effect, 228

ICP, see Inductively coupled plasma
Indeterminate errors, 47–48, see also Error analysis
distribution of, 53
Indicators, 82, 83, see also specific indicators
acid/base, 125–127
recipes for solution preparation, 487
Indicator electrodes, see Electrodes
Indirect titration, 117–119, 164–165
Inductively coupled plasma (ICP), 289, 297, 307, 310
Infrared (IR) spectra
fingerprint region of, 260–261
peak ID region of, 261
Infrared (IR) spectrometry, see also Gas chromatography-infrared spectrometry (GC-IR); Liquid chromatography-infrared spectrometry (LC-IR)
applications of, 462
correlation chart for, 264
FTIR, see Fourier Transform Infrared Spectrometry
general discussion, 249–265
instruments for, 256–259
liquid sampling with, 250–254
qualitative analysis, 249–250, 259–261, 265–266
quantitative analysis, 261–265, 267
solid sampling with, 254–256
diffuse reflective in, 255–256
KBr pellet in, 254–255
Nujol mull in, 255
Injection port, in gas chromatography, 357, 358
Instrumental chromatography, 341
Instrumental methods
applications of, 461–464
calculations in, sample pretreatment effect on, 188–190
computers in, 190–193
data and readout from, 177–184
electrical connections for, 193–195
general discussion, 2–3, 177
for quantitative analysis, 184–186
vs. wet methods, 2–3, 177
Instruments, see specific instrument; specific method
Integration,
of chromatography peaks, 376–377
in NMR spectroscopy, see Nuclear magnetic resonance spectroscopy
Interface, 190
Interferences
in atomic absorption, 302–303
in UV/vis spectrophotometry, 230–231
Interferogram, 259
Interferometer, 258
Internal standard method, 184–185
Iodine, 164–166
Iodometry, 164–165
Ion chromatography, see Ion-exchange chromatography

Ion-electron equation balancing method, see Oxidation-reduction reactions
Ion-exchange chromatography, 334–335, see also High-performance liquid chromatography (HPLC)
  columns for, 400–401
  conductivity detectors in, 407–408
Ion-selective electrodes
  determination of fluoride using, 451–452
  general discussion, 435–437
IR, see Infrared spectra
IR light, see Infrared spectrometry
Iron
  atomic absorption determination of
    in soil, 313
    in water, 315
  colorimetric determination of, 235–236
  gravimetric determination of, 25–26
Isocratic elution, 397
Isoelectric focusing, 344
Isoelectric point, 344

Jones reductor, 166

K_a, see Acid/base dissociation constants
K_b, see Acid/base dissociation constants
K_{sp}, see Solubility product constant
Karl Fischer titration, 464
KBr pellet technique, see Infrared spectrometry, solid sampling
KHP, see Potassium hydrogen phthalate
Kjeldahl method, 117–119, 131–132, 460
Kuderna-Danish evaporative concentrator, see Concentrators

Lambda pipet, 92, 93
LC, see Liquid chromatography
LC-IR, see Liquid chromatography-infrared spectrometry
LC-MS, see High-performance liquid chromatography-mass spectroscopy
Least squares method, 182–183
Le Chatelier's Principle, 64, 67, 71, 72, 144
Ligands, 139
  examples of, 139–142
  properties of, 140–141
Light
  absorption of, 207–212
  dual nature of, 201
  emission of, 207–212
  monochromatic, 211
  nature of, 201–202
  parameters associated with (energy, frequency, speed, wavelength, wave number), 201–207
  particle theory, 201
  polychromatic, 212
  source of, UV/vis spectrophotometers and, 218
  wave theory of, 201–202
Limestone, determination of calcium in, 170
Line spectrum, 210, 292–293, 294–295, 299
Liquid chromatography (LC), 393, see also High-performance liquid chromatography (HPLC)
Liquid-liquid chromatography (LLC), 331, 332, 393
Liquid-solid chromatography (LSC), 331
Liquid sampling, in IR spectrometry, see Infrared spectrometry
Liquid-solid extraction, see also Extraction
Loop injector, see High-performance liquid chromatography, sample injection

Magnetic resonance, see Nuclear magnetic resonance (NMR) spectrometry
Magnetic sector mass spectrometry, 280–282
Manganese, in steel, spectrophotometric determination of, 238–239
Manifold, for automated segmented flow analysis, 466
Masking, 141
Mass spectra, 282–283
Mass spectrometers, see Mass spectrometry, instruments for
Mass spectrometry, see also Gas chromatography-mass spectrometry (GC-MS); Liquid chromatography-mass spectrometry
  general discussion, 280–283
  instruments for, 280–282
  mass spectra and, 282–283
Matched cuvettes, see Cuvette
Matrix matching, see Atomic absorption
McReynolds Constants, 363
Mean, defined, 48

Measuring pipet, see Pipets
Median, defined, 48
Mercury, safety hazards of, 464
Methanol, 38, 400
Methylene chloride, 37, see also Dichloromethane IR spectrum of, 253
Methyl Paraben, determination of, by HPLC, 413–414
Metric system, 205–206, 493
Milliequivalents, 119–120
Millimoles, 119–120
Mineral oil, IR spectrum of, 256
Mobile phase, see Chromatography
Mohr method for chloride, see Chloride
Moisture, determination of
Mode, defined, 49
Mohr method for chloride, 167, 170–171
Mohr pipet, 91–93
Mohr's salt, see Ferrous ammonium sulfate
Molar absorptivity, 228–229
Molar extinction coefficient, see Molar absorptivity
Molar solutions, preparation of, 84
  by dilution, 85, 486
  by dissolving pure solid chemicals, 84–85, 485
Molarity, 84
  of some concentrated acids and bases, 483
Molecular absorption, 209–212
Molecular absorption spectrum,
  double-beam spectrophotometry and, 223, 226
  examples of, 227
  qualitative analysis and, 226–228
Molecular spectroscopy, see specific types
Monochromatic light, 212
Monochromator
  flame atomic absorption, 298, 301
  flame photometry, 295
  UV/vis spectrophotometry, 218–220
Multielement hollow cathode lamp, 299

Neat liquids, IR spectrometry and, 250, 252
Nebulizer, 291
Nernst Equation, 427–429, 431, 432, 433–434
Neutralization, acid/base, 105–106
N.F. grade of chemical, 40
Nitrate in water, determination of, 236–237
Nitric acid, concentrated
  properties of, 35, 36, 483
  for sample preparation, 35, 36
Nitrogen/phosphorus detector (NPD), 372–373
NMR spectroscopy, see Nuclear magnetic resonance spectroscopy
Non-Faradaic current, see Current
Normal distribution curve, 53
Normality
  in acid/base systems, 109–112
Normal phase HPLC, see High-performance liquid chromatography
Normal solutions, preparation of
  by dilution, 112, 486
  by dissolving pure solid chemicals, 111, 485
Notebook, laboratory, 3–4,
NPD, see Nitrogen/phosphorus detector, 372–373
Nuclear magnetic resonance (NMR) spectroscopy
  applications of, 463
  chemical shifts in, 278–279
  Fourier transform instrument for, 278
  general discussion, 275–280
  peak splitting and integration in, 279–280
  traditional instrument for, 277–278
Nujol mull, see Infrared spectrometry, solid sampling

Open-column chromatography, see Chromatography
Open tubular columns, in gas chromatography, see Capillary columns
Ostwald-Folin pipet, see Pipets
Oxidants, in flame atomic absorption, 290
Oxidation, 157, 160
Oxidation number, 158–159
  rules for assigning, 158
Oxidation-reduction reactions, 157–167
  method for equation balancing, 160–162
Oxidizing agent, 160
Ozone in air, determination of, 237–238

Packed columns, in gas chromatography, 361
Paper chromatography, see Chromatography
Paper electrophoresis, 342–343
Particle theory of light, 201
Partition chromatography, 332–334, see also Bonded-phase chromatography, Gas-liquid chromatography; Liquid-liquid chromatography, High-performance liquid chromatography
  columns for, 399–400
Partition coefficient, 326
Parts per million (ppm), 145–148, 188–190
  of calcium carbonate in water, calculation of, 148–149
Path length, 228
  in IR cells, 250–252
Peak height method, in gas chromatography, 376
Peak ID region, IR spectra, see Infrared spectra
Peak size measurement, in instrumental chromatography, 376–377
Peak splitting, integration and in NMR spectroscopy, see Nuclear magnetic resonance spectrometry
Peptization, 26
Percent
  in gravimetric analysis, 11–12, 15
  in titrimetric analysis, 115, 116, 120, 486
Percent transmittance, in UV/vis spectrophotometry, 222, see also Transmittance
Perchloric acid, concentrated
  properties of, 35, 36, 483
  for sample preparation, 35, 36
Peristaltic proportioning pump, see Pumps
pH, see also Acids and bases; Buffer solutions
  determination of, in soil, 451
  measurement of, 433–435
pH electrode, 433–435
pH meter, 120, 121
1,10 Phenanthroline (*o*-phenanthroline), 141, 235
Phosphate buffer solutions, 74, 75
Phosphoric acid, concentrated, properties of, 483
Phosphorus, spectrophotometric determination of, in water and soil, 239–240
Photoionization detector (PID), 374–375
Photometry, flame, see Flame photometry
Photomultiplier tube, 212, 221
Photon, 201, 204
PID, see Photoionization detector
Pipets
  blow-out, 92
  cleaning of, 95–97
  disposable, 92
  duopette, 92, 93
  lambda, 92, 93
  measuring, 91
  Mohr, 91–93
  Ostwald-Folin, 92, 93
  serological, 91–93
  transfer, see Volumetric pipet
  volumetric, see Volumetric pipet
Pipetting
  basics of technique, 91, 99–100
  temperature effects, 99
Pipetting devices, 93–94
Planck's constant, 204
Plasma source, 307
Plating solutions, 460–461
Polarogram, 443
Polarographic maxima, 445
Polarography, 408, 439, 440–448, see also specific methods (Differential pulse, Pulse, Sampled dc)
  applications of, 448, 464
  classical, 442–445
  modern techniques of, 445–446, 447
  quantitative analysis using, 446–447
Polychromatic light, 212
Potassium acid phthalate, see Potassium hydrogen phthalate
Potassium biphthalate, see Potassium hydrogen phthalate
Potassium, determination of, in soil, 313
Potassium dichromate, 96, 165, 166
Potassium hydrogen phthalate (KHP), 113
  standardization of sodium hydroxide using, 128–130
  titrimetric analysis of unknown containing, 130–131
Potassium permanganate
  absorption spectrum of, 227
  standardization of, 167–168
  use as oxidizing agent, 163–164, 165
Potential, 490–491, see also Potentiometry; specific type
Potentiometric titrations, see Titration

Potentiometry, 429–438
  indicator electrodes in, 433–437, see also Indicator electrodes
  Nernst Equation and, 429
  reference electrodes in, 430–433, see also Reference electrodes
Potentiostat, 439
ppm, see Parts per million
Practical grade of chemical, 40
Precipitation titrations, 167
Precision, see Accuracy
Premix burner, 291–292
Preparation of solutions, see Solution preparation
Preparative columns, in gas chromatography, 361
Primary lines
  flame atomic absorption and, 296, 298, 303
Primary standard
  calcium carbonate, for EDTA standardization, 149–150
  examples, acid/base, 113–115
  examples, redox, 164, 165
  grade of chemical, 39
  requirements for, 97
  standardization using, 97, 107–108, 113, 114, 120, 150–151
Prism, see Dispersing element
Proportioning pump, peristaltic, see Pumps
Proton magnetic resonance, see Nuclear magnetic resonance (NMR) spectroscopy
Pulse polarography, 446, 447
Pumps
  proportioning, peristaltic, 466
  HPLC, 397
Purified grade of chemical, 40

Q test, 54–55
Quadratic equation and formula, 69
Quadrupole mass spectrometer, 280–282
Qualitative analysis
  defined, 2
  in gas chromatography, see Gas Chromatography
  in HPLC, see High-performance liquid chromatography
  in IR spectrometry, see Infrared spectrometry
  in emission spectrography, 308
  in UV/vis spectrophotometry, see UV/vis spectrophotometry
Quality assurance, for reagents, 39–40
Quality assurance laboratory, 459
Quality assurance program, 39
Quality control, statistics in, 57–58
Quality control charts, 57–58
Quantitative analysis, see also other specific techniques
  general discussion, 2–3
  in fluorometry, 233
  in gas chromatography, see Gas Chromatography
  in HPLC, see High-performance liquid chromatography
  in instrumental analysis, see Instrumental methods
  in IR spectrometry, see Infrared spectrometry
  in polarography, see Polarography
  in UV/vis spectrophotometry, see UV/vis spectrophotometry
Quantitative transfer, defined, 239

Radial chromatography, see Chromatography
Random errors, see indeterminate errors
Readout
  flame atomic absorption, 301
  instrumental, 178–181
  UV/vis spectrophotometry, 221–222
Reagent grade of chemical, 39–40
Reciprocating piston pump, in HPLC, see Pumps
Recorders, 177–178, 179, 193–195
Recrystallization, 322
Redox reactions, see Oxidation-reduction reactions
Reducing agent, 160, 166
Reduction, 157, 158, 160
Reference electrodes, see also specific types (Saturated calomel (SCE); Silver-silver chloride; Standard hydrogen
  in amperometric detector, 408
  in three-electrode systems, 439
Reference sources, 464, 523–525
Reflectance, diffuse, see Infrared spectrometry, solid sampling
Refractive index, 203, 405

Refractive index detector, 405–406
Refractometer, 406
Refractometry, 203
Relative deviation, defined, 50
Relative error, 58
Relative retention, in gas chromatography, 375
Relative standard deviation, defined, 51–52
Representative sample, 32–33
Residual current, see Charging current
Resistance, electrical, 490
Resolution
  in gas chromatography, 368
  in HPLC, 402
Retention time
  in gas chromatography, 366, 375
    altered, 382
  in HPLC, 402, 410
    decreased, 412
Reverse-phase HPLC, see High-performance liquid chromatography
R_f factors, in chromatography, 338–339
Riboflavin
  fluorometric analysis for, 241–242
  structure of, 234
Right angle configuration
  in atomic fluorescence, 309
  in fluorometry, 233
  in HPLC fluorescence detector, 405
Robotics, 470
Rotary evaporator, see Concentrators
Rotational energy transitions, 208–209

Safety, xiii–xiv
  atomic absorption and, 303
  cleaning solutions and, 96
Salt bridge, 422
  in SCE reference electrodes, 430
Sample
  control, see Control sample
  obtainment of, see Sampling
  pretreatment of, 34–39
    effect of, on calculations, 188–190
    HPLC and, 395–396
  representative, see Representative sample
Sample "clean-up", open-column chromatography and, 340
Sample compartment, in UV/vis spectrophotometer, 220–221
Sampled dc polarography, 445, 446
Sample handling, see Sample preservation
Sample injection, see Gas chromatography; High-performance liquid chromatography
Sample preparation
  concentrated acids in, 35–36
  by extraction, 37
  by fusion, 39
  reagents for, 35–38
  water in, 34
Sample preservation, 33–34
Samplers, automatic, see Automatic samplers
Sampling, see also Infrared spectrometry, liquid sampling; Infrared spectrometry, solid sampling
  general discussion, 32–33
Saturated calomel reference electrode (SCE), 430–431
SCE, see Saturated calomel reference electrode
Sealed cell, 250
Sealed demountable cell, 250
Secondary lines, in atomic absorption, 296, 303
Segmented flow methods, 466–469
Selectivity
  in gas chromatography, 369
  in HPLC, 403
Separations, see Analytical separations
Separatory funnel, 37, 324–326
Serial dilution method, 184
Serological pipet, 91–93
Significant figures, rules for, 475–478
Silver-silver chloride electrode, 431–432
Single-beam spectrophotometer, 218, 223
Single-pan balance, see Balances
Size exclusion chromatography, see also High-performance liquid chromatography Slit, monochromator, 218, 219
Snyder column, 328–329
Soda ash, 115, 130
Sodium benzoate, determination of, in soda pop, 414–415
Sodium carbonate
  determination of, in soda ash, 130
  as primary standard, 113–114
  titration of, 113–114, 127–128, 130
Sodium, determination of
  in snack chips, 314–315
  in soda pop, 315–316

Sodium oxalate, determination of, with potassium permanganate, 168
Sodium thiosulfate
  standardization of, 168–169
  uses of, 165, 166
Sodium hydroxide, 97
  preparation of solution of, 128
  reaction with acids, 109
  standardization of solution of, 128–130
Soil
  pH of, see pH
  sampling of, 32–33
Solid sampling, see Infrared spectrometry
Solubility product constant, 64–67
Solubility rules, 34
Solution preparation
  buffer, 74–75, 150
  dilution in, 85, 486
  molarity and, 84–87, 120, 485–486
  normality and, 111–112, 485–486
  parts per million and, 146–148
Solvent extraction, see Liquid-liquid extraction
Solvents, see also Sample preparation; specific solvents
Solvent strength, 398
Soxhlet extractor, 330, 331
Spark emission spectrography, 307–308, 310
Spectator ions, 161
Spectro grade of chemical, 40
Spectrometer, 212, see also specific type
Spectrometry, 212, see also Spectrophotometry; Infrared spectrometry; Mass spectrometry
Spectrophotometer, 212
  double-beam, see Double–beam instruments
  single-beam, see Single-beam spectrophotometer
Spectrophotometry, 212, see also Spectrometry; UV/vis spectrophotometry
  applications of, 461
Spectroscopy, 212
  atomic, see Atomic spectroscopy
  emission, see Emission spectroscopy
  molecular, see specific type
Speed of light, see Light
Standard, primary
Standard additions method, 185–186
Standard curve, 181, 229
Standard deviation, defined, 51
Standard hydrogen electrode, 432
Standard reduction potentials, 423–426
Standard solutions, 97
  series of, 181
Standardization, 97–98
  calculations, 106–108, 112–115, 120, 486
  of EDTA, 148–151
  of hydrochloric acid, 128–130
  of potassium permanganate, 167–168
  of sodium hydroxide, 128–130
  of sodium thiosulfate, 165, 168–169
Starch indicator, 16
Stationary phase, see Chromatography
Statistics, 47–58, see also Error analysis
Steel, manganese in, spectrophotometric determination of, 238–239
Stoichiometry, 12–13
Stopcock, 82, 83
Strip-chart recorder, see Recorders
Stripping voltammetry, 449
Student's *t*, 55–56
Sulfate, gravimetric determination of, 18–25
Sulfuric acid, concentrated
  properties of, 35, 36, 483
  for sample preparation, 35, 36, 117
Supporting electrolyte, 439–440
Suppressors, conductivity detectors and, 407–408
Suspended solids in wastewater samples, 460
Systematic errors, see Determinate errors

Tailing, in gas chromatography, 381–382
Taring, 7
TCD, see Thermal conductivity detector
TC imprint
  on duopettes, 93
  on lambda pipets, 93
  on volumetric flasks, 86
TD imprint
  on pipets, 88, 90, 93
Technical grade of chemical, 40
Technicon AutoAnalyzer, see AutoAnalyzer
Temperature programming, in gas chromatography, 364–365, 366
Tesla coil, in ICP, 307
THAM, see Tris-(hydroxymethyl)amine methane

Theoretical plates
  in chromatography, 368, 369, 402
  in distillation, 324
Thermal conductivity detector (TCD), 369–371
Thin-layer chromatography (TLC), 336–339
Three-electrode system, 439
Titrand, 82, 83
Titrant, 82, 83
Titration, 82–83, see also Titrimetric analysis
  amperometric, see Amperometric titration
  automatic, see also Titrators, automatic
  back, 116–119
  indirect, 117–119
  potentiometric, 437–438
Titration curves, acid/base, 120–125
  measurement by computer, 195–196
Titrators, automatic, see Automatic titrators
Titrimetric analysis, 81–84, see also Titration
  applications of, 460
  comparison to gravimetric analysis, 82
  Kjeldahl method in, see Kjeldahl method
  standardization in, see Standardization
  of water hardness, see Water hardness
TLC, see Thin-layer chromatography
Toluene, 38, 400
  IR spectrum of, 260
  molecular absorption spectrum of, 227
  spectrophotometric analysis for, 236
Top-loading balances, see Balances
Total alkalinity, see Alkalinity
Total consumption burner, 290–291
Total solids, in wastewater, 460
Transfer pipet, see Volumetric pipet
Transmission (transmittance) spectra, 232
  examples of, 227
Transmittance, in UV/vis spectrophotometry, 221, 224
Triangulation method, in measuring chromatography peak size, 376
Trichloroacetic acid, in buffer solutions, 74
TRIS, see Tris-(hydroxymethyl)amine methane
Tris-(hydroxymethyl)amine methane (THAM)
  in buffer solutions, 74
  as primary standard, 113, 114
Tungsten filament bulb, 218, 219

U.S.P. grade of chemical, 40
UV absorption detector, for HPLC, 404
UV/vis spectrophotometry, 217–235, see also Beer's Law
  applications of, 461
  instruments for
    calibration of, 222–223
    design of, 218–224
    detector/readout, 221–222
    light sources for, 218
    monochromator, 218–220
    sample compartment (cuvette), 220–221
  interferences in, 230–231
  qualitative analysis with, 226–228
  quantitative analysis with, 228–230

Vapor generation methods, for atomization, 306–307, 310
Vapor pressure, gas chromatography, 356
Variance, 51
Vibrational energy transitions, 208–212
Visible spectrophotometry, see Colorimetry; UV/vis spectrophotometry
Volhard method for silver, 167
Voltage, electrical, 490
Voltammetry, 408, 439, 448–451, see also specific techniques (Amperometric titration, Cyclic voltammetry, Stripping voltammetry)
Volumetric analysis, see Titrimetric analysis
Volumetric flask, 86–90
Volumetric glassware
  cleaning of, 95–97
  general discussion, 86–97
Volumetric pipet, 90–91

Walden reductor, 166
Water
  deionized, see Deionized water
  distillation of, 322
  for sample preparation, 34

Water hardness, titrimetric determination of, 144–145, 151–152, 460
Water of hydration, determination of, in hydrated barium chloride, 15–18
Wavelength of light, see Light
Wavelength of maximum absorbance, 229, 232
Wave number of light, see Light
Weak acids, 68–71
Weak bases, 71
Weighing by difference, 20–21
Weighing devices, 4–10
Wet chemical analysis, 2
  applications of, 460
Winkler method for dissolved oxygen, 167
Working electrode, 439

X-Y recorders, see Recorders

PERIODIC TABLE OF THE ELEMENTS

New notation	1	2	3	4	5	6	7	8	9	10	11	12	13	14	15	16	17	18	
Previous IUPAC form	Group IA	IIA	IIIA	IVA	VA	VIA	VIIA		VIIIA		IB	IIB	IIIB	IVB	VB	VIB	VIIB	VIIIA	
CAS version			IIIB	IVB	VB	VIB	VIIB		VIII				IIIA	IVA	VA	VIA	VIIA		Shell
	1 H +1 −1 1.00794 1																	2 He 0 4.0020602 2	K
	3 Li +1 6.941 2-1	4 Be +2 9.012182 2-2											5 B +3 10.811 2-3	6 C +2 +4 −4 12.011 2-4	7 N +1 +2 +3 +4 +5 −1 −2 −3 14.00674 2-5	8 O −2 15.9994 2-6	9 F −1 18.9984032 2-7	10 Ne 0 20.1797 2-8	K-L
	11 Na +1 22.989768 2-8-1	12 Mg +2 24.3050 2-8-2											13 Al +3 26.981539 2-8-3	14 Si +2 +4 −4 28.0855 2-8-4	15 P +3 +5 −3 30.97362 2-8-5	16 S +4 +6 −2 32.066 2-8-6	17 Cl +1 +5 +7 −1 35.4527 2-8-7	18 Ar 0 39.948 2-8-8	K-L-M
	19 K +1 39.0983 -8-8-1	20 Ca +2 40.078 -8-8-2	21 Sc +3 44.955910 -8-9-2	22 Ti +2 +3 +4 47.867 -8-10-2	23 V +2 +3 +4 +5 50.9415 -8-11-2	24 Cr +2 +3 +6 51.9961 -8-13-1	25 Mn +2 +3 +4 +7 54.93085 -8-13-2	26 Fe +2 +3 55.845 -8-14-2	27 Co +2 +3 58.93320 -8-15-2	28 Ni +2 +3 58.6934 -8-16-2	29 Cu +1 +2 63.546 -8-18-1	30 Zn +2 65.39 -8-18-2	31 Ga +3 69.723 -8-18-3	32 Ge +2 +4 72.61 -8-18-4	33 As +3 +5 −3 74.92159 -8-18-5	34 Se +4 +6 −2 78.96 -8-18-6	35 Br +1 +5 −1 79.904 -8-18-7	36 Kr 0 83.80 -8-18-8	-L-M-N
	37 Rb +1 85.4678 -18-8-1	38 Sr +2 87.62 -18-8-2	39 Y +3 88.90585 -18-9-2	40 Zr +4 91.224 -18-10-2	41 Nb +3 +5 92.90638 -18-12-1	42 Mo +6 95.94 -18-13-1	43 Tc +4 +6 +7 (98) -18-13-2	44 Ru +3 101.07 -18-15-1	45 Rh +3 102.90550 -18-16-1	46 Pd +2 +4 106.42 -18-18-0	47 Ag +1 107.8682 -18-18-1	48 Cd +2 112.411 -18-18-2	49 In +3 114.818 -18-18-3	50 Sn +2 +4 118.710 -18-18-4	51 Sb +3 +5 −3 121.760 -18-18-5	52 Te +4 +6 −2 127.60 -18-18-6	53 I +1 +5 +7 −1 126.90447 -18-18-7	54 Xe 0 131.29 -18-18-8	-M-N-O
	55 Cs +1 132.90543 -18-8-1	56 Ba +2 137.327 -18-8-2	57* La +3 138.9055 -18-9-2	72 Hf +4 178.49 -32-10-2	73 Ta +5 180.9479 -32-11-2	74 W +6 183.84 -32-12-2	75 Re +4 +6 +7 186.207 -32-13-2	76 Os +3 +4 190.23 -32-14-2	77 Ir +3 +4 192.217 -32-15-2	78 Pt +2 +4 195.08 -32-16-2	79 Au +1 +3 196.96654 -32-18-1	80 Hg +1 +2 200.59 -32-18-2	81 Tl +1 +3 204.3833 -32-18-3	82 Pb +2 +4 207.2 -32-18-4	83 Bi +3 +5 208.98037 -32-18-5	84 Po +2 +4 (209) -32-18-6	85 At (210) -32-18-7	86 Rn 0 (222) -32-18-8	-N-O-P
	87 Fr +1 (223) -18-8-1	88 Ra +2 226.025 -18-8-2	89** Ac +3 227.028 -18-9-2	104 Unq +4 (261) -32-10-2	105 Unp (262) -32-11-2	106 Unh (263) -32-12-2	107 Uns (262) -32-13-2												O P Q

*Lanthanides	58 Ce +3 +4 140.115 -19-9-2	59 Pr +3 140.90765 -21-8-2	60 Nd +3 144.24 -22-8-2	61 Pm +3 (145) -23-8-2	62 Sm +2 +3 150.36 -24-8-2	63 Eu +2 +3 151.965 -25-8-2	64 Gd +3 157.25 -25-9-2	65 Tb +3 158.92534 -27-8-2	66 Dy +3 162.50 -28-8-2	67 Ho +3 164.93032 -29-8-2	68 Er +3 167.26 -30-8-2	69 Tm +3 168.93421 -31-8-2	70 Yb +2 +3 173.04 -32-8-2	71 Lu +3 174.967 -32-9-2	N O P
**Actinides	90 Th +4 232.0381 -18-10-2	91 Pa +5 +4 231.03588 -20-9-2	92 U +3 +4 +5 +6 238.0289 -21-9-2	93 Np +3 +4 +5 +6 237.048 -22-9-2	94 Pu +3 +4 +5 +6 (244) -24-8-2	95 Am +3 +4 +5 +6 (243) -25-8-2	96 Cm +3 (247) -25-9-2	97 Bk +3 +4 (247) -27-8-2	98 Cf +3 (251) -28-8-2	99 Es +3 (252) -29-8-2	100 Fm +3 (257) -30-8-2	101 Md +2 +3 (258) -31-8-2	102 No +2 +3 (259) -32-8-2	103 Lr +3 (260) -32-9-2	O P Q

KEY TO CHART

Atomic Number → 50 ← +2 +4 Oxidation States
Symbol → Sn
1989 Atomic Weight → 118.71
18 18 4 ← Electron Configuration

The new IUPAC format numbers the groups from 1 to 18. The previous IUPAC numbering system and the system used by Chemical Abstracts Service (CAS) are also shown. For radioactive elements that do not occur in nature, the mass number of the most stable isotope is given in parentheses.

FORMULA WEIGHTS

Formula	Weight
$AgBr$	187.78
$AgCl$	143.32
Ag_2CrO_4	331.73
AgI	234.77
$AgNO_3$	169.87
$AgSCN$	165.95
Al_2O_3	101.96
$Al(OH)_3$	78.00
$Al_2(SO_4)_3$	342.14
As_2O_3	197.85
$BaCO_3$	197.35
$BaCl_2$	208.25
$BaCl_2 \cdot 2H_2O$	244.27
$BaCrO_4$	253.33
BaO	153.34
$Ba(OH)_2$	171.36
$BaSO_4$	233.40
Bi_2O_3	466.0
$C_6H_{12}O_6$ (glucose)	180.16
$C_{12}H_{22}O_{11}$ (sucrose)	342.30
$CHCl_3$	119.38
CO_2	44.01
$CaCl_2$	110.99
$CaCO_3$	100.09
CaC_2O_4	128.10
CaF_2	78.08
CaO	56.08
$Ca(OH)_2$	74.10
$CaSO_4$	136.14
CeO_2	172.12
$Ce(SO_4)_2$	332.25
Cr_2O_3	151.99
CuO	79.54
Cu_2O	143.08
$CuSO_4$	159.60
$Fe(NH_4)_2(SO_4)_2 \cdot 6H_2O$	392.14
FeO	71.85
Fe_2O_3	159.69
Fe_3O_4	231.54
HBr	80.92
$HC_2H_3O_2$ (acetic acid)	60.05
$HCO_2C_6H_5$ (benzoic acid)	122.12
HCl	36.46
$HClO_4$	100.46
$H_2C_2O_4$	90.04
$H_2C_2O_4 \cdot 2H_2O$	126.07
HNO_3	63.01
H_2O	18.015
H_2O_2	34.01
H_3PO_4	98.00
H_2S	34.08
H_2SO_3	82.08
H_2SO_4	98.08
HSO_3NH_2 (sulfamic acid)	97.09
HgO	216.59
Hg_2Cl_2	472.09
$HgCl_2$	271.50
$Hg(NO_3)_2$	324.61
KBr	119.01
$KBrO_3$	167.01
KCl	74.56
$KClO_3$	122.55

Formula	Weight
KCN	65.12
K_2CO_3	138.22
K_2CrO_4	194.20
$K_2Cr_2O_7$	294.19
KHC_2O_4	128.13
$KHC_8H_4O_4$ (KHP)	204.23
$KH(IO_3)_2$	389.92
K_2HPO_4	174.18
KH_2PO_4	136.09
$KHSO_4$	136.17
KI	166.01
KIO_3	214.00
KIO_4	230.00
$KMnO_4$	158.04
KNO_3	101.11
KOH	56.11
$KSCN$	97.18
K_2SO_4	174.27
$MgCl_2$	95.22
MgO	40.31
$Mg(OH)_2$	58.33
$Mg_2P_2O_7$	222.57
$MgSO_4$	120.37
MnO_2	86.94
Mn_2O_3	157.88
Mn_3O_4	228.81
$Na_2B_4O_7 \cdot 10H_2O$	381.37
$NaBr$	102.90
$NaC_2H_3O_2$	82.03
$Na_2C_2O_4$	134.00
$NaCl$	58.44
$NaClO$	74.44
$NaCN$	49.01
Na_2CO_3	105.99
NaF	41.99
$NaHCO_3$	84.01
$Na_2H_2EDTA \cdot 2H_2O$	372.23
NaH_2PO_4	119.99
Na_2HPO_4	141.98
$NaOH$	40.00
Na_3PO_4	163.95
$NaSCN$	81.07
Na_2SO_4	142.04
$Na_2S_2O_3 \cdot 5H_2O$	248.18
NH_3	17.03
NH_4Cl	53.49
$NH_2(HOCH_2)_3$ (THAM)	121.14
$NH_2CO_2NH_2$ (Urea)	60.05
$(NH_4)_2C_2O_4 \cdot H_2O$	142.11
NH_4NO_3	80.04
$(NH_4)_2SO_4$	132.14
$(NH_4)_2S_2O_8$	228.18
$PbCrO_4$	323.18
$PbSO_4$	303.25
P_2O_5	141.94
Sb_2O_3	291.50
SiO_2	60.08
$SnCl_2$	189.60
SnO_2	150.69
SrO_4	183.68
SO_2	64.06
SO_3	80.06
TiO_2	79.90